THIRD EDITION

WHITTINGTON'S DICTIONARY OF PLASTICS

Edited by James F. Carley, Ph.D., PE

LANCASTER · BASEL

Whittington's Dictionary of Plastics
a **TECHNOMIC**® publication

Published in the Western Hemisphere by
Technomic Publishing Company, Inc.
851 New Holland Avenue, Box 3535
Lancaster, Pennsylvania 17604 U.S.A.

Distributed in the Rest of the World by
Technomic Publishing AG
Missionsstrasse 44
CH-4055 Basel, Switzerland

Printed in the United States of America

10 9 8 7 6 5 4 3 2 1

Main entry under title:
Whittington's Dictionary of Plastics—Third Edition

A Technomic Publishing Company book
Bibliography: p.

Library of Congress Catalog Card No. 93-60943
ISBN No. 1-56676-090-9

To the memory of the late Mrs. B. L. (Molly) Carley, my early-widowed mom with two acquired kids, whose dedication of her labor, love, and a large fraction of her hard-won income to supporting my education has made it possible for me to write the book.

PREFACE

Since 1978, when the second edition of this Dictionary appeared, many of its terms have changed in meaning or taken on additional new meanings, a few have become obsolete, and many new terms have entered the plastics literature. New materials and new processing techniques developed during that interval have given rise to new terminology. There are now about three times the number of technical journals dealing with plastics, resin composites, and polymers as there were in the early 1970s. The number of trade journals serving the industry has not increased much, yet they still introduce much new terminology and develop its usage.

Toxicity, safety, flammability, and environmental concerns have continued to grow in importance and to influence decisions on choices of materials and processing methods. This edition, even more than the second, reflects those concerns in many of its entries. The push toward constant improvement of quality has caused us to add definitions of terms basic to process control, statistics and quality management even though they are not specific to plastics.

There has been a huge growth in the number of new polymers and blends since 1978, the vast majority of which are mainly of interest to the scientific community. However, technologists are finding that they need to be more conversant with what their scientific *compadres* are doing because it often impinges on their own everyday work. More emphasis has been given in this edition to basic polymer science.

In 1978, the use of the International System of units (SI) in the US was still half-hearted, even though several national engineering and scientific societies were then requiring that SI units be used in their journals. Since then, acceptance of SI in the science and engineering communities has been rapid but in the plastics industry and its trade journals, disappointingly slow. The good news is that the old cgs metric system is fading fast, giving way to SI. Also, the fact that most of the rest of the world, with whom we must do business, is using SI is putting pressure on our industry and commerce to adopt and actively use the system, to make it part of our everyday thinking. It is a much simpler and more coherent system than the jumble we have been using. In this work, all major SI units that do or could have application in plastics technology and commerce, and some that are mainly of interest to science, are defined. Also, in entries containing quantitative entries, SI units are the ones mainly used, though some on-the-spot equivalents in English units are given. To facilitate the inevitable changeover, I have included an Appendix of conversion factors useful in our industry.

Anticipating that this Dictionary will be used not only by experienced plastics people, but also by newcomers to plastics and by readers for whom English is a second language, I have given the full names of materials mentioned in entries, with the single exception of "PVC" (polyvinyl chloride), which is almost universally known, even among lay persons. Most of such abbreviations in use are listed as separate entries. Nearly all of the entries of the Second Edition were retained, but every one was examined for accuracy, corrected or updated where necessary, and reread for clarity. I am hugely indebted and grateful to Lloyd Whittington for the extensive groundwork he laid in the first two editions, particularly in the identification of additives used with plastics. This edition has been strengthened in the areas of processing, rheology, plastics properties, and materials tests. ASTM (formerly American Society for Testing Materials) publishes four volumes of tests, specifications, and recommended procedures for plastics in section 08 of its *Annual Book of ASTM Standards.* Some others of its 15 sections (57 volumes in all) also contain tests relevant to plastics products. In the Dictionary's entries, a test number, e g, ASTM D 638, will be found in one of the four volumes of section 08 unless another section is designated. ASTM data is from the 1991 edition of the *Standards.* A few leading trade names are included. For a comprehensive list of trade names, see the works by M. Ash and I. Ash in the references. More structural formulas for chemicals and monomers are given, as are many of the important equations of polymer science and plastics engineering. To save the user the inconvenience of having to refer repeatedly to a distant table of symbols, the symbols used in each equation are defined on the spot. Some definitions have been quoted from ISO 472 (see references), some from lists of the International Union of Pure and Applied Chemistry (IUPAC), and some from ASTM D 883. At many entries, the reader is referred to other, related entries printed in SMALL CAPITALS.

While this work is called a dictionary, we have filled out many definitions beyond the concise definitions typical of the conventional dictionary to provide greater breadth and depth of insight, yet have stopped well short of what one would find in an encyclopedia of much greater size. Wherever possible, we have tried to use simple language understandable to students and workers without technical education. Hundreds of recently published books and technical and trade journals were searched for new and relevant terms. As an employee and retiree of Lawrence Livermore Laboratory, I had access to its fine library (and those of the University of California at Berkeley), where I found most of the books and the much consulted ASTM volumes. My publisher very generously furnished his suite of sixteen scientific journals and newsletters, some of which I might not otherwise have come by. I am in debt, also, to my friend James P. Harrington, Editor of the *Journal of Plastic Film & Sheeting*, for sending a list of packaging terms and suggesting many new words for this Dictionary.

Finally, strenuous efforts were made to free the Dictionary from errors of fact, writing, and typography. Abbreviations, except for personal initials, have no periods. The abbreviation "w s" stands for "which see", replacing its old Latin alias "q v". Typos that eluded my triple rereadings were spotted by hawk-eyed Ms Kimberly Martin of Technomic Publishing Co, to whom I give many thanks. However, perfection, though diligently sought, is rarely achieved. Since I prepared camera-ready copy, responsibility for errors is totally mine. We will be grateful to readers who notify us of errors and omissions, so that we may correct and improve future printings and editions.

References

Aldrich Chemical Co, *Catalog Handbook of Fine Chemicals* (1990). Brief descriptions and basic data for 19,000 chemicals with correct names and alternates, including structural formulas where needed.

Alger, M. S. M., *Polymer Science Dictionary*, Elsevier Science Publishers, Barking, Essex, England (1989). This excellent volume is "...restricted to polymer science and deliberately excludes technological areas. Specifically, polymer processing terms are not included. Within the fields of polymer chemistry and physics, an attempt has been made to be as comprehensive as possible...the entries are longer than is usual in a dictionary strictly concerned with definitions...and [it] could perhaps be considered more as an encyclopaedic dictionary."

Annual Book of ASTM Standards, chiefly section 08, vols 08.01, 08.02, 08.03, and 08.04, ASTM, Philadelphia, PA (1991). Section 08 contains 2953 pages of analytical methods, material specifications, tests of mechanical, optical, permanence, and thermal properties, and general procedures for plastics. At the end of each volume is a list, for all four volumes, coordinating section 08 tests with tests of ISO (International Organization for Standardization). Sections 6, 7, 9, 10, 13, and 15, occasionally cited in this Dictionary, are titled, in that order, *Paints, Related Coatings, and Aromatics*; *Textiles*; *Rubber*; *Electrical Insulation and Electronics*; *Medical Devices and Services*; *General Products, Chemical Specialties, and End Use Products.*

Ash, M. and I. Ash, *Encyclopedia of Plastics, Polymers, and Resins*, vols I-III, Chemical Publishing Co, New York, (1982-3); also, the supplement, vol IV (1988). "This encyclopedia is an attempt to coordinate and unify practical information on plastic, polymer, and resin trademarks." The most complete source of information on trade names.

Billmeyer, F. W. Jr., *Textbook of Polymer Science*, 3rd Ed, Wiley-Interscience, New York (1984).

Bittance, J. C., (Ed), *Engineering Plastics and Composites*, ASM International, Materials Park, OH 44073 (1990). Another source of information on plastics trade names (old and new).

Brandrup, J. and E. H. Immergut, (Eds), *Polymer Handbook*, 3rd Ed, Wiley-Interscience, New York (1989).

Elias, H.-G. et al, "Abbreviations for Thermoplastics, Thermosets, Fibers, Elastomers, and Additives," *Polymer News* (9) 101-110 (1983); "—Part II," *Polymer News* (10) 169-172 (1985).

Gove, P. B. (Ed-in-Chief), *Webster's Third New International Dictionary*, G. & C. Merriam Company, Springfield, MA (1961). My authority on word division; enlightening on some items in this Dictionary, as well.

ISO 472 Plastics—Vocabulary, 2nd Ed, International Organization for Standardization, Geneva, Switzerland (1988). Brief definitions of some 1300 terms agreed to by international committee and relating to plastics and polymer science, ordered alphabetically by the English versions, with French alongside, and a French-ordered index list at the end.

Kroschwitz, J. I., (Exec Ed), *Concise Encyclopedia of Polymer Science and Engineering*, John Wiley & Sons, New York (1990). An excellent, 1341-page condensation of the 19-volume *Encyclopedia of Polymer Science and Engineering* from the same publisher.

Lee, S. M., *Dictionary of Composite Materials Technology*, Technomic Publishing Co, Inc, Lancaster, PA (1989). Contains definitions of some 6400 terms relating to very broadly defined composites.

Modern Plastics Encyclopedia, an annual by the editors of *Modern Plastics* magazine, included with subscription. Content and format change from year to year. The 1992 issue was the most recent that contained the short summary articles on commercially important resins; 1990 was the latest containing the valuable "Designing with Plastics". Properties charts for resins, films, plasticizers, and machinery and equipment charts are strong features.

"Practice for the Use of the International System of Units (SI)," ASTM E 380-89a, given in full in *1991 Book of ASTM Standards*, vol 14.02, excerpted in many other volumes, and available separately from ASTM, Philadelphia, PA.

James F. Carley
Livermore, CA
June, 1993

ABOUT THE EDITOR

It seems unlikely that Alice Grannis Harakal supposed, as she led James F. Carley through the rigors and joys of sophomore and junior English in Eastchester, New York, High School, that he would someday write a dictionary. Her no-nonsense nurturing, though, is where his preparation for this work began. A few years later, as a chemical engineering student and Navy V-12er at Cornell University, under the demanding tutelage of Fred H. ("Dusty") Rhodes, he wrote (and, alas! often rewrote) numerous formal reports on Unit Ops experiments, training that subsequently served him well in his five years with Du Pont's Polychemicals (later, Plastics) Department. It was at Cornell, too, that he first became acquainted with plastics in a survey course with Charles C. Winding, and glimpsed the enormous future of plastics in our society.

After three years in naval training and communications, Jim Carley returned to Cornell, completed his delayed fifth year, earned his second degree, and decided to continue at Cornell to do graduate work. His PhD in chemical engineering was granted in 1951, by which time he was already at work with Du Pont, studying single- and twin-screw extrusion, injection molding, and performance properties of plastics. He visited many processors with Du Pont's technical-service people, and was an in-house consultant on extrusion to several of Polychemicals' plants. His mentor at Du Pont was William L. Gore, who later founded W. L. Gore Associates. Bill vigorously supported Jim's interest in statistics and experiment design and established an almost collegial—and highly productive—atmosphere of lively debate in his team of processing researchers.

Late in 1955, when *Modern Plastics* magazine, then the dominant trade publication in the field, needed a new Engineering Editor, a letter that Carley had earlier written to Editor Hiram McCann brought him an invitation and, soon after, an offer of the position, which he accepted and fulfilled for three and a half years. Here his knowledge of plastics, reinforced plastics, and the plastics industry were greatly broadened. He visited many plants and laboratories and his writing skills developed. Once a year, his Engineering Section contained a critical, comprehensive review of significant worldwide advances in plastics processing.

In 1960, Carley went to the University of Arizona, Tucson, where he taught chemical engineering, including report writing, and conducted research in plastics processing. In 1962, he became Technical Director of Prodex Corporation of Fords, NJ, manufacturers of extruders and extrusion lines. There he did more extrusion R&D, exercised the mechanical-engineering training that the Navy had required of him at Cornell, solved customers' problems in the field, helped with product design, and engaged in many a hot debate with his dynamic bosses, the late Albert A. Kaufman and Frank R. Nissel, now President of Welex.

After a short stint with Celanese Plastics in 1964, Carley joined the University of Colorado, Boulder, where, for twelve years, he taught in two departments, Chemical Engineering and Engineering Design, teaching courses at all levels and continuing research in rheology, plastics processing and other topics. He also served as an extrusion and product-design consultant to several industrial companies. By the time he left CU to join the Oil Shale Project at Lawrence Livermore National Laboratory, Carley had published about forty papers in plastics engineering.

During his six years with Oil Shale, during which time he wrote many reports and papers on oil-shale technology, he also published papers on corona-discharge treatment of films, viscosities of polymer blends, blend morphology, pressure drop through screen packs, and flow in wire-coating dies. Carley then joined the Polymers Section of LLNL, where he worked on the development of low-density materials for the "star wars" project. During the years 1982-88 he was Technical Editor of *Modern Plastics*, writing a monthly page of literature abstracts, "Technical Review". In 1985 he was elected a Fellow of the Society of Plastics Engineers. In 1989 he was Sidney Levy's coauthor for the second edition of *Plastics Extrusion Technology Handbook*. Retired from LLNL at the end of 1990, he has continued part-time work there on unfinished projects while working on this Dictionary at home.

WHITTINGTON'S DICTIONARY OF PLASTICS

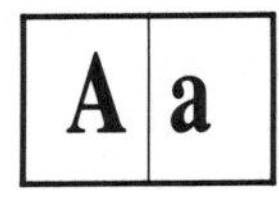

a (1) SI abbreviation for prefix ATTO-, w s. (2) Symbol for acceleration.

A Abbreviation for AMPERE, w s.

Å Abbreviation for deprecated ANGSTROM UNIT, w s.

Ab (= absolute) A prefix attached to the names of practical electrical units to indicate the corresponding unit in the old cgs system (emu), e g, abampere, abvolt.

ABFA See AZOBISFORMAMIDE.

Abherent (adhesive) A coating or film applied to one surface to prevent or reduce its adhesion to another surface brought into intimate contact with it. Abherents applied to plastic films are often called *anti-blocking agents*. Those applied to molds, calender rolls, etc are sometimes called *release agents* or PARTING AGENTs, w s.

Abietic Acid $C_{19}H_{30}COOH$. A monocarboxylic acid derived from rosin. Plasticizers derived from it include HYDROABIETYL ALCOHOL, HYDROGENATED METHYL ABIETATE, and METHYL ABIETATE, w s.

Ablation Derived from the Latin *ablatio*, meaning "a carrying away", this term has been used by astrophysicists to describe the erosion and disintegration of meteors entering the atmosphere, and more recently by space scientists and engineers for the layer-by-layer decomposition of a plastic surface when heated quickly to a very high temperature. Usually, the decomposition is highly endothermic and the absorption of energy at the surface slows penetration of high temperature to the interior.

ABL Bottle A filament-wound test vessel about 46 cm in diameter and 61 cm long, subjected to rising internal hydrostatic pressure to determine the quality and strength of the composite from which it was made.

Abrasion Cycle The number of repetitive abrading motions to which a specimen is subjected in a test of abrasion resistance.

Abrasion Resistance The ability of a material to withstand mechanical actions such as rubbing, scraping, grinding, sanding, or erosion that tend progressively to remove material from its surface. ASTM tests for abrasion resistance are D 1044, "Resistance of Transparent Plastics to Surface Abrasion", and D 1242, "Resistance to Abrasion of Plastic Materials".

Abrasive Finishing A method of removing flash, gate marks and rough edges from plastics articles by means of grit-containing belts or wheels. The process is usually employed on large rigid or semi-rigid products with intricate surfaces that cannot be treated by tumbling or other more efficient methods of finishing.

ABS Abbreviation for acrylonitrile-butadiene-styrene. See ABS RESIN.

Absolute Alcohol Ethyl alcohol that has been refined by azeotropic distillation to 99.9+% purity (200 proof). Other commercial ethanols contain about 5% water and may contain denaturants that make the alcohol undrinkable.

Absolute Error The difference between a measurement and the true value of the quantity measured.

Absolute Humidity See HUMIDITY.

Absolute Temperature Temperature measured from the absolute zero, at which all molecular motions cease; 0.00 K (kelvin) = –273.15°C.

Absolute Viscosity See VISCOSITY.

Absorbance See LIGHT ABSORBANCE.

Absorption (1) The penetration of a substance into the mass of another substance by chemical or physical action. See, for example, WATER ABSORPTION. (2) The process by which energy is dissipated within a specimen placed in a field of radiant energy. Since some part of the impinging energy may be transmitted through the specimen and another part be reflected, the energy absorbed will nearly always be less than that impinging.

Absorption Spectrophotometry (spectrophotometry) An analytical technique utilizing the absorption of electromagnetic radiation by a specimen (or solution) as a property related to the composition and quantity of a given material in the specimen. The radiation is usually in the ultraviolet, the visible, or the near-infrared portions of the electromagnetic spectrum. When the absorbing medium is in the gaseous state, the absorption spectrum consists of dark lines or bands, being the reverse of the emission spectrum of the absorbing gas. When the absorbing medium is in the solid or liquid state, the spectrum of the transmitted light shows broad dark regions which are not resolvable into lines and have no sharp or distinct edges. In quantitative spectrophotometry, the intensity of the radiation passing through a specimen or solution is compared with the intensity of the incident radiation and with radiation passing through a nonabsorbing solvent (*blank*). The percent absorbed by the solution is exponentially related to the solute concentration (Beer's law). Modern spectrophotometers are capable of generating nearly monochromatic radiation, so they can develop plots of percent absorption vs wavelength—*absorption spectra*—for the test compound. See INFRARED SPECTROPHOTOMETRY.

ABS Resin Any of a family of thermoplastics based on acrylonitrile, butadiene and styrene combined by a variety of methods involving polymerization, graft copolymerization, physical mixtures, and combinations thereof. Hundreds of standard grades of ABS resins are available, plus many special grades, alloyed or otherwise modified to yield unusual properties. The standard grades are rigid, hard and tough, and possess good impact strength, heat resistance, low-temperature properties, chemical resistance, and electrical properties. ABS compounds in pellet form can be extruded, blow molded, calendered, and injection molded. ABS powders are used as modifiers for other resins, for example PVC. Typical applications for ABS resins are household appliances, automotive parts, business-machine and telephone components, pipe and pipe fittings, packaging and shoe heels.

Accelerated Test A test procedure in which conditions such as temperature, humidity and ultraviolet radiation are intensified to reduce the time required to obtain a deteriorating effect similar to one resulting from exposure to normal service conditions for much longer times.

Accelerated Weathering See ARTIFICIAL WEATHERING.

Accelerator (promoter) A substance that hastens a reaction, usually by acting in concert with a CATALYST or a CURING AGENT, w s. Accelerators are sometimes used in the polymerization of thermoplastics, but are used most widely in curing systems for thermosets and natural and synthetic rubbers. Sometimes called *cocatalyst*.

Acceptable Quality Level (AQL) The maximum lot fraction (percent) defective, in sampling inspection of attributes, that can be considered satisfactory.

Accumulator (1) In blow molding and injection molding, an auxiliary ram extruder providing fast parison delivery or fast mold filling. The accumulator cylinder is filled with plasticated melt from the main extruder between parison deliveries shots, and stores this melt until the plunger is called upon to deliver the next parison or shot. (2) A pressurized gas reservoir that stores energy in hydraulic systems.

Acetaldehyde (ethanal, ethyl aldehyde, acetic aldehyde) CH_3CHO A colorless, flammable liquid made by the hydration of acetylene, the oxidation or dehydrogenation of ethyl alcohol, or the oxidation of saturated hydrocarbons or ethylene.

Acetal Resin (polyformaldehyde, polyoxymethylene, polycarboxane) A thermoplastic produced by the addition polymerization of an aldehyde through the carbonyl function, yielding unbranched polyoxymethylene $(\text{-O-CH}_2\text{-})_n$ chains of great length. Examples are Du Pont's "Delrin" and Hoechst-Celanese's "Celcon" (acetal copolymer based on trioxane). The acetal resins are among the strongest and stiffest of all thermoplastics, and are characterized by good fatigue life, resilience, low moisture sensitivity, high resistance to solvents and chemicals, and good electrical properties. They may be processed by conventional injection molding and extrusion techniques, and fabricated by welding methods used for other thermoplastics. Their main arena of application is industrial and mechanical products, e g, gears, rollers, and many automotive parts.

Acetate (1) A salt or ester of acetic acid. (2) A generic name for cellulose acetate plastics, particularly for their fibers. Where at least 92% of the hydroxyl groups have been acetylated, the term *triacetate* may be used as the generic name of the fiber. (3) A compound containing the acetate group, CH_3COO–.

Acetate Fibers (see ACETATE) Fibers made by partly acetylating cellulose. At one time, acetate fibers were in third place among synthetic fibers, surpassed only by rayon and nylon. Their major application was for ladies' apparel and home-furnishing fabrics. The triacetates were developed later and became more widely used due to their similarity to acrylics, nylons and polyesters. According to the FTC, *triacetate* must have at least 92% of the cellulose –OH groups acetylated. At one time, acetate fibers were called rayon, usage that has been deprecated because RAYONs (w s) have a completely different structure.

Acetic Acid (ethanoic acid, methanecarboxylic acid, vinegar acid) CH_3COOH. A colorless liquid with the familiar taste and odor of vinegar, of which it is the chief constituent in dilute form. Acetic acid was originally derived by souring wine and beer, but is synthesized today by oxidation of acetaldehyde in the presence of a catalyst. Among the uses of acetic acid in the plastics industry is the manufacture of cellulosic plastics such as cellulose acetate (CA), CA butyrate and CA propionate, vinyl acetate, and acetate esters for plasticizing thermoplastics.

Acetic Aldehyde See ACETALDEHYDE.

Acetic Anhydride $(CH_3CO)_2O$. This pungent liquid may be thought of as the condensation product of two molecules of acetic acid by removal of one molecule of water, though in fact it is made by reaction of acetic acid with ketene, $CH_2{=}C{=}O$. It is a strong acetylating agent, used for many of the same purposes as its parent acid.

Acetic Ether See ETHYL ACETATE.

Acetone (dimethyl ketone, 2-propanone) CH_3COCH_3. The simplest and most important member of the ketone family of solvents. All the cellulosic plastics and polyvinyl chloride, polyvinyl acetate, polymethyl methacrylate, epoxies, and some thermosetting resins are soluble in acetone. It is also an intermediate in the production of bisphenol A, antioxidants, and resins such as polymethyl methacrylate, cellulose acetate, and epoxies.

Acetone Extraction In molded phenolic products, the amount of acetone-soluble material that can be extracted from the material is an indication of the degree of cure. A test for determining such material is

ASTM D 494, which can be found in versions of ASTM Standards published before 1985, when D 494 was discontinued by ASTM.

Acetone Resin A synthetic resin produced by the reaction of acetone with materials such as phenol or formaldehyde.

Acetylation The introduction of an acetyl group (CH_3CO–) into the molecule of an organic compound having an –OH or –NH_2 group, by treatment with acetic anhydride or acetyl chloride.

Acetyl Chloride CH_3OCl. Acetic acid in which the –OH group has been replaced by –Cl; a very active acetylating agent.

Acetyl Cyclohexane Sulfonyl Peroxide A polymerization initiator, often used in conjunction with a dicarbonate such as di-sec-butyl peroxydicarbonate in vinyl chloride polymerization. These initiators have largely replaced benzoÿl and lauroÿl peroxides, the principal initiators in the early years of PVC production.

Acetylene (ethyne) HC≡CH. A colorless (but not odorless) gas obtained by reacting water with calcium carbide, CaC_2, or by cracking petroleum hydrocarbons. In the plastics industry, it is an important intermediate in the production of vinyl chloride, neoprene, acrylonitrile, and trichloroethylene. See also POLYACETYLENE.

Acetylene Black Extremely finely divided carbon produced by incomplete combustion of acetylene. It is used as a filler in plastics, imparting electrical conductivity. See also CARBON BLACK.

Acetylene Polymers See POLYACETYLENE.

***N*-Acetyl Ethanolamine** (hydroxyethyl acetamide) $CH_3CONHC_2H_4$-OH. A plasticizer for polyvinyl alcohol and cellulosic plastics.

Acetyl Peroxide (diacetyl peroxide) $(CH_3CO)_2O_2$. A polymerization catalyst.

4-Acetyl Resorcinol (2,4-dihydroxyacetophenone) $C_6H_3(OH)_2CO$-CH_3. A light stabilizer for plastics.

Acetyl Ricinoleates A generic term for a family of important plasticizers.

Acetyl Triallyl Citrate $CH_3COOC_3H_4(COOCH_2CH{=}CH_2)_3$. A cross-linking agent for polyesters, and a polymerizable monomer. Easily polymerized with peroxide catalysts, it forms a clear, hard thermosetting resin.

Acetyl Tributyl Citrate $CH_3COOC_3H_4(COOC_4H_9)_3$. A plasticizer commonly used in vinyl plastics, derived from the esterification and acetylation of citric acid. It has been FDA-approved for food-contact use.

Acetyl Triethyl Citrate $CH_3COOC_3H_4(COOC_2H_5)_3$. A plasticizer produced by esterifying and acetylating citric acid, used in cellulose nitrate, cellulose acetate, and certain vinyls, e g, polyvinyl acetate. It has been FDA-approved for food-contact use.

Acetyl Tri-2-Ethylhexyl Citrate $CH_3COOC_3H_4(COOC_8H_{17})_3$. A plasticizer for vinyls, with limited compatibility for cellulose nitrate and ethyl cellulose.

Acetyl Value The number of milligrams of potassium hydroxide (KOH) necessary to neutralize the acetic acid liberated by hydrolysis of one gram of an acetylated compound.

Acid Acceptor A compound that acts as a stabilizer by chemically combining with acid that may be initially present in minute quantities in a plastic, or that may be formed during the decomposition of the resin. See also STABILIZER.

Acid Number See ACID VALUE.

Acidolysis A chemical reaction analogous to hydrolysis in which an acid plays a role similar to that of water. See also ESTER INTERCHANGE.

Acid Resistance The ability of a plastic to withstand attack by acids, specifically strong mineral acids. Most plastics have excellent acid resistance. Tests for resistance of plastics to some acids are included in ASTM 543.

Acid Value The measure of the free-acid content of a substance. It is expressed as the number of milligrams of potassium hydroxide (KOH) neutralized by the free acid present in one gram of the substance. This value, also called *acid number*, is sometimes used in connection with the end-group method of determining molecular weights of polyesters. It is also used in evaluating plasticizers, in which acid values should be as low as possible.

Acoustic-Emissions Testing A nondestructive test for determining material or structural integrity by detecting and recording location, amplitude, and frequency of sound emissions as test loads are applied.

Acrolein (propenal, acrylic or allyl aldehyde) $CH_2{=}CHCHO$. A liquid derived from the oxidation of allyl alcohol or propylene, used as an intermediate in the production of polyester resins and polyurethanes.

Acrylamide $CH_2{=}CHCONH_2$. A crystalline solid produced by hydrolysis of acrylonitrile; the monomer of POLYACRYLAMIDE, w s, and a useful comonomer.

2-Acrylamido-2-Methylpropanesulfonic Acid (AMPS) A solid aliphatic sulfonic-acid monomer produced by Lubrizol Corp. Its homopolymers are water-soluble and hydrolytically stable. It can be incorporated into other polymers by crosslinking.

Acrylate Resin See ACRYLIC RESIN.

Acrylic Acid (propenoic acid) $CH_2{=}CHCOOH$. A colorless, unsaturated acid that polymerizes readily. The homopolymer is used as a thickener and textile-sizing agent and, crosslinked, as a cation-exchange resin. Acrylic-acid esters are widely used as monomers for acrylic resins.

Acrylic Aldehyde See ACROLEIN.

Acrylic Ester (acryl ester) An ester of acrylic or methacrylic acid or of structural derivatives thereof. Polymers derived from these monomers range from soft, elastic, film-forming materials to hard plastics. They are readily polymerized as homopolymers or copolymers with many other monomers, contributing improved resistance to heat, light and weathering. Some members of the acrylic-ester family, e g, butylene dimethacrylate and trimethylolpropane trimethacrylate, function as reactive plasticizers in PVC and elastomers. They serve as plasticizers during processing, then polymerize while curing to impart hardness to the finished article. See also ACRYLIC RESIN.

Acrylic Fiber Generic name for any manufactured fiber in which the fiber-forming material is a long-chain synthetic polymer composed of at least 85% by weight of acrylonitrile units, $-CH_2CHCN-$ (Federal Trade Commission).

Acrylic Foam A cellular polymer used for lining drapes and made by mixing an emulsified acrylic resin with compressed air in the ratio of one part emulsion to four or five parts air, spreading the foam on a substrate, then drying in an oven. The emulsion may contain fillers and pigments to provide opacity, and a foaming aid such as ammonium stearate. When the coated fabric must have abrasion resistance for washing and cleaning, the acrylic foam can be crushed between rollers to partly collapse the cell structure.

Acrylic Resin A polymer of acrylic or methacrylic esters, sometimes modified with nonacrylic monomers such as the ABS group. The acrylates may be methyl, ethyl, butyl, or 2-ethylhexyl. Usual methacrylates are the methyl, ethyl, butyl, lauryl, and stearyl. The resins may be in the form of molding powders or casting syrups, and are noted for their exceptional clarity and optical properties. Acrylics are widely used in lighting fixtures because they are slow-burning or even, with additives, self-extinguishing, and do not produce harmful smoke or gases in the presence of flame.

Acrylic Rubber (AR) A synthetic rubber made at least partly from acrylonitrile, or from ethyl acrylate copolymerized with many of the monomers or block polymers of the synthetic-rubber family.

Acrylonitrile (propenenitrile, vinyl cyanide) A monomer with the structure $CH_2=CHCN$. It is most useful in copolymers. Its copolymer with butadiene is nitrile rubber, and several copolymers with styrene exist that are tougher than polystyrene. It is also used as a synthetic fiber and as a chemical intermediate.

Acrylonitrile-Butadiene Copolymer (NBR) Any of a family of copolymers ranging from about 18 to 50% acrylonitrile, and sometimes including small amounts of a third monomer. The family includes the German materials Perbunan and Buna-N, and the nitrile rubbers. The outstanding property of this nitrile-rubber family is excellent resistance to oils, fats, and hydrocarbons such as motor fuels, making them useful for motor gaskets, abrasion linings, conveyor belts, and hoses for oils and fuels.

Acrylonitrile-Butadiene-Styrene Resin See ABS RESIN and NITRILE BARRIER RESIN.

Acrylonitrile-Styrene Copolymer Any of a group of copolymers that have the transparency of polystyrene, but with improved resistance to solvents and stress cracking.

ACS Abbreviation for the American Chemical Society, headquartered at 1155 Sixteenth St, NW, Washington, DC 20036. The Society's Polymer Chemistry Division holds national meetings and publishes several journals.

Activation (1) Inducing radioactivity in a specimen by bombardment with neutrons or other types of radiation. (2) Rendering a thermoplastic surface more receptive to printing inks, paints and adhesives by chemical treatment, corona discharge or flame treatment. (3) The energetic elevation of a molecule to a state in which it becomes ready to react with another molecule. (4) The creation of "holes" within a liquid enabling the "jumping" of molecules or, in the case of polymers, "flow segments", and thus enabling flow or creep.

Activation Energy (E, E_A) The energy required to facilitate reaction between two molecules or, in the context of the Eyring theory of flow, the energy required to cause a molecule of liquid or chain segment of a polymer to "jump" from its present position to a nearby hole (i e, an empty volume of molecular or chain-segment size) in the liquid. Activation energies are usually expressed per mole of substance (SI: J/mol) and are evaluated by fitting reaction-rate or flow data at several temperatures to an equation of the Arrhenius form. See ARRHENIUS EQUATION.

Activator (1) An agent added to the accelerator in natural or synthetic resins to enhance the action of the accelerator in the vulcanizing process. (2) A chemical additive used to initiate the chemical reaction in a specific mixture.

Actuator A device that controls the movement or mechanical action of a machine indirectly rather than directly or by hand. Actuators can perform linear or rotary motions, and are usually driven by pneumatic or hydraulic cylinders, or solenoids.

Adapter (die adapter) In an extrusion setup, the portion of the die assembly that attaches the die to the extruder and provides, inside, a flow channel for the molten plastic between the extruder and the die.

Adapter Plate In injection molding, the plate holding the mold to the press frame or platen.

Adapter Ring An annular retaining part for extrusion and injection apparatus.

ADC See ALLYL DIGLYCOL CARBONATE.

Addition Polymerization A reaction in which unsaturated monomer molecules join together to form a polymer in which the molecular formula of the repeating unit is identical (except for the double bond) with that of the monomer. The molecular weight of the polymer so formed is thus the total of the molecular weights of all of the combined monomer units. Example: $n\ CH_2{=}CH_2 \rightarrow (-CH_2CH_2)_n$, with molecular weight = $n \times 28.03$.

Additive Any substance that is added to a resin, usually in a relatively small percentage, to alter properties. Examples are slip additives, pigments, stabilizers, and flame retardants.

Adduct (1) The cyclic product of an addition reaction between one unsaturated compound, such as a diene, and another. (2) A crystalline mixture, not a true compound, in which molecules of one of the components are contained within the crystal-lattice framework of the other component. Such complexes are stable at room temperature but the entrapped component can escape when the mixture is melted or dissolved.

Adherend (n) A body that is held to another by an adhesive.

Adherometer An instrument that measures the strength of an adhesive bond.

Adhesion The state in which two surfaces are held together by interfacial forces that may consist of valence forces or interlocking action or both. One method for testing the strength of adhesive bonds is ASTM D 952. Three others, in Section 15.06, are D 3163, D 3164, and D 3807.

Adhesion, Mechanical Adhesion between surfaces in which the adhesive holds the parts together by interlocking action. See ADHESION, SPECIFIC.

Adhesion Promoter A coating that is applied to a substrate before it is coated with a plastic, to improve the adhesion of the plastic to the substrate. Typical adhesion promoters are based on silanes and silicones with hydrolyzable groups on one end of their molecules that react with moisture to yield silanol groups, which in turn react with or adsorb to inorganic surfaces to enable strong bonds to be made. At the other ends of the molecules are reactive, but nonhydrolyzable groups that are compatible with resins or elastomers in adhesive formulations. Adhesion promoters are added to the adhesive as water or ethanol solutions.

Adhesion, Specific Adhesion between surfaces that are held together by valence forces of the same type as those that give rise to cohesion. See ADHESION, MECHANICAL.

Adhesive (n) A material capable of bonding one surface to another. Adhesives are used in the plastics industry to join a plastic article to another article of (a) the same plastic, (b) a different plastic, or (c) a nonplastic material. Adhesives based on plastics are also used in other industries to join nonplastic materials, for example plywood, glass, cloths, and metals. Adhesives used in all of these applications can be classified into five types. A *monomeric cement* contains a monomer of at least one of the polymers to be joined and is catalyzed so that a bond is produced by polymerization. A *solvent cement* is one that dissolves the plastics being joined, forming strong intermolecular bonds, then evaporates. *Bonded adhesives* are solvent solutions of resins, sometimes containing plasticizers, that dry at room temperature. *Elastomeric adhesives* contain natural or synthetic rubbers either dissolved in solvents or suspended in water or other liquid, and are cured at room or elevated temperatures. *Reactive adhesives* are those containing partly polymerized resins, e g, epoxies, polyesters, or phenolics, that cure with the aid of hardeners to form a bond. See also ELECTROMAGNETIC ADHESIVE and HOT-MELT ADHESIVE.

Adhesive Assembly The process of joining two or more plastic parts other than flat sheets (for which the term *laminating* is used) by means of an adhesive. A related term is SOLVENT CEMENTING, w s.

Adhesive Film A thin film of dry resin, usually a thermoset, used as an interleaf in the production of laminates such as plywood. Heat and pressure applied in the laminating process cause the film to bond the layers together.

Adiabatic Denoting a process or system in which no heat is added or removed, while work may be delivered to or by the system. The term has been used somewhat loosely to describe the manner of extrusion in which the enthalpy increase of the plastic multiplied by the throughput is about equal to the power delivered by the screw.

Adiabatic Extrusion See AUTOTHERMAL EXTRUSION.

Adiabatic Temperature Rise (1) In autothermal extrusion, the rise in the temperature of the plastic, over a section of the extruder, that would occur if no heat flowed into, or out of the polymer. (2) In resin casting, the exothermic rise in temperature, assumed to be equal in all parts of the casting, that would occur if the reacting resin were perfectly insulated from the surroundings.

Adipate Plasticizer For plasticizers derived from adipic acid, see BENZYLOCTYL–, BIS(2,2,4-TRIMETHYL-1,3-PENTANEDIOL) MONOISOBUTYRATE–, DIBUTOXYETHOXY ETHYL–, DIBUTOXYETHYL–, DIBUTYL–, DICAPRYL–, DIDECYL–, DIETHOXYETHYL–, DIETHYL–, DI(2-

ETHYLHEXYL)–, DI-*n*-HEXYL–, DIISOBUTYL–, DIISODECYL–, DIISOOCTYL–, DIMETHOXYETHYL–, DI(METHYLCYCLOHEXYL)–, DINONYL–, DITETRAHYDROFURFURYL–, *n*-OCTYL-*n*-DECYL–, and POLYPROPYLENE ADIPATE.

Adipic Acid (hexanedioic acid, 1,4-butanedicarboxylic acid) HOOC–$(CH_2)_4COOH$. A dicarboxylic acid used in the production of polyamides, alkyd resins, and urethane foams. Esters of adipic acid are used as plasticizers and lubricants. A list of some plasticizers derived from adipic acid is given under ADIPATE PLASTICIZER.

Adiponitrile Carbonate (ADNC) See 5,5'-TETRAMETHYLENE DI-(1,3,4-DIOXAZOL-2-ONE).

ADNC Abbreviation for ADIPONITRILE CARBONATE, w s.

Adsorption The molecular adhesion of gases, dissolved substances, or liquids to the surfaces of solids with which they are in contact; a concentration of a substance at a surface or interface of another substance.

Advanced Composite Composites made up of ADVANCED FIBER and/or ADVANCED RESIN (w s) and having one or more service properties well exceeding those of epoxy-glass and polyester-glass laminates. Their costs, both for material and processing, are also usually higher, in some cases much higher.

Advanced Fiber Any of a class of reinforcing fibers characterized by either very high strength and modulus or high operating temperature, beyond those of more familiar fibers such as glass, nylon, and polyester, that have been in use for decades. Examples are aramid, fibers of some metals, carbon, boron, silicon carbide and silicon nitride, and whiskers of metals and inorganics.

Advanced Resin Any of a new multiclass of thermoplastics of various chemical natures, distinguished from the established and more common ENGINEERING PLASTICs by one or more outstanding properties, such as higher strength or modulus, or serviceability at 200+°C. These materials also command premium prices and some require special processing. Examples are polyimides, polyetheretherketone, liquid-crystal polymers, polytetrafluoroethylene, and polybenzimidazole.

Aerogel A porous, foam-like network with very small "cells", whose substance may be silica or a polymer or a carbonized polymer. Densities range from half of solid values to as low as 3 mg/cm^3. Applications for these new and still costly materials, developed by SDI researchers, are being explored

Aerosol A suspension of liquid or solid particles in a gas, typically under pressure. In the packaging industry, the term means a self-contained sprayable product in which the propellant force is supplied by a compressed or liquefied gas, e g, isopentane.

Affine Deformation A deformation in which each element in the volume distorts in the same way as does the volume as a whole.

Affinity With respect to an adhesive, affinity is attraction or polar similarity between the adhesive and an ADHEREND.

After-Bake A technique used with phenolic and amino resins to increase the output of a molding press by ejecting moldings before they are fully cured, subsequently completing the cure by baking them. After-baking may also be done with fully cured parts to improve their electrical properties and heat resistance.

After-Flame In an ignition test, persistence of flame after removal of the ignition source.

After-Flame Time The duration of AFTER-FLAME, w s.

Afterglow Persistence of visible glowing in a material being ignition-tested after removal of the ignition source.

Ag Chemical symbol for the element silver.

Aging (ageing) The process of, or the results of, exposure of plastics to natural or artificial environmental conditions for a prolonged period of time. See also ARTIFICIAL AGING, ARTIFICIAL WEATHERING.

Aggregate In the reinforced-plastics industry, a mixture of a hard, fragmented material with an epoxy binder, used as a flooring or surfacing medium, or in epoxy tooling.

AI Abbreviation of *amide-imide* (polymer). See POLYAMIDE-IMIDE RESIN.

Air-Assist Vacuum Forming A modification of the process of SHEET THERMOFORMING, w s, in which partial preforming of the sheet is effected by air flow or air pressure before vacuum pull-down.

Air Brush A small artist's spraygun designed for manually spraying paint onto small areas and in relatively fine lines, for example in the decoration of toys and doll faces.

Air-Bubble Viscometer An instrument used to measure the viscosities of oils, varnishes and resin solutions by matching the rate of rise of an air bubble in the sample liquid with the rate of rise in one of a series of standard liquids.whose viscosities are known. The Gardner-Holt bubble viscometer is such an instrument.

Air Gap (1) In extrusion of film, sheet or a coating, the distance from the die opening to the nip formed by the pressure roll and the chill roll. (2) In the radio-frequency heating of plastics and corona treatment of films, the space between the electrode and the surface of the material.

Air-Knife Coating A coating technique especially suitable for thin coatings such as adhesives, wherein a high-pressure jet of air is forced through orifices in a knife to meter and control the thickness of the coating. See also SPREAD COATING.

Airless Blast Deflashing The process of removing flash from molded parts by bombarding them with tiny nonabrading pellets that break off flash by impact. See also BLAST FINISHING.

Airless Spraying A spraying process used for decorating plastics or other materials with coatings, in which the coating material is forced through a small orifice at such a high velocity that atomization occurs without need for carrier air.

Air Lock A surface depression on a molded part, caused by air trapped between the surfaces of melt and mold.

Air Ring (air-cooling ring) In the process of blowing tubular film, a circular manifold with one or more annular openings concentric with and just above the die lip, that blows a uniform stream of air on or along the plastic tube. The air may be refrigerated.

Air Shot (air purge) In injection molding, a shot made with the nozzle withdrawn from the mold, so that the expelled melt may fall freely and be caught for measurement of CUP TEMPERATURE or for weighing.

Airslip Forming (airslip vacuum forming) A variation of SNAP-BACK FORMING in which a male mold is enclosed in a box so that, as the mold advances toward the softened sheet, air is trapped between mold and sheet. The ballooning sheet is thus kept from touching the mold during most of the latter's advance. At the end of the stroke, vacuum is applied, destroying the air cushion, and the sheet is sucked against the plug. See also SHEET THERMOFORMING.

Air Vent A passageway, typically a fine groove or scratch, between a mold cavity and the outside edge of the mold face, that allows air to escape as melt is injected into the cavity. See BURN MARK.

Al Chemical symbol for the element aluminum.

Albedo The fraction of incident light or other electromagnetic radiation that is reflected by a surface.

Alcohol (1) A generic term for organic compounds having the general structure ROH. In aliphatic alcohols, R has the formula C_nH_{2n+1}, as in methyl alcohol, CH_3OH, or n-butanol, C_4H_9OH. In more complex alcohols R may be other alkyl, acyclic or alkaryl groups. Alcohols are classified according to the number of –OH groups they contain—monohydric, dihydric, trihydric, or polyhydric. Dihydric alcohols are called glycols; trihydric alcohols are also known as glycerols; and the term polyol is used for any polyhydric alcohol. Alcohols have many important applications in the plastics industry. They are used directly as solvents and diluents. Many esters of alcohols with organic acids are plasticizers. As intermediates, alcohols are used in the production of resins such as acrylics, alkyds, aminos, polyurethanes, and epoxies. (2) Specifically, ethanol, C_2H_5OH.

Alcohol, Denatured Ethyl alcohol that has been adulterated with a toxic material such as acetaldehyde or benzene so as to render it unfit for human consumption but still useful as an industrial solvent or, occasionally, as a reactant.

Alcoholysis By analogy to *hydrolysis*, any chemical reaction in which an alcohol acts in a way similar to that of water. See also ESTER INTERCHANGE.

Aldehyde A generic term for organic compounds containing a double-bonded oxygen and hydrogen bonded to the same terminal carbon atom of the molecule, i e, the –CHO group. Thus they may be represented by the general formula RCHO. The simplest one is formaldehyde, HCHO, in which R is hydrogen. For all others, R represents a hydrocarbon radical.

Alfin Catalyst A catalyst obtained from alkali alcoholates derived from a secondary alcohol, used for polymerizing olefins.

Alginate Any derivative of alginic acid; alginates are used as emulsifying agents, as thickeners, and in films.

Aliphatic Designating a large class of organic compounds (and their radicals) having open-chain structures and consisting of the paraffin, olefin, and acetylene hydrocarbons and their derivatives. Examples are butane, isopropyl alcohol, many fats and oils, adipic acid, amyl acetate, ethyl amines.

Alizarin 1,2-dihydroxyanthraquinone, the raw material in making many pigments.

Alkali Any of the hydroxides and carbonates of the alkali metals (lithium, sodium, potassium,) and the radical ammonium. The term is also used more generally for any strong base in aqueous solution capable of forming salts.

Alkali Cellulose See REGENERATED CELLULOSE.

Alkali Resistance The ability of a plastic material to withstand the action of an alkali. ASTM D 543 lists several alkalis as reagents for testing the chemical resistance of plastics.

Alkane The generic term for any saturated, aliphatic hydrocarbon, i e, a compound consisting of carbon and hydrogen only and containing no double or triple bonds. Linear alkanes are representable as C_nH_{2n+2} while cyclic alkanes have the general formula C_nH_{2n}. Examples are propane, C_3H_8, and cyclohexane, C_6H_{12}.

Alkane-Imide Resin A thermoplastic introduced by Raychem Corp under the trade name Polyimidal. The polymer retains high strength up to 200°C and melts at 302°C. It has good electrical properties, low water absorption, and high solvent resistance.

Alkene Same as OLEFIN, w s.

Alkyd Molding Compound A compound based on an ALKYD RESIN containing fillers, pigments, lubricants and other additives. Alkyd molding compounds are chemically similar to polyesters, but the term alkyd is usually applied to those polyester formulations that use lesser quantities of monomers of the high-viscosity or dry types, resulting in free-flowing granular and nodular types. The compounds are used for applications requiring good electrical properties and long-term dimensional stability such as automotive distributor caps, rotors, and coil caps. Alkyds can be compression molded at low pressure, they cure rapidly, and they present no venting problems because no volatiles are liberated during cure.

Alkyd Resin A polyester resin resulting from the condensation of a polyfunctional alcohol and acid, typically glycerine and phthalic anhydride. (See GLYPTAL.) Today the term is mostly used for (1) resins modified with drying oils and used as vehicles for varnishes and paints; and (2) for crosslinking resins in ALKYD MOLDING COMPOUNDS, w s. The word *alkyd* is an acronym, from *al–* for alcohol, and *–cid* (changed to *kyd*) for acid.

Alkyl A general term for a monovalent aliphatic hydrocarbon radical, which may be represented as having been derived from an alkane by dropping one hydrogen from the formula, C_nH_{2n+2}. Examples of alkyl groups are C_2H_5– (ethyl), and $(CH_3)_2CH_2CH$– (isobutyl).

Alkyl Aluminum Compound Any of a family of organo-aluminum compounds widely used as catalysts in the Ziegler-process polymerization of olefins. Members include trialkyl compounds such as triethyl–, tripropyl–, and triisobutyl aluminums; alkyl aluminum hydrides such as diisobutyl aluminum hydride and diethyl aluminum hydride; and alkyl aluminum halides such as diethyl aluminum chloride.

Alkylaryl Phosphate (octylphenyl phosphate) $(C_8H_{17}O)(C_6H_5O)PO$. A phosphate diester, a plasticizer for cellulose acetate butyrate, ethyl cellulose, polystyrene, and vinyl resins.

Alkylaryl Phthalate Any of a family of diesters of phthalic acid containing two alkoxy groups, one aliphatic and one aromatic, used as plasticizers for cellulosic plastics, polymethyl methacrylate, polystyrene, and vinyl resins.

Alkylthio Cadmium A stabilizer for PVC.

Alkyne (n) A hydrocarbon containing at least one pair of carbon atoms linked by a triple bond (–C≡C–). The simplest alkyne is acetylene, HC≡CH. (adj) Signifying the presence in a compound of the triple bond.

Allomerism A similarity of crystalline form with a difference in chemical composition. See POLYALLOMER.

Allotropy The existence of a substance in two or more solid, liquid or gaseous forms due to differences in the arrangement of atoms of molecules. Examples are amorphous, graphite, and diamond forms of carbon; NO_2 and N_2O_4.

Allowable Stress In engineering design, the maximum stress to which a structure or structural element may be subjected under the expected operating conditions. The allowable stress is normally less, by a sizeable FACTOR OF SAFETY, w s, than the stress of the same type that would cause the member to fail under the same conditions.

Alloy A blend of a polymer or copolymer with other polymers or elastomers. An important example is a blend of of styrene-acrylonitrile copolymer with butadiene-acrylonitrile rubber. The term *polyblend* is sometimes used for such mixtures. Some writers restrict the term alloy to mixtures of polymers that form a single phase, reserving the term *blend* for nonhomogeneous mixtures. The sales of plastics blends and alloys worldwide was 1.3 billion pounds (0.59 Tg) in 1987 and was predicted to have increased by more than 50% by 1992.

Allyl The unsaturated radical, $CH_2{=}CHCH_2–$, which upon liberation forms biallyl (1,5-hexadiene), a pungent, volatile liquid.

Allyl Alcohol (propenyl alcohol, AA, 2-propene-1-ol) $CH_2{=}CHCH_2$-OH. A colorless liquid with a characteristic pungent odor, obtained by the hydrolysis of allyl chloride (from propylene) with dilute caustic, or by the dehydration of propylene glycol. It is a basic material for all allyl resins, and its esters are used as plasticizers.

Allyl Aldehyde See ACROLEIN.

Allyl Chloride (3-chloropropene, α-chloropropylene, AC) $CH_2{=}CH$-CH_2Cl. Used in the preparation of allyl alcohol and various thermosetting resins.

Allyl Cyanide (3-butenoic acid nitrile, allyl carbylamine, vinylacetonitrile) $CH_2{=}CHCH_2CN$. Used as a crosslinking agent.

Allyl Diglycol Carbonate (ADC) A colorless, water-clear monomer that can be polymerized and cast into a variety of transparent, optical-grade products. It can be copolymerized with other unsaturated monomers such as vinyl acetate, maleic anhydride, and methyl methacrylate to produce polymers with a wide spectrum of properties.

Allyl Diglycol Carbonate Resin A thermosetting-resin group with outstanding optical clarity, good mechanical properties, and the highest scratch resistance of all transparent plastics. The resins are made by polymerizing the monomer of the same name with catalysts such as benzoÿl peroxide or, preferably, diisopropyl peroxy dicarbonate.

Allyl Esters Esters of allyl alcohol, used in the production of plasticizers and resins.

Allyl Resins Resins formed by the addition polymerization of compounds containing the group $CH_2{=}CHCH_2$, such as esters of allyl alcohol with dibasic acids. They are commercially available as monomers, as partly polymerized prepolymers, and as molding compounds. The dominant compound in the family is diallyl phthalate (DAP). Others are diallyl isophthalate (DAIP), diallyl maleate (DAM), and diallyl chlorendate (DAC). The monomers and partial polymers may be cured with peroxide catalysts to thermosetting resins that are stable at high temperatures and have good solvent and chemical resistance. The molding compounds may be reinforced with glass fibers or other reinforcements, and are easily molded by compression- and transfer-molding methods.

Alpha- A prefix, usually ignored in alphabetizing compound names, and usually abbreviated by the Greek letter α, signifying that the substitution is on the carbon atom immediately adjacent to the main functional group of the compound. An example is α-aminobutanol, CH_3-$CH_2CH(NH_2)CH_2OH$. Similarly, substituents on the second and third carbon atoms distant from the main functional group are designated β and γ, respectively.

Alpha Cellulose A colorless filler obtained by treating wood pulp with alkali, used in light-colored thermosetting resins such as urea formaldehyde and melamine formaldehyde. The material is sometimes treated with resinous agents to coat the individual particles and reduce water absorption of the finished articles.

Alpha Olefins Olefins having 5 to 20 carbon atoms. See OLEFIN.

Alpha Paper Paper made from purified wood cellulose, often beautifully preprinted when used as surfacing sheets of decorative laminates.

Alpha Particle (alpha ray) A charged particle, essentially a helium nucleus, emitted during the radioactive decay of certain elements. Alpha particles have little penetrating power, typically dissipating their energy in passing through a few centimeters of air, but they can do harm if released within the human body.

Alternating Copolymer A polymer in which two different mer units alternate along the chain in a regular pattern of –A–B–A–B–....

Alternating-Strain Amplitude Related through the complex modulus to the ALTERNATING STRESS AMPLITUDE, w s.

Alternating Stress A stress mode typical of fatigue tests in which the specimen is subjected to stress that varies sinusoidally between tension and compression, the two maximum stresses being equal in magnitude. In some tests, the stress cycles between zero and a tensile maximum, or other unsymmetrical limits. The term applies also to other modes of loading, such as bending and torsion.

Alternating-Stress Amplitude A test parameter of a dynamic fatigue test, others being frequency and environment. It is one half the algebraic difference between the highest and lowest stress in one full cycle.

Alumina (corundum) The oxide of aluminum, Al_2O_3, very refractory and next to diamond and boron nitride in hardness, obtained by the calcination of bauxite. Alumina powder is used as a fire-retardant filler in plastics and, over the past two decades, ALUMINA FIBERs, w s, have enjoyed increasing use as reinforcements for plastics, metals and even ceramics. Its density is 3.965 g/cm^3.

Alumina Fiber A class of reinforcing fibers available as whiskers or continuous filaments, with quite different properties. Whiskers are almost pure Al_2O_3 (corundum) and are grown by passing a stream of moist hydrogen over aluminum powder heated to 1300–1500°C. Their strength ranges from 4 to 24 GPa, modulus ranges from 400 to 1000 GPa, and they cost about $15/gram. Continuous filaments are lower in crystallinity and/or alumina content (densities range from 2.7 to 3.7 g/cm^3), tensile strengths range from 1.3 to 2.1 GPa, and moduli from 105 to 380 GPa, depending on the manufacturer.

Alumina Trihydrate (aluminum hydroxide, aluminum hydrate, hydrated aluminum oxide) $Al_2O_3 \cdot 3H_2O$ or $Al(OH)_3$. A white crystalline powder, alumina trihydrate accounts for about half of all flame retardants used in plastics. When heated above about 220°C, it releases water endothermically.

Aluminum Alkyl (aluminum trialkyl) See ALKYL ALUMINUM COMPOUND.

Aluminum Chelate Chemically modified aluminum secondary butoxide, used as a curing agent for epoxy, phenolic and alkyd resins.

Aluminum Distearate $Al(OH)[OOC(CH_2)_{16}CH_3]_2$. A white powder used as a lubricant for plastics.

Aluminum Dihydroxy Stearate $Al(OH)[OOC(CH_2)_{10}CHOH(CH_2)_5$-$CH_3]_2$. A white powder used as a plastics lubricant.

Aluminum Isopropylate (aluminum isopropoxide) $Al[OCH(CH_3)_2]_3$. A white solid, a crosslinking agent.

Aluminum Monostearate $Al(OH)_2[OOC(CH_2)_{16}CH_3]$. A white or yellowish-white powder, a stabilizer.

Aluminum Oleate $Al[OOC(CH_2)_7CH{=}CH(CH_2)_7CH_3]_3$. A plastics lubricant.

Aluminum Palmitate $Al_2(OH)_2[OOC(CH_2)_{14}CH_3]$. A plastics lubricant.

Aluminum Silicate Any of a large group of minerals with various proportions of Al_2O_3 and SiO_2, occurring naturally in clays. They are used as pigments and fillers in plastics.

Amber A natural fossil resin formed during the Oligocene age by exudation from a species of pine now extinct. Its empirical formula is $C_{10}H_{16}O$, it softens at about 150°C, can be fabricated and polished. It has been used in jewelry, cigarette holders, and pipe mouthpieces.

Ambient Completely surrounding; indicative of the surrounding environmental conditions such as temperature, pressure, atmosphere, etc. When no values are given, the temperature is presumed to be room temperature (18–23°C) and the atmosphere to be air at standard pressure (101.3 kPa).

Ambient Temperature (1) The temperature of the medium immersing an object. (2) The prevailing room temperature.

American National Standards Institute (ANSI) address: 1430 Broadway, New York, NY 10018. Clearinghouse and national coordinator for voluntary standards of engineering, equipment, industrial processing, and safety. Formerly known as American Standards Association (ASA). In the plastics field, ANSI works closely with the SOCIETY OF THE PLASTICS INDUSTRY (SPI) and the SOCIETY OF AUTOMOTIVE ENGINEERS (SAE) to develop and publish standards for plastics materials, processing equipment, operations, and operating safety.

Amide A compound containing the $-CONH_2$ group, formed by the reaction of an organic acid or an ester with ammonia. Except for formamide, all amides are crystalline solids at room temperature. Examples are acetamide, CH_3CONH_2, and urea, H_2NCONH_2.

Amide-Imide Resin See AMINO RESIN.

Amine A compound derived (in concept) from ammonia by substitution of one or more hydrogen atoms by a hydrocarbon radical. Amines in which one, two or all three of the ammonia hydrogens have been substituted are termed *primary, secondary,* and *tertiary amines.*

Amine Resin See AMINO RESIN.

Amino- A prefix signifying the presence in a compound or resin of an $-NH_2$ or $=NH$ group.

Amino Acid An organic acid containing an amino group attached to the carbon atom adjacent to the $-COOH$ group, obtained by the hydrolysis of a protein or by synthesis. Examples are glycine, $CH_2(NH_2)$-COOH, and cysteine, $HSCH_2CH(NH_2)COOH$.

***N-β*(Aminoethyl)-*γ*-Aminopropyltrimethoxy Silane** A silane coupling agent used in reinforced epoxy, phenolic, melamine and polypropylene resins.

Aminoplasts Thermosetting resins made by the polycondensation of formaldehyde with a nitrogen compound and a higher aliphatic alcohol. The two general types are *urea-formaldehyde* and *triazine-formaldehyde.* Melamine is the triazine most often used. See also AMINO RESIN.

γ-Aminopropyltriethoxy Silane $NH_2(CH_2)_3Si(OC_2H_5)_3$ A silane coupling agent used in reinforced epoxy, phenolic, melamine and many thermoplastic resins.

Amino Resin (polyalkene amide, aminoplast) A generic term for a group of nitrogen-rich polymers containing amino nitrogen or its derivatives. The starting amino-bearing material is usually reacted with formaldehyde to form a reactive monomer that is condensation-polymerized to a thermosetting resin. Included amino compounds are urea, melamine, copolymers of both with formaldehyde, and, of limited use, thiourea, aniline, dicyandiamide, toluenesulfonamide, benzoguanidine, ethylene urea, and acrylamide. Not included, because properties warrant separate classification, are polyamides of the nylon type, polyurethanes, polyacrylamide, and acrylamide copolymers. The most important members of the amino-resin family are melamine-formaldehyde and urea-formaldehyde resins. The basic resins are clear, water-white syrups or white powdered materials that can be dispersed in water to form colorless syrups. They cure at high temperatures with appropriate catalysts. Molding powders are made by adding fillers to the uncured syrups, forming a consistency suitable for compression and transfer molding.

AMMA Abbreviation for copolymers of acrylonitrile and methyl methacrylate. In Europe, written A/MMA.

Amorphous Devoid of crystallinity or stratification. Most plastics are amorphous at processing temperatures, many retaining this state under all normal conditions.

Ampere (A) The primary electrical unit of the SI system, upon which all other electrical units are based. The ampere itself is defined as that current which, if maintained in two long, parallel, fine wires located 1 meter apart in a vacuum, will produce between these conductors a force of 2×10^{-7} newton per meter of length. Practically, an ampere is the current that flows between two points connected with an electric resistance of one ohm when their potential difference is one volt.

Amphoteric Designating an element or a compound that can behave either as an acid or a base, i e, as an electron donor or an electron acceptor. Polymerization emulsifiers having both anionic and cationic groups are called *amphoteric emulsifiers.*

Amyl The radical $C_5H_{11}-$, also known as *pentyl.* The amyl radical occurs in six isomeric forms, and the term amyl usually refers to any mixture of the isomers.

Amyl Acetate (banana oil, pear oil, amylacetic ester) $CH_3COOC_5H_{11}$. A commercial solvent for several resins, including the cellulosics, vinyls, acrylics, polystyrene, and uncured alkyds and phenolics. It has a strong, fruity odor (hence its nicknames), and its main constituent is isoamyl acetate, but other isomers such as normal- and secondary-amyl acetates are present in amounts determined by the grade and origin.

Amyl Formate $HCOOC_5H_{11}$. A solvent for resins and cellulose derivatives.

Amyl Oleate Solvent and plasticizer for cellulosic and vinyl resins.

***p-tert*-Amyl Phenol** $(CH_3)_2C_2H_5CC_6H_4OH$. A white crystalline material made by alkylating phenol with amyl chlorides or amylenes, then separating by distillation. Resins made by reacting *p-tert*-amyl phenol with formaldehyde or paraformaldehyde are used in varnishes for wood, wire coating and coil insulation. They are also used as plasticizers and/or stabilizers in hot-melt adhesives based on ethyl cellulose.

Amyl Salicylate See ISOAMYL SALICYLATE.

Anaerobic Adhesive An adhesive that cures only in the absence of air after being confined between assembled parts. An example is dimethacrylate adhesive, used for bonding assembly parts, locking screws and bolts, retaining gears and other shaft-mounted parts, and sealing threads and flanges.

Analysis of Variance (ANOVA, AOV) A statistical method whose central principle is the partition of the total sum of squares of a set of measurements about their overall mean into additive components, and the parallel partition of the total degrees of freedom in the sample, so that, when the several sums of squares are divided by their assigned degrees of freedom, the resulting *mean squares* for effects (experimental factors) or sources of variation may be rigorously compared with an error mean square (error variance) to judge their statistical significance and relative importance. AOV is a powerful technique, particularly when qualitative factors, such as different machines, materials, or operators are involved, or when no credible chemical or physical models are known for the system under study.

Anatase (octahedrite) A crystalline ore of TITANIUM DIOXIDE, w s.

Anchorage Part of an insert that is molded inside of a plastic part and held fast by shrinkage of the plastic onto the insert's knurled surfaces.

Andrade Creep A type of creep behavior in which the compliance of the sample is proportional to the cube root of the time under stress. A variety of polymers exhibit this behavior.

Anelasticity The dependence of elastic strain on both stress and time, resulting in a lag of strain behind stress. In materials subjected to cyclic stress, the anelastic effect causes DAMPING, w s.

Angel's Hair Fibrous strands of material pulled away from thermoplastic films, particularly polypropylene, in heat-sealing and cutting operations that employ hot knives or wires. The angel's hair accumulates on the cutting mechanism, eventually affecting performance and requiring removal.

Angle Head (offset head, crosshead) An extruder head so designed that the principal direction of the extrudate makes an angle with the (extended) axis of the screw. See also CROSSHEAD.

Angle of Repose (angle of rest) The maximum angle that a conical pile of particles makes with the horizontal surface on which it rests. No ASTM test is listed for this important property of plastic powders and pellets. The smaller the angle of repose, the more easily does the material flow through hoppers and constrictions.

Angle-Ply Laminate A laminate in which equal numbers of plies are oriented at equal plus and minus angles from the plies in the length direction, making the laminate *orthotropic*. The most commonly chosen angles are ±60°, giving nearly equal strengths in all directions in the plane of the laminate. See also CROSS LAMINATE.

Angle Press A hydraulic molding press equipped with horizontal and vertical rams, used in the production of complex moldings containing deep undercuts or side cavities.

Angstrom Unit (Å) A now deprecated unit of optical wavelength, 10^{-8} cm. It has been replaced by the nanometer (nm). 1 Å = 0.1 nm.

Angular Welding See FRICTION WELDING.

Anhydride (1) A compound from which water has been extracted. (2) An oxide of a metal (basic anhydride) or of a nonmetal (acidic anhydride) that forms a base or an acid, respectively, when united with water. (3) An organic compound made (conceptually) by the union of two acid molecules with the elimination of a molecule of water. In practice, organic anhydrides are usually produced by other reactions.

Anhydrous Perfectly dry; containing no water.

Aniline (phenylamine, aminobenzene) $C_6H_5NH_2$. A colorless, oily liquid made by the reduction of nitrobenzene with iron chips and an acid catalyst. It is used in the production of aniline-formaldehyde resins and certain catalysts and antioxidants.

Aniline-Formaldehyde Resin An aminoplastic that is made by condensing formaldehyde and aniline in an acid solution. The resins are thermoplastic and are used in making molded and laminated insulating materials with high dielectric strength and good chemical resistance. See also AMINO RESIN.

Aniline Ink A fast-drying ink used for printing on cellophane, polyethylene, etc. Aniline inks were first made from solutions of coal-tar dyes in organic solvents, hence the name. Modern inks generally employ pigments rather than dyes.

Animal Black (animal char, animal charcoal, boneblack) A form of charcoal derived from animal bones, used as a pigment. See also CARBON BLACK.

Anion An atom, molecule or radical that has gained an electron to become positively charged.

Anion-Exchange Resin See ION-EXCHANGE RESIN.

Anionic Pertaining to a negatively charged atom, radical, or molecule, or to any compound or mixture having negatively charged groups.

Anionic Polymerization See IONIC POLYMERIZATION.

Anisotropic Said of materials whose properties, e g, strength, refractive index, thermal conductivity, are unequal in different directions. Oriented thermoplastics and unidirectionally fiber-reinforced resins are typically anisotropic.

Anisotropy The quality of being anisotropic; having directionally dependent properties.

Annealing The process of relieving stresses in molded plastics by heat-ing to a predetermined temperature, maintaining this temperature for a set period of time, and slowly cooling the articles. Sometimes the articles are placed in jigs to prevent distortion as internal stresses are relieved during annealing.

ANSI Acronym: AMERICAN NATIONAL STANDARDS INSTITUTE, w s.

Anthophyllite A type of ASBESTOS, w s, the major source of which is in Finland. Anthophyllite is a natural magnesium iron silicate, formerly used as a filler in polypropylene to provide heat stability.

Antiblocking Agent (antiblock) An additive that is incorporated into resins and compounds to prevent surfaces of products (mainly films) from sticking to each other or to other surfaces. The term is not generally used for coatings, dusts, or sprays applied to surfaces for the same purpose, or as SLIP AGENTs, w s, after products have been formed, Antiblocking agents usually are finely divided, solid, infusible materials, such as silica, but can be minerals or waxes. They function by forming minute, protruding asperities that maintain separating air spaces that interfere with adhesion.

Antifoaming Agent An additive that reduces the surface tension of a solution or emulsion, thus inhibiting or modifying the formation of bubbles and foam. Commonly used are insoluble oils, dimethyl polysiloxanes and other silicones, certain alcohols, stearates and glycols. In many polymerizations, these agents prevent foaming altogether. They are also used to delay foaming when producing cellular plastics.

Antifogging Agent An additive that prevents or reduces the condensation of fine droplets of water on a shiny surface. Such additives function as mild wetting agents that exude to the surface and lower the surface tension of water, thereby causing it to spread into a continuous film. Antifogging agents are much used in PVC wrapping films for meats and other moist foods. Examples are alkylphenol ethoxylates, complex polyol monoesters, polyoxyethylene esters of oleic acid, polyoxyethylene sorbitan esters of oleic acid, and sorbitan esters of fatty acids.

Antigelling Agent An additive that prevents a solution from forming a gel.

Antimicrobial Agent See BIOCIDE.

Antimony Trioxide (antimony white, flowers of antimony, antimony oxide) Sb_2O_3. A very fine white powder made by vaporizing antimony metal in an oxidizing atmosphere, then cooling and collecting the oxide dust. Available in several ranges of particle size, it is used as a flame retardant and pigment in plastics, usually in synergistic combination with an organo-halogen compound. PVC is the biggest consumer.

Antioxidant A substance incorporated in a material to inhibit oxidation at normal or elevated temperatures. Antioxidants are used mainly with natural and synthetic rubbers, petroleum-based resins, and other such polymers that oxidize readily due to structural unsaturation. However, some thermoplastics, namely polypropylene, ABS, rubber-modified polystyrene, acrylic and vinyl resins, also require protection by antioxidants for some uses. There are two main classes: (1) Those that inhibit oxidation by reacting with chain-propagating radicals, such as hindered phenols that intercept free radicals. These are called primary antioxidants or free-radical scavengers. (2) Those that decompose peroxide into nonradical and stable products; examples are phosphites and various sulfur compounds, e g, esters of thiodipropionic acid. These are referred to as secondary antioxidants, or peroxide decomposers.

Antiozonant A substance added to elastomers to retard or prevent deterioration caused by exposure to air containing ozone.

Antisag Agent See THICKENING AGENT.

Antistatic Agent (antistat) A chemical that imparts slight electrical conductivity to plastics compounds, thus preventing the accumulation of electrostatic charges on finished articles. The agent may be incorporated in the materials before molding or applied to their surfaces afterward. Antistats function either by being inherently conductive or by absorbing moisture from the air. Examples are long-chain aliphatic amines and amides, phosphate esters, quaternary ammonium salts, polyethylene glycols, polyethylene-glycol esters, and ethoxylated long-chain amines. See also STATIC ELIMINATOR, SOOT-CHAMBER TEST.

Antistat See ANTISTATIC AGENT.

Ant Oil See FURFURAL.

Apertured Nonwoven Fabric A NONWOVEN FABRIC (w s) having many small through-holes made by laying the fabric on a perforated plate or screen and applying fluid pressure.

Apparent Density The mass per unit volume of material including voids inherent in the material as tested, such as pellets, powders, foams, chopped film or fiber scrap. The term BULK DENSITY, w s, is synonymous for particulate materials. ASTM tests are D 1895 for pellets and powders, except PTFE powders, for which D 1457 applies; D 1622 for rigid foams. See also DENSITY and BULK FACTOR.

Apparent Viscosity At any point in a fluid undergoing laminar shear, the nominal shear stress divided by apparent shear rate. In simple fluids, viscosity is a state property, depending only on composition, temperature and pressure. In polymer melts and solutions, it is, nearly always, also dependent on the shear rate (or stress), hence the term *apparent viscosity*. The term is also applied to the quotient of the shear stress at the tube wall of a capillary viscometer ÷ the Newtonian shear rate at the wall, which reduces to $\pi \cdot R^4 \cdot \Delta P/(8 \cdot Q \cdot L)$, where R and L are the radius and length of the tube, ΔP is the pressure drop through the tube, and Q is the volumetric flow rate. More at VISCOSITY.

Aprotic Solvent An organic solvent that neither donates protons to nor accepts them from a substance dissolved in it. Benzene, C_6H_6, is such a solvent.

Aprotic Substance A substance that can act neither as an acid nor as a base.

Aqueous Acrylic See LATEX.

AR Abbreviation for ACRYLIC RUBBER, w s.

Aragonite See CALCIUM CARBONATE.

Aramid Acronym for *ar*omatic poly*amid*e, currently available in fiber form only, having at least 85% of the amide groups bonded to two aromatic rings. Du Pont's Kevlar® is poly(*p*-phenylene terephthalamide).

Aramid Fiber Any of a family of high-strength, high-modulus fibers made from aramid resin. Du Pont's Kevlar®-49 and -29 are the best known. K-49's ultimate strength is 3.4 GPa (500 kpsi), modulus is 131 GPa (19 Mpsi), ultimate elongation is 2.4%. K-29 has about equal strength, but half the modulus and twice the elongation. Density of either is 1.44 g/cm^3. Strength/density is higher for either of these fibers than for any others except some whiskers. They are available in the same forms as glass reinforcements, and produce composites with high moduli and fatigue resistance, low thermal expansion, and good electricals. K-29 is used in auto tires and bullet-stopping garments.

Arc Resistance The ability of a plastic material to maintain low conductivity along the path of exposure to a high-voltage electrical arc, usually stated in terms of the time required to render the material electrically conductive. Failure of the specimen may be caused by heating to incandescence, burning, tracking or carbonization of the surface.

Arc Tracking See TRACKING.

Arithmetic Mean (arithmetic average, mean, $\bar{x}$) (1) In statistics, the average of a set of measurements found by summing the measurements and dividing the sum by the number of measurements. (NUMBER-AVERAGE MOLECULAR WEIGHT is an arithmetic average.) (2) The conceptual mean, μ, of the population from which a set of measurements was drawn, rarely known exactly. The sample mean, $\bar{x}$, is the most efficient estimator of the population mean, μ.

Aromatic Compounds A class of organic compounds containing a resonant, unsaturated ring of carbon atoms. Included are benzene, naphthalene, anthracene, and their derivatives. The term *aromatic* stems from the fact that many of these compounds have an agreeable odor.

Aromatic Polyester A POLYESTER that has aromatic rings in its chain, e g, POLYETHYLENE TEREPHTHALATE. (See both.)

Aromatic Hydrocarbon A compound of carbon and hydrogen whose molecular structure contains one or more rings of six carbon atoms, with at least one of the rings containing alternating, resonant single and double bonds. Benzene, which is the simplest of the aromatic hydrocarbons, has the molecular formula C_6H_6. The family includes many solvents for plastics.

Aromatic Polyamide See ARAMID.

Arrhenius Equation (1) A classical equation describing how rates of chemical reactions increase with rising absolute temperature:

$$r = A\ e^{-E/(R \cdot T)}$$

in which r = reaction rate (in appropriate units), A = the collision factor (in the same units as r), e = 2.71828..., E = the activation energy of the reaction, J/mol, R = the universal molar-energy constant, 8.3144 J/(mol·K), and T = absolute temperature, K. (2) An almost identical form, differing only in that the sign of the exponent is positive (viscosities *decrease* with rising temperature), has been used with good results to represent the temperature dependence of liquid viscosities, including those of polymer solutions and melts. It, too, is often referred to as an Arrhenius equation. This form has also been successful in modeling the temperature dependence of creep failure and property retention during heat aging. For viscosity work, the logarithmic form of this equation is more convenient.

$$\ln (\mu/\mu_0) = E(T - T_0)/(R \cdot T \cdot T_0).$$

Here E, R and T have the same meanings as above; μ_0 represents the viscosity at a reference temperature, T_0, which is usually chosen to be within the temperature range over which the viscosities have been measured.

Artificial Aging The accelerated testing of plastics to determine their changes in properties such as dimensional stability, water resistance,

resistance to chemicals and solvents, light stability, and fatigue resistance.

Artificial Weathering The process of exposing plastics to continuous or repeated environmental conditions generated in the laboratory and designed to simulate conditions encountered in actual outdoor exposure. Such conditions include temperature, humidity, light in the ultraviolet range, and direct water spray. The laboratory conditions are usually intensified to a degree greater than those normally encountered outdoors in order to decrease the time required to achieve significant results. Pertinent ASTM tests are D 756, D 1435, D 1499, D 2565, and G 23.

ASA See AMERICAN NATIONAL STANDARDS INSTITUTE.

Asbestos The commercial term for a family of fibrous mineral silicates comprising some 30 known varieties, of which six were, for many years, commercially important. They are of two general types, *serpentine* and *amphibole*. The serpentine type contains CHRYSOTILE, w s, and was the most widely used as a reinforcement in thermosetting resins and laminates and, in finer form, as a filler in polyethylene, polypropylene, nylons, and vinyls. A great deal was used in flooring sheet and tiles. The amphiboles provided better chemical resistance and lower water absorption, but the outstanding properties imparted by all asbestos types were resistance to heat, fire retardance, and resistance to chemicals. Because the cancers of many long-time asbestos workers are now believed to have been caused, in part at least, by inhaled asbestos fibers, asbestos is now used much less in the US for reinforcing plastics, though brake linings for trucks still contain it. Extreme precautions must be taken in handling asbestos and asbestos-filled materials. The OSHA limit for such workers is 2 fibers per m^3 of air, averaged over an 8-h shift.

Ascaridole (1,4-peroxide-*p*-menthene-2) $C_{10}H_{16}O_2$. A naturally occurring peroxide with uses as a polymerization initiator.

Aseptic Packaging A package that has been sterilized with gas, heat or radiation after it was sealed.

Ash Content The solid residue remaining after a substance has been incinerated or heated to a temperature sufficient to drive off all volatile or combustible substances.

Ashing A finishing process used to produce a satin-like finish on plastic articles, or to remove cold spots or teardrops from irregular surfaces which cannot be reached by wet sanding. The part is applied to a loose muslin disk loaded with wet ground pumice, rotating at a lineal speed of about 20 m/s.

ASM International A technical society for 51,000 engineered-material executives and engineers, headquartered in Materials Park, OH 44073, publishers of periodicals, handbooks, etc, and sponsors of an annual Materials Week conference.

fiber. In flaky materials such as mica, the ratio of the equivalent diameter to the thickness.

Asphalt (bitumen) A dark brown or black, bituminous, viscous material found in some natural deposits and produced also as a residue of petroleum refining.

Assembly of Plastics Plastic parts may be joined to others by many methods. Self-tapping screws are made with special thread designs to suit specific resins. Threaded inserts to receive mounting screws may be molded in or installed by press-fitting or by means of self-tapping external threads. Press-fitting may be employed to join plastics to similar or dissimilar materials. Snap-fit joints are made by molding or machining an undercut in one part, and providing a lip to engage this undercut in the mating component. Other methods are BUTT FUSION, CEMENTING, HEAT SEALING, HOT-PLATE WELDING, LASER, STAKING, THERMOBAND WELDING, ULTRASONIC INSERTING, ULTRASONIC STAKING, and WELDING, w s.

A-Stage An early stage in the preparation of certain thermosetting resins in which the material is still fusible and soluble in certain liquids. Sometimes referred to as a *resol*. See also B-STAGE and C-STAGE.

ASTM Originally an abbreviation for American Society for Testing Materials, now its official name. ASTM, located at 1916 Race St, Philadelphia, PA 19103, may be the largest nongovernmental, standards-writing body in the world, with 33,000 members. ASTM Committee D-20 has, with changing membership through the years, been responsible for thousands of standards and test methods for plastics. Committee D-30 performs the same functions for high-modulus fibers and their composites. The 1990 set of ASTM Standards contains 15 Sections comprising 2 to 9 volumes each, plus Section 00, the master index. Most plastics and plastics products and test methods are covered in Section 8 (four volumes), but Sections 6, 7, 9, 10, 13, 14, and 15 also contain information on polymeric materials and products.

Asymmetric Of such a form that no point, line or plane exists about which opposite portions are congruent. The opposite of symmetrical.

Asymmetry A molecular arrangement in which a particular carbon atom is joined to four different groups.

Atactic Pertaining to a polymer in which the pendant side groups, as $-CH_3$ in polypropylene, are randomly located around the main chain.

Atactic Block A block of chain units in a polymer or copolymer that has a random distribution of equal numbers of the possible configurational base units (UPAC).

Atactic Polymer A polymer with molecules in which substituent groups or atoms are arranged at random around the backbone chain of atoms. The opposite of a *stereospecific polymer*.

ATE Abbreviation for aluminum triethyl (triethylaluminum), a polymerization catalyst for olefins.

Atomic Unit (a u, Dalton) A mass equal to 1/12 the mass of an atom of carbon-12. The unit is deprecated by the SI system.

Atomic Weight (atomic mass) The mass of an elemental isotope relative to that of the C-12 isotope of carbon, whose mass has been set at exactly 12.0000 atomic units (Daltons). For most elements, the tabulated atomic weight is the average, weighted by natural mass abundance, over all the element's isotopes, so is never an integer. The actual mass—fraction of a gram—of one a u is the reciprocal of Avogadro's number, $1/(6.02283 \times 10^{23})$.

Attenuation (1) The process for making slim and slender, for example, the formation of fibers from molten glass. (2) The gradual diminution of intensity or amplitude of a damped vibration with distance or time.

atto- (a) The SI prefix meaning $\times 10^{-18}$.

a u Abbreviation for ATOMIC UNIT, w s.

AU Abbreviation for POLYURETHANE ELASTOMERs, w s, with polyester segments.

Autoacceleration In some vinyl polymerizations, as the reaction approaches completion and the viscosity of the reaction medium rises, there is a rising rate of increase of molecular weight of the chains that have not yet been terminated. This rising increase is called autoacceleration, or the *Trommsdorff effect*, or *gel effect*.

Autoadhesion (tackiness) The ability of two contiguous surfaces of the same material, when pressed together, to form a strong bond that prevents their separation at the place of contact.

Autocatalytic Degradation A type of breakdown in which the initially generated products accelerate the rate at which later degradation proceeds.

Autoclave A strong pressure vessel with a quick-opening door and means for heating and applying pressure to its contents. Autoclaves are widely used for bonding and curing reinforced-plastic laminates such as polyesters, epoxies, and phenolics.

Autoclaveable Capable of being sterilized in steam at two to three times standard atmospheric pressure with no change in properties.

Autoclave Molding See BAG MOLDING.

Auto-Flex Die Trade name for a type of sheet-extrusion die with a flexible lip in which each lip-adjusting bolt, which can either push against the lip or pull it, is paired with a nearby cartridge heater. When a signal from the BETA-RAY GAUGE that is constantly traversing the width of the sheet indicates that the sheet is too thick or too thin at a given point, the heater voltage at the relevant bolt is raised or lowered, expanding or contracting the bolt and decreasing or increasing the lip opening at that point. See FLEXIBLE-LIP DIE.

Autogenous Extrusion Equivalent to autothermal extrusion. See EXTRUSION, AUTOTHERMAL.

Autoignition Temperature The temperature at which a combustible material will ignite and burn spontaneously under specified conditions. See FLASH POINT and FLAMMABILITY TESTS, ASTM D 1929.

Automatic Control In processing, control achieved by instruments that sense the state of the process and adjust process inputs such as feed rate, heater voltage, screw speed, or hydraulic pressure to bring the process to the desired, or target state. Opposed to manual control, in which operators read sensing instruments and turn knobs to make process changes.

Automatic Mold A mold for compression, transfer, or injection molding, that is equipped to perform all operations of the molding cycle, including ejection of the molded parts, in a completely automatic manner without human assistance.

Automatic Profile Control In film and sheet extrusion, a system for controlling the uniformity of thickness across the sheet. The main components are a traversing thickness sensor such as a BETA-RAY GAUGE, w s, a computer and program that uses the sensor's signals to direct a mechanism that rotates the die-lip-adjusting bolts.

Automatic Screen Changer A SCREEN CHANGER, w s, that operates when signaled by a control instrument that monitors the head pressure in the extruder.

Automatic Unscrewing Mold A mold for making threaded products—bottle caps are typical—that incorporates a mechanism for unscrewing the product from the mold core (or *vice versa*) as the mold opens, thereby releasing the product.

Autooxidation After polyolefins have been exposed to an oxidizing process such as CORONA-DISCHARGE TREATMENT to render them receptive to inks or adhesives, the oxidation may continue for a time after exposure to the oxidizing agent has been terminated. Such self-sustaining oxidation is called autooxidation.

Autothermal Extrusion (adiabatic or autogenous extrusion) In screw extruders, a steady state of operation in which the increase in enthalpy of the plastic from feed throat to die entry is equal to the net energy furnished by the drive to the screw.

Average Molecular Weight See NUMBER-AVERAGE..., VISCOSITY-AVERAGE..., and WEIGHT-AVERAGE MOLECULAR WEIGHT.

Average Outgoing Quality (AOQ) The average fraction defective in lots subjected to sampling inspection. Random samples of size *n* are routinely inspected. If the sample contains *c* or fewer defective items, the lot is accepted and the sample, *sans* its defectives, is returned to the lot. If more than *c* defective items are found, the entire lot is inspected and, presumably, all defective items are culled out. For a given incoming lot fraction defective, AOQ is the long-term average fraction defective in outgoing (inspected) lots. More elaborate plans may involve double or multiple sampling, but AOQ has the same meaning for all. Also called *average quality level* (AQL).

Average Outgoing Quality Limit (AOQL) See AVERAGE OUTGOING QUALITY. In a sampling-inspection plan, AOQL is the maximum value that the AOQ can reach over the whole range of incoming fraction defective.

Axial Winding A method of FILAMENT WINDING, w s, in which the filaments are wound in a direction parallel to the axis of rotation (0° helix angle).

Azelaic Acid (nonanedioic acid, 1,7-heptanedicarboxylic acid) $HOOC(CH_2)_7COOH$. A yellowish to white crystalline powder, derived from a fatty acid such as oleic acid by oxidation with ozone. It is an intermediate used in the production of plasticizers, polyamides and alkyd resins. For plasticizers derived from azelaic acid, see DICYCLOHEXYL–, DI(2-ETHYLBUTYL)–, DI(2-ETHYLHEXYL)–, DI(2-ETHYLHEXYL)-4-THIO–, DI-*n*-HEXYL–, DIISOBUTYL–, and DIISOOCTYL AZELATEs.

Azeotropic Copolymer A copolymer in which the relative numbers of the different mer units are the same as in the mixture of monomers from which the copolymer was obtained.

2(1-Aziridinyl)ethyl Methacrylate A vinyl monomer that combines a reactive vinyl group with an aziridinyl functional group. It can be polymerized alone or with other vinyl monomers to yield polymers with pendant aziridinyl groups that promote adhesion of coatings to the polymer. As little as 0.5–2.0% of this monomer is effective for achieving good coating adhesion.

Azobisformamide (ABFA, azodicarbonamide) $H_2NCON{=}NCONH_2$. An aliphatic azo compound widely used as a chemical blowing agent in PVC, polystyrene, polyolefins, many other plastics, and in natural and synthetic rubbers. It is nontoxic, odorless, nonstaining, and, unlike other organic blowing agents, it is self-extinguishing and does not support combustion. Since ABFA in the pure state decomposes at temperatures above 216°C, when used with heat-sensitive plastics such as PVC, an activator that lowers its decomposition temperature is added to the compound. Such activators are compounds of cadmium, zinc, and lead, which also act as heat stabilizers, either directly or synergistically with other stabilizers.

Azobis(isobutyronitrile) A blowing agent developed in Germany for use in rubber and PVC. It is nonstaining and yields white PVC foam of fine uniform cell structure. However, its decomposition product, tetramethyl succinonitrile, *is* toxic and must be eliminated from the expanded product. For this reason the material is not used commercially in the US as a blowing agent. It *is* used as a polymerization initiator.

Azodicarbonamide See AZOBISFORMAMIDE.

Azo Dye Any of an important family of dyes containing the –N=N– group, produced from amino compounds by the processes of diazotization and coupling. By varying the composition it is possible to produce acidic, basic, triazo, and tetrazo types, depending on the number of –N=N– groups in the molecule.

Azo Group The structural grouping, –N=N–.

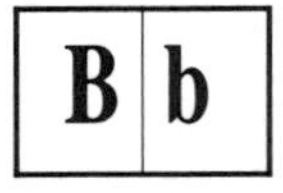

b SI abbreviation for BARN, w s.

B (1) Chemical symbol for the element boron. (2) Symbol for magnetic induction.

Ba Chemical symbol for the element barium.

Back Draft (back taper, counterdraft) A slight undercut or tapered area in a mold tending to prevent removal of the molded part. See also UNDERCUT.

Backing Plate In injection molding, a plate used as a support for the cavity blocks, guide pins, bushing, etc. Sometimes called *support plate*.

Back Pressure In extrusion, the head pressure. In screw-injection molding, the head pressure just before the valve opens to make the shot.

Back-Pressure Relief Port A side channel in the head of an extruder, usually leading to a RUPTURE DISK, w s, through which the melt can escape if the pressure exceeds a safe limit.

Back Taper See BACK DRAFT.

Bacteriocide An agent capable of destroying bacteria. See also BIOCIDE.

Bacteriostat An agent that, when incorporated in a plastics compound, will prevent the growth of bacteria on surfaces of articles made from the compound.

Baffle A plug or other device inserted in a flow channel to restrict the flow or change its direction.

Bagasse (megass) A tough fiber derived from sugar cane, remaining after the sugar juice has been extracted. It is used as a reinforcement in some laminates and molding powders.

Bag Molding A method of forming and curing reinforced-plastic laminates employing a flexible bag or mattress to apply pressure uniformly over one surface of the laminate. A preform comprising a fibrous sheet impregnated with an A- or B-stage resin is placed over or in a rigid mold forming one surface of the article. The bag is applied to the upper surface, then pressure is applied by vacuum, in an autoclave, in a press, or by inflating the bag. Heat may be applied by steam in the autoclave, or through the rigid half of the mold. When an autoclave is used, the process is sometimes called *autoclave molding*.

Bakelite A tradename derived from the name of Leo H. Baekeland, a pioneering Belgian chemist who developed phenolic resins in the early 1900s. The tradename was long used by the Bakelite Corporation, later absorbed by Union Carbide, who still uses the name for some of its resins.

Balanced Construction In plywood, a laminate with an odd number of plies, symmetrical around its center plane.

Balanced Design In reinforced plastics, a winding pattern so designed that the stresses in all filaments are equal.

Balanced Gating In multicavity injection molds, the ideal is to have all the cavities begin to fill simultaneously, to fill to the same final pressure, then to have their gates freeze at the same time. Where BALANCED RUNNERS (w s) are not practical, the imbalance of the runners can be offset by providing more resistance to flow in the gates nearest the sprue and less in those further from it. In molds where the cavities differ in volume, such as FAMILY MOLDS, balanced runners often will not be feasible. With such a mold the designer will attempt to balance the gates by having the flow rate through each gate be in proportion to the volume of its cavity. Modern computer programs can, by repeated simulation and design changes, find a combination of runner and gate dimensions that will let the mold approach the ideal performance.

Balanced Laminate A composite structure in which fiber layers laid at angles to the main axis occur in pairs, at equal ± angles, that may or may not be adjacent.

Balanced Runners In a multicavity injection mold, the runners are balanced when the injected melt reaches all the cavity gates at the same instant after the start of injection. In practice, with identical cavities whose shape, size, number, and layout permit, all runner branches are given equal cross sections and corresponding branch lengths are made equal. Uniform metal temperature throughout is assumed. Modern computer programs for flow of plastics in molds can equalize the total resistance to flow through runners (or cavity-fill time) with more flexible geometries. See also BALANCED GATING.

Balata, natural A material identical in properties and composition to GUTTA-PERCHA, w s, obtained from trees in South America.

Balata, synthetic A stereospecific rubber, the trans isomer of polyisoprene, made by catalyzed addition polymerization of isoprene.

Ball-and-Ring Test (ring-and-ball test) A method of determining the softening temperature of resins, ASTM E 38 (in Section 06.03). A specimen is cast or molded in a metal ring of 16-mm inside diameter and 6.4-mm depth. This ring is placed upon a metal plate in a liquid bath heating at a controlled rate, and a steel ball 9.5 mm in diameter weighing 3.5 g is placed in the center of the specimen. The softening point is considered to be the temperature of the liquid when the ball penetrates the specimen and touches the lower plate.

Ball Mill (pebble mill) A cylindrical or conical shell rotating horizontally about its axis, partly filled with a grinding medium such as natural flint pebbles, ceramic pellets, or hard metal balls. The material to be ground is added to just fill, or slightly more than fill, the voids between the balls. Water may or may not be added. The shell is rotated at a speed that causes the balls to cascade, thus reducing the particle sizes by repeated impacts. The operation may be batchwise or continuous. In the plastics industry, the term *ball mill* is reserved by some persons for mills containing metallic grinding media, and the term *pebble mill* for nonmetallic media. See also JAR MILL.

Ball-Rebound Test A method for measuring the resilience of polymeric materials by dropping a steel ball on a specimen from a fixed height and observing the height of rebound. The difference between the two heights is proportional to the energy absorbed. By conducting tests over a range of temperature, results can indicate temperature of first- and second-order transitions, and effects of additives and plasticizers. Section G of ASTM 3574 (in Section 09.02) describes such a test for flexible urethane foams.

Ball-up A term used in adhesive circles to describe the tendency of an adhesive to stick to itself.

Ball Viscometer See FALLING-BALL VISCOMETER.

Balsa Wood from the tree *Ochroma lagopus*, grown mainly in Ecuador. Its density is only 0.12–0.2 g/cm^3, yet it has good strength, especially end-grain compressive strength, so it has found application as an interlayer in reinforced-plastics SANDWICH STRUCTUREs, w s.

Balsam See OLEORESIN.

Banana Liquid A solution of nitrocellulose in amyl acetate or similar solvent.

Banana Oil See AMYL ACETATE.

Banbury Mixer An intensive mixer originally used for rubber, and for many years used for mixing plastics such as cellulosics, vinyls, polyethylene and others. It consists of two counterrotating, spiral-shaped blades encased in segments of cylindrical housing, the housing halves joined along internal ridges between the blades. Blades and housing may be cored for circulating heating or cooling liquids. A recent adaptation of the design, with connections to feed and discharge screws, permits continuous operation.

Band Heater See HEATER BAND.

B & S Gauge See WIRE GAUGE.

Bank In calendering and roll-milling, a cylindrical accumulation of working material in the nip of the rolls at the feed point.

Bar A deprecated unit of pressure, long used in meteorological work as approximately one atmosphere, and actually equal to 0.987 standard atmosphere, i e, 100 kPa.

Barcol Hardness The resistance of a material to penetration by a sharp steel point under a known load with an instrument called the Barcol Impressor. Direct readings are obtained on a scale from 0 to 100. The instrument has often been used as a way of judging the degree of cure of thermosetting resins. The ASTM test is D 2583, "Indentation Hardness of Rigid Plastics". See INDENTATION HARDNESS.

Barefoot Resin See NEAT RESIN.

Barite See BARIUM SULFATE.

Barium-Cadmium Stabilizer Any of a family of stabilizers based on salts of the title metals with organic acids, often in combination with a zinc salt of such acids, phosphites, and epoxides. These provide moderate to good heat stability at low cost, but cannot be used in compounds that will be in contact with foods or drinking water.

Barium Ferrite See FERRITE.

Barium Hydroxide Monohydrate (barium monohydrate) $Ba(OH)_2 \cdot H_2O$. A white powder used in the production of phenol-formaldehyde resins and barium soaps.

Barium Peroxide BaO_2 or $BaO_2 \cdot 8H_2O$. An oxidizing catalyst used in some polymerization reactions.

Barium Ricinoleate $Ba(OOCC_7H_4CH{=}CHCH_2CHOHC_5H_{10}CH_3)_2$. A heat stabilizer imparting good clarity, used most often in vinyl plastisols and organosols.

Barium Stearate $Ba(OOCC_{17}H_{35})_2$. A heat stabilizer, used particularly when sulfur staining is to be avoided. Also used as a lubricant where high temperatures are to be encountered.

Barium Sulfate (barytes, blanc fixe, heavy spar, permanent white, terra ponderosa) $BaSO_4$. A white powder obtained from the mineral barite or synthesized chemically. One of the synthetic varieties, *blanc fixe*, is made by mixing aqueous solutions containing sulfate and barium ions. As a filler in plastics and rubbers, barium sulfate imparts opacity to X rays but only a low order of optical opacity. Thus it is useful as a filler when it is desired to increase specific gravity without adversely affecting the tinctorial power of pigments.

Bar Mold A mold in which the cavities are arranged in rows on separate bars that may be individually removed to facilitate stripping.

Barn (b) A miniscule area unit commensurate with the cross sections of atomic nuclei. 1 barn = 10^{-28} m^2. Nuclear cross sections range from about 0.01 to 1.5 b.

Barrel (1) The tubular main cylinder of an extruder, within which the screw rotates. See EXTRUDER BARREL. (2) A container, agitated by rotation or vibration, used for tumbling moldings to remove flash and sharp edges. Also used for mixing of dry solids, e g, pigments with resin pellets.

Barrel Mixing See TUMBLING.

Barrier Layer In multilayer films, coextruded sheet, and blow-molded containers, a layer of polymer having very low permeability to the gases and/or vapors of interest for the application of the film, sheet, or container.

Barrier Plastics Thermoplastics with low permeability to gases and/or vapors. Most important commercially are NITRILE BARRIER RESINs, w s. Several others, however, are based on various copolymers, some of which are more permeable than nitrile resins but are easier to process. A major application is bottles for carbonated beverages.

Barrier Screw See SOLIDS-DRAINING SCREW.

Barrier Sheet An inner layer of a laminate, placed between the core and an outer layer.

Bar Stock Standard lengths of plastics extrusions having simple cross-sectional shapes such as circular (*rod stock*), square, hexagonal, and rectangular of low aspect ratio, used in fabricating plastics parts by machining, welding, fastening, and adhesive bonding.

Barytes A filler material made from the naturally occurring form of BARIUM SULFATE, w s.

Basebox In the metal-coating trade, a unit of area equal to 0.4861 m^2.

Basic Lead Carbonate $2PbCO_3 \cdot Pb(OH)_2$. A very effective heat stabilizer, used where toxicity is of no concern as in electrical-insulating compounds. Its use is limited because of its tendency to form blisters during processing and to cause spew when exposed to weather, also by rising concern about lead in the environment.

Bast Fiber Any of a group of fibers taken from the inner barks of plants that run the length of the stem, are surrounded by enveloping tissue, and are cemented together by pectic gums. Included in the group are jute, flax, sunn, hemp, and ramie, some of which are used to reinforce plastics.

Batting Layers of loose fibers, less compacted and less dense than FELT, w s.

Baumé (Bé) A dual transformation of specific gravity (S) for liquids devised by the French chemist Antoine Baumé for the graduation of hydrometers and, like Twaddell and API, deservedly becoming obsolete. Letting S equal the ratio of the density of the subject liquid at 15.6°C to that of water at the same temperature, the Baumé transformations are:

for liquids more dense than water, $°Bé = 145\,[1 - (1/S)]$

and for liquids less dense than water, $°Bé = (140/S) - 130$.

BBP See BUTYL BENZYL PHTHALATE.

Be Chemical symbol for the element beryllium.

Beader A device for rolling beads on the edges of thermoplastic sheets or cylinders.

Bead Polymer A polymer in the form of nearly spherical particles about 1 mm in diameter.

Bead Polymerization A type of polymerization identical to SUSPENSION POLYMERIZATION, w s, except that the monomer is dispersed as relatively large droplets in water or other suitable inert diluent by vigorous agitation.

Bearing Strength The ability of plastics sheets to sustain edgewise loads that are applied by pins, rods, or rivets used to assemble the sheets to other articles.

Becquerel (Bq) The SI unit for rate of disintegration of a radioactive element, equal to 1 disintegration per second.

Bending Moment The resultant moment about the neutral axis of a beam or column, at any point along its span, of the system of forces that produce bending.

Bending Strength See FLEXURAL STRENGTH.

Bentonite A type of clay, used as a filler, resulting from the weathering of volcanic ash and consisting essentially of montmorillonite, and anhydrous silicate of alumina. The name is derived from Ft. Benton, Wyoming, where it was discovered. The material has the unique quality of absorbing much water per unit mass.

Benzaldehyde (benzoic aldehyde, oil of bitter almonds, benzoÿl hydride, benzene carbonal) C_6H_5CHO. A solvent, particularly for polyester and cellulosic plastics.

Benzene (benzol, phene) C_6H_6. The fundamental compound and building block of all aromatic organic chemistry. It took almost a century of investigation after its discovery by Faraday in 1823 to establish the structure of this extraordinarily stable ring: its system of resonant, alternating single and double bonds. Benzene is a solvent and intermediate in the production of phenolics, epoxies, STYRENE, w s, and nylon. Hydrogenation of benzene yields cyclohexane, a solvent and raw material for preparing adipic acid, from which nylon is derived. As a solvent, benzene will dissolve ethyl cellulose, polyvinyl acetate, polymethyl methacrylate, polystyrene, coumarone-indene resins, and certain alkyds. Benzene is toxic and has been declared to be a carcinogen, so it requires very careful handling.

Benzene Ring (phenyl ring) The six carbon atoms, diagramed as a hexagon, joined by alternating single and double bonds, each carbon with an attached hydrogen in the case of benzene itself, or with one or more hydrogens replaced by other atoms or radicals. The alternating double bonds, in either of two possible arrangements, may or may not be shown, depending on the expected audience's knowledge of organic chemistry. The Greek letter, φ, is also used as a symbol for *phenyl–*.

Benzenesulfonylbutylamide $C_6H_5SO_2NHC_4H_9$. A plasticizer for cellulosics and polyvinyl acetate.

Benzenesulfonylhydrazide [4,4'-oxybis(benzenesulfonylhydrazide), OBSH] A blowing agent, a white crystalline solid that melts and begins to decompose near 105°C. It produces a white unicellular foam when incorporated in PVC plastisol but has a strong residual odor that does not evolve when it is used in rubbers. It is also used in epoxy and phenolic foams and serves as a cross-linking agent in rubber compositions and rubber/resin blends.

Benzidine Orange, Benzidine Yellow Metal-free diazo pigments based on dichlorobenzidine. They are highly transparent, bright in color, and low in cost due to their high tinctorial strength, but tend to bleed and fade upon exposure to light.

Benzine See LIGROIN.

Benzofuran See COUMARONE.

Benzofuran Resin See COUMARONE-INDENE RESIN.

Benzoguanamine (2,4-diamino-6-phenyl-1,3,5-triazine) $C_6H_5N_3(NH_2)_2$. A crystalline compound that reacts with formaldehyde to give thermosetting resins with resistance to heat and alkalies, and gloss generally superior to those of melamine-formaldehyde resins. Benzoguanamine resins are used for protective coatings, paper additives and finishes, laminating agents, textile finishes, and adhesives.

Benzoic Acid (carboxybenzene, benzene carboxylic acid, phenylformic acid) C_6H_5COOH. A white, crystalline compound occurring naturally in benzoin gum and some berries, also synthesized from phthalic acid or toluene. It is used in making plasticizers such as 2-ethylhexyl-*p*-oxybenzoate, diethyleneglycol dibenzoate, dipropyleneglycol dibenzoate, ethyleneglycol dibenzoate, triethyleneglycol dibenzoate, polyethyleneglycol-(200)- and -(600)-dibenzoate, and benzophenone.

Benzoic Ether See ETHYL BENZOATE.

Benzophenone (1) (diphenylketone) $(C_6H_5)_2C{=}O$. An involatile solvent and chemical intermediate. (2) Any of a family of UV stabilizers based on substituted 2-hydroxybenzophenone ("B"). Typical members are 4-methoxy-B, 4-octyloxy-B, 4-dodecyloxy-B, 2,2'-dihydroxy-4-methoxybenzophenone, and 2,2'-dihydroxy-4,4'-dimethoxybenzophenone. They function both as direct UV absorbers and, in the case of polyolefins, also as energy-transfer agents and radical scavengers.

***p*-Benzoquinone** (1,4-benzoquinone, chinone) $O{=}C_6H_4{=}O$. A yellow crystalline compound used, along with many of its derivatives, as an inhibitor in unsaturated polyester resins to prevent premature gelation during storage.

Benzotriazole (1) $C_6H_5N_3$. A double-ring compound, parent to many derivatives. (2) Any of a family of UV stabilizers, derivatives of 2-(2'-hydroxyphenyl)benzotriazole, that function primarily as UV absorbers. Typical examples are 2-(2'-hydroxy-5'-methylphenyl)benzotriazole and the corresponding 5'-*t*-octylphenyl analog. The benzotriazoles offer intense and broad UV absorption with a fairly sharp wavelength cutoff close to the visible region. The higher alkyl derivatives are less volatile and therefore more suitable for processing at higher temperatures.

Benzoÿl Peroxide (dibenzoÿl peroxide, DBP) $(C_6H_5CO)_2O_2$. A catalyst employed in the polymerization of styrene, vinyl and acrylic resins. It is also a curing agent for polyester and silicone resins, usually used together with an accelerator such as dimethylaniline. It can be dispersed in diluents or plasticizers to diminish the explosion hazard associated with the dry product.

Benzyl (α-tolyl) The radical $C_6H_5CH_2$–, which exists only in combination.

Benzyl Acetate (phenylmethyl acetate) $C_6H_5CH_2OOCCH_3$. A colorless liquid with a pleasant aroma, a solvent for cellulosic resins.

Benzyl Alcohol (α-hydroxytoluene, phenylcarbinol) $C_6H_5CH_2OH$. A water-white solvent for cellulosics and some other resins.

Benzyl Benzoate $C_6H_5CH_2OOCC_6H_5$. A water-white liquid that freezes at room temperature, a plasticizer.

Benzyl Butyrate $C_6H_5COOC_3H_7$. A liquid with a heavy, fruity odor, a plasticizer.

Benzyl Cellulose A benzyl ether of cellulose, it is a cellulosic plastic used in lacquers. It also may be formulated for making films and compounds for molding and extrusion.

Benzyl Formate $C_6H_5CH_2OOCH$. A solvent for cellulosic resins.

Benzyloctyl Adipate (BOA) $C_6H_5CH_2OOC(CH_2)_4COOC_8H_{17}$. A plasticizer for polystyrene, vinyl and cellulosic resins.

Benzyltrimethylammonium Chloride $C_6H_5CH_2N(CH_3)_3 \cdot Cl$. A quaternary ammonium salt, a solvent for cellulosics and a catalyst for phenolic resins.

Berlin Blue A term used for any of the variety of iron-based blue pigments; Prussian blue.

Berlin Red A pigment consisting essentially of red iron oxide.

Beryllium Copper Copper containing about 2.7% beryllium and 0.5% cobalt, used for blow molds and insertable injection-mold cavities. The small percentages of Be and Co greatly increase the strength and hardness of the copper while preserving its high thermal conductivity and corrosion resistance. Beryllium copper is easily pressure cast and hobbed into mold cavities.

Beta- A prefix, usually abbreviated as the Greek letter β and usually ignored in alphabetizing compound names, signifying that the so-labeled substitution is on the second carbon atom away from the main functional group of the molecule. See ALPHA-.

Beta Particle A subatomic particle created at the instant of emission from a decaying radioactive atomic nucleus, having a mass (at rest) of 9.1095×10^{-28} g. A negatively charged beta particle is identical to an ordinary electron, and a positively charged one is identical to a positron. A stream of beta particles is called a *beta ray*. Such rays are used in equipment for measuring and controlling the thickness of plastics films, sheets, and other extrudates.

Beta-Ray Gauge (beta gauge) A device for measuring the thickness of plastics films, sheets, or extruded shapes, consisting of a source of beta rays and a detecting element. When material is passed between the source and the detector, some of the rays are absorbed, the percent absorbed being a measure of the thickness of the material. Signals from the detecting element can be used to control equipment that automatically regulates the thickness. The most usual sources are krypton 85 and strontium 90. Also used for particular applications are cesium 137, promethium 147, and ruthenium 106. See also THICKNESS GAUGING.

Betatron An accelerator that uses an electrostatic field to impart high velocities to electrons. Energies of 5 to 6 MeV will produce X rays equivalent in energy to gamma radiation of 12 to 20 g of radium.

BHT Abbreviation for butylated hydroxytoluene. See DI-*tert*-BUTYL-*p*-CRESOL.

Bias Ply A layer of reinforcing fiber, cloth, or sheet oriented at an angle, less than 90° and typically 45°, to the fiber direction in the main reinforcing layers.

Biaxial Laminate See BIDIRECTIONAL LAMINATE.

Biaxial Orientation (1) The process of stretching hot plastic film or other article in two perpendicular directions, resulting in molecular alignment. See also ORIENTATION and TENTERING. (2) The state of the material that has been subjected to such stretching.

Biaxial Winding A method of FILAMENT WINDING in which the helical bands are laid in sequence, side-by-side, with no crossover of fibers.

Bicarbureted Hydrogen See ETHYLENE.

Bidirectional Laminate A fiber-reinforced material in which the fibers are laid in two different directions, typically in the length and width directions. In particular such a laminate in which equal volumes of reinforcing fibers are laid in the two directions.

Bierbaum Scratch Hardness See SCRATCH HARDNESS.

Bikerman Boundary-Layer Theory The theory that adhesive bonding occurs through the formation of an adsorbed, strong boundary layer on the surface of the adherend.

Billow Forming A variant of THERMOFORMING, w s, in which the hot plastic sheet is clamped in a frame and expanded upwards with mild air pressure against a male plug or female die as the plug or die descends into the frame. The process is suitable for thin-walled containers with high draw ratios.

Bimetallic Cylinder In most modern extruders and injection machines, the barrel is lined, by centrifugal casting from the melt, with any of several white irons containing chromium and boron carbides, and having hardnesses near Rockwell C65. After finish-grinding and polishing, the liner, about 1 mm thick, provides excellent resistance to wear or corrosion or both, depending on the formulation. The best known trade name is XALOY®, w s.

Bin Activator A device that promotes the steady flow of granular or powdered plastics from storage bins or hoppers. Among the many types of equipment are vibrators or mallets acting upon the outside of the container, prodding devices or air jets acting directly on the material, inverted-cone baffles with vibrating means located at the bottom of the hopper, and other "live bottom" devices such as scrapers, rolls and chains.

Binder An adhesive material holding particles of dry material together. For example, resinous adhesives are used to bind foundry sands. The term is also sometimes used for the continuous phase—a thermoplastic or thermosetting resin—in reinforced plastics.

Bingham Plastic A model for flow behavior, sometimes realized in the flow of pastes, in which no flow occurs until the shear stress exceeds a critical level called the YIELD VALUE, above which shear rate is proportional to the excess of stress over the yield value. When a Bingham plastic flows through a circular tube, there is a critical radius r_C at which the shear stress, $\Delta P \cdot r_C/(2 \cdot L)$, equals the yield stress. All actual flow occurs between that radius and the wall radius, R, while from the center to r_C there is a solid plug carried along by the stream.

Biocide An agent incorporated in or applied to the surfaces of plastics to destroy bacteria, fungi, marine organisms, etc. Some plastics, e g, acetals, acrylics, epoxies, phenoxies, ABS, nylons, polycarbonate, polyesters, fluorocarbons and polystyrene, are normally resistant to attack by bacteria or fungi. Others, e g, alkyds, phenolics, low-density polyethylene, urethanes, and flexible vinyls can under some circumstances be affected by growth of these organisms on their surfaces. Even though the resins themselves might be resistant, additives such as plasticizers, stabilizers, fillers, and lubricants can serve as food for fungi and bacteria. Examples of biocides are organotins, brominated salicylanilides, mercaptans, quaternary ammonium compounds, and compounds of mercury, copper, and arsenic.

Biodegradation The gradual breakdown of plastics by living organisms such as bacteria, fungi, and yeasts. Most of the commonly used plastics are essentially not biodegradable, exhibiting limited susceptibility to assimilation by microorganisms. An exception is POLYCAPROLACTAM, w s. However, the growing emphasis on environmental aspects of discarded plastics has stimulated research in ways of attaining biodegradability after a predetermined time period. One method is to add a UV-light sensitizer that causes photodegradation after a period of exposure to light, followed by breakup after prolonged exposure to the elements, after which bacteria will finish the job. A third method is the deliberate incorporation of weak links in the polymer chain, temporarily protected by a degradable stabilizer. See also PHOTODEGRADATION.

Biopolymer A polymer produced by living organisms, such as cellulose, natural rubber, silk, rosin, and leather.

Biot Number A dimensionless group, $h\ t/k$, important in convective heating and cooling of sheets. h = the heat-transfer coefficient at the sheet's surface, t = the sheet thickness or half-thickness, and k = the thermal conductivity of the sheet material.

Biphenyl (diphenyl, phenyl benzene) $(C_6H_5)_2$. A stable, high-boiling (256°C) liquid long used as a heat-transfer medium.

Bipolymer A polymer derived from two species of monomer (IUPAC). The more commonly used synonym is *copolymer*.

Birefringence (double refraction) The difference between any two refractive indices in a single material. When the refractive indices measured along three mutually perpendicular axes are identical, the material is said to be optically isotropic. Orientation of a polymer by drawing may alter the refractive index in the direction of draw so that it is no longer equal to that in the perpendicular directions, in which event the material is said to display birefringence. Crystalline polymers, normally birefringent, may become optically isotropic at their melting points. Studies of birefringence provide useful information regarding the shapes of molecules, degrees of orientation and crystal habits.

Bis(4-*t*-Butylcyclohexyl)peroxy Dicarbonate A catalyst of the organic-peroxide family, used in reinforced plastics and vinyl polymerization. Unlike other percarbonates, it does not require refrigeration for storage or handling.

Biscuit See PREFORM.

Bis(ethoxyethoxyethyl) Phthalate $C_6H_4(COOC_2H_4OC_2H_4OC_2H_5)$. A good primary plasticizer for polyvinyl acetate, nitrocellulose, cellulose acetate, and many other polymers.

Bis(β-Hydroxyethyl)-γ-Aminopropyltriethoxy Silane A silane coupling agent used in reinforced epoxy resins, also in many reinforced thermoplastics such as PVC, polycarbonates, nylon, polypropylene, and polysulfones.

Bismaleimide Resin (BMI) Any of a family of heat- and amine-curing resins based on the structure diagramed below.

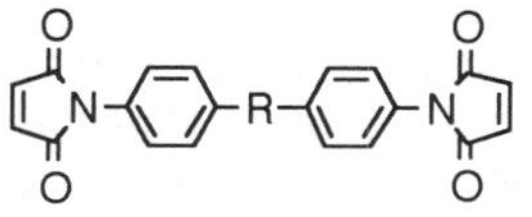

These resins are comparable to epoxies but have much higher temperature resistance (T_g = 400°C), and are finding application in aircraft composites. "R" may be CH_2, O, SO_2, or other groups. BMIs are marketed in the US by BASF under the tradename Narmco. See also POLYAMINOBISMALEIMIDE RESIN.

Bisphenol A (4,4'-isopropylidenediphenol) An intermediate used in the production of epoxy, polycarbonate, and phenolic resins. The name was coined after the condensation reaction by which it may be formed—two (bis) molecules of phenol with one of acetone (A). It has the structure:

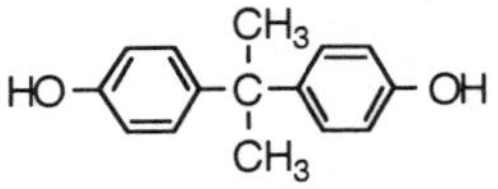

CH_3
HO–C–OH
CH_3

Bis(Tri-n-Butyltin) Oxide A liquid derived by the hydrolysis of tributyltin chloride, used to control the growth of most fungi, bacteria and marine organisms on plastics used in boat construction and in urethane foams.

Bis(Tri-n-Butyltin) Sulfosalicylate An antimicrobial agent used in flexible PVC film and urethanes.

Bis(2,2,4-Trimethyl-1,3-Pentanediol) Monoisobutyrate Adipate A plasticizer for cellulosic resins and polystyrene.

Bite (1) The ability of an adhesive to penetrate surfaces and thereby produce an adhesive bond. (2) Synonym for NIP, w s.

Bitter-Almond Oil, synthetic See BENZALDEHYDE.

Biuret (allophanamide, carbamylurea) $NH_2CONHCONH_2 \cdot H_2O$. A white crystalline material derived from urea by heat or by reaction with an isocyanate. It is used primarily in analytical chemistry, but the biuret group is formed during some polymerization reactions, such as primary bonding in urethane elastomers.

Bivinyl See BUTADIENE.

Blanc Fixe A synthetic form of barium sulfate prepared by reacting aqueous solutions containing barium ions with others containing sulfate ions, and precipitating the reaction product. It is used as a special-purpose filler to impart X ray opacity and high specific gravity.

Blank (1) A piece punched or die-cut from a sheet and intended for further forming into its final shape. (2) In chemical analysis, a dummy sample or solution of the same matrix as the sample to be analyzed but containing none of the analyte sought.

Blanking (die cutting) The cutting of flat sheet stock to shape by striking it sharply with a punch while it is supported on a mating die. Punch presses are often used for the operation. An alternate method is to make the cut with a thin, sharp-edged, shaped steel blade called a STEEL-RULE DIE. See also DIE CUTTING.

Blanking Die A metal die used in the BLANKING process.

Blast Finishing The removal of flash from molded objects (and/or dulling their surfaces) by impinging media such as steel balls, crushed apricot pits, walnut shells or plastic pellets upon them with sufficient force to fracture the flash. When the material being deflashed is not sufficiently brittle at room temperature, the articles are first chilled to a temperature below their brittleness temperature. Typical blast-finishing machines consist of wheels rotating at high speeds, fed at their centers with the media, which are thrown out at high velocities against the objects.

Bleed (1) (n) An escape passage at the parting line of a mold, similar to an AIR VENT but deeper, serving to allow material to escape or bleed out. (2) See BLEEDING.

Bleeding The diffusion of color from a plastic article into a surrounding surface or part, caused by inherent solubility of the pigment in one or more ingredients of the composition. The terms *migration, crocking, blooming,* and *bronzing* are sometimes used loosely to describe the same phenomenon.

Bleedout In filament winding, the excess liquid resin that migrates to the surface of a winding.

Blend See ALLOY.

Blending (1) Any process in which two or more components or ingredients are physically intermingled without significant change of the physical states of the components. (2) The bringing together of two or more polymers, using whatever means may be needed, such that the final SCALE OF SEGREGATION is microscopic or finer.

Blending Resin (extender resin) With respect to vinyl plastisols and organosols, a blending resin is one of larger particle size and lower cost than the dispersion resins normally used, a partial replacement for the primary resin. Blending resins are sometimes used to achieve a better balance of properties other than cost.

Blister (1) An imperfection on the surface of a plastic article caused by a pocket of air or gas beneath the surface. (2) A thermoformed canopy or pocket roughly hemispherical, for example an aircraft cockpit cover or a shape used in BLISTER PACKAGING, w s.

Blister Packaging The enclosing of articles in thermoformed, transparent "blisters" shaped to more or less fit the contours of the articles. The preformed blisters, usually slightly oversized to provide ample room, are made of thermoplastics such as vinyl, polystyrene, or cellulosic plastics. They are placed inverted in fixtures and loaded with the articles, then cards coated with an adhesive are applied and sealed to the flanges between and around the blisters by means of heat and pressure.

Block A portion of a polymer molecule comprising many mer units that has at least one constitutional or configurational feature not present in the adjacent portions.

Block Copolymer A copolymer with chains composed of shorter homopolymeric chains that are linked together. These "blocks" can alternate regularly or randomly. Such copolymers usually have some higher properties than either of the homopolymers or their physical blends.

Blocked Curing Agent A CURING AGENT, w s, that is temporarily rendered unreactive but that can, when it is needed, be reactivated by physical or chemical means.

Blocking An undesirable adhesion between layers of a plastic, particularly rolled-up film, that may develop during storage. Blocking can be prevented by agents added to the compound prior to extrusion, or applied to the surfaces of the extrudate. Such agents are called ANTI-BLOCKING AGENTS, w s.

Block Polymer A polymer whose molecules consist of blocks connected linearly. The blocks are connected directly or through a constitutional unit that is not part of the blocks, and the individual blocks are regular and of the same species. When blocks are of different monomer species, the term *block copolymer* is used.

Block Press (1) A press used to agglomerate laminate squares under heat. The squares, which have been cut from laminated sheet, are crossed to combat the anisotropy that normally occurs during laminating. (2) A press used to mold very large blocks of polystyrene foam.

Blood Red A pigment consisting essentially of red iron oxide.

Bloom An undesirable cloudy effect or whitish powdery deposit on the surface of a plastic article caused by the exudation of a compounding ingredient such as a lubricant, stabilizer, pigment, plasticizer, etc. The term is also used to identify a discoloration of a metal mold.

Blowing Agent (foaming agent) A substance that, alone or in combination with other substances, is capable of producing a cellular structure in a plastic or rubber mass. Thus, the term includes compressed gases that expand when pressure is released, soluble solids that leave pores when leached out, liquids that develop cells when they vaporize, and chemical agents that decompose or react under the influence of heat to form a gas. Liquid foaming agents include certain aliphatic and halogenated hydrocarbons, low-boiling alcohols, ethers, ketones and aromatic hydrocarbons. The chemical blowing agents range from simple salts such as ammonium or sodium bicarbonate to complex nitrogen-releasing agents, of which azobisformamide (ABFA) is an important example.

Blow Molding The process of forming hollow articles by expanding a hot plastic element against the internal surfaces of a mold. In its most common form, the process comprises extruding a tube (*parison*) downward between the opened halves of a metal mold, closing the mold to pinch off and seal the parison at top and bottom, injecting air through a needle inserted through the parison wall, cooling the mass by contact with a chilled mold, opening the mold and removing the formed article. Many variations of the process exist. In the earliest use of the process, two sheets of cellulose nitrate were used instead of a parison, a method still in use today. The parison is sometimes formed by injection molding, and sometimes extruded in advance, cut into lengths, and reheated when needed. Another variation, *cold-parison blow molding*, employs a preformed parison made by extruding a tube and forming one of its ends into a closed hemisphere so that the preform resembles a test tube. These preformed parisons are later reheated by infrared radiation, then blow molded in the usual manner. A more recent variation called *stretch blow molding* achieves biaxial orientation by stretching the parison longitudinally as well as radially within a narrow temperature range wherein the stretching produces molecular orientation. The advantages are better clarity, reduced creep, higher impact strength, reduced rates of gas and water-vapor transmission, and lower weight. The technique is limited to simple bottle shapes and is used mostly for larger bottles in the 0.5- to 2-liter sizes, and up to 5 gallons.

Blown Film See FILM BLOWING.

Blow-up Ratio (BUR) (1) In blow molding, the ratio of the largest diameter of a cavity into which a parison is to be blown to the outside diameter of the parison. (2) In blown-film extrusion, the ratio of the film diameter, before collapsing or gusseting, to the mean diameter of the die opening.

Blue Asbestos (crocidolite) An iron-rich form of ASBESTOS, w s, fibers of which were long used in reinforced plastics when good chemical resistance was essential.

Bluing A mold blemish in the form of a blue oxide film on the polished surface of a mold, caused by overheating.

Bluing Off A term used by mold makers for the process of checking the accuracy of two mating surfaces by applying a thin coating of Prussian blue to one surface, pressing the coated surface against the other surface, and observing the areas of intimate contact where the blue color has transferred.

Blushing The formation of a whitish discoloration on a freshly applied solution coating or lacquer that occurs when fast evaporation of a solvent cools the film below the ambient dew point, causing moisture to condense on the wet surface. It is encountered most often at times of high humidity, and can sometimes be avoided by using slower-drying solvents in the formulation. The term "blushing" is sometimes used for the tendency of a plastic article to turn white or chalky in areas that are highly stressed, but this use of the term is not approved by ASTM. See also GATE BLUSH.

BMC Abbreviation for *bulk molding compound.* See PREMIX.

BOA See BENZYL OCTYL ADIPATE.

Bob-and-Cup Viscometer See BROOKFIELD VISCOMETER.

Body (1) A term used loosely in the paint and adhesives industries to denote overall consistency, i e, a combination of viscosity, density, pastiness, tackiness, etc. (2) An aspect of fabric quality, akin to DRAPE and HAND, w s.

Body Putty A paste-like mixture of resin, often a polyester, and a filler such as talc, used to smooth and repair metal surfaces such as auto bodies.

Boil-in-Bag A type of food packaging—foil, plastic, or laminate—intended to be dropped into boiling water in order to cook the contents.

Bolster A spacer or filler in a mold.

Boltzmann Superposition Principle Strain is presumed to be a linear function of stress, so the total effect of applying several stresses is the sum of the effects of applying each one separately.

Bolus Alba See KAOLIN.

Bon-Arylamide Red Any of a group of metal-free monazo pigments based on substituted 2-hydroxy-3-naphthoic acid.

Bonded Adhesive See ADHESIVE.

Bonded Fabric A web of fibers held together by an adhesive medium that does not form a continuous film.

Bonding Resin Any resin used for bonding aggregates such as foundry sands, grinding wheels, abrasive papers, asbestos brake linings, and concrete masses. Plywood adhesives are sometimes called bonding resins.

Bond Strength (1) Of an adhesive bond between joined substrates, the force required to break the bond divided by the bond area (Pa). See TENSILE-SHEAR STRENGTH. (2) Of fiber-reinforced laminates, the strength of the bond between fiber and matrix. (3) The degree of attraction between adjacent atoms within a molecule, usually expressed in J/mol.

BON Pigment (1) Any of several brilliant reds and maroons widely used in plastics and rubbers, resistant to bleeding, migration, and crocking. The initials BON stand for ***β*-oxyn**aphthoic acid, the basic raw material. (2) A related azo pigment made by coupling *β*-hydroxynaphthoic acid to an amine and forming the barium, calcium, strontium, or manganese salt thereof. The colors range from yellowish red to deep maroon.

Booster Ram A hydraulic ram used as an auxiliary to main ram of a molding press.

BOP Abbreviation for butyl octyl phthalate. See BUTYL ETHYLHEXYL PHTHALATE.

BOPP Abbreviation for biaxially oriented polypropylene.

Boric-Acid Ester One of a family of flame retardants for plastics, etc, and plasticizers. Examples are the trimethyl, tri-n-butyl, tricyclohexyl, and tri-*p*-cresyl borates.

Bornyl Acetate $C_{10}H_{17}OOCCH_3$. A solvent and plasticizer for nitrocellulose.

Boron-Epoxy Composite A composite in which BORON FIBERS (w s) are embedded in an epoxy matrix. Modulus is about 200 GPa, tensile strength about 1.6 GPa, rising with fiber content. The combination of high specific modulus and high specific strength has made these composites attractive for aerospace vehicles in spite of their high cost.

Boron Fiber An advanced reinforcing fiber produced by passing 10-μm, resistively heated tungsten wire through an atmosphere of boron trichloride and hydrogen. Hydrogen reduces the BCl_3 and the boron deposits on the wire to make a filament from 120 to 140 μm in diameter. Density is low, 2.4 to 2.6 g/cm^3. Boron fiber is extremely strong and stiff with strength near 3.1 GPa (450 kpsi) and modulus near 400 GPa (58 Mpsi). The very high cost of these fibers (ca $1/g) has limited their use.

Boron Hydride See DIBORANE.

Boss A protuberance provided on an article to add strength, facilitate alignment during assembly, or for attaching the article to another part.

Boston Round A family of variously sized, blown bottles, either glass or plastic, of circular-cylindrical shape with a short curved shoulder and length-to-diameter ratio in the body of about 1.7:1.

Boyer-Beaman Rule A rule of thumb stating that the ratio of a polymer's glass-transition temperature to its melting temperature (T_g/T_m), both in kelvin, usually lies between 0.5 and 0.7. For symmetrical polymers, such as polyethylene, it is close to 0.5; for unsymmetrical ones, such as polystyrene and polyisoprene, it is near 0.7.

Boyle's Law The part of the ideal-gas law, due to Robert Boyle, stating that the volume of a gas is inversely proportional to its pressure.

BPF Abbreviation for British Plastics Federation.

Br Chemical symbol for the element bromine.

BR Abbreviation for butadiene rubber (British Standards Institution). See POLYBUTADIENE.

Brabender Plastograph (PlastiCorder®) An instrument that continuously measures the torque exerted in shearing a softened polymer or compound specimen over a wider range of shear rates and temperatures, including those conditions anticipated in actual processing. The instrument records torque, time and temperature on a graph called a *plastigram*, from which one can infer processability. It shows the effects of additives and fillers, measures and records lubricity, plasticity, scorch, cure, shear and heat stability, and polymer consistency. The modern instrument is equipped with interchangeable heads for bench-scale studies of mixing and extrusion, and with computer control and data acquisition.

Branched Polymer A polymer that contains either short or long side chains, or both, unlike a *linear* polymer.

Branching The growth of a new polymer chain from an active site on an established chain, in a direction different from that of the original chain. Branching occurs as a result of chain-transfer processes or from the polymerization of difunctional monomers, and is an important factor influencing polymer properties.

Breakdown Voltage See DIELECTRIC BREAKDOWN VOLTAGE and DIELECTRIC STRENGTH.

Breaker Plate In extrusion, a strong metal disk containing many closely spaced, smoothly faired holes and located immediately forward of the screw tip, supporting the screen pack (if one is used). The holes are usually circular, about 4 mm in diameter, but some designs have slot-shaped holes. Breaker plates have sometimes been used in injection molders in the hope of improving color uniformity in the moldings.

Breaking Extension See ELONGATION.

Breaking Strength See ULTIMATE STRENGTH.

Breathable Film A film that is at least somewhat permeable to gases due to the presence of open cells throughout its mass or to minute perforations.

Breathing (1) The passage of air through a plastic film due to a degree of porosity. (2) In injection and compression molding, the momentary opening and closing of a mold during the early part of the cycle to permit the escape of air from the cavities, or other gases or vapors from the compound.

Bridging (1) In the flow by gravity of powders or pellets in feed hoppers and the throats of extruders and injection molders, the stoppage of flow caused by the formation of an arch across the flow path. Bridging is related to particle shape and properties such as density and coefficient of friction, also to hopper angles and to the ratio of particle size to throat diameter. (2) A related but even more complicated phenomenon that sometimes occurs in the transition zone between the feed and metering zones of an extruder, and that can cause surging, or even interruption, of extrudate flow.

Brightening Agents (optical brighteners, fluorescent bleaches, optical whiteners) Chemicals used primarily in fibers, but also to some extent in molded and extruded products, to overcome yellow casts and to enhance clarity or brightness. In contrast to *bluing agents*, which act by removing yellow light, the optical brighteners absorb ultraviolet rays and convert their energy into visible blue-violet light. Thus, they cannot be used in compounds containing UV-absorbing agents. Optical brighteners are used in PVC sheet and film, fluorescent lighting fixtures, vinyl flooring, nylon fishing line, polyethylene bottles, etc. A few examples of optical brighteners are coumarins, naphthotriazolyl stilbenes, benzoxazolyl-, benzimidazolyl-, naphthylimide-, and diaminostilbene disulfonates.

Brinell Hardness The hardness of a material as determined by pressing a hardened steel or carbide ball, 10 mm in diameter, into the specimen under a fixed load. The Brinell number is the load in kilograms divided by the area in mm^2 of the spherical impression formed by the ball. For nonferrous materials, the prescribed load is 500 kg applied for 30 seconds. The Brinell test is used mostly for ductile metals, for which the Brinell number is related to yield strength in a simple way.

Brinkman Number A dimensionless group relevant for heat transfer in flowing viscous liquids such as polymer melts. It is defined by $N_{Br} = \mu \cdot V^2/(k \cdot \Delta T)$, in which μ = viscosity, V = velocity, k = thermal conductivity, and ΔT = the difference in temperature between the stream and the confining wall. The number represents the ratio of the rates of heat generation and heat conduction.

Bristle (1) A generic term for a short, stiff, coarse fiber. (2) Any grade of nylon monofilament used in toothbrushes, hairbrushes, paintbrushes (*tapered bristle*), etc.

British Thermal Unit (Btu) Before the introduction of the SI system, the Btu was variously defined as the quantity of heat required to raise the temperature of one pound (avoirdupois) of water 1°F, either at or near 39°F (the temperature of maximum density), at 59°F, or 60°F, or averaged from 32 to 212°F ("mean Btu"). The ASTM Standard for Metric Practice lists conversions for six slightly different Btus, all nearly equal to 1055 J. It seems likely that, in future, the "Btu (International Table)", defined as 1055.056 J, will become the dominant conversion equivalent as the Btu fades out of general use.

Brittle Fracture An abrupt breaking of a material in which there is very little or no elongation or distortion of the part before failure. In a tensile test, the stress-strain graph is nearly linear to the breaking point.

Brittleness Temperature The temperature at which a plastic or elastomer breaks in cantilever-bending impact under specified conditions. One method, described in ASTM D 746, consists of determining the temperature at which 50% of a group of specimens fail by a single impact. The brittleness temperature is related to that of the GLASS TRANSITION, w s, for those plastics with T_gs below room temperature, such as flexible PVC.

Broadgoods A fabrics-industry term for woven materials, including glass fabrics, that are over 46 cm in width.

Bronze Pigment Simulated bronze- or gold-colored pigments made by staining aluminum flakes with brown or yellow colorants.

Bronzing A term sometimes used for BLEEDING, w s, but more specifically referring to the appearance of an iridescent metallic luster caused by a film of dry pigment on a glossy surface.

Brookfield Viscometer The Brookfield "Synchrolectric" viscometer is the most widely used instrument for measuring the viscosities of plastisols and other liquids, both Newtonian and nonNewtonian. About a dozen models are available to accommodate subranges of viscosity in the overall range from 0.01 to 1000 Pa·s. It is portable and can be hand held. A synchronous motor provides 4 or 8 spindle speeds by shifting gears. Near the tip of the spindle, and concentric with it, is a horizontal disk whose drag torque in the liquid is detected by a torsion spring; a pointer indicates viscosity. By taking readings at different rotational speeds, one can estimate the pseudoplasticity of the liquid. For accurate work, the spindle guard should be removed and the diameter of the vessel containing the liquid should be at least five times that of the disk.

Bruceton Staircase Method See UP-AND-DOWN METHOD.

BS Abbreviation for British Standard.

BSI Abbreviation for British Standards Institute.

B-Stage An intermediate product in the formation of certain thermosetting resins in which the material swells when in contact with certain liquids and softens when heated, but may not entirely dissolve or fuse (ASTM D 883). The resin in an uncured thermosetting molding compound is usually in this stage. Sometimes referred to as a *resistol.* See also A-STAGE AND C-STAGE.

B-10 Life A statistical term used mostly with reference to thrust bearings in extruders and screw-injection molders, B-10 life is the expected operating time in hours, at a head pressure of 35 MPa (5000 psi) and a constant speed of 100 rpm, at which 10% of bearings will fail. A typical design life is 100,000 h. Life drops sharply with increasing screw speed and head pressure.

Btu (BTU) See BRITISH THERMAL UNIT.

BTX Abbreviation for the group of aromatic solvents comprising benzene, toluene, and xylenes.

Bubble (1) In blown-film extrusion, the expanding tube moving from the die to the collapsing rolls at the top of the tower. (2) A void within a molding or an extrusion.

Bubble Forming A variant of SHEET THERMOFORMING, w s, in which the plastic sheet is clamped in a frame suspended above a mold, heated, expanded into a blister shape with air pressure, then molded to its final shape by means of a descending plug applied to the blister and forcing it downward into the mold.

Bubbler A device inserted into a mold force, cavity, or core that delivers water to the bottom of the hole into which it is inserted. The water then flows back around the bubbler to the discharge line. Efficient cooling of isolated mold sections can be achieved in this way.

Buckling (1) A crimping of the fibers in a composite material that may occur in glass-reinforced thermosets due to shrinkage of the resin during cure. (2) The principal mode of failure of axially loaded, slender structural members such as columns and panels.

Buffing A surface-finishing method used with plastics and other materials in which the object is rubbed with cloths or cloth wheels that may contain fine, mild abrasives or waxes or both. Also see TUMBLING.

Bulk Density The density of a particulate material (granules, powder, flakes, chopped fiber, etc) expressed as the ratio of weight to total volume, voids included. ASTM D 1895 tells how to measure bulk densities of plastics powders and pellets.

Bulk Factor The ratio of the volume of a given mass of plastic particles to the volume of the same mass of material after molding or forming. The bulk factor is also equal to the ratio of the density after molding or forming to the apparent density (BULK DENSITY) of the material as received.

Bulking Value A synonym for SPECIFIC VOLUME, w s.

Bulk Modulus The modulus of volume elasticity, i e, the resistance of a solid or liquid to change in volume with change in pressure, at constant temperature. The thermodynamic definition is:

$$B = -v(\partial P/\partial v)_T = -(\partial P/\partial \ln v)_T$$

in which v = specific volume, P = pressure, and T = the specified temperature. This may be approximated over an interval of pressure as $0.5 \cdot (v_2 + v_1) \cdot (P_2 - P_1)/(v_1 - v_2)$. With plastics, B gradually diminishes as temperature rises and increases with rising pressure.

Bulk Molding Compound (BMC) See PREMIX.

Bulk Polymerization (mass polymerization) The polymerization of a monomer in the absence of any medium other than a catalyst or accelerator. The monomer is usually a liquid, but the term also applies to the polymerization of gases and solids in the absence of solvents or any other dispersing medium. Polystyrene, polymethyl methacrylate, low-density polyethylene, and styrene-acrylonitrile copolymers are examples of polymers most frequently produced by bulk polymerization. Acrylic monomers may be simultaneously polymerized and formed into products by conducting the polymerization in molds such as those for rods and sheets. Other bulk polymerizations are conducted in heated kettles, usually equipped with agitators and means for controlling the atmosphere.

Bulk Specific Gravity The specific gravity of a porous solid when the volume of the solid used in the calculation includes both the permeable and impermeable voids. Compare SPECIFIC GRAVITY and BULK DENSITY.

Bulk Yarn Yarn of glass (or other) fiber in bulk form, as opposed to roving, mat, or woven forms.

Buna-N See ACRYLONITRILE-BUTADIENE COPOLYMER.

Buna-S (GR-S) A synthetic elastomer produced by the copolymerization of butadiene and styrene. Also known as STYRENE-BUTADIENE RUBBER, w s.

BUR Abbreviation for BLOW-UP RATIO, w s.

Burned Showing evidence of excessive heating during processing or use of a plastic, as evidenced by blistering, discoloration, charring, or distortion.

Burning Rate See FLAMMABILITY, OXYGEN-INDEX FLAMMABILITY TEST, SELF-EXTINGUISHING.

Burn Mark Any visual sign of burning or charring at a particular spot on a part. In injection molding with poorly vented molds, burn marks can be caused by severe compression-heating of the air trapped in the mold cavity and consequent ignition of the molten plastic in contact with the hot air pocket (*dieseling*).

Bursting Strength Of rigid plastic tubing, the internal liquid pressure required to cause rupture of a test specimen in ASTM D 1180 or D 4021. Tubes with internal diameters between 3.2 and 152 mm (1/8 to 6 in) may be tested and diameter and wall thickness must be reported. The term has much the same meaning for filament-wound pressure vessels.

Bushing (1) In an extruder, the outer ring of any type of a die for circular tubing or pipe that forms the outer surface of the tube or pipe. (2) See SPRUE BUSHING.

Butadiene (1,3-butadiene, erythrene, vinylethylene, bivinyl, divinyl) $CH_2{=}CHCH{=}CH_2$. A gas, insoluble in water but soluble in alcohol and ether, obtained from cracking of petroleum, from coal-tar benzene, or from acetylene. It is widely used in the formation of copolymers with styrene, acrylonitrile, vinyl chloride and other monomers, imparting flexibility to the products made from them. Its homopolymer is a synthetic rubber.

Butadiene-Acrylonitrile Copolymer (NBR) See ACRYLONITRILE-BUTADIENE COPOLYMER.

Butadiene Rubber (BR) See POLYBUTADIENE.

Butadiene-Styrene Thermoplastics See STYRENE-BUTADIENE THERMOPLASTIC.

Butaldehyde See BUTYRALDEHYDE.

Butanal See BUTYRALDEHYDE.

1,4-Butanedicarboxylic Acid See ADIPIC ACID.

1,3- or 1,4-Butanediol See 1,3- OR 1,4-BUTYLENE GLYCOL.

1,2,4-Butanetriol $HOCH_2CHOHCH_2CH_2OH$. A nearly colorless liquid, an intermediate for alkyd resins and a plasticizer for cellulosics.

Butanoic Acid See BUTYRIC ACID.

n-Butanol See n-BUTYL ALCOHOL.

Butene Any of the monounsaturated C_4 hydrocarbons listed below:

IUPAC Name	Alternative Names
1-butene	α-butylene
cis-2-butene	*cis*-β-butylene
trans-2-butene	*trans*-β-butylene
methylpropene	isobutylene, isobutene

The term *butenes* refers to the first three compounds above as a group. The butylenes are used as monomers for rubbery homopolymers and copolymers with styrene, acrylics, other olefins, and vinyls. They are also used in adhesives for many plastics, and in making plasticizers.

2-Butene-1,4-diol $HOCH_2CH{=}CHCH_2OH$. A nearly colorless, odorless liquid, an intermediate for alkyd resins, plasticizers, nylon, and a cross-linking agent for resins.

2-Butoxyethanol See ETHYLENE GLYCOL MONOBUTYL ETHER.

2-Butoxyethyl Pelargonate $CH_3(CH_2)_7COOC_2H_4OC_4H_9$. A plasticizer for polystyrene, vinyl chloride polymers and copolymers, and cellulosics.

Butoxyethyl Stearate $CH_3(CH_2)_{16}COOC_2H_4OC_4H_9$. A high-boiling, plasticizer for nitrocellulose, polystyrene, ethyl cellulose, and polyvinyl acetate.

Butt Fusion A method of joining pipe, sheet or other forms of a thermoplastic resin wherein the ends of the two pieces to be joined are heated into the lower end of the polymer's melting range, then rapidly pressed together and allowed to cool, forming a homogeneous bond. ASTM D 2657 describes a recommended practice for butt-joining polyolefin pipe by heat fusion.

Butt Joint A joint made by fastening and/or bonding two surfaces that are perpendicular to the main surfaces of the parts being joined.

Butyl (1) The radical C_4H_9–, occurring only in combination. (2) Abbreviation used by British Standards Institution for BUTYL RUBBER, w s.

Butyl Acetate $CH_3COOC_4H_9$. A pleasantly aromatic solvent of moderate strength for ethyl cellulose, cellulose nitrate, vinyls, polymethyl methacrylate, polystyrene, coumarone-indene resins, and certain alkyds and phenolics.

***sec*-Butyl Acetate** (2-butyl acetate) $CH_3COOCH(CH_3)C_2H_5$. A solvent for nitrocellulose, ethyl cellulose, PVC, acrylics, polystyrene, phenolics and alkyd resins.

Butyl Acetoxystearate $CH_3(CH_2)_5CH(CH_3COO)(CH_2)_{10}COOC_4H_9$. A plasticizer similar to BUTYL ACETYL RICINOLEATE, but with the double bond saturated. It is compatible with cellulosic and vinyl resins.

Butyl Acetyl Ricinoleate $CH_3(CH_2)_5CH(CH_3COO)CH_2CH{=}CH(CH_2)_7$-$COOC_4H_9$. A yellow, oily liquid derived from castor oil, butyl alcohol and acetic anhydride, used as a plasticizer, compatible with cellulosics and vinyls.

***n*-Butyl Acrylate** $CH_2{=}CHCOOC_4H_9$. A colorless liquid that polymerizes readily on heating.

***n*-Butyl Alcohol** (1-butanol) $CH_3(CH_2)_2CH_2OH$. A medium-boiling alcohol, liquid above 35°C, used as a solvent for cellulosic, phenolic, and urea-formaldehyde resins. It is also used as a diluent/reactant in the manufacture of urea-formaldehyde and phenol-formaldehyde resins, and as an intermediate in the production of butyl acetate, dibutyl phthalate, and dibutyl sebaçate.

***n*-Butyl Aldehyde** See BUTYRALDEHYDE.

Butylated Hydroxytoluene (di-*tert*-*p*-cresol, BHT) A white, crystalline solid, the most widely used antioxidant for plastics such as ABS and LDPE. It has been approved by the FDA for use in foods and food-packaging materials.

Butylated Resin A resin containing the butyl radical, C_4H_9–.

Butyl Benzenesulfonamide (*N*-*n*-butyl benzenesulfonamide) $C_6H_5SO_2$-NHC_4H_9. A plasticizer for some synthetic resins and an intermediate for resin manufacture.

Butyl Benzoate (*n*-butyl benzoate) $C_6H_5COOC_4H_9$. A plasticizer and solvent for cellulosics.

Butyl Benzyl Phthalate (BBP) $(C_4H_9OOC)C_6H_4(COOCH_2C_6H_5)$. A clear, oily liquid used as a plasticizer for cellulosic and vinyl resins. It imparts good stain resistance, low volatility at calendering and extruding temperatures, low oil extraction and good heat and light stability.

Butyl Benzyl Sebaçate $C_4H_9OOC(CH_2)_8COOCH_2C_6H_5$. An ester-type plasticizer with a light straw color. It combines the desirable properties of dibenzyl sebaçate and dibutyl sebaçate.

Butyl Borate See TRIBUTYL BORATE.

Butyl Cyclohexyl Phthalate $(C_4H_9OOC)C_6H_4(COOC_6H_{11})$. A plasticizer for PVC, other vinyls, cellulosics, and polystyrene.

Butyl Decyl Phthalate $(C_4H_9OOC)C_6H_4(COOC_{10}H_{21})$. A plasticizer for polystyrene, PVC, and vinyl chloride-acetate copolymers.

Butyl Diglycol Carbonate [diethyleneglycol bis(n-butylcarbonate)] $C_{14}H_{26}O_7$. A colorless liquid of low volatility, used as a plasticizer with many resins.

1,3-Butylene Dimethacrylate A polymerizable monomer used in PVC and rubber systems to obtain rigid or semirigid products from materials that are normally flexible. The monomer acts as a plasticizer at room temperature and crosslinks at processing temperatures.

1,3-Butylene Glycol (1,3-butanediol) $CH_3CHOHCH_2CH_2OH$. A colorless liquid made by catalytic hydrogenation of aldol (3-hydroxy-n-butyraldehyde). Its most important use is as an intermediate in the manufacture of polyester plasticizers.

1,4-Butylene Glycol (1,4-butanediol, tetramethylene glycol) $HOCH_2$-$CH_2CH_2CH_2OH$. A stable, hygroscopic, colorless liquid used in the production of polyesters by reaction with dibasic acids, and in the production of polyurethanes by reaction with diisocyanates.

Butylene See BUTENE.

1,3-Butylene Glycol Adipate Polyester (Santicizer® 334F) A polymeric plasticizer for PVC.

Butyl Epoxy Stearate A plasticizer for PVC, imparting low-temperature flexibility.

Butyl Ethylhexyl Phthalate (butyl octyl phthalate, BOP) A mixed ester of butanol and 2-ethylhexanol, widely used as a primary plasticizer for PVC compounds and plastisols, in which it performs like dioctyl phthalate in most respects. It is also compatible with vinyl chloride-acetate copolymers, cellulose nitrate, ethyl cellulose, polystyrene, chlorinated rubber and, at lower concentrations, with polymethyl methacrylate.

Butyl Formate $HCOOC_4H_9$. A solvent for several resins, including cellulose acetate.

***tert*-Butyl Hydroperoxide** $(CH_3)_3COOH$. A highly reactive peroxy compound used as a polymerization catalyst.

Butyl Isodecyl Phthalate (decyl butyl phthalate) $(C_4H_9OOC)C_6H_4$-$(COOC_{10}H_{21})$. A plasticizer for PVC and polystyrene.

Butyl Isohexyl Phthalate $(C_4H_9OOC)C_6H_4(COOC_6H_{13})$. A plasticizer for cellulosics, acrylic resins, polystyrene, PVC, and other vinyl resins.

Butyl Lactate $CH_3CHOHCOOC_4H_9$. A solvent for nitrocellulose, ethyl cellulose, and many synthetic resins.

Butyl Laurate $C_{11}H_{23}COOC_4H_9$. A plasticizer for cellulosic resins, polystyrene, and vinyl resins.

***n*-Butyl Methacrylate** $H_2C{=}CHCH_3COOC_4H_9$. A polymerizable monomer used in the production of acrylic resins and potting compounds.

***n*-Butyl Myristate** $CH_3(CH_2)_{12}COOC_4H_9$. The butyl ester of myristic acid, an oily liquid used as a plasticizer for cellulosic plastics.

Butyl Octadecanoate See BUTYL STEARATE.

Butyl Octyl Phthalate See BUTYL ETHYLHEXYL PHTHALATE.

Butyl Oleate $CH_3(CH_2)_7CH{=}CH(CH_2)_7COOC_4H_9$. A solvent, plasticizer and lubricant, used mainly with neoprene and other synthetic rubbers, chlorinated rubber, and ethyl cellulose; also a mold lubricant.

***n*-Butyl Palmitate** $C_{15}H_{31}COOC_4H_9$. A plasticizer for polystyrene and cellulosic plastics.

***tert*-Butyl Perbenzoate** $C_6H_5{\cdot}O_2{\cdot}C(CH_3)_3$. A catalyst for the polymerization of acrylic and styrene monomers, and the curing of polyesters. Also used in the compounding of silicones and polyethylene. This peroxide has long been the workhorse in sheet molding compounds because it is stable enough for all practical purposes but is slow-reacting, requiring activation temperatures of 121–127°C unless boosted by a less stable peroxide.

***tert*-Butyl Permaleic Acid** $(CH_3)_3CC{\cdot}O_2{\cdot}COCH{=}CHCOOH$. A polymerization catalyst.

***t*-Butyl Peroxy Neodecanoate** A polymerization initiator for vinyl chloride.

***t*-Butyl Peroxy Pentanoate** A peroxyester catalyst.

***t*-Butyl Perphthalic acid** $(CH_3)_3C{\cdot}O_2{\cdot}COC_6H_4COOH$. A polymerization catalyst.

***p-t*-Butyl Phenol** $(CH_3)_3CC_6H_4OH$. A white crystalline solid used as a plasticizer for cellulose acetate.

***p-t*-Butylphenyl Salicylate** A plasticizer approved by FDA for contact with foods, also used as a light-absorbing agent.

***n*-Butylphosphoric Acid** $C_4H_9H_2PO_4$. A reddish amber liquid used as a catalyst, e g, in urea-resin production.

Butyl Phthalyl Butyl Glycollate $C_4H_9OOCC_6H_4COOCH_2COOC_4H_9$. A plasticizer with good light stability, used mainly with PVC and polystyrene, but compatible with most other thermoplastics. It has been approved by the FDA for contact with foods.

***n*-Butyl Propionate** $C_2H_5COOC_4H_9$. A colorless liquid with an apple-like odor, used as a solvent for nitrocellulose.

Butyl Ricinoleate $CH_3(CH_2)_5CHOHCH_2CH{=}CH(CH_2)_7COOC_4H_9$. A plasticizer for vinyl resins and cellulose acetate butyrate, derived from castor oil and butyl alcohol.

Butyl Rubber (Butyl, formerly GR-1) A synthetic elastomer produced by copolymerizing isobutylene with a small amount (ca 2%) of isoprene or butadiene. It has good resistance to heat, oxygen, and ozone, and low gas permeability. Thus, it is widely used in inner tubes and to line tubeless tires.

Butyl Stearate (butyl octadecanoate) $C_{17}H_{35}COOC_4H_9$. A mold lubricant and plasticizer, compatible with natural and synthetic rubbers, chlorinated rubber, and ethyl cellulose. It can be used in vinyls in very low concentrations as a nontoxic, secondary plasticizer and lubricant. In the production of polystyrene, butyl stearate is added to the emulsion polymerization to impart good flow properties to the resin.

Butyraldehyde (butaldehyde, n-butanal, n-butyl aldehyde, butyric aldehyde) $CH_3(CH_2)_2CHO$. An aldehyde sometimes used in place of formaldehyde in the production of resins. Butyraldehyde reacts with polyvinyl alcohol to form polyvinyl butyral.

Butyrate (1) A salt or ester of BUTYRIC ACID, w s. (2) The industry nickname for CELLULOSE ACETATE BUTYRATE (CAB), w s.

Butyric Acid (n-butyric acid, butanoic acid, ethylacetic acid, propylformic acid) $CH_3(CH_2)_2COOH$. A liquid with a strong rancid-butter odor, used in the production of CELLULOSE ACETATE BUTYRATE, w s. Derivatives of butyric acid are used in the production of plasticizers for cellulosic plastics.

γ-Butyrolactone (butyrolactone) A hygroscopic, colorless liquid obtained by the dehydrogenation of 1,4-butanediol with the structure:

H_2C—CH_2
H_2C $C{=}O$
O

It is a solvent for cellulosics, epoxy resins, and vinyl copolymers.

C c

c (1) Abbreviation for SI prefix, CENTI-, w s. (2) Abbreviation for cubic.

C (1) Chemical symbol for the element carbon. (2) Abbreviation for celsius or centigrade. (3) Symbol for electrical capacitance.

Ca Chemical symbol for the element calcium.

CA See CELLULOSE ACETATE.

CAD/CAM Acronym for COMPUTER-AIDED DESIGN and COMPUTER-AIDED MANUFACTURING, w s.

Cadmium Ethylhexanoate A metallic soap used as a stabilizer for vinyls, especially to avoid plate-out in calendering compounds.

Cadmium Pigment Any inorganic pigment based on cadmium sulfide or cadmium sulfoselenide, used widely in PVC, polystyrene, and polyolefins. Included are cadmium-maroon, -orange, -red, and -yellow. The cadmium pigments have good resistance to heat (up to 500°C) and to alkalis, and do not bleed. Light stability is good in solid colors, but may be poor when used for tints with white pigments. Resistance to acids is poor.

Cadmium Ricinoleate $[CH_3(CH_2)_5CHOHCH_2CH{=}CH(CH_2)_7COO]_2$-Cd. A white powder derived from castor oil, used as a heat stabilizer for vinyl chloride polymers and copolymers.

Cadmium Stearate $(C_{17}H_{35}COO)_2Cd$. A heat- and light-stabilizer, used when good clarity in transparent compositions is desired.

Calcined Clay See CLAY.

Calcium Acetate (vinegar salts, gray acetate, lime acetate, brown acetate) $(CH_3COO)_2Ca \cdot H_2O$. A stabilizer.

Calcium Carbonate (aragonite, calcite, chalk, limestone, lithographic stone, marble marl, travertine, whiting) $CaCO_3$. Grades of calcium carbonate suitable as fillers for plastics are obtained from naturally occurring deposits as well as by chemical precipitation. The natural types are prepared by dry grinding, yielding particles usually over 20 μm, used in stiff products such as floor tiles; or by wet grinding, yielding particles under 16 μm, used in flexible products. The chemically precipitated types range from 0.05 to 11 μm in size, and are most often used in plastisols and highly flexible products. Both the wet-ground and precipitated types are available with coating such as resins, fatty acids, and calcium stearate. These coated grades have low oil absorption, of particular value in compounding plastisols. The calcium stearate coatings provide improved electrical properties, heat stability and lubricity during processing, beneficial in extrusion compounds.

Calcium Glycerophosphate (calcium glycerinophosphate) $CaC_3H_7O_2$-PO_4. A white, crystalline powder, odorless and nearly tasteless, used as a stabilizer for plastics.

Calcium Oxide (calx, lime, quicklime, burnt lime) CaO. A white powder with affinity for water, with which it combines to form calcium hydroxide. It has been used to remove traces of water in vinyl plastisols.

Calcium Phosphate (calcium orthophosphate, tricalcium phosphate, tricalcic phosphate, tertiary calcium phosphate) $Ca_3(PO_4)_2$. A stabilizer.

Calcium Phosphate, Dibasic (calcium biphosphate, acid calcium phosphate, primary calcium phosphate) $CaH_4(PO_4)_2 \cdot H_2O$. A stabilizer.

Calcium Phosphate, Monobasic (dicalcium orthophosphate, bicalcic phosphate, secondary calcium phosphate) $CaHPO_4$ or $CaHPO_4 \cdot H_2O$. A stabilizer.

Calcium Ricinoleate $[CH_3(CH_2)_5CHOHCH_2CH{=}CH(CH_2)_7COO]_2Ca$. A white powder derived from castor oil, used as a nontoxic stabilizer for PVC.

Calcium Silicate (wollastonite) $CaSiO_3$. A naturally occurring mineral found in metamorphic rocks, used as a reinforcing filler in low-density polyethylene, polyester and other thermosetting molding compounds. It imparts smooth molded surfaces and low water absorption.

Calcium Stearate $(C_{17}H_{35}COO)_2Ca$. A nontoxic stabilizer and lubricant. It is not often used alone because of its early color development, but is used in combination with zinc and magnesium derivatives and epoxides in manufacturing other nontoxic stabilizers.

Calcium Sulfate (anhydrite) $CaSO_4$. A filler and white pigment. The hydrated forms are known as gypsum, *terra alba*, and plaster of paris.

Calcium Thiocyanate (calcium sulfocyanate, calcium rhodanate) $Ca(SCN)_2 \cdot 3H_2O$. In water solution, a solubilizer for acrylic and cellulosic resins.

Calcium-Zinc Stabilizer Any of a family of stabilizers based on compounds and mixtures of compounds of calcium and zinc. Their effectiveness is limited, but they are among the few that have been approved by the FDA for materials to be contacted by foods.

Calender The machine performing the process of CALENDERING, w s.

Calender Coating The process of applying plastics to substrates such as paper or fabric by passing both the substrate and a plastic film through calender rolls.

Calendering The process of forming thermoplastics sheeting and films by passing the material through a series of rigid, heated rolls. Four rolls are typical. The gap between the last pair of heated rolls determines the thickness of the sheet. Subsequent chilled rolls cool the sheet. The plastic compound is usually premixed and plasticated on separate eqiupment, then fed continuously into the nip of the first pair of calender rolls.

Calibrate (1) To determine the exact relationship between the indicated or recorded readings of a measuring instrument or method and the true values of the quantities measured, over the instrument's range. (2) (mainly European usage) To bring the dimensions of an extruded tube or pipe within their specified ranges.

Calibration (1) The process of ascertaining the errors of a measuring technique or instrument by comparing its readings with the corresponding values of known standards or the readings of a more accurate technique or instrument. Calibration standards for the whole gamut of measurements are available from the (US) National Institute of Standards and Technology (NIST, formerly the National Bureau of Standards), Gaithersburg, MD 20899. They can also furnish standard materials, some of them polymeric. See NIST special publication 260, "Standard Reference Materials Catalog". NIST also calibrates instruments on request; see special publication 250, "NIST Calibration Services Users' Guide". Both publications are available from the US Government Printing Office in Washington, DC. (2) See CALIBRATE (2).

Caliper (1) The thickness of film or sheet, typically stated in thousandths of an inch (mils). (2) Any of several types of precise instruments used to measure thickness, as *micrometer caliper*.

Calorie (small calorie) A deprecated, small unit of heat energy: the amount of heat required to raise the temperature of one gram of water, at or near 4°C, one degree C. The ASTM "Standard for Metric Practice" (E 380) lists five slightly different calories, all nearly equal to 4.185 joules. In future, most of these will fade away, leaving only the "calorie (International Table)", equal to 4.186800 J. The term calorie is also used loosely, especially in the nutrition field, to mean 1000 calories, the kilocalorie, or "large calorie".

Calorimeter A device for measuring the heat liberated (or absorbed) during chemical reactions or physical changes of state.

CAM Acronym for COMPUTER-AIDED MANUFACTURING, w s.

Camphor (*d*-2-camphanone, 2-keto-1,7,7-trimethylnorcamphane) $C_{10}H_{16}O$. A bridged-ring, naturally occurring ketone with the structural formula:

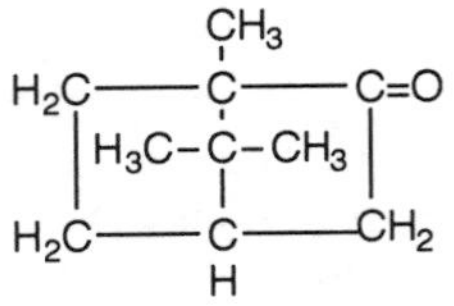

```
            CH3
            |
  H2C ------C------ C=O
   |        |        |
   |  H3C - C - CH3  |
   |        |        |
  H2C ------C------ CH2
            H
```

It is a colorless, aromatic, crystalline material originally derived from camphor oil but now mostly synthesized from pinene. In 1870, the Hyatt brothers were awarded a US patent for a horn-like compound of camphor and cellulose nitrate, which they called Celluloid. The event marked the start of the US plastics industry.

Camphoric Acid $C_8H_{14}(COOH)_2$. A dibasic-acid plasticizer for cellulose nitrate, derived from CAMPHOR, w s, by oxidizing the =C=O and adjacent $=CH_2$ groups and opening the right-side ring.

CAN Abbreviation for cellulose acetate nitrate.

Candela (cd) One of the six basic units of the SI system, the unit of luminous intensity, defined as the luminous intensity normal to the surface of 1/60 of a square centimeter of a black body at the temperature of freezing platinum (1772°C) under a pressure of 101.325 kPa. In older literature, the same abbreviation has been used for both *candlepower* and *candle*, which have different values.

Cantilever-Beam Stiffness A method of determining stiffness of plastics by measuring the force and angle of bend of a cantilever beam made of the specimen material. The ASTM test is D 747. See also FLEXURAL MODULUS.

Caoutchouc An early name for natural rubber, still in use in the French literature.

CAP See CELLULOSE ACETATE PROPIONATE.

Capacitance (C, electric capacity) The property of a system of two conducting surfaces (plates or foils, typically) separated by a nonconductor (*dielectric*) that permits storage of electric charge in proportion to the voltage difference between the conductors. The SI unit is the farad (F). A capacitor storing one coulomb of charge at a potential difference of one volt has a capacitance of one farad. Plastics are much used as capacitor dielectrics.

Capillary Rheometer An instrument for measuring the flow properties of polymer melts, comprising a capillary tube of known diameter and length, means for applying pressure to force the molten polymer from a reservoir through the capillary, means for closely maintaining the desired temperatures of the apparatus, and means for measuring differential pressures and flow rates. Many designs, varying in sophistication, are in use. The data obtained from capillary rheometers is usually presented as logarithmic graphs, for several temperatures, of nominal shear stress vs apparent Newtonian shear rate. Some instruments and recommended procedures are described in ASTM D 1238, D 1823, and D 3835. See also VISCOSITY.

Capillary Viscometer This term is frequently used for two types of capillary instruments—one used for concentrated solutions or polymer melts and described just above, the other, described here, used for measuring dilute-solution viscosities. The most widely used of the latter types employ a glass capillary tube and means for timing the flow of a measured volume of the solution through the tube under the force of gravity. This time is then compared with the time taken for the same volume of pure solvent, or of another liquid of known viscosity, to flow through the same capillary. Relevant ASTM tests are D 1243, D 1601, D 2587, S 4603, and D 4889. See also DILUTE-SOLUTION VISCOSITY.

Capric Acid (decanoic acid, decylic acid, decoic acid) $CH_3(CH_2)_8COOH$. A plasticizer and an intermediate for resins.

Caprolactam (ε-caprolactam, hexanoic acid-ε-amino lactam) A cyclic amide having the structure shown below.

$H_2C—CH_2$, H_2C, CH_2, H_2C, N, C=O, H

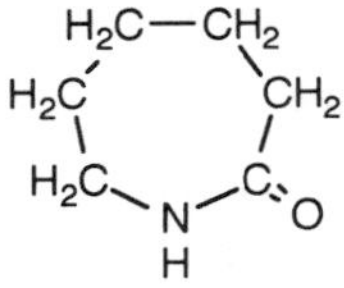

When the ring is opened, caprolactam is polymerizable to a nylon resin known as NYLON 6, w s, or *polycaprolactam.* It is also used as a cross-linking agent for polyurethanes, and as a plasticizer. In the late 1960s it was found that caprolactam could be rotationally cast by heating the solid monomer to its melting point (ca 71°C) and introducing the molten monomer into a mold along with a catalyst, then heating and rotating the mold in the usual manner. The liquid gradually thickens and gels against the mold in the manner of a plastisol, and conversion to nylon 6 is accomplished in a few minutes.

Caprylic Acid (octanoic acid, octoic acid, octylic acid, caprilic acid) $CH_3(CH_2)_6COOH$. A plasticizer and organic intermediate.

Capstan In the plastics industry, a drum or pulley that controls the speed of a filament, wire or web between production stages.

Captive Production Production of materials or components by a manufacturer for its own use or for later incorporation in its products. Compare CUSTOM MOLDER.

Carbamide See UREA.

Carbamide Phosphoric Acid (urea phosphoric acid) $CO(NH_2)_2 \cdot H_3PO_4$. A catalyst for acid-setting resins.

Carbamyl Urea See BIURET.

Carbazole (dibenzopyrrole, diphenylenimine) Extracted from coal tar, this unsaturated compound has the structure shown below.

N, H

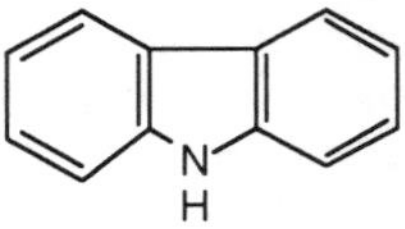

It is used in the manufacture of POLY(N-VINYLCARBAZOLE), w s.

Carbitol [2,(2-ethoxyethoxy)ethanol, ethyl cellosolve] $C_2H_5OCH_2$-

$CH_2OCH_2CH_2OH$. Trade name of an active, water-soluble solvent, the monoethyl ether of diethylene glycol.

Carbolic Acid A synonym for PHENOL, w s.

Carbon 14 (radiocarbon) Radioactive carbon of mass number 14, naturally occurring but usually made by irradiating calcium nitrate. It has been used as a beta-ray source in gages for measuring the thickness of plastics films.

Carbon Black A generic term for the family of colloidal carbons. More specifically, carbon black is made by the partial combustion and/or thermal cracking of natural gas, oil, or another hydrocarbon. *Acetylene black* is the carbon black derived from burning acetylene. *Animal black* is derived from bones of animals. *Channel blacks* are made by impinging gas flames against steel plates or channel irons (hence the name), from which the deposit is scraped at intervals. *Furnace black* is the term sometimes applied to carbon blacks made in a refractory-lined furnace. *Lamp black*, the properties of which are markedly different from other carbon blacks, is made by burning heavy oils or other carbonaceous materials in closed systems equipped with settling chambers for collecting the soot. *Thermal black* is produced by passing natural gas through a heated brick checkerwork where it thermally cracks to form a relatively coarse carbon black. Carbon blacks are widely used as fillers and pigments in PVC, phenolics, polyolefins, and several other resins, also imparting resistance to ultraviolet rays. In polyethylene, carbon black acts as a crosslinking agent; in rubbers, as a reinforcement.

Carbon-Carbon Composite A combination of carbon or graphite fibers in a carbon or graphite matrix, produced by impregnating a carbon- or graphite-fiber cloth or mat structure with a carbonizable binder such as pitch.

Carbon Fiber (graphite fiber) Any fiber consisting mainly of elemental carbon and increasingly used in reinforced-plastics products. They may be prepared by growing single crystals in a carbon electric arc under high-pressure inert gas; by growth from a vapor state via pyrolysis of a hydrocarbon gas; or by pyrolysis of organic fibers, the most widely used method. Polyacrylonitrile and rayon fibers are most commonly used as starting materials. The terms "carbon fibers" and "graphite fibers" are used somewhat interchangeably. However, PAN-based carbon fibers are 93–95% C by elemental analysis, whereas graphite fibers are usually 99+% C. The difference is due mainly to the temperature of formation, 1315°C for fibers formed from PAN, while the high-modulus graphite fibers are graphitized at 3450°C. The higher the graphite content, the higher the elastic modulus but the lower the strength. Properties hold to very high temperatures in inert atmospheres. (Properties transverse to the fiber length are much lower than along the length.) In recent years, carbon fibers have become the leading reinforcement for high-performance composites. The less expensive carbon fibers produced from pitch have broadened the markets for these reinforcements. Strength and modulus range considerably depending on the supplier and grade, from 1.7 to 3.5 GPa (250 to 500 kpsi) for strength and from 230 to 830 GPa (34 to 120 Mpsi) for modulus. Rayon-based fiber has lower density (1.6 g/cm^3) than PAN- or pitch-based fiber (1.9 g/cm^3).

Carbon-Fiber-Reinforced Plastics (CRP) Plastics, either thermosetting or thermoplastic—most commonly epoxies or high-performance resins—that contain carbon or graphite fibers.

Carbon Tetrachloride (tetrachloromethane) CCl_4. A clear, dense, pungent-smelling liquid, a powerful solvent for many resins and miscible with most other organic solvents, and a starting compound in the synthesis of NYLON 7. Its use as a solvent has now been severely curtailed because of its high toxicity and known carcenogenicity.

Carbonyl (carbonyl group) The divalent organic radical =C=O, found only in combination, as in aldehydes, ketones and organic acids.

Carborundum See SILICON CARBIDE.

Carborane (dicarbadodecaborane) $C_2B_{10}H_{12}$. An icosahedral cage compound containing mostly boron, with two active hydrogens on the carbon atoms, and available in several isomers. It is polymerizable, but only the silicone copolymers have been manufactured.

Carboxylic Term for the –COOH group, the radical occurring in organic acids.

Carboxymethyl Cellulose A glycolic ether of cellulose, usually sold as its sodium salt, and used as a soluble thickener.

Carboxy Nitroso Rubber (CNR) A fluorocarbon elastomer, synthesized as a terpolymer from tetrafluoroethylene, trifluoronitrosomethane, and nitrosoperfluorobutyric acid. CNR has unique resistance to strong oxidizers and is nonflammable in pure oxygen, hence has found applications in the aerospace field. The gum can be processed on standard rubber-mixing equipment for molding, or dissolved for application by spraying, dipping or brushing.

Carburizing A process for case-hardening steels in which the objects to be hardened are heated with carbonates and charcoal in the absence of air for about 24 h at 840–950°C, then quenched in oil. The depth of hardening is about 1.7 mm and the surface hardness is from 50 to 55 on the Rockwell C scale.

Carcinogen Any material that has been tested and found to cause cancer in laboratory animals or that, through statistical studies, is correlated with the incidence of cancers in humans. An example from the plastics industry is vinyl chloride monomer, which is believed to have caused human liver cancer. PVC polymer, on the other hand, is noncarcenogenic. A list of known carcinogens is available from OSHA (US Office of Health and Safety Administration).

Cartridge Heater A rod-shaped electrical heating element, consisting of a metal outer shell, sealed within which is a Nichrome-wire coil embedded in a thermally stable, electrically insulating powder, such as magnesium oxide. Cartridge heaters come in a wide range of physical sizes and wattages and are used in heating and controlling the temperatures of dies and molds, injection nozzles, hot stampers, etc.

Cascade Coating A process for applying epoxy and other thermosets to objects such as electrical resistors and capacitors, in which finely powdered resin is poured over the preheated object. The article is usually rotated as the powder is applied.

Cascade Control (piggy-back control) In automatic control, a system in which the output of one unit is the input of the next, the goal being to obtain closer control of an important, final variable by controlling more sensitive, linked variables. A cascade system may be of the open-loop or closed-loop type.

Case Hardening Any of several processes by which the working surfaces of steel tools and molds are hardened after being machined in their softer, original states. See CARBURIZING, FLAME HARDENING, NITRIDING.

Casein The protein substance occurring in milk and cheese. It can be obtained by treating skim milk with a dilute acid, but the type used mainly for plastics (rennet casein or paracasein) is made by treating warm skim milk with a rennet extract. See also CASEIN PLASTIC.

Casein Plastic A family of thermosetting plastics derived from CASEIN, used widely in the early years of the plastics industry but less important now. Casein plastics have poor water resistance and dimensional stability, which limits their applications.

Cashew Resin A thermosetting resin produced from the phenolic fraction of the oil extracted from cashew-nut shells.

Casing A term coined by Bell Telephone Laboratories, an acronym for the process of **C**rosslinking by **A**ctivated **S**pecies of **IN**ert **G**ases, developed to impart printability and adhesive receptivity to polymers such as PTFE and polyethylene. In this process, articles are exposed to a flow of activated inert gases in a glow-discharge tube, forming a shell of highly crosslinked molecules having high SURFACE ENERGY on the article surfaces.

Cast (v) See CASTING.

Cast Embossing A process of casting films against an embossed temporary carrier. Vinyl plastisols, organosols, solutions, or latices are used as film formers, which may be backed up with layers of foam or fabric. The temporary carrier is often paper, embossed with the desired pattern and treated so as to be easily stripped from the fused film or laminate.

Cast Film Film produced by pouring or spreading a solution, hot melt, or dispersion of plastic material onto a temporary carrier—typically a polished metal roll, hardening the material by suitable means, and stripping the solidified film from the surface. Cellulosic, polystyrene, and vinyl films are often produced in this manner.

Casting (1,v) The process of forming solid or hollow articles from fluid plastic mixtures or resins by pouring or injecting the fluid into a mold or against a substrate with little or no pressure, followed by solidification and removal of the formed object. See also CAST EMBOSSSING, CENTRIFUGAL-, FILM-, SLUSH-, SOLID-, ROTATIONAL-, and SOLVENT CASTING; and EMBEDDING, ENCAPSULATION and POTTING. (2,n) The finished product of a casting operation.

Casting Syrup (casting resin) Liquid monomers or partially polymerized polymers, usually containing catalysts or curing agents, capable of polymerizing to the solid state after they have been cast in molds. The materials most generally used are the acrylics, styrenes, polyesters, epoxies, silicones, and nylons. Also called *potting syrups* when used for encapsulating articles such as electrical components or assemblies.

Cast-in Heater A type of heater used on cylindrical surfaces, such as extruder barrels, in which a rod-type heating element is bent to a semicylindrical shape and cast within an aluminum channel shape whose inner surface has the cylinder's radius. Pairs of such cast-in elements are then strapped tightly to the barrel with Monel bands and high-strength bolts, leaving the heater terminals exposed for electrical connections. Copper tubing may also be cast into the same elements, permitting circulation of water or, better, an involatile heat-transfer liquid, for barrel cooling.

Cast-in Lining See BIMETALLIC CYLINDER.

Castor Oil (ricinus oil) A pale-yellowish oil derived from the seeds of the castor bean, *Ricinus communis*, and consisting essentially of ricinolein. It is an important starting material for plasticizers, certain nylons, and alkyd resins; and an ingredient in certain urethane foams.

Catalyst A substance that causes or accelerates a chemical reaction when added to the reactants in a minor amount, and that is not consumed in the reaction. A negative catalyst (INHIBITOR) decreases the rate of reaction or prevents it altogether. See also ACCELERATOR, CURING AGENT, INITIATOR.

Caterpillar (caterpillar puller) A device used downstream of the extruder in extrusion of pipe and profiles, consisting of two driven and counterrotating belts, having an elongated oval shape about 0.8 m long, with pads attached to the outsides of the belts. One of the belts is elevatable to adjust the clearance between them so as to firmly grip the extrudate being pulled away from the cooling tank, yet not so strongly as to deform it. Belt speed is adjustable over a wide range to accommodate different rates of extrusion.

Cathode (1) In an electrolytic cell through which current is being forced by an external emf, the cathode is the negative electrode, giving up electrons to cations in the electrolyte. In a cell or battery *delivering current*, the cathode is the positive terminal. (2) In a vacuum tube, the cathode is the electrode from which electrons are emitted.

Cathode Sputtering See VACUUM METALIZING.

Cation An atom, molecule or radical, usually in aqueous solution, that has lost an electron and has become positively charged.

Cation-Exchange Resin See ION-EXCHANGE RESIN.

Cationic Pertaining to any positively charged atom, radical, or molecule; or to any compound or mixture containing positively charged groups.

Cationic Polymerization See IONIC POLYMERIZATION.

Caul A sheet of metal, wood, or other material used in laminating to apply and equalize pressure.

Cause-and-Effect Diagram (fishbone diagram) A graphical way of analyzing a process, based on ideas and experiences of workers and engineers concerning the materials, machines, and methods of the process, in order to identify possible causes of product defects.

Cavity A depression, or sometimes the set of matching or associated depressions, in a plastics mold that forms the outer surfaces of the cast or molded article(s). The cavity may surround a CORE, the portion of the mold that forms the inner surfaces of a hollow article.

Cavity-Retainer Plate A plate in a mold that holds the cavities and forces. Such plates are at the mold parting line and usually contain the guide pins and bushings. Also called *force-retainer plate.*

Cavity Side (British) The side of an injection mold that is adjacent to the nozzle.

Cavity-Side Part (US) The stationary part of an injection mold.

Cavity-Transfer Mixer A two-piece device installed at the end of an extruder screw to accomplish both distributive and dispersive mixing. The *stator* is a barrel extension into whose inside surface is machined an array of many hemispherical cavities. The *rotor* is a screw extension whose exterior is similarly contoured. The lands of rotor and stator have the usual close clearance of screw and barrel. As the melt stream passes through, it is smeared between the lands and is repeatedly cut into small globs and recombined, passing from rotor to stator, stator to rotor, until it emerges.

CBA In the plastics-foam industry, abbreviation for *chemical blowing agent*.

cd SI abbreviation for CANDELA, w s.

Cd Chemical symbol for the element cadmium.

CDP Abbreviation for CRESYL DIPHENYL PHOSPHATE, w s.

Cell In the cellular-plastics industry, a single void produced by a blowing agent, by mechanically entrained gas, or by the evaporation of a volatile constituent. When the void is completely surrounded by polymer, the cell is said to be *closed*. A *completely open* cell has no wall membranes but is part of a three-dimensional network of connected fibers or rods.

Cell Collapse A defect in foamed plastics characterized by slumping and cratered surfaces, with the internal cells resembling a stack of leaflets when viewed in cross section under a microscope. The condition is caused by tearing of the cell walls, weakened by plasticization or other mechanism.

Celloidin (celluidine, photoxylin) A form of cellulose nitrate made by precipitation from an ether-alcohol solution of collodion cotton. See CELLULOSE NITRATE and COLLODION.

Cellophane Regenerated cellulose film, chemically similar to RAYON, made by mixing cellulose xanthate with dilute sodium hydroxide solution to form a viscose, then extruding the viscose into an acid bath for regeneration. (The term *rayon* is used when the regenerated material is in fibrous form.) Cellophane is widely used for packaging, most often with coatings of other polymers to overcome its tendency to absorb moisture and to improve the film's heat-sealability.

Cellosolve See ETHYLENE GLYCOL MONOETHYL ETHER.

Cellular Mortar See SYNTACTIC FOAM.

Cellular Plastic (expanded plastic, foamed plastic) A plastic with numerous cells of gas distributed throughout its mass. The terms *cellular-*, *expanded-*, and *foamed plastic* are used synonymously. A cellular plastic may be produced by (1) incorporating a blowing agent that decomposes to liberate a gas; (2) mechanically whipping in a gas or vaporizable liquid; (3) by adding a water-soluble salt or a solvent-extractable agent to the mix prior to forming, then leaching out the agent after forming to leave voids; or (4) other techniques described under EPOXY FOAM, PHENOLIC FOAM, POLYSTYRENE FOAM, SYNTACTIC FOAM, URETHANE FOAM. Cellular plastics range in density from some only slightly less than that of the parent resin to less than 0.01 g/cm^3. The cells may be open or closed, depending on the process and density. See also STRUCTURAL FOAM.

Cellular Striation In a cellular plastic, a layer of cells differing in size or nature from the majority of cells in the same mass.

Celluloid An old trade name, now generic, for CELLULOSE NITRATE, w s, compounded with camphor and ethanol. The ethanol is removed after processing by heating, leaving behind the camphor which toughens the compound.

Cellulose A natural carbohydrate polymer of high molecular weight, having the structure shown below:

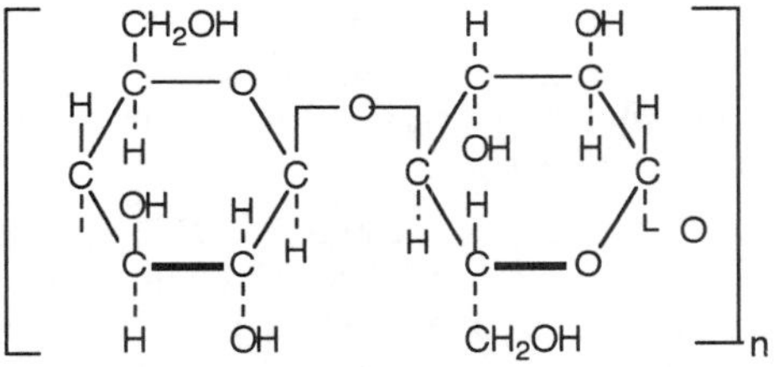

Cellulose is a constituent of most higher plants (*spermatophyta).* Cotton fiber is almost pure cellulose. Cotton linters and wood pulp are the major sources of cellulose for CELLULOSIC PLASTICS, w s.

Cellulose Acetate (CA) An acetic-acid ester of cellulose, forming a tough, transparent thermoplastic material when compounded with plasticizers. It is obtained by the action, under rigidly controlled conditions, of acetic acid and acetic anhydride on purified cellulose, usually obtained from cotton linters. All three available hydroxyl groups in each glucose unit of the cellulose can be acetylated, but in the material normally used for plastics it is usual to acetylate fully, then to lower the acetyl value by partial hydrolysis, leaving, on average, 2.4 acetate groups per C_6 unit. Cellulose acetate compounds are used when toughness, permanence, flame resistance, and transparency are required at moderate cost. However, they absorb up to 2.5% of atmospheric moisture, making them unsuitable for long-term outdoor exposure.

Cellulose Acetate Butyrate (CAB) A mixed ester produced by treating fibrous cellulose with butyric and acetic acids and anhydrides in the presence of sulfuric acid. CAB is generally supplied in the form of pellets prepared by mixing the molten ester with a plasticizer, then extruding and pelletizing. It is one of the toughest of the cellulosic plastics, and has good transparency, colorability, weatherability, electrical properties and resistance to inorganic chemicals. It can be processed by extrusion, injection molding, blow molding, rotational molding, and thermoforming. Applications include pipe, tool handles, instrument housings, lighting, packaging film, and marine hardware.

Cellulose Acetate Propionate (CAP, cellulose propionate) A thermoplastic formed by treating fibrous cellulose with propionic and acetic acids and anhydrides in the presence of sulfuric acid. CAP is easily extruded and injection molded, forming tough, flexible products with shock resistance close to that of ethyl cellulose.

Cellulose Acetobutyrate See CELLULOSE ACETATE BUTYRATE.

Cellulose Ester Any derivative of cellulose in which the free hydroxyl groups attached to the cellulose chain have been replaced wholly or in part by acidic groups, e g, nitrate, acetate, propionate, butyrate, or stearate groups. Esterification is effected by the use of a mixture of an acid with its anhydride in the presence of a catalyst such as sulfuric acid. Mixed esters of cellulose, e g, cellulose acetate butyrate, are prepared by using mixed acids and mixed anhydrides.

Cellulose Ether A cellulose derivative based on the etherification products of cellulose, such as ethyl cellulose, methyl cellulose, and sodium carboxymethyl cellulose.

Cellulose Nitrate (CN, nitrocellulose, NC, pyroxylin) Cellulose nitrate, dating back to the work of French chemist Braconnet in 1833, is the oldest of the synthetic plastics. It is made by treating fibrous cellulose with a mixture of nitric and sulfuric acids, and was first used in the form of a lacquer (see COLLODION). In 1870, John Wesley Hyatt and his brother patented the use of plasticized cellulose nitrate as a solid, moldable material, the first commercial thermoplastic (celluloid). Camphor was the first (and is still the best) plasticizer for CN, although many camphor substitutes have been developed. Alcohol is normally used as a volatile solvent to assist in plasticization, after which it is removed. Molded products of CN are extremely tough, but highly flammable and subject to discoloration in sunlight. CN is amenable to many decorative variations. Its principal uses today are in knife handles, table-tennis balls, and eyeglass frames.

Cellulose Propionate See CELLULOSE ACETATE PROPIONATE.

Cellulose Triacetate A member of the cellulosic plastics family made by reacting purified cellulose with acetic anhydride in the presence of a catalyst in such a manner that at least 92% of the hydroxyl groups are replaced by acetyl groups. Because of its high softening point, this material cannot be molded or extruded. Its major use is for casting films or spinning fibers from solutions, such as in a mixture of methylene chloride and methanol.

Cellulosic Plastic (cellulosic resin) Any of a family of thermoplastics made by substituting various chemical groups for the hydroxy groups in the cellulose molecules of cotton and purified wood pulp. See the following: CELLULOSE ACETATE, CELLULOSE ACETATE BUTYRATE, CELLULOSE ACETATE PROPIONATE, CELLULOSE ESTERS, CELLULOSE NITRATE, CELLULOSE TRIACETATE, ETHYL CELLULOSE, HYDROXYETHYL CELLULOSE, REGENERATED CELLULOSE.

Cementing Joining plastics to themselves or dissimilar materials by means of solvents (see SOLVENT CEMENTING), dopes, or chemical cements. *Dope adhesives* comprise a solvent solution of a plastic similar to the plastic to be joined. *Chemical cements*, the only type suitable for thermosetting plastics, are based on monomers or semi-polymers that polymerize in the joint to form a strong bond. See ADHESIVE.

Cenospheres Hollow microspheres in fly ash formed during combustion of coal in electric-power plants. They have had some use as a lower-cost substitute for glass microspheres in SYNTACTIC FOAMs.

Center-Gated Mold In injection molding, a mold in which each cavity is fed through an orifice at the center of the cavity. This type of gating is employed for items such as cups and bowls.

Centi- (c) The SI-approved prefix signifying multiplication by 10^{-2}.

Centipoise (cp) A deprecated, but still widely used viscosity unit, 0.01 POISE, w s. Water at 20°C has a viscosity of 1.002 cp. The SI equivalent is: 1 cp = 0.001 Pa·s.

Centistoke (cs) A deprecated, but still used unit of kinematic viscosity, 0.01 STOKE, w s, the approximate kinematic viscosity of water at 20°C. The SI equivalent is: 1 cs = 10^{-6} m^2/s.

Centrifugal Casting The process of forming tubes or other hollow cylindrical objects by introducing a measured amount of fluid resin dispersion into a rotatable container or mold, rotating the mold about the cylinder's axis at a speed high enough to force the fluid against all parts of the mold by centrifugal force, maintaining such rotation while solidifying the plastic by applicable means such as heating, then cooling if necessary, and removing the formed part. The fluid resin may be a dispersion such as a plastisol, or an A-stage thermoset with or without reinforcing strands. This process should not be confused with ROTATIONAL CASTING, which involves rotation at low speeds about one or more axes of rotation and gravity flow. See also CENTRIFUGAL MOLDING. Centrifugal casting is also used with metals, in particular, to manufacture BIMETALLIC CYLINDERs, w s.

Centrifugal Impact Mixer A device used for continuously mixing free-flowing dry blends, comprising a conical hopper in which are rotated at high speeds a rotor disk and a peripheral impactor. The material is fed to the center of the rotor which throws it against the impactor blades, which in turn throws the material against fixed impactors at the extremities of the cone. From there, the material flows downward to a discharge orifice. Compare: HIGH-INTENSITY MIXER.

Centrifugal Molding A process similar to CENTRIFUGAL CASTING, w s, except that the materials employed are dry, sinterable powders such as polyethylene. The powders are fused by heating the mold, then solidified by cooling it.

Ceramic Fiber A term embracing all reinforcing fibers made of refractory oxides such as Al_2O_3, BeO, MgO, $MgO \cdot Al_2O_3$, ThO_2, and ZrO_2. Although glasses are also ceramic materials, glass fibers are not generally included. Ceramic fibers are produced by chemical vapor deposition, melt drawing, spinning, and extrusion. Their main advantages are high strength and modulus, and resistance to high temperatures.

Ceraplast Any reinforced thermoplastic, particularly polyethylene, containing ceramic or mineral particles that have been dispersed in the polymer melt to their ultimate size (no agglomerates) and completely enveloped in resin. Bonding of the envelope to the filler particles and the matrix polymer is aided by the addition of a small percentage of reactive monomer or resin precursor. It is believed that, in the extremely thin transition envelope, there is a smooth gradient of modulus from that of the particulate material to that of the polymer. The mechanical properties of ceraplasts are superior to those in which the same fillers have been conventionally incorporated.

Cermet Any refractory composition made by bonding grains of ceramics, metal carbides, nitrides, etc, with a metal. Codeposition of cermets with nickel in the electroless-nickel process provides excellent wear resistance and chemical resistance to molds, dies, extruder screws and other tooling components used in the plastics industry.

C Glass A type of glass fiber not quite as strong and stiff as E GLASS, w s, but having better chemical resistance. Its major constituents are SiO_2 65%, CaO 13%, Na_2O 8%, B_2O_3 5%, Al_2O_3 3%, and MgO 2%. Fiber density is 2.49 g/cm^3, tensile modulus is 71 GPa, and tensile strength is 3.2 GPa.

CFRP See CARBON-FIBER-REINFORCED PLASTICS.

CGPM Abbreviation for *Conférence Générale des Poids et Mesures*, the international group that developed the system of weights and measures intended for worldwide use. The name *Système Internationale des Unités* and the abbreviation *SI* were adopted by the 11th CGPM in 1960. For information on SI units see SI (2) and the Appendix.

Chain Flexibility The ability of polymer molecules to assume a variety of configurations, arising from freedom of segments to rotate around C–C bonds. Polar side groups generally hinder rotation, making chains stiffer, while alkyl side chains tend to increase flexibility.

Chain Length The number of monomeric or structural units in a linear polymer, or the main chain in a branched polymer. See also DEGREE OF POLYMERIZATION.

Chain Stiffness See CHAIN FLEXIBILITY.

Chain-Transfer Agent A substance, used in polymerization, that has the ability to stop the growth of a molecular chain by yielding an atom to the active radical at the end of the growing chain, but in turn being converted to another radical that can initiate the growth of a new chain. Examples are thiols and carbon tetrachloride. Such agents are useful for preventing the occurrence of too long chains and too high WEIGHT-AVERAGE MOLECULAR WEIGHTs.

Chalk A soft white mineral consisting essentially of CALCIUM CARBONATE, w s, occurring as the remains of sea shells and minute marine organisms.

Chalking A specific type of BLOOM, w s, characterized by a dry, chalk-like appearance of the surface of a plastic article.

Change-Can Mixer (pony mixer) A type of planetary mixer comprising several paddle blades mounted on a vertical shaft rotating in one direction while the can or container counter-rotates. The paddle shaft is usually mounted on a hinged structure so that it can be swung out of the can, permitting the can to be removed, emptied, and replaced easily. This type of mixer is employed for relatively small batches (12 to 480 L) of fluid dispersions and dry materials.

Channel Black A type of CARBON BLACK, w s, made by impingement of a natural-gas flame against a metal plate, from which the deposit is scraped at intervals.

Channel Depth (h or H) Of an extruder screw, at any point along its length, the radial distance between the flight-tip surface and the screw-root surface. In a screw section of constant depth, half the difference between the outer (major) diameter of the screw and its root diameter.

Channel-Depth Ratio In an extruder screw, the ratio of the depth in the first turn of the screw at the feed end to the depth in the last turn at the delivery end. If the LEAD (w s) of the screw is constant, the channel-depth ratio is slightly larger than the CHANNEL-VOLUME RATIO (which follows).

Channel-Volume Ratio In an extruder screw, the ratio of the volume of the first turn of the screw at the feed end to the volume of the last turn at the delivery end. The term COMPRESSION RATIO is commonly used as a synonym in the extrusion industry.

Char (1, n) Animal or vegetable CARBON BLACK, w s, used as a decolorant in the process industries. (2, v) To partly burn and blacken, especially the outside surface of, a carbonaceous material.

Charge (n) The amount of material used to load a mold at one time or for one cycle. The amount may be expressed in either mass or volume units. In injection and transfer molding, the charge includes sprues and runners.

Charles' Law (Gay-Lussac's Law) The temperature part of the IDEAL-GAS LAW, w s, as follows: at constant pressure, the volume of any gas is directly proportional to its absolute temperature (*T*, K).

Charpy Impact Test A destructive test (ASTM D 256B) of impact resistance using a centrally notched test specimen 126 mm long and typically 12.7 mm square. The specimen is supported horizontally near its ends and struck on the side opposite the notch by a pendulum having sufficient kinetic energy to break the specimen with one blow. The result is expressed as the quotient of the energy absorbed from the pendulum divided by the specimen width (J/cm or ft-lb/in).

Chase (shoe) An enclosure of any shape used to (1) shrink-fit parts of a mold cavity in place, (2) prevent spreading or distortion in hobbing, or (3) enclose an assembly of two or more parts of a split-cavity block.

Checking Equals CRAZING, w s.

Cheese A supply of glass fiber wound into a cylindrical mass.

Chelate A compound comprised of metallic ions bound by a CHELATING AGENT, w s.

Chelating Agent A term derived from the Greek word *chele*, meaning claw. Thus, a chelating agent is a substance whose molecules are capable of seizing and holding metallic ions in a clawlike grip. The "claw" is usually a ring or cage structure of nitrogen, oxygen, or sulfur ("ligand atoms"), each of which donates two electrons to form a coordinate bond with the ion. See also SEQUESTERING AGENT.

Chemically Foamed Plastic A cellular plastic in which the cells are formed by thermal decomposition of a blowing agent or by the reaction of gas-liberating constituents. See also CELLULAR PLASTIC.

Chemical Resistance (reagent resistance) The ability of a plastic to maintain structural and esthetic integrity when exposed to acids, alkalis, solvents, and other chemicals. ASTM tests for chemical resistance of plastics include: C 581, Chemical Resistance of Thermosetting Resins Used in Glass-Fiber-Reinforced Structures; D 543, Resistance of Plastics to Chemical Reagents; D 1239, Resistance of Plastic Films to Extraction by Chemicals; D 1712, Resistance of Plastics to Sulfide Staining; D 2151, Test Method for Staining of Polyvinyl Chloride Compositions by Rubber-Compounding Ingredients; D 2299, Determining Relative Stain Resistance of Plastics; D 3615, Chemical Resistance of Thermoset Molding Compounds Used in the Manufacture of Molded Fittings; D 3681, Chemical Resistance of Reinforced-Thermosetting-Resin Pipe in a Deflected Condition; D 3753, Specification for Glass-Fiber-Reinforced Polyester Manholes; and D 4398, Determining the Chemical Resistance of Fiberglass-Reinforced Thermosetting Resins by One-Side Panel Exposure.

Chemisorption ADSORPTION, w s, particularly when irreversible, by chemical action rather than physical action.

Chill Roll A metal roll—a shell within a shell with a relatively small clearance between the two—with water or other heat-transfer medium circulating through the so-formed annular space, used to cool an extruded or cast film or sheet prior to winding or cutting. The surface of the roll may be polished or textured to impart a finish to the plastic.

Chill-Roll Extrusion Film extrusion in which the molten film is drawn over cooled, polished rollers, imparting high gloss to the film.

China Clay Synonym for KAOLIN, w s.

Chinese White Synonym for ZINC OXIDE, w s.

Chirality The property of an organic molecule of not being identical with its mirror image. All asymmetric molecules are chiral; however, not all chiral molecules are asymmetric since some having axes of rotational symmetry are chiral. Chiral and prochiral atoms are sites or potential sites, respectively, of STEREOISOMERISM, w s.

Chlorendic Anhydride (1,4,5,6,7,7-hexachloro-5-norbornene-2,3-dicarboxylic anhydride, HET anhydride) A difunctional acid anhydride having the structure:

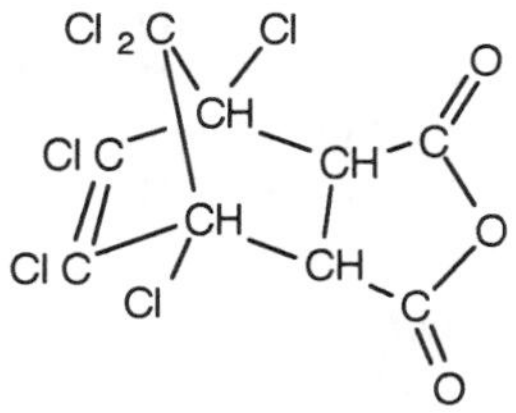

HET anhydride is a white, crystalline powder used as a hardening agent and flame retardant in epoxy, alkyd, and polyester resins.

Chlorinated Biphenyl (chlorinated diphenyl) Any of a group of plasticizers ranging from liquids to hard solids, used with polyvinylidene chloride and polystyrene. They are also used in conjunction with DOP (w s) as coplasticizers for PVC, and in conjunction with polyvinyl acetate, ethyl cellulose and other thermoplastics, as adhesives.

Chlorinated Hydrocarbon Any of a wide variety of liquids and solids resulting from the substitution or addition of chlorine in hydrocarbons such as methane, ethylene, and benzene. They are employed as solvents, plasticizers, and monomers for plastics manufacture.

Chlorinated Paraffin (chlorocosane) Any of a family of yellow to light amber liquids produced by chlorinating a paraffin oil, with uses as secondary plasticizers for vinyls, polystyrene, polymethyl methacrylate, and coumarone-indene resins. Chlorinated paraffins also impart flame resistance to polyolefins, polystyrene, PVC, natural rubber, and unsaturated polyester resins.

Chlorinated Polyether [poly-3,3-bis-(chloromethyl)oxacyclobutane, Penton®] A corrosion-resistant thermoplastic obtained by polymerization of chlorinated oxetane monomer, the oxetane being derived from PENTAERYTHRITOL, to a high molecular weight (250,000 to 350,000). The polymer is linear, crystalline, and extremely resistant to degradation at processing temperatures. It may be injection molded, extruded, or applied as a coating by the fluidized-bed method. The resin is widely used in valves, pumps, flowmeters, etc, for chemical plants.

Chlorinated Polyethylene (CPE) Any polyethylene modified by simple chemical substitution of chlorine on the linear backbone chain. CPEs range from rubbery amorphous elastomers at 35–40% Cl to hard, semicrystalline materials at 68–75% Cl. They are sometimes included with chlorinated natural and butyl rubbers under the term *chlorinated rubbers*. Certain CPEs are used as modifiers in PVC compounds to obtain better flexibility and toughness, particularly low-temperature toughness, greater latitude in compounding, and ease of processing.

Chlorinated Polyvinyl Chloride (CPVC) A PVC resin modified by post-chlorination. A series of such resins, known as "Hi-Temp Geon", is available from B F Goodrich Chemical Co. Compared to conventional rigid PVC, CPVC withstands service temperatures 20° to 30°C higher, is stronger, and has better chemical resistance. CPVC is mildly hygroscopic, so requires predrying before processing. With that proviso, it can be processed by all the methods used for rigid PVC with few modifications.

Chlorinated Rubber (rubber chloride) Natural rubber in which about two-thirds of the hydrogen atoms have been replaced by chlorine atoms. It has adhesive properties and, because of its good fire resistance, is used in paints.

Chlorobenzene (chlorbenzene, chlorbenzol, chlorobenzol, phenyl chloride) C_6H_5Cl. A solvent, and an intermediate in the production of phenol.

Chlorobutanol (chlorbutanol, 1,1,1-trichloro-2-methyl-2-propanol, acetone chloroform, trichloro-*tert*-butyl alcohol) $Cl_3CC(CH_3)_2OH$. A plasticizer for esters and ethers of cellulose.

Chlorodiphenyl Resin (chlorobiphenyl resin) Any resin made from chlorinated biphenyl, rosin or rosin ester, and the higher fatty acids. These resins are used as plasticizers and modifying resins in plastics, and in lacquers and varnishes.

Chloroethane Synonym for ETHYL CHLORIDE, w s.

Chloroethene (chloroethylene) Synonym for VINYL CHLORIDE, w s.

Chlorofluorocarbon Resin Any resin made by the polymerization of monomer(s) containing only carbon, chlorine, and fluorine. The principal member is POLYCHLOROTRIFLUOROETHYLENE (PCTFE), w s.

Chlorofluorohydrocarbon Resin A resin made by polymerization of monomer(s) containing only carbon, chlorine, fluorine, and hydrogen.

Chloroform (trichloromethane) $CHCl_3$. A pungent, toxic, dense liquid, useful as a solvent for epoxy resins and others, less used nowadays than formerly because of its toxicity.

Chlorohydrin (α-chlorohydrin, 3-chloropropane-1,2-diol, glyceryl α-chlorohydrin) $ClCH_2CHOHCH_2OH$. A solvent, mainly for cellulosics.

Chlorohydrin Rubber See EPICHLOROHYDRIN RUBBER.

Chloronaphthalene Oil Any of several nearly colorless oils derived by chlorinating naphthalene, used as plasticizers and flame retardants.

α-Chloro-*meta*-Nitroacetophenone $O_2NC_6H_4COOH_2Cl$. A bacteriostat and fungistat for plastics.

Chloroprene Polymer See NEOPRENE.

Chloropropylene Oxide Synonym for EPICHLOROHYDRIN, w s.

Chlorostyrenated Polyester An unsaturated polyester resin made by reacting a fluid polyester with monochlorostyrene in place of styrene. (See POLYESTER, UNSATURATED.) Monochlorostyrene is less volatile and more reactive than styrene, providing faster cure rates and increased flexural strength and modulus in glass-fiber laminates.

Chlorosulfonated Polyethylene (CSM, CSPR) An elastomer resistant to solvents, chemicals, and ozone. It is produced by simultaneous treatment, with sulfur dioxide and chlorine, of dissolved, radicalized polyethylene. A commercial product (Hypalon®, Du Pont) contains 22–26% Cl and 1.3–1.7% S.

Chlorotrifluoroethylene $ClFC{=}CF_2$. A colorless gas, the monomer for POLYCHLOROTRIFLUOROETHYLENE, w s. It is obtained by either dehalogenation or dehydrohalogenation of saturated chlorofluorocarbons or chlorohydrocarbons, e g, by reacting 1,1,2-trichlorotrifluoroethane with zinc.

Choker Bar (restrictor bar) A bendable metal bar incorporated in a sheet-extrusion die for controlling flow distribution and lateral sheet thickness, and for reducing stagnation in the melt. The shape of a flow passage between the choker bar and the lower die body is altered by turning bolts connected to the bar.

Cholesteric See LIQUID-CRYSTAL POLYMER.

Chopped Strand A type of GLASS-FIBER REINFORCEMENT, w s, consisting of strands of individual glass fibers that have been chopped into lengths from 1 to 12 mm. The individual fibers are bonded together within the strands so that they remain in bundles after being cut. See also ROVING.

Chopped-Strand Mat A mat formed from randomly oriented, chopped strands of glass and held together, just strongly enough for handling, by a binder.

Chromatography A process in which a gas or liquid solution moves through a calibrated column containing a subdivided solid phase into which some components of the solution are absorbed, smaller molecules more quickly and thoroughly than larger ones. This is followed by pure carrier gas or solvent, the stream being monitored by a differential detector. Larger molecules emerge first, smaller ones later. The detector signal is proportional to the concentration of each species in the effluent. The process is mainly used for analysis of organic mixtures, but also for their separation. The name "chromatography" derives from the work of the Russian botanist N. Tswett, who first used the process to separate chloroplast pigments, obtaining colored bands on filter paper. Some variations of the process are *gas chromatography* (the gas mixture is passed through a porous bed, or through a capillary tube lined with an absorbent liquid or solid phase); *paper chromatography* (a drop of specimen is placed near one end of a porous paper); *ion-exchange chromatography*; *thin-layer chromatography* (the sample is placed on an absorbent cake spread on a smooth glass plate); and, important for plastics, SIZE-EXCLUSION CHROMATOGRAPHY, w s.

Chrome Green (brunswick green) Any of a family of pigments ranging from light yellow-green through dark green, based on physical mixtures of chrome yellow and iron blue (a complex ammonium iron hexacyanoferrate). The amount of iron blue determines the shade, about 2% being used for light yellow-green, and up to 64% being used for dark greens.

Chrome-Orange Pigment Any pigment based on basic lead chromate, $PbO \cdot PbCrO_4$, that is of a deep orange color.

Chrome-Oxide Green A stable pigment based on anhydrous chromium oxide, Cr_2O_3. The form based on hydrated chromium oxide is called *Guignet's green*.

Chromel and Alumel Special high-nickel alloys that, when their wires are joined, develop high thermoelectric power, and are therefore useful for temperature measurement. "Type K" thermocouples are usually chosen for high-temperature work in oxidizing atmospheres.

Chrome-Yellow Pigment (primrose chrome, permanent yellow) Any pigment based on normal lead chromate, $PbCrO_4$, that is characterized by a medium yellow color. Other shades ranging from light greenish-yellow to medium reddish-yellow are made by coprecipitating lead chromate with other insoluble compounds such as lead sulfate or lead phosphate.

Chromic Chloride (chromium chloride, chromium chloride hexahydrate) $CrCl_3$ or $[Cr(4H_2O)Cl_2]Cl \cdot 2H_2O$. A catalyst for polymerizing olefins.

Chromium Plating (chrome plating) An electrolytic process that deposits a hard, inert, smooth layer of chromium onto working surfaces of other metals for resistance to corrosion and wear. Extruder screws, chill rolls for sheet and film production, calendering rolls, dies, and molds are commonly chromium plated

Chromophore A group such as –NO, $-NO_2$, or –N=N– that, when present in a molecule, enables the molecule to be transformed into a dye upon the introduction of an acid group.

Chrysotile (serpentine) $Mg_3Si_2O_5(OH)_4$ or $3MgO\cdot 2SiO_2\cdot 2H_2O$. A hydrated magnesium orthosilicate, the chief constituent of *serpentine* asbestos. Chrysotile-bearing asbestos has been the most used type, once accounting for over 90% of the world production. Its fine and silky fibers, and mats and felts made therefrom, were widely used as fillers and reinforcements for plastics, providing excellent resistance to chemicals and fire. Many of its former uses are now prohibited because of the carcenogenicity of some types of asbestos.

CIL Flow Test A capillary-rheometer test developed at Canadian Industries Ltd for characterizing the flow of thermoplastics. The reported flow unit is the amount of melt that is forced through a specified orifice per unit time when a suitably chosen force is applied. Similar to MELT-FLOW INDEX, w s.

CIM Acronym for **c**omputer-**i**ntegrated **m**anufacturing.

Circuit In filament winding, one complete traverse of the fiber-feed mechanism of the winding machine; or a complete traverse of a winding band from one arbitrary point along the winding path to another point on a plane through the starting point and perpendicular to the axis.

cis- A chemical prefix (Latin: "on this side"), usually ignored in alphabetizing lists, denoting an isomer in which certain atoms or groups are on the same side of a plane. Opposite of *TRANS-*, w s.

Citrate Plasticizer Any of a family of plasticizers derived from citric acid, $HO_2CCH_2C(OH)(CO_2H)CH_2CO_2H$, noted for their low order of toxicity. Included citrates are: triethyl, tri(2-ethylhexyl), tricyclohexyl, tri-n-butyl, acetyl triethyl, acetyl tri-n-butyl, acetyl tri-n-octyl, n-decyl, and acetyl tri(2-ethylhexyl).

Cl Chemical symbol for the element chlorine.

Clamping Capacity The largest rated projected area of cavities and runners that an injection or transfer press can safely hold closed at full molding pressure.

Clamping Force In injection and transfer molding, the force applied to the mold to keep it closed, and opposing the pressure exerted by the injected plastic acting upon the projected area of cavities and runners. Per square centimeter of cavities and runners, at least 3.5 kN of clamping force is required.

Clamping Plate A plate, fitted to a mold, that secures the mold to the frame of the molding machine.

Clamping Pressure In injection and transfer molding with a hydraulically operated mold, the hydraulic-fluid pressure applied to the mold ram to keep the mold closed during the molding cycle. Compare CLAMPING FORCE.

Clamshell Molding A term applied to the modern version of the oldest form of blow molding—preheating two sheets of plastic, placing them between halves of a split mold, closing the mold, drawing the sheets against their respective mold surfaces by means of vacuum, then completing the forming with air pressure between the sheets. The modern process, mechanized and conveyorized, is superior to blow molding from a parison for very large parts and for those in which uniformity of wall thickness is important.

Clarifier An additive that increases the transparency of a plastic material.

Clash-Berg Point The rising temperature at which the apparent modulus of rigidity of a specimen falls to 931 MPa, the end point of "flexibility" as defined by Clash and Berg in their studies of low-temperature flexibility. In a similar test described in ASTM D 1043, the deciding shear modulus is one-third the C-B value.

Clay Any naturally occurring sediment rich in hydrated silicates of aluminum, predominating in particles of colloidal or near-colloidal size. There are many types of clays and clay-like minerals. Those of particular interest to the plastics industry are varieties refined by nature and man to a state of good color and particle-size distribution, such as KAOLIN (china clay), w s. They are used as fillers in epoxy and polyester resins, PVC compounds, and urethane foams. *Calcined clays* are those that have been heated to a high temperature to drive off the chemically bound water, sometimes also surface-treated to improve their chemical inertness and moisture resistance. They are used primarily in vinyl insulation.

Clearance The space or gap between two adjacent surfaces particularly those in relative motion, such as a shaft and its journal bearing, or mating parts in an assembly.

Clear Point With regard to vinyl plastisols, clear point is the rising temperature at which an unpigmented plastisol suddenly becomes transparent, signifying that the resin particles have completely dissolved in the warm plasticizer. This test is useful for determining the relative fusion temperatures of different plastisols.

Clicker Die A cutting die for stamping out blanks from plastic sheet.

Clicker Press A stamping press used with CLICKER DIES to cut shapes from plastic sheet. Compare DIE CUTTING.

Clicking See DIE CUTTING.

Closed-Cell Foamed Plastic (unicellular foam) A CELLULAR PLASTIC, w s, in which interconnecting cells are too few to permit the bulk flow of fluids through the mass.

Closed-Loop Control See FEEDBACK CONTROL.

Cloud Point (1) In condensation polymerization, the temperature at which the first turbidity appears, caused by water separation when a reaction mixture is cooled. (2) In petroleum and other oils, the falling temperature at which the oil becomes cloudy, from precipitation of wax or other solid.

CMC Abbreviation for CARBOXYMETHYL CELLULOSE, w s, or for ceramic-metal composite.

CN Abbreviation for CELLULOSE NITRATE, w s.

Co Chemical symbol for the element cobalt.

Coacervation The separation of a polymer solution into two or more liquid phases, one of which is a polymer-rich liquid. The term was introduced to distinguish this phenomenon from the precipitation of a polymer solute in solid form. The process is used in MICROENCAPSULATION, w s, by emulsifying or dispersing the material to be encapsulated with a solution of the polymer. By changing the temperature or concentration of the mixture, or by adding another polymer or solvent, a phase separation may be induced and the polymeric portion forms a thin coating on the external surfaces of the particles. After further treatment to solidify the polymeric wall, the capsules can be isolated in powder form by filtration.

Coagulant A substance that (1) initiates the formation of relatively large particles in a finely divided suspension. or (2) assists in the formation of a gel, thus accelerating settling of the particles or their deposition on a substrate.

Coagulation A physical or chemical action inducing transition from a fluid to a semi-solid or gel-like state, or the bringing together of small, individual particles into clumps.

Coal-Tar Resin See COUMARONE-INDENE RESIN.

Coated Fabric A cloth that has been impregnated and/or coated with a plastic material in the form of a solution, dispersion, hot melt, or powder. The term is sometimes used when a preformed film is applied to the fabric by calendering, although such products are more properly termed *laminates*.

Coathanger Die A sheet- or film-extrusion die whose melt-distribution manifold has the obtuse-isoceles outline of a coathanger. This popular die design is said to yield more uniform distribution of material across the full width of the extruded web, thus producing sheet of laterally more uniform thickness. Side-fed blow-molding dies and spiral-type dies for blown film may also be spoken of as coathanger dies.

Coating Methods See:

AIR-KNIFE COATING
CALENDER COATING
CASCADE COATING
CURTAIN COATING
DECORATING
DIP COATING
ELECTROPHORETIC DEPOSITION
ELECTROPLATING ON PLASTICS
ELECTROSTATIC FLUIDIZED-BED COATING
ELECTROSTATIC SPRAY COATING
EXTRUSION COATING
FLAME-SPRAY COATING
FLOCKING
FLOW COATING
FLUIDIZED-BED COATING
FRICTION CALENDERING
GRAVURE COATING
INTUMESCENT COATING
KISS-ROLL COATING
PAINTING OF PLASTICS
PLASMA-SPRAY COATING
PRINTING ON PLASTICS
REVERSE-ROLL COATING
ROLLER COATING
SILVER-SPRAY PROCESS
SINTER COATING
SOLUTION COATING
SPRAY COATING
SPREAD COATING
STRIPPABLE COATING
TRANSFER COATING
URETHANE COATINGS
VACUUM METALIZING

Cobalt Naphthenate See NAPHTHENIC ACID.

Cocatalyst (promoter) A chemical that alone is a feeble catalyst, but that greatly increases the activity of a given catalyst.

Co-Cure To cure two or more different materials in one step.

Coefficient of Determination (R^2) In statistical analysis by multiple, polynomial, or nonlinear regression, the fraction of the total variance of the dependent variable accounted for or "explained" by the regression. In most cases, R^2 is a useful measure of the goodness of fit of the regression model to the data and of its ability to predict new results. It is defined by the equation

$$R^2 = 1 - [\Sigma(y_i - y)^2/\Sigma(y_i - y^2]$$

where y_i is any one of the n measurements of the dependent variable, y is the corresponding value estimated from the regression equation, y is the mean of all the observed y-values, and the sums are performed over all n data points.

Coefficient of Elasticity A rarely seen synonym for MODULUS OF ELASTICITY, w s.

Coefficient of Friction See FRICTION.

Coefficient of Thermal Conductivity See THERMAL CONDUCTIVITY.

Coefficient of Thermal Expansion The fractional change in length (or sometimes in volume, when specified) of a material for a unit change in temperature, as given by the equation

$$CE = (1/L)\cdot dL/dT = d\ln L/dT \approx \Delta L/\Delta T$$

See also VOLUME COEFFICIENT OF THERMAL EXPANSION.

Coefficient of Twist Contraction The shortening of a yarn, due to twist, per unit length of untwisted yarn, usually expressed in percent.

Coefficient of Variation (CV) A statistical term, the quotient of the STANDARD DEVIATION (w s) divided by the MEAN (w s), either of a set of measurements or of the population from which they are conceived to have been drawn. CV is usually expressed as a percentage. Analytical chemists often use the synonym *relative standard deviation* for this statistic.

Coextrusion The process by which the outputs of two or more extruders are brought smoothly together in a FEED BLOCK, w s, to form a single multilayer stream that is fed to a die to produce a layered extrudate. The extruder streams may be split within the feed block to form dual layers, usually in a symmetrical arrangement about the center plane of the final sheet. Sheeting containing up to nine layers is commercially produced. Coextrusion is employed in film blowing, sheet and flat-film extrusion, blow molding, and extrusion coating. The advantage of coextrusion is that each ply imparts a desired characteristic property, such as stiffness, heat-sealability, impermeability, or resistance to some environment, all of which properties would be impossible to obtain with any single material. Layers of poorly compatible plastics can be coextruded by including a thin adhesive layer between them.

Coextrusion Blow Molding A variant of extrusion BLOW MOLDING, w s, in which the parison contains two or more layers of at least two materials. See COEXTRUSION.

Cogswell Rheometer See EXTENSIOMETER.

Cohesion The holding together of the atoms or molecules of a single substance by primary or secondary valence forces.

Cohesive Energy Density For liquids, the heat of vaporization per unit mass, divided by the specific volume, or the same quantity based on molar properties. Cohesive energy density is also equal to the square of the SOLUBILITY PARAMETER, w s.

Coining A term borrowed from the metal-stamping industry for a process of forming integral hinges from plastics. In the case of a polypropylene article, the hinge is produced by molding a thin section between the two parts of the article to be hinged. Such a thin section cannot be molded easily in articles of nylon or acetal resin because of the difficulty of filling the half of the mold cavity opposite the gated half through the thin section. In the coining process, the area to be formed into a hinge is molded in a thickness suitable for the molding process. Subsequently, the article is placed between bars in a press that quickly squeezes the plastic to the desired thickness. The material must be deformed beyond its compressive yield point but short of failure, so that it remains essentially stable with little recovery from the deformation. This rapid cold pressing produces a high degree of orientation that imparts high strength and flexibility to the integral hinge area.

Coinjection A process similar in its results to COEXTRUSION, w s, but accomplished by modifications of the injection-molding process. By means of various nozzle and valving arrangements, two or more materials can be injected either simultaneously or sequentially to form an article with an outer shell of one material with certain desired properties, the shell filled with another material to attain other desired properties such as reduced cost.

Cold-Bend Test A test for measuring the flexibility of a plastic material at low temperatures. A specimen in a series is chilled to one of several specified low temperatures, then bent to a predetermined radius until the temperature at which half the specimens tested do not survive the bend has been identified.

Cold Drawing (cold stretching) A stretching process performed at a temperature below a thermoplastic's melting range to orient the material and improve the tensile modulus and strength.

Cold Flow See CREEP.

Cold Forming A group of processes by which sheets or billets of thermoplastics are formed into three-dimensional shapes at room temperature by processes used in the metal-working industry such as forging, brake-press bending, deep drawing, rolling, stamping, heading, and coining. The materials used, generally in relatively thick sections, include ABS, polycarbonate, polyolefins, and rigid PVC. When either the material or the forming dies are preheated, the preferred term is SOLID-PHASE FORMING, w s.

Cold Heading A process for forming short plastic rods into rivets by uniformly loading the projecting shaft end in compression while holding and containing the shaft trunk. All thermoplastics can be cold-headed but acetal and nylon are particularly suitable.

Cold Molding A process similar to COMPRESSION MOLDING, w s, except that no heat is applied during the molding cycle. The formed part is subsequently cured by heating and cooling. A-stage phenolic resins and bituminous plastics are sometimes molded by this process.

Cold-Parison Blow Molding See BLOW MOLDING.

Cold Pressing A bonding operation in which an assembly is subjected to pressure without application of heat.

Cold-Runner Injection Molding (runnerless injection molding)

Whereas in injection molding thermoplastics the runners are sometimes kept hot to reduce scrap (see HOT-RUNNER MOLD), with thermosets the runners are kept *cooler* than the cavities to prevent material from curing within the runner system. In cold-runner injection molding of thermosets the mold is divided into two sections: a heated section containing the cavities, and an insulated manifold section containing the injection sprue and runners. Material is fed from runners to cavities through very short gates or sub-sprues.The insulated manifold is maintained at a temperature high enough to soften the uncured material, generally in the vicinity of 90°C, but well below the curing temperature prevailing in the cavity section.

Cold Slug The first material to enter an injection mold, so called because in passing through the sprue orifice it is cooled below the effective molding temperature. In some molds, a small well in the mold opposite the sprue catches the cold slug and thereby prevents it from entering the runner system.

Cold-Slug Well A small circular cavity, directly opposite the sprue opening in an injection mold, that traps the COLD SLUG, w s.

Collapse (1) Inadvertent densification of cellular material during manufacture resulting from breakdown of cell structure (ASTM D 883). (2) Inward contraction of the walls of a molded container, e g, while cooling, resulting in permanent indentation.

Colligative Property A property of a material or solution that depends mainly on the number, rather than the nature, of the molecules, atoms or ions present. Examples are gas pressure and osmotic pressure of solutions.

Collimated Roving Roving with strands that are more nearly parallel than those in standard roving, usually made by parallel winding.

Collodion A solution of cellulose nitrate in alcohol and ether used as a coating lacquer or to cast very thin films of cellulose nitrate on water ("microfilm").

Collodion Cotton See CELLULOSE NITRATE.

Colloid An emulsion or suspension of extremely small particles that is stable (will not settle out) and will not diffuse readily through vegetable or animal membranes. Colloidal particles are usually aggregates of molecules ranging in diameter from about 0.1 to 50 μm.

Colloidal Clay See BENTONITE.

Colloid Mill A device for preparing emulsions and reducing particle size, consisting of a high-speed rotor and a fixed or counter-rotating element in close proximity to the rotor. The liquid is conveyed continuously from a hopper to the space between the shearing elements, then discharged into a receiver. See also HOMOGENIZER.

Colmonoy See HARD-FACING ALLOY.

Colorant Any dye or pigment that can impart color to a plastic. The dyes are natural or synthetic compounds of submicroscopic or molecular size, soluble in most common solvents, yielding transparent colors. Their generally poor heat resistance and tendency to migrate limit their use as additives to a few families that are superior in heat resistance. However, dyes are sometimes used to post-color finished parts such as buttons and fibers. The pigments are organic and inorganic substances with larger particle sizes, rarely below 1 μm, and usually insoluble in the common solvents. Organic pigments produce translucent and nearly transparent colors, resist migration better than dyes, and are somewhat more heat-resistant. Inorganic pigments are, with few exceptions, opaque and superior to organics in light-fastness, heat resistance and resistance to migration. Colorants are added to plastics by dry-coloring (simply tumbling the colorant with the base or compounded resin powder or pellets); by extrusion coloring (extruding a dry-colored mixture and chopping it into pellets to be reprocessed); by masterbatching (see COLOR CONCENTRATE); or by stirring colorants or dispersions thereof into liquid plastisols or resin systems. See also:

BON PIGMENT
FLUORESCENT PIGMENT
FLUSHED PIGMENT
GLITTER
INORGANIC PIGMENT
LIQUID COLORANT
LUMINESCENT PIGMENT
METALLIC-FLAKE PIGMENT
ORGANIC PIGMENT
PEARLESCENT PIGMENT
PERYLENE PIGMENT
PHOSPHORESCENT PIGMENT
PHTHALOCYANINE PIGMENT
QUINACRIDONE PIGMENT
RHODAMINE
ULTRAMARINE-BLUE PIGMENT

Color Bleeding See BLEEDING.

Color Concentrate A plastic compound that contains a high percentage of pigment, to be blended in precise amounts with the base resin or compound so that the correct final color will be achieved. The concentrate provides a clean and convenient method of obtaining accurate color shades in extruded and molded products. The term *masterbatch* is sometimes used for color concentrate, as well as for concentrates of other additives. See also MULTIFUNCTIONAL CONCENTRATE.

Colorfastness See LIGHT RESISTANCE.

Colorimeter An instrument for matching colors with results approximately equal to those of visual inspection, but more consistent. The sample is illuminated by light from three primary-color filters, and scanned by an electronic detecting system. A colorimeter is sometimes used together with a SPECTROPHOTOMETER, w s, for close control of color in production.

Colorimetry (color-identification testing) A method of analysis based on the fact that certain plastics undergo characteristic color changes when exposed to certain chemicals.

Color Migration The movement of dyes or pigments through or out of a material.

Color Stability The constancy of the characteristics of color in a plastic compound—hue, intensity, and saturation—in its products over their service lives in their design environments. See LIGHT RESISTANCE.

Combination Mold Synonym for FAMILY MOLD, w s.

Combining Weight Synonym for EQUIVALENT WEIGHT, w s.

Combustible Liquid A liquid that emits flammable vapors in the temperature range from 27 to 66°C. Most solvents used in plastics manufacture and processing are combustible by this definition.

Combustion Analysis Any of several methods for quantitatively determining, by burning, the elemental composition of organic compounds, including plastics. First introduced in the 1830s by J. von Liebig, it was refined to permit accurate analysis of small samples (10–50 mg) by F. Pregl, who led the development of the microbalance. Modern combustion analysis is highly automated, but still relies on the microbalance.

Comminute (v) To pulverize or reduce to very small sizes, as by grinding.

Comoforming A fabrication process that combines vacuum-formed thermoplastic shapes with cold-molded fiberglass-reinforced resin to produce parts having excellent surface appearance and weatherability.

Comonomer A monomer that is mixed with one or more other monomers for a polymerization reaction, to make a COPOLYMER, w s.

Compact (n) See POWDER COMPACT.

Compatibility The ability of two or more substances to mix together or be joined without objectionable separation. In plastics technology the term is most often used in connection with plasticizers, but is also applied to resin pairs or to a resin and prospective compounding ingredients. ASTM tests for compatibility of plasticizers and PVC resins are D 2383 and D 3291. See also LOOP TEST.

Compatibilizer A material that, added to a blend of ordinarily incompatible polymers, suppresses phase separation.

Complexing Agent See CHELATING AGENT, SEQUESTERING AGENT.

Complex Modulus (complex dynamic modulus) A property of viscoelastic materials subjected to periodic variation or reversal of stress. The stress may be any of the three principal types; the material may be in the solid or liquid state (molten or concentrated solution). In such materials, the strain lags the applied stress in time and when the stress is periodic, the time lag is characterized by a phase angle, θ. The modulus—the ratio of stress to strain—is resolved into two parts, a "real" or in-phase part and an "imaginary" part lagging the real part by $\pi/2$ radians (90°). For example, in shear, the complex modulus is stated as $G = G' + iG''$, where $i = (-1)^{0.5}$. The vector sum of the two components is called the absolute (dynamic) modulus. The real part is often referred to as the "modulus" while the imaginary part is called the "dynamic viscosity". Both parts vary with frequency, both diminish with rising temperature, and, like static moduli, they are different for different modes of stress.

Compliance The degree to which a material deforms under stress; the reciprocal of the modulus. Thus, in each mode of stress, the material is characterized by three moduli and their reciprocals, three compliances. However, when the stress is varying, the "real" and "imaginary" parts of the complex compliances are *not* equal to the reciprocals of their counterparts in the COMPLEX MODULUS, w s.

Composite An article or substance made up of two or more distinct phases of different substances. In the plastics industry the term applies broadly to structures of reinforcing members (*dispersed phase*) incorporated in compatible resinous binders (*continuous phase*). Such composites are subdivided into classes on the basis of the reinforcing constituents: LAMINATE, w s; *particulate* (the dispersed phase consists of small particles); *fibrous* (the dispersed phase consists of unlayered fibers); *flake* (flat flakes forming the dispersed phase); and *skeletal* (composed of a continuous skeletal matrix filled by a resin).

Composite Laminate A term sometimes applied to a laminated plastic bonded to a nonplastic material such as copper, vulcanized fiber, rubber, asbestos, lead, aluminum, etc. An example is the copper-clad laminated plastic used for printed-circuit boards.

Composite Mold A mold in which different shapes are produced in one cycle from the several cavities. See also FAMILY MOLD.

Composite Molding The process of molding two or more materials in the same cavity in the same shot, by a combination of transfer and compression molding. For example, in making a ring gear, a loose nylon-fiber-filled material is loaded into an open mold around the tooth circle, the mold is closed, then molten nylon is injected by transfer molding.

Composites Manufacturing Association (CMA/SME) Formerly a subgroup of the Society of Manufacturing Engineers, CMA is now a separate entity with 3000 members to promote composites, publish books and tapes, and hold conferences. Its office is at 1 SME Dr, P O Box 390, Dearborn, MI 48121.

Composition (1) A synthetic material containing resins or/and elastomers, and perhaps other components, in specified percentages. (2) The list of constituents and their percentages in such a material.

Compound (1, n) A mixture of resin and the ingredients necessary to modify the resin to a form suitable for processing into finished articles having the desired performance properties. (2, n) In chemistry, a combination of atoms ionically or covalently bonded in fixed ratios to form a molecule. (3, v) To produce a plastic compound by blending ingredients with resins in intensive mixers, extruders, or dry-blenders.

Compound Curve A surface having curvature in two principal directions. Simply curved surfaces, such as cylinders and cones, having only one direction of curvature, may be cut along an element and laid flat. Compound curves, such as spheres and hyperbolic paraboloids, cannot be laid flat without distortion no matter how they are cut. Structures having compound curvature (reinforced-plastic roofs, for example) have high stiffness for their mass.

Compounding Mixing basic resins with additives such as plasticizers, stabilizers, fillers, and pigments in a form suitable for processing into finished articles. In some areas of the industry the term includes fusion of the polymer, for example in the production of molding powders by extrusion and pelletizing. In the plastisol industry, the compounding step ends with the preparation of the dispersion, fusion being part of the molding step.

Compreg A contraction of "**com**pressed im**preg**nated wood", usually referring to an assembly of veneer layers impregnated with a liquid resin and bonded under high pressure.

Compregnate To impregnate and simultaneously or subsequently compress, as in the production of compregs.

Compressed-Air Ejection The removal of a molding from its mold by means of a jet of compressed air.

Compressibility The relative change in volume per unit change in pressure; the reciprocal of BULK MODULUS, w s.

Compression Mold A mold used in the process of COMPRESSION MOLDING, w s.

Compression Molding A method of molding in which a thermosetting molding material, generally preheated, is placed in an open, heated mold cavity, the mold is closed with a top force or plug member, pressure is applied to force the material into intimate contact with all mold surfaces, and heat and pressure are maintained until the material has cured and solidified. The process most often employs thermosetting resins in a partly cured stage, either in the form of granules or putty-like masses, or sometimes in preformed shapes roughly conforming to the shape of the mold. Compression molding has also been used with thermoplastics, most notably phonograph records. In this process, the mold is cooled following the compression-flow stage.

Compression-Molding Pressure (1) The force per unit of projected area applied to the molding material in a compression mold. The area is projected from all parts of the material under pressure during the complete closing of the mold. (2) The *hydraulic* pressure applied to the compression ram during molding.

Compression Ratio In an extruder, the ratio of the volume of the first turn of the feed section to that of the last turn of the metering section. This ratio is a rough indication of the total compaction performed on the feedstock. More precisely called the *channel-volume ratio* or, for a screw of constant pitch, *channel-depth ratio*.

Compression Set A permanent deformation remaining after release of a compressive stress. It is a property of interest in elastomers and cushioning materials, such as plastic foams. See, for example, ASTM tests D 395 and D 1565, in Section 9.

Compression Zone (compression section) The part of an extruder screw, connecting the feed and metering sections, in which the volume per turn is decreasing because of decreasing channel depth, or lead, or both. In some older designs, rarely made today, there were no distinct feed and metering sections and the volume per turn decreased over the entire screw length, so the compression zone was the entire screw. In two-stage screws used for vented operation, the rear metering zone is normally followed by a deep zone of *decompression*, then a second, short compression zone and second metering zone. Where decreasing channel depth is the means of volume reduction, it may be done with a conical screw-root profile or, more usual today, a helical root profile.

Compressive Modulus The ratio of compressive stress to compressive strain below the proportional limit. While it is theoretically equal to Young's modulus determined from tensile testing, compressive modulus is usually somewhat greater in plastics.

Compressive Strain The reduction in thickness (length), in the direction of compression, of a specimen under test, divided by the original, unstressed thickness, usually expressed as a percent.

Compressive Strength The load at which a test specimen fails in compression, divided by the original cross-sectional area perpendicular to the load. For rigid plastics, ASTM test D 695 is used; for rigid foams, D 1621. These tests also prescribe procedures for estimating compressive moduli. The actual mode of failure of a stiff material in a test of compressive strength is usually by diagonal shear.

Compressive Stress The compressive load per unit area of perpendicular cross section carried by the specimen during a compression test.

Computer-Aided Design (CAD) Using a computer, and appropriate programs (software), in the engineering design—even to the production of finished working drawings—of parts, tools, molds, and assemblies.

Computer-Aided Manufacturing (CAM) Using a computer—usually dedicated and typically a minicomputer—with appropriate programs (software), to control parts or all of a manufacturing operation.

Computer Numerical Control (CNC, numerical control) The use of a dedicated small computer to implement the control of a process or, typically, a machining task.

Concavity Factor The entire stress-strain curves of rubbers and elastomers that have no elastic limit are typically concave toward the stress axis (and convex to the strain axis). The concavity factor is the ratio (less than 1) between the energy beneath the extension curve to that beneath the straight line to the same final point.

Concentricity The characteristic of circles or circular cylindrical surfaces of different radii having a common center. More loosely, the property, in *any* annular shape, of constant radial wall thickness. Concentricity is important in blown film, pipe and tubing, wire-coating, and many noncircular extrusions.

Condensation (1) The process of reducing a gas or vapor to liquid or solid form. (2) A chemical reaction in which two or more molecules combine with the separation of water or some other simple molecule. If a polymer is formed the process is called *polycondensation.*

Condensation Agent A chemical compound that acts as a catalyst and also furnishes a complement of material necessary for a polycondensation reaction to proceed.

Condensation Polymerization A polymerization reaction in which water or some other simple molecule is eliminated from two or more monomer molecules as they combine to form the polymer or crosslinks between polymer chains. Examples of resins so made are alkyds, phenol-formaldehyde and urea-formaldehyde, polyesters, polyamides, acetals, and polyphenylene oxide.

Condensation Resin A resin formed by CONDENSATION POLYMERIZATION, w s.

Conditioning Subjecting a test specimen to standard environmental and/or stress history prior to testing. Several "ambient" conditioning atmospheres, approved by ISO, are listed in ASTM specification E 171, and ASTM D 618 provides detailed information on conditioning. Plastics test specimens are usually conditioned at 23°C and 50% RH for several days or more. Test D 638 for tensile properties of plastics specifies "...not less than 40 hours..." at these conditions.

Conductance (1) The electrical term for the reciprocal of resistance, measured by the ratio of current flowing through a conductor to the difference of potential between its ends. The SI unit is the *siemens* (S), replacing the now-deprecated mho, to which it is exactly equal. ASTM D 257 prescribes tests for the resistance and conductance of electrical-insulating materials. (2) In heat transfer, the ratio of heat *flux* through a solid wall or a stagnant fluid film divided by the difference in temperature through the wall or film. The SI unit is $J/(m^2 \cdot K)$.

Conductimetric Analysis A method of analysis for certain ions in solution based on measurement of the solution's electrical conductivity.

Conducting Polymer One of a class of polymers that, unlike most organic polymers, have high electrical conductivity. Polyacetylene is the best known member and at least one type has been prepared by doping that has conductivity almost equal to that of copper. While these polymers are expected to be useful in a wide range of devices, only a few commercial applications exist. Also, see POLYELECTROLYTE.

Conductive Composite A composite material having a VOLUME RESISTIVITY less than 500 ohm·cm. Such composites may be created by adding metal or graphite powders to ordinary resins. Others are made from inherently CONDUCTING POLYMERs, w s. They are useful for static elimination, RF shielding, and in storage-battery components.

Conductive Heat Transfer See HEAT TRANSFER.

Conductivity (Electrical) The reciprocal of volume resistivity; the CONDUCTANCE, w s, of a unit cube of material. The SI unit is siemens per meter (S/m).

Configuration See CONFORMATION.

Configurational Base Unit A molecular repeating unit or mer whose configuration is defined at least at one site of stereoisomerism in the main chain of a polymer molecule. Note: in a regular polymer, a configurational base unit corresponds to the mer. For example, in regular polypropylene, the mer is $-CH(CH_3)CH_2-$ and the configurational base units are

```
      H                    CH3
      |                     |
    —C—CH2—      and      —C—CH2—
      |                     |
     CH3                    H
```

These two configurational base units are enantiomeric to each other (mirror images).

Configuration Repeating Unit The smallest set of one, two, or more successive configurational base units that prescribes configurational repetition at one or more sites of stereoisomerism in the main chain of a polymer molecule (IUPAC).

Configurational Unit A molecular unit having one or more sites of defined stereoisomerism (IUPAC).

Conformal Coating See ENCAPSULATION.

Conformation (configuration) In polymer technology, the overall spatial arrangement of the atoms and groups in a polymer molecule, or the general shape of the molecule.

Conical Dry-Blender A device consisting of two hollow cones joined at their bases by a short cylindrical section, mounted on a central shaft perpendicular to the conicylindrical axis. Material is charged and discharged at openings in the apexes of the cones. Mixing is accomplished by cascading, rolling, and tumbling of the charge as the chamber rotates about the shaft.

Conical Transition In a metering-type extruder screw, the root surface of the screw between the feed section and metering section having the shape of a cone whose diameter increases from that of the deeper feed section to that of the shallower metering section.

Conjugated In organic chemistry, referring to the regular alternation of single and double bonds between carbon atoms. For example, in the conventional representation of the benzene molecule, below,

each single bond represents one pair of shared electrons, each double bond, two pairs.

Conjugated Double Bonds A chemical term denoting double bonds separated from each other by a single bond. An example is the bonding in 1,3-butadiene, $CH_2{=}CH{-}CH{=}CH_2$.

Consistency The density, firmness, viscosity, or resistance to flow of a substance, slurry, or aggregate. See VISCOSITY.

Consistometer An instrument for measuring the flow characteristics of a viscous or plastic material. See VISCOMETER and RHEOMETER.

Constantan An alloy containing about 55% copper and 45% nickel and having a low thermal coefficient of resistivity. Its main use in the plastics industry is in thermocouple wire with either iron or copper as the mating element. Iron-constantan, Type J, and chromel-alumel, Type K, are widely used to sense temperatures in plastics-processing equipment.

Constitutional Repeating Unit The smallest molecular unit whose repetition describes a regular polymer (IUPAC).

Constitutive Equation In material science, an equation that relates stress in a material to strain or strain rate. Simple examples are (1) Hooke's law, which states that, in elastic solids, strain is directly proportional to stress, and (2) Newton's law of flow, which states that, in laminar shear flow, the shear rate is equal to the shear stress divided by the viscosity. Few plastic solids and liquids obey either of these laws.

Consumer's Risk In quality control and acceptance sampling, the risk of making a Type II error, i e, of accepting, under a given sampling plan, a lot that is of definitely *un*acceptable quality.

Contact Adhesive A liquid adhesive that dries to a film that is not sticky to other materials but very sticky to itself. A typical contact adhesive is a neoprene elastomer mixed with either an organic-solvent vehicle or an aqueous dispersion medium. The adhesive is applied to both surfaces to be joined and dried at least partly. When pressed together with light to moderate pressure, a bond of high initial strength results. Some definitions of "contact adhesive" stipulate that, for satisfactory bonding, the surfaces to be joined shall be no further apart than about 0.1 mm.

Contact Angle The angle between the edge of a liquid meniscus or drop and the solid surface with which it is in contact. A droplet placed on a horizontal solid surface may remain spherical or spread to a degree that is related to the surface energies of the two materials. The angle between the solid surface and the tangent to the droplet at the curve of contact with the surface is the contact angle. By measuring such droplet-contact angles for droplets of several liquids whose surface energies are known, one may estimate the surface energy of the solid. See, for example, ASTM Test D 2578.

Contact Laminating See contact-pressure molding.

Contact-Pressure Molding (contact molding) This term encompasses processes for forming shapes of reinforced plastics in which little or no pressure is applied during the forming and curing steps. It is usually employed in connection with the processes of SPRAYUP and HAND-LAYUP MOLDING, w s, when such processes do not include the application of pressure during curing.

Contact-Pressure Resin (contact resin, impression resin, low-pressure resin) A liquid resin that thickens or crosslinks on heating and, when used for bonding laminates, requires little or no pressure. Typical components are an unsaturated monomer such as an allyl ester, or a mixture of styrene or other vinyl monomer with an unsaturated polyester or alkyd.

Contact Resin See CONTACT-PRESSURE RESIN.

Continuous Filament A single, small-diameter fiber of indefinite length.

Continuous-Filament Yarn A yarn formed by twisting together two or more—typically scores—of continuous filaments.

Continuous Phase (1) In a SUSPENSION or EMULSION, w s, the continuous phase refers to the liquid medium in which the solid or second-liquid particles are dispersed. The solid particles or droplets are called the *disperse phase.* (2) In a plastic filled or reinforced with solid particles, flakes or fibers, the binding resin is the continuous phase.

Continuous Polymerization A polymerization process in which monomer is steadily fed to the reactor and polymer is steadily removed.

Continuous Roving See ROVING.

Contraction Allowance See SHRINKAGE ALLOWANCE.

Controlled-Atmosphere Packaging The packaging of a product in a gas other than air, typically an inert gas such as nitrogen.

Convection Any process by which heat energy or material exchange is effected by flow of the medium. Convection may be *natural,* driven by gravity and density or concentration differences at different points in the medium; or *forced,* driven by pumps, blowers, or vibrating devices. Forced convection is almost always faster and more efficient because of the higher velocities and greater turbulence produced.

Convergent Die An extrusion die in which the internal channels of the die leading to the die orifice are decreasing in cross section in the direction of flow.

Conversion (1) In a chemical process, the molar percentage of any reactant, often the primary or most costly reactant, that is changed into product. (2) In the packaging industry, the intermediate processing and fabrication of plastic film or sheeting into useful forms by slitting, die cutting, heat-sealing into bags, etc, for resale to packagers.

Cooling Channel A passageway provided in a mold, platen or die for circulating water or other cooling medium, in order to control the temperature of the metal surfaces in contact with the plastic being molded or extruded. Proper sizing and placement of cooling channels can do much to speed processing and optimize properties.

Cooling Fixture (shrink fixture) A structure of wood or metal shaped to receive and restrain a part after its removal from a mold, so as to prevent distortion of the part while it is cooling.

Coordination Catalyst A catalyst comprising a mixture of an organometallic compound, e g, triethylaluminum, and a transition-metal compound, e g, titanium tetrachloride. Often called Ziegler or Ziegler-Natta catalysts, they are used in polymerizing olefins and dienes.

Coordination Compound (Werner complex) A complex compound whose molecular structure contains a central atom bonded to other atoms by coordinate covalent bonds based on a shared pair of electrons, both of which are from a single atom or ion. A CHELATE, w s, is a special type of coordination compound.

Copolycondensation The copolymerization of two or more monomers by CONDENSATION POLYMERIZATION, w s.

Copolymer This term usually, but not always, denotes a polymer of two chemically distinct monomers. It is sometimes used for terpolymers, etc, containing more than two types of mer units. Three common types of copolymers are BLOCK COPOLYMER, GRAFT COPOLYMER, and RANDOM COPOLYMER, w s. The IUPAC term for a polymer derived from two species of monomer, *bipolymer*, eschews the foregoing ambiguity but is nevertheless rarely seen or heard.

Copolymerization The simultaneous reaction of two or more monomers to form a COPOLYMER, w s.

Copper-Clad Laminate A laminated plastic surfaced with copper foil or plating, used for preparing printed circuits.

Core (1) The central member of a laminate to which the faces of the sandwich are bonded. (2) A channel in a mold, extruder screw, or cast-in heating element for circulation of heat-transfer media. (3) Part of a complex mold that forms undercut sections of a part, usually withdrawn to one side before the main members of the mold are opened. (4) The central member of a die for extruding pipe, tubing, wire-coating, or a parison to be blow-molded. (5) The central conductor in a coaxial cable.

Core and Separator See CORE (4), above.

Cored Screw An extruder screw bored centrally from the rear to permit circulation of temperature-controlled liquid within all or part of the screw's length.

Corfam® See POROMERIC.

Cork The outer bark of *Quercus suber*, a species of oak native to Mediterranean countries, having density in the range 0.22 to 0.26 g/cm^3. Cork is used as a core material (See CORE 1) in sandwich structures and, in ground form, as a density-lowering filler in thermoplastic and thermosetting compounds for special applications such as flooring, ablative plastics, insulating compositions, and shoe inner soles.

Cork Composite A compound consisting of ground CORK, w s, resins and possible other additives and reinforcements, formed into rods, sheets, etc. Cork composites have relatively low density and are used in sporting goods, for thermal insulation, and in ablative materials.

Corona Discharge The flow of electrical energy from a conductor to or through the surrounding air or gas. The phenomenon occurs when the voltage difference is sufficient ($\geq$ 5000 V) to cause partial ionization of the gas. The discharge is characterized by a pale violet glow, a hissing noise, and the odor of ozone formed when the surrounding gas contains oxygen. Corona discharge occurs around high-voltage cables, thus making ozone resistance an important factor in compounding plastics for insulation of electrical wires and cables. See also CASING and CORONA-DISCHARGE TREATMENT.

Corona-Discharge Treatment A method of rendering inert plastics, primarily polyolefins, more receptive to inks, adhesives, and decorative coatings by subjecting their surfaces to a corona discharge. A typical method of treating film is to pass the film over a grounded metal cylinder above which is located a sharp-edged, high-voltage electrode spaced so as to leave a small air gap between the film and the electrode. The corona discharge oxidizes the film, forming polar groups on vulnerable sites, increasing the surface energy and making the film receptive to inks, etc. See also FLAME TREATING.

Corona Resistance In an insulated, high-voltage electrical conductor, the resistance of the insulation to breakdown caused by ionized air in voids existing in the insulation.

Correlation Coefficient (r, product-moment correlation coefficient) The measure of the closeness of *linear* relationship between two variables x_i and y_i, defined by the equation

$$r = \sum_{i=1}^{n}\left[\left(x_i - \bar{x}\right)\left(y_i - \bar{y}\right)\right] \div \left[\sum_{i=1}^{n}\left(x_i - \bar{x}\right)^2 \sum_{i=1}^{n}\left(y_i - \bar{y}\right)^2\right]^{0.5}$$

in which $\bar{x}$ and $\bar{y}$ are the ARITHMETIC MEANs of x_i and y_i, The possible range of values for r is from -1 to $+1$, either extreme indicating perfect relationship, while absolute values below 0.5 indicate weak relationship and 0 indicates none at all. Correlation does not imply cause-and-effect between x and y, but does indicate that both are subject to common causes to the fractional extent of r. Where the relationship between x and y is *non*linear, a more meaningful measure of correlation is the square root of the COEFFICIENT OF DETERMINATION, w s.

Corrosion Resistance A broad term applying to the ability of plastics to resist many environments, but in particular, attack by acids, bases, and oxidants. See ACID RESISTANCE, ALKALI RESISTANCE, ARTIFICIAL WEATHERING, CHEMICAL RESISTANCE, DETERIORATION, PERMANENCE, SOLVENT RESISTANCE, STAIN RESISTANCE, SULFIDE STAINING, LIGHT RESISTANCE, VOLATILE LOSS, and WEATHERING.

Corundum Natural aluminum oxide (including many gemstones), extremely hard, used as a filler in plastics to impart hardness, heat resistance, and abrasion resistance.

Cotton Linters See LINTERS.

Couette Flow Shear flow in the annulus between two concentric cylinders, one of which is usually stationary while the other turns. By measuring the relative rotational velocity and the torque required to maintain steady flow, one can infer the viscosity of the liquid. See ROTATIONAL VISCOMETER. Flow in the metering section of a single-screw extruder resembles Couette flow, modified by the presence of the flight and, normally, by the pressure rise along the screw.

Coulomb (C) (1) A quantity of electricity defined in the SI system as equal to a current of one ampere flowing for one second, i e, 1 C = 1 A·s. (2) Before SI, the quantity of electricity that must pass through a circuit to deposit 0.0011180 gram of silver from a solution of silver nitrate. (3) The quantity of electricity on the positive plate of a one-farad capacitor when the potential difference between the plates is one volt.

Coumarin (cumarin) A dual-ring aromatic ketone, $C_9H_6O_2$, the sweet-smelling constituent of white clover, also produced synthetically. It is sometimes copolymerized with styrene to increase the deflection temperature above that of polystyrene.

Coumarone (2,3-benzofuran, cumarone) A colorless liquid derived from the fraction of coal-tar naphtha boiling between 150 and 200°C, and having the structure

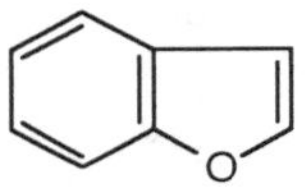

It is used in making COUMARONE-INDENE RESINs, w s.

Coumarone-Indene Resin Any of a family of resins produced by polymerizing a coal-tar naphtha containing coumarone and indene. The naphtha is first washed with sulfuric acid to remove some impurities, then is polymerized in the presence of sulfuric acid or stannic chloride as a catalyst. Remaining impurities determine the quality of the resin, which can range from a clear, viscous liquid to a dark, brittle solid. Coumarone-indene resins have no commercial applications when used alone, but are used primarily as processing aids, extenders, and plasticizers with other resins and with rubbers.

Coupling Agent A chemical capable of reacting with both the reinforcement and the resin matrix of a composite material to form or promote a stronger bond at the interface. The agent may be applied from the gas phase or a solution to the reinforcing fiber, or added to the resin, or both. See also SILANE COUPLING AGENT, TITANATE COUPLER, ADHESION PROMOTER.

Coupon A representative specimen of a material or sheet product, cut from the product and set aside for testing.

Cp Abbreviation for the deprecated but still widely used viscosity unit, CENTIPOISE, w s.

CP (1) Abbreviation for Cellulose Propionate. See CELLULOSE ACETATE PROPIONATE. (2) Abbreviation for *chemically pure*, a designation for laboratory chemicals now largely superseded by *analytical reagent*.

CPE Abbreviation for CHLORINATED POLYETHYLENE, w s.

CPET Abbreviation for CRYSTALLINE POLYETHYLENE TEREPHTHALATE, w s.

CPVC (1) Abbreviation for CHLORINATED POLYVINYL CHLORIDE, w s. (2) In the paint industry, abbreviation for *critical pigment concentration*, a source of confusion often encountered in the paint and color literature.

Cr Chemical symbol for the element chromium.

CR Abbreviation for chloroprene rubber (British Standards Institution). See NEOPRENE.

Cracking (1) Formation of a narrow separation in a surface or bulk of a structure, usually by rupture of chemical bonds or reinforcing fibers. (2) Breakdown of an organic compound, typically a hydrocarbon, at high temperatures or by catalytic action, to form smaller compounds, even free carbon, with release of hydrogen.

Crack Stopper A method or material used or applied to delay the propagation of a potential or existing crack. Techniques include drilling of holes, installing a load-spreading doubler, or including in the design an interruption in part continuity.

Crammer-Feeder (force feeder) A device fitted to the inlet port of an extruder that precompacts a low-density feedstock and propels it into the feed section of the extruder screw. As originally produced by Prodex Corp (now a division of HPM), the crammer-feeder consisted of a conical shell within which turned a decreasing-diameter screw driven independently of the extruder.

Crater A small, shallow surface imperfection (ASTM D 883).

Crazing An undesirable defect in plastics articles characterized by distinct surface cracks or minute frost-like internal cracks, resulting from stresses within the article that exceed the tensile strength of the plastic. Such stresses may result from molding shrinkage, or machining, flexing, impact shocks, temperature changes, or the action of chemicals and solvents. See also STRESS CRACKING.

Creel The spool and its supporting structure on which continuous strands or rovings of reinforcing material are wound for use in the filament-winding process.

Creep (n) Due to its viscoelastic nature, a plastic subjected to a load for a period of time tends to deform more than it would from the same load released immediately after application. The degree of this deformation increases with the duration of the load and with rising temperature. Creep is the permanent deformation resulting from prolonged application of a stress below the elastic limit. This deformation, after any time under stress, is partly recoverable (*primary creep*) upon the release of the load and partly unrecoverable (*secondary creep*). Creep at room temperature is sometimes called *cold flow*. See also ANDRADE CREEP.

Creep Modulus The total deformation measured at constant load over a period of time divided into the applied stress. Creep modulus, if known, simplifies some plastic part designs by allowing the designer to use standard formulas for deformation in which the creep modulus, for the expected service life under load, replaces the conventional, short-time modulus.

Creep Rupture The rupture of a plastic under a continuously applied stress that is less than the short-time strength. This phenomenon is caused by the viscoelastic nature of plastics. Creep-rupture tests are generally conducted over a series of loads ranging from those causing rupture within a few minutes to those requiring several years or more.

Creep Strength The initial stress at which failure occurs after a measured time under load. Thus, creep strength (at any temperature) must be labeled with the time to failure. Like CREEP MODULUS (w s), creep strength is useful to designers for applications in which plastic articles or members will carry sustained loads.

Crepe Rubber Natural rubber of a pale- to dark amber color prepared by coagulating natural-rubber latex with acid, then milling this coagulum into sheets. The other basic form of solid natural rubber (i e, ribbed sheet) is prepared by drying the latex on rolls in the presence of smoke.

Cresol (hydroxytoluene, methylphenol) $H_3CC_6H_4OH$. An important family of coal-tar derivatives, occurring in ortho, meta, and para isomers, and used in the production of phenol-formaldehyde resins and tricresyl phosphate, an important plasticizer for PVC.

Cresol Resin A phenolic-type resin obtained by condensing a cresol with an aldehyde.

Cresyl Diphenyl Phosphate (CDP) $(H_3CC_6H_4O)PO(C_6H_5O)_2$. A plasticizer for cellulosics, vinyl chloride polymers and copolymers, with a high degree of flame resistance and good low-temperature properties. It is also acceptable for use in food-packaging films. It is most often used, in low percentages of the total plasticizer, as a flame retardant.

Cresylic Acid A term sometimes applied to mixtures of *o*-, *m*-, and *p*-cresol, which are mildly acidic, but also including wider fractions of phenolic compounds derived from coal tar or petroleum that contain xylenols and other higher-boiling phenols in addition to the cresols. It is used in the production of phenolic resins and tricresyl phosphate.

Crimp The waviness of a fiber. It determines the capacity of fibers to cohere under light pressure. Crimp is measured by either the number of crimps or waves per unit length or by the decrease in length upon crimping divided by the uncrimped length, expressed as a percent.

Critical Damping In a damped vibrating system, damping so strong that the system, when displaced from rest, returns to rest in one-half cycle. Compare LOGARITHMIC DECREMENT.

Critical Shear Stress In extrusion, the shear stress in the melt at the die wall that signals the onset of MELT FRACTURE, w s. The stress is on the order of 0.1 to 0.4 MPa.

Critical Strain In a strength test, the strain at the yield point.

Critical Surface Tension That value of the surface tension of a liquid, γ_C, below which a drop of the liquid will wet and spread, forming a zero contact angle on a substrate whose surface energy is characterized by γ_C. This property is positively correlated with a polymer's solubility parameter. See also SURFACE TENSION and SOLUBILITY PARAMETER.

Crocidolite (blue asbestos) See ASBESTOS.

Crocking (1) See BLEEDING. (2) Physical transfer of color from one material to another. In ASTM D 1593, a test for resistance to crocking of flexible films subjects the test film to rubbing with a 5-cm square of white cotton cloth.

Crosshead (1) A device that receives a molten stream of plastic emerging from an extruder, diverts the flow to a direction usually 45° or 90° from the axis of the extruder, and forms the extrudate to a shape such as a parison for blow molding or a jacket around a wire. An essential element of the crosshead is the mandrel, a tubular core with grooves of various shapes and held in place by a perforated plate, web, or spider legs. Material emerging from the space between the mandrel and the crosshead housing is given its final shape by means of a die mounted on the end of the crosshead. (2) The moving member of a testing machine.

Crosshead Die See CROSSHEAD (1).

Cross Laminate A laminate in which the direction of greatest strength in some layers is perpendicular to that direction in other layers. A simple example is a laminate in which alternate layers contain unidirectional reinforcing fibers laid in perpendicular directions.

Crosslinking Applied to polymer molecules, the setting up of chemical links among the molecular chains. When extensive, as in most cured thermosetting resins, crosslinking creates one infusible super-molecule from all the chains. Crosslinking can also occur between polymer molecules and other substances. For example, polyethylene can be crosslinked with carbon-black particles, which have sites to which polyethylene chains can link in the presence of a catalyst. The mixture of resin, filler, and catalyst can be molded as a thermoplastic, then transformed to a thermoset by crosslinking in the curing cycle. Crosslinking can be achieved by irradiation with high-energy electron beams, or by means of chemical crosslinking agents such as organic peroxides.

Crosslinking Agent A substance that promotes or regulates intermolecular covalent bonding between polymer chains.

Crosslinking Index The number of crosslinked units per primary polymer molecule, averaged over the whole specimen.

Crossply A layer containing reinforcing fibers mainly running in one direction and perpendicular to the main fiber direction in adjacent layers. See also CROSS LAMINATE.

Crotonaldehyde $CH_3CH{=}CHCHO$. A colorless liquid synthesized by the aldol condensation of acetaldehyde, accompanied or followed by dehydration. It can be polymerized by triethylamine to a resin with film-forming properties, or copolymerized with many compounds. Other uses include solvent for PVC, short-stopper in the polymerization of vinyl chloride, and plasticizer synthesis.

Crotonic Acid (*trans*-2-butenoic acid, *trans*-β-methacrylic acid) CH_3-$CH{=}CHCOOH$. A white crystalline solid prepared by the oxidation of crotonaldehyde. It forms copolymers with vinyl acetate, used as hot-melt adhesives. Esters of crotonic acid are used as plasticizers for acrylic and cellulosic plastics.

Crotonyl Peroxide A catalyst for the polymerization of vinyl and vinylidene halides.

Crown (1) Of a calender roll, a gradual, small increase in the diameter of a roll toward the center to compensate for the slight deflection due to bending of the roll under pressure. (2) The much more convex surface of a transmission-belt pulley designed to keep the belt running in the center of the pulley.

CR-39 Abbreviation for "Carbonate Resin 39". See DIETHYLENE GLYCOL BIS-(ALLYL CARBONATE).

Cryogenic Pertaining to very low temperatures, usually temperatures below about –150°C (123 K). Evaluations of plastics at cryogenic temperatures are conducted for potential space applications.

Cryogenic Grinding (freeze grinding) Thermoplastics are difficult to grind to small particle sizes at ambient temperatures because they soften, cohere in lumpy masses, and clog screens. When chilled by dry ice, liquid carbon dioxide, or liquid nitrogen, thermoplastics can be finely ground to powders suitable for electrostatic spraying and other powder processes.

Cryptometer An instrument for measuring the opacity of surface coatings.

Crystal A homogeneous solid having an orderly and repetitive three-dimensional arrangement of its atoms. Crystalline polymers are never wholly crystalline but contain some amorphous material and many CRYSTALLITES, w s.

Crystal Lattice The spatial arrangement of atoms or radicals in a crystal.

Crystallinity (1) A state of molecular structure in some resins attributed to the existence of solid crystals with a definite geometric form. Such structures are characterized by uniformity and compactness. See LIQUID-CRYSTAL POLYMER. (2) The percentage of a polymer sample that has formed crystals.

Crystallite A perfect portion of an ordinary crystal; that is, a portion with its atoms and molecules arranged in a lattice free of defects. Ordinary crystals are composed of large numbers of crystallites, which may or may not be perfectly aligned with one another.

Crystallized Polyethylene Terephthalate (CPET) PET resin to which a fractional percentage of nucleating agent has been added to encourage the development of crystallinity in extruded or molded products. The percent crystalline material is typically between 15 and 35. Modulus and strength increase with crystalline content. See also POLYETHYLENE TEREPHTHALATE.

Crystal Polystyrene Styrene homopolymer, which, though actually 100% amorphous, was so called because of its excellent clarity and the glitter of the early cube-cut pellets. See POLYSTYRENE.

CS Abbreviation for CASEIN PLASTIC, w s.

C-Stage (resite) The final, fully cured state of a thermosetting resin. See also A-STAGE and B-STAGE.

CTA See CELLULOSE TRIACETATE.

CTFE Resin See POLYCHLOROTRIFLUOROETHYLENE.

Cu Chemical symbol for the element copper (Latin: cuprum).

Cubic (1) Having the appearance of a cube. (2) Of algebraic equations, containing the unknown to the power 3 and not higher. (3) Characterizing volume measure, as *cubic meter*. (4) The simplest of the six crystal systems, in which the three principal axes are mutually perpendicular and the atomic spacing is the same along all three.

Cull (n) (1) A rejected material or product. (2) In transfer molding, the material remaining in the transfer pot after the mold has been filled. A certain amount of cull is usually necessary for the operator to be confident that the cavity has been properly filled.

Cultured Stone A term applied to decorative embedments of natural stones such as marble, granite, terrazzo, and slate in thermosetting resins. They are made by casting the resin, usually a polyester, in molds containing the stones. The embedments are used for counter tops, window sills, wall facings, flooring, giftware, etc.

Cumarone See COUMARONE.

Cumene (isopropylbenzene, isopropylbenzol, cumol) $C_6H_5CH(CH_3)_2$. A volatile liquid in the alkyl-aromatic family of hydrocarbons. It is used as a solvent and intermediate for the production of phenol, acetone, and α-methyl styrene; and as a catalyst for acrylic and polyester resins.

Cumene Hydroperoxide $C_6H_5C(CH_3)_2OOH$. A colorless liquid derived from an oxidized solution or emulsion of cumene, used as a polymerization catalyst.

Cumylphenol Derivative One of a group of polymer intermediates based on cumylphenol that offer higher performance at lower cost than nonylphenol competitors. The free phenol is an accelerator for amine hardeners of epoxy resins. Cumylphenyl acetate and the glycidyl ether are reactive in epoxy systems, giving enhanced strength. The benzoate of cumylphenol aids extrusion of PVC compounds.

Cup-Flow Test A British Standard Test (B S 771) for measuring the flow properties of phenolic resins. A standard mold is charged with the specimen material and then closed under preset pressure. The time in seconds for the mold to close completely is the cup-flow index.

Cuprammonium Rayon A regenerated cellulose formed by dissolving cotton or wood-pulp linters in a solution of ammonia and copper oxide (from sulfate), then extruding the solution through spinnerets into warm water, where the filaments harden. The finest filaments (lowest denier) are made this way.

Cup Temperature (1) In injection molding or extrusion, the measured temperature of a glob of melt collected at the injection nozzle or extrusion die in an insulated vessel and presumed to be equal to, or slightly less than, the average melt temperature leaving the machine. (2) (cup-mixing temperature, flow-average temperature) In a flowing stream of fluid, the local product of velocity times temperature integrated over the stream cross section, said integral divided by the integral of the local velocity over the cross section.

Cure (v) To change the molecular structure and properties of a plastic or resin by chemical reaction, which may be condensation, polymerization, or addition; usually accomplished by the action of either heat or a catalyst or both, with or without pressure. The term cure is used almost exclusively in connection with thermosetting plastics, vulcanizable elastomers and rubbers. It is not ordinarily used for the hardening of thermoplastics by physical methods such as heating, cooling, or evaporation of solvents. However, in the early years of the art, some authors used the term cure in connection with the fusion of plastisols before the term fusion came into general use. See also POST-CURING.

Cure Stress Internal stress in cast or molded thermosetting parts, caused by unequal shrinkage in different sections of the parts. Depending on the directions of applied stress in service relative to the principal cure stress, parts may be considerably weaker than their designers expected them to be.

Curing Agent (hardener) A substance or mixture of substances added to a plastic or rubber composition to promote or control the curing reaction. An agent that does not enter into the reaction is known as a *catalytic hardener* or *catalyst*. A *reactive curing agent* or *hardener* is generally used in much greater proportions than a catalyst, and is actually converted in the reaction. See also ACCELERATOR.

Curing Temperature The temperature to which a thermosetting or elastomeric material is brought in order to commence and complete its final stage of cure.

Curing Time (molding time) In the molding of thermosets, the time elapsing between the moment relative movement between the mold parts ceases, and the instant that pressure is released.

Curtain Coating A process in which the substrate to be coated is conveyed quickly through a free-falling liquid "curtain" of a low-viscosity resin, solution, suspension, or emulsion. The coating thickness is governed by the rates of fluid delivery and substrate travel.

Custom Molder (Brit: trade moulder) A firm specializing in the molding of items or components to the specifications of another firm that handles the sale and distribution of the item, or incorporates the custom-molded component into one of its own products.

Cut In the fiber industry, including glass and asbestos, the number of 100-yd lengths of fiber per pound. A now deprecated unit, 1 cut corresponds to a lineal density of 0.0049604 kg/m or 4960.4 texes.

Cut-Layers As applied to laminated plastics, a condition of the surface of machined or ground rods and tubes and of sanded sheets in which cut edges of the surface layer or lower laminations are revealed (ASTM D 883).

Cut-off (flash groove, pinch-off) In compression molding, the line where the two halves of a mold come together, often along a sharp mating ridge and groove.

Cut-off Saw (traveling cut-off) In extrusion of pipe, rod, and profiles, a circular saw that periodically swings forward, while moving downline at the same rate as the extruded product, to cut it into desired lengths. When the saw has completed its cut, it swings back, at the same time quickly reversing travel to return to its starting point, poised for the next cut.

Cyanoguanidine See DICYANDIAMIDE.

Cyanuric Acid (1,3,4-triazine-2,4,6-triol, tricyanic acid, tricarbimide) An acid evolved from the blowing agent, azodicarbonamide, when it decomposes. The acid is corrosive, and is the chief cause of plate-out on components of extruders in the structural-foam process when azodicarbonamide is used as the blowing agent.

Cycle The series of sequential operations entering into a repeating batch process or part of the process. In a molding operation, cycle time is the average time elapsing, over several normal cycles, between a particular occurrence in one cycle and the same occurrence in the next cycle.

Cyclized Rubber A thermoplastic resin produced by reacting natural rubber with stannic chloride or chlorostannic acid. It is claimed (US Patent 3,205,093) that a solution of cyclized rubber in toluene is one of the few lacquers known to adhere to polyolefins without their being pretreated or the lacquered surface being post-treated with radiation. Other uses include films and hot-melt coatings.

Cyclohexane (hexamethylene, hexahydrobenzene) C_6H_{12}. A saturated hydrocarbon with a six-membered ring, cyclohexane is derived from the catalytic hydrogenation of benzene, and is used as a solvent for cellulosics and as an intermediate in the production of nylon.

Cyclohexanol (hexahydrophenol) $CH_2(CH_2)_4CHOH$. A colorless, viscous liquid prepared by the oxidation of cyclohexane or by the hydrogenation of phenol. It is used as an intermediate in the production of nylon 6/6, it is a solvent for cellulosic resins, and it is an intermediate in the manufacture of phthalate-ester plasticizers.

Cyclohexanol Acetate $CH_3COOC_6H_{11}$. A nonflammable solvent for cellulosics and many other resins.

Cyclohexanone (pimelic ketone, ketohexamethylene) $CH_2(CH_2)_4C{=}O$. A colorless liquid produced by the oxidation of cyclohexane or cyclohexanol. Its most important use is for the manufacture of adipic acid for nylon 6/6, and caprolactam for nylon 6. It is also an excellent high-boiling, slowly evaporating solvent for many resins including cellulosics, acrylics and vinyls. It is one of the most powerful solvents for PVC, and is often used in lacquers to improve their adhesion to PVC.

Cyclohexene Oxide $C_6H_{10}O$. This epoxide is a highly reactive, colorless liquid that resembles ethylene oxide in most of its reactions. It is useful as an intermediate in the production of many organic chemicals used in plastics. Its epoxide structure is especially useful in applications where an HCl scavenger is required.

Cyclohexyl Methacrylate $H_2C{=}C(CH_3)COOC_6H_{11}$. A colorless monomer, polymerizable to resins for optical lenses and dental parts, and useful for potting electrical components.

Cyclohexyl Stearate $C_6H_{11}OOCC_{17}H_{35}$. A plasticizer for polystyrene, ethyl cellulose, and cellulose nitrate.

Cyclopentane (pentamethylene) C_5H_{10}. A solvent for cellulose ethers.

Cylinder (barrel) In extruders and injection molders, the heavy-walled tube within which the screw rotates. Friction between feedstock and cylinder is the force that propels the feed forward, as directed by the screw flight.

Cylindrical Weave A type of woven, knitted, or braided sleeve generated as reinforcement in tubular reinforced-plastic structures such as pipe. Typically, the reinforcement will constitute about half the volume of the finished product; but it can be much less, as in knit-reinforced garden hose.

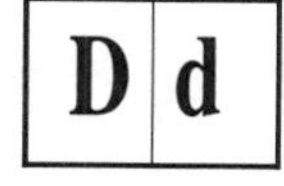

d (1) SI abbreviation for prefix DECI-, w s. (2) Abbreviation in use with SI system for time interval of one day (= 86,400 s).

D (1) Symbol for diameter. (2) Chemical symbol for the hydrogen isotope of atomic weight 2, deuterium.

da SI abbreviation for prefix DECA-, w s.

DABCO (former trade name, now generic) Abbreviation for 1,4-diazabicyclo-2,2,2-octane. See TRIETHYLENEDIAMINE.

DAC Abbreviation for DIALLYL CHLORENDATE, w s.

DAF Abbreviation for diallyl fumarate, a polymerizable monomer.

DAIP Abbreviation for diallyl isophthalate resin. See ALLYL RESIN.

Dam A ridge around the perimeter of a mold that retains excess resin during pressing and curing.

DAM Abbreviation for diallyl maleate. See ALLYL RESIN.

Damping Reduction of vibration amplitude due to viscous or frictional resistance within a material or to drag of its environment upon a structure. The vibration may be mechanical, sonic, or electronic. Energy dissipated per cycle is called HYSTERESIS loss, w s. See also CRITICAL DAMPING.

Dancer Roll A roller mounted on an axis which is movable with respect to axes of other rollers in an apparatus, used to control or measure tension of a continous web or strand as it passes through a series of rollers. Dancer rolls are used as tension-sensing devices in the extrusion coating of wire, and as tension-maintaining devices in film winding.

DAP Abbreviation for DIALLYL PHTHALATE, w s.

Dart Impact Test See FREE-FALLING-DART TEST.

Dashpot (1) A device used for damping vibration and cushioning shock in hydraulic systems. Typically, a dashpot is a liquid- or gas-filled cylinder with a piston that is attached to a moving machine part. (2) A modeling concept useful in visualizing the mechanical behavior of viscoelastic materials, a purely viscous element that may operate alone or connected in series and/or parallel with springs and sliders.

Daylight Fluorescent Pigment See FLUORESCENT PIGMENT.

Daylight Opening The clearance between two platens of a molding press when in the open position.

dB Abbreviation for DECIBEL, w s.

DBEP Abbreviation for DIBUTOXYETHYL PHTHALATE, w s.

DBP Abbreviation for DIBUTYL PHTHALATE, w s.

DBPC Abbreviation for DI-*tert*-BUTYL-*p*-CRESOL, w s.

DBS Abbreviation for DIBUTYL SEBAÇATE, w s.

DBTDL Abbreviation for DIBUTYLTIN DILAURATE, w s.

DC Drive (direct-current drive) A machine drive, particularly that of an extruder, powered by a direct-current motor. The availability of economical solid-state rectifiers, and the good torque-vs-speed characteristic and tight speed regulation of DC drives have made them the most popular choice today for variable-speed service.

DCHP Abbreviation for DICYCLOHEXYL PHTHALATE, w s.

DCP Abbreviation for DICAPRYL PHTHALATE, w s.

DDM Abbreviation for 4,4'-DIAMINODIPHENYL METHANE, w s.

DDP Abbreviation for DIDECYL PHTHALATE, w s.

DE Abbreviation for diatomaceous earth. See DIATOMITE.

Dead Flat Of a finish, the quality of having no luster or gloss.

Dead Fold A fold that does not spontaneously unfold; a crease.

Dead Zone In control work, a range of the quantity sensed in which a small change in the quantity causes no change in the indication of it, nor any control action. Compare INDUCTION PERIOD.

Deaerate To remove air from a substance. Deaeration is an important step in the production of vinyl plastisols and most casting operations, accomplished by subjecting the liquid to a moderate vacuum with or without agitation, to remove air that would cause objectionable bubbles or blisters in finished products.

Debonding In a bonded joint, separation of the bonded surfaces. In a laminate, separation of layers or fibers from the matrix.

Deborah Number (De, N_{De}) The dimensionless ratio of the relaxation time of a viscoelastic fluid to a characteristic time for the flow process being considered. Relaxation time is often taken as the ratio of viscosity to modulus, while the characteristic time is a significant length divided by average velocity. The Deborah number has been useful in studying such phenomena as die-exit swell in extrusion. If *De* is high, elastic effects probably dominate; if *De* is low, the flow is essentially viscous.

Deburr To remove from a machined part rough edges or corners left by the machining operations. Compare DEFLASHING.

Deca- (da) SI-acceptable prefix meaning × 10.

Decabromodiphenyl Oxide (decabromodiphenyl ether) Diphenyl ether in which all ten phenyl hydrogens have been replaced by bromine, containing 83% bromine. The commercial product, a free-flowing white powder, is much used as a flame retardant in HIGH-IMPACT POLYSTYRENE and STRUCTURAL FOAM, usually with synergistic ANTIMONY OXIDE, w s. Typical would be 12% decabromodiphenyl oxide and 4% antimony oxide in the resin.

Decahydronaphthalene (Decalin®) A saturated bicyclic hydrocarbon, $C_{10}H_{18}$, essentially two CYCLOHEXANE rings fused together, sharing two hydrogen atoms. The commercial product is a mixture of *cis-* and *trans-* isomers. It is a colorless liquid with an aromatic odor, derived by treating molten naphthalene with hydrogen in the presence of a catalyst. It is a solvent for many resins.

Decalcomania (decal) A printed design on a temporary carrier such as paper or film, and subsequently transferred to the item to be decorated. Decals are widely used for decorating many materials including plastics. The imprint is adhered to the plastic surface by means of a pressure-sensitive adhesive, solvent welding, or heat and pressure.

Decanedioic Acid Synonym for SEBACIC ACID, w s.

Decarboxylate To remove from an organic acid its carboxylic-acid groups.

Deci- (d) The SI-approved prefix meaning × 0.1.

Decibel (dB) In acoustics and electronic circuits, a change in sound intensity or circuit power by the factor $10^{0.1} = 1.259$. These changes correspond to changes in sound *pressure* or circuit *voltage* by the square root of the above factor, 1.122. In acoustics most measurements are referred to a sound-power level of 10^{-12} W, corresponding to a sound-pressure level of 0.00002 Pa, the presumed threshold of human audibility. 1 dB is the least *change* in intensity detectable by normal ears. Thus, because of the exponential definition of the decibel, an 80-dB sound is ten times as intense as a 70-dB sound. OSHA regulations prescribe maximum sound levels permitted for employee exposure. In many plants, scrap grinders, which used to be major noise generators, now run much more quietly or are isolated by enclosures, or both.

Deckle (deckle rod, cut-off plate) In extrusion of film or sheet, or extrusion coating, a small rod or plate attached to each end of the die that symmetrically shortens the length of the die opening, thus facilitating the production of a web of less than the maximum width.

Decomposition Breakdown of molecular structure by chemical or thermal action. With polymers, depending on the severity of conditions, decomposition products can range from subpolymers and oligomers down to monomers and even atoms and ions.

Decompression Zone In a vented extruder, the zone of deep flights immediately forward of the first metering zone and beneath the vent, where unwanted gases and vapors are released from the (typically) foaming melt. In some vented single-screw machines and most twin-screw machines, the two zones are separated by a DYNAMIC VALVE, w s, whose resistance to flow helps to increase the superheating of the melt prior to decompression.

Decorating The processes used for decorating plastics are defined under the following headings:

AIRLESS SPRAYING
ASHING
DECALCOMANIA
DOUBLE-SHOT MOLDING
ELECTROLESS PLATING
ELECTROPHORETIC DEPOSITION
ELECTROPLATING ON PLASTICS
ELECTROSTATIC PRINTING
ELECTROSTATIC SPRAY COATING
EMBEDMENT DECORATING
EMBOSSING
FILL-AND-WIPE
FLEXOGRAPHIC PRINTING
FLOCKING
FLOW COATING
GRAVURE COATING
GRAVURE PRINTING
HOT STAMPING
IN-MOLD DECORATING
LETTERPRESS PRINTING
METALLIZING
OFFSET PRINTING
PAINTING OF PLASTICS
PRINTING ON PLASTICS
ROLLER COATING
RUBBER-PLATE PRINTING
SCREEN PRINTING
SECOND-SURFACE DECORATING
SPRAY-AND-WIPE PAINTING
SPRAY COATING
THERMOGRAPHIC-TRANSFER PROCESS
VACUUM METALIZING
VALLEY PRINTING

Decorative Board (decorative laminate) A special term for laminates used in the furniture and cabinetry industries, which are defined by the Decorative Board Section of NEMA as "...a product resulting from the impregnation or coating of a decorative web of paper, cloth, or other carrying medium with a thermosetting type of resin and consolidation of one or more of these webs with a cellulosic substrate under heat and pressure of less than 500 pounds per square inch." This includes all boards that were formerly called low-pressure melamine and polyester laminates, but does not include vinyls. "Cellulosic" here means impregnated paper, wood, or plywood.

Decyl Butyl Phthalate See BUTYL ISODECYL PHTHALATE.

Decyl-Octyl Methacrylate $H_2C{=}C(CH_3)COO(CH_2)_nCH_3$, with n = 7, 8 and 9. A polymerizable mixed monomer for acrylic plastics.

***n*-Decyl-*n*-Octyl Phthalate** (NDOP) See *n*-OCTYL-*n*-DECYL PHTHALATE.

Decyl Tridecyl Phthalate $C_{10}H_{21}COOC_6H_4OOCC_{13}H_{27}$. A plasticizer for vinyls, cellulosics, and polystyrene.

Deep Drawing The process of forming a thermoplastics sheet in a mold in which the ratio of depth to the shortest lateral opening is 1:1 or greater.

Deflashing The removal of flash or rind left on molded plastic articles by spaces between the mold-cavity edges. Methods include TUMBLING, BLAST FINISHING (w s), use of dry or wet abrasive belts, and hand methods using knives, scrapers, broaching tools, and files. Soft thermoplastic parts are sometimes deflashed by tumbling them while immersed in a severe coolant such as liquid nitrogen. See also ABRASIVE FINISHING, AIRLESS BLAST DEFLASHING.

Deflection Temperature According to ASTM D 648, the stress-labeled temperature at which a standard test bar, centrally loaded as a simple beam to develop a (theoretical) maximum stress of 455 or 1820 kPa (66 or 264 psi), and warmed at 2°C/min, deflects 0.25 mm. For many years this temperature was called the "heat-distortion point", a term now deprecated and fading from use.

Deflocculating Agent A substance that breaks down agglomerates into primary particles or prevents the latter from combining into agglomerates (ISO).

Defoamer (defoaming agent) A substance that, when added in small percentages to a liquid containing gas bubbles, causes the small bubbles to coalesce into larger ones that rise to the surface and break.

Deformation (1) Change in dimension or shape of a plastic product, particularly a test specimen. (2) In a tensile or compression test, the change in length of the specimen in the direction of applied force.

Deformation divided by original length equals *elongation*, often expressed as a percent. (3) In a shear mode, the angle of shear in radians.

Degassing (breathing) In injection or transfer molding, the momentary opening and closing of a mold during the early stages of the cycle to permit the escape of air or gas from the heated compound. See also DEAERATE.

Degating The removal of material left on a plastic part formed by the passage between the runner and the cavity, i e, the gate. The operation is sometimes performed automatically by a mold element. Otherwise, the gate may be removed by manual breaking or cutting, sometimes followed by sanding or burnishing.

Degradable Plastic See BIODEGRADATION.

Degradation A deleterious change in chemical structure, physical properties, or appearance of a plastic caused by exposure to heat (*thermal degradation*), light (*photodegradation*), oxygen (*oxidative degradation*), or weathering. The ability of plastics to withstand such degradation is called *stability*. See also ARTIFICIAL WEATHERING, AUTOCATALYTIC DEGRADATION, BIODEGRADATION, CORROSION RESISTANCE, DETERIORATION, DEW-CYCLE WEATHERING TEST, DISCOLORATION, PINK STAINING, XENON-ARC AGING.

Degree of Crosslinking The fraction of mer units of a polymer that are crosslinked, equal to the quotient of the mer weight and the average molecular weight of segments between crosslinks.

Degree of Cure The extent to which curing or hardening of a thermosetting resin has progressed. See also A-STAGE, B-STAGE, AND C-STAGE.

Degree of Polymerization (DP, chain length) The average number of monomer units per polymer molecule. $DP = M_n/M_{mer}$, where M_n = the polymer's NUMBER-AVERAGE MOLECULAR WEIGHT, w s, and M_{mer} = the molecular weight of the mer (which may differ from monomer molecular weight). In a polymer having worthwhile mechanical properties, *DP* exceeds 500, and many commercial thermoplastics have *DP*s between 1000 and 10,000. See also MOLECULAR WEIGHT.

DEHP (DOP) Abbreviation for DI(2-ETHYLHEXYL) PHTHALATE, w s.

Dehydration The removal of water from a substance either by ordinary drying or heating, or by absorption, chemical action, condensation of water vapor, or by centrifugal force or filtration.

Dehydroacetic Acid (DHA) $CH_3C{=}CHC(O)CH(COCH_3)C(O)O$. A heterocyclic compound used as a plasticizer, fungicide, and bactericide.

Dehydrogenation The removal of hydrogen from a compound by chemical means. See also CRACKING (2).

Dehydrohalogenation The splitting of hydrogen chloride or other hydrohalide from polymers such as PVC, by action of excessive heat or light.

Deka- (da) A prefix permitted by SI and meaning × 10.

Delamination The separation of one or more layers in a laminate caused by the failure of the adhesive bond.

Deliquescent Said of some salts that are so strongly attractive of water that they can absorb enough from room air to liquefy.

Delustrant A chemical agent used either before or after spinning to produce dull surfaces on synthetic fibers to obtain a more natural, silk-like appearance.

Denatured Alcohol Ethyl alcohol to which odorous and/or nauseating substances have been added in small quantities to make it unpotable without diminishing its solvent power. Typical denaturants are benzene, acetaldehyde, and pyridine.

Denier The weight in grams of 9000 meters of a fiber in the form of continuous filament. Although deprecated in SI, this is still the most widely used unit of lineal density in the textile industry to indicate fineness of natural or synthetic fibers. 1 denier = 1.1111×10^{-7} kg/m. See also CUT, GREX NUMBER, and TEX.

Densitometer An instrument for measuring the optical density of, or transmission of light through, a partly transparent medium, specifically, photographic film that has been exposed.

Density (absolute density) Mass per unit volume of a substance. The SI unit is kg/m^3, but others in common use are g/cm^3, lb/ft^3, and lb/gallon. It is an important criterion in specifying some plastics, e g, polyethylene, which can vary in density from 0.92 to 0.98 g/cm^3, with significant associated variation in properties such as modulus. Since density decreases with rising temperature, the temperature of measurement should be stated. A common one is 23°C and, when temperature is *not* stated, 23°C may be assumed. ASTM tests for densities of (non-cellular) plastics include D 792, measurement by liquid displacement; D 1505, using a density-gradient tube; and, the newest one, D 4883, density of polyethylene by the ultrasound technique. For cellular plastics, other methods are prescribed: D 1565, D 1622, and D 1667. See also APPARENT DENSITY, BULK DENSITY, GRADIENT-TUBE DENSITY METHOD, RELATIVE DENSITY, and SPECIFIC GRAVITY.

Deoxy- (desoxy-) A prefix denoting replacement of a hydroxyl group with hydrogen.

DEP Abbreviation for DIETHYL PHTHALATE, w s.

Depolymerization The reversion of a polymer to its monomer, or to a polymer of lower molecular weight. Such reversion occurs in most plastics when they are exposed to very high temperatures in the absence of air.

Deposition Applying a coating to a substrate by vacuum evaporation, sputtering, electrolysis, chemical reaction, etc.

Depth of Draw In thermoforming, the depth of the lowest point in the formed object relative to the clamped edge of the sheet. Also see DRAW RATIO (2).

Depth of Screw Synonym for FLIGHT DEPTH, w s.

Desiccant A substance capable of absorbing water vapor from air and other materials enclosed with it; a drying agent. Typically used to maintain low humidity in a storage or test vessel or package.

Desiccator A plastic or heavy-glass bowl with a tight-fitting lid, containing a drying agent, in which a sample can be kept in a controlled, dry atmosphere, typically under a mild vacuum.

Design Life The period through which a part or product is expected to perform satisfactorily, remaining within preset performance tolerances.

Desorption The escape or loss of a substance previously absorbed into, or adsorbed onto, another.

Destaticization The treating of a plastic to minimize its tendency to accumulate static electric charge, or removal of charge. See also ANTI-STATIC AGENT, STATIC ELIMINATOR, SOOT-CHAMBER TEST.

Destructive Distillation Decomposition of an organic material by heating in the absence of air. Destructive distillation of wood, which produces charcoal and several small organic molecules, was once the major source of methanol.

DETA Abbreviation for DIETHYLENETRIAMINE, w s.

Detection Limit Of chemical quantitative analyses, the least quantity or concentration of the substance being sought that can be detected with a stated level of confidence. This widely used term may be giving way to MINIMUM DETECTABLE AMOUNT, w s.

Deterioration A usually gradual process of permanent impairment in the appearance or application properties of a plastic article.

Devolatilization The removal from a resin of a substantial percentage of some unwanted volatile matter such as water, solvent, or monomer. Vented extruders have been widely used for this task. See EXTRACTION EXTRUSION.

Dew-Cycle Weathering Test An accelerated-weathering test, ASTM D 4329, in which specimens mounted on panels are alternatively exposed to unfiltered light, such as from a carbon arc, and dew caused by spraying cold water on the backs of the panels. This test is thought to be a fast and reliable means of predicting the weatherability of plastic coatings, but its validity for acrylic coatings has been disputed.

DGEBA Abbreviation for DIGLYCIDYL ETHER OF BISPHENOL A, w s.

D-Glass Glass with a high boron content, used for fibers in laminates that require a precisely controlled dielectric constant.

DHA Abbreviation for DEHYDROACETIC ACID, w s.

DHP Abbreviation for dihexyl phthalate. See DI(2-ETHYLBUTYL) PHTHALATE.

DHXP Same as DHP, preceding.

Di- A prefix meaning two or twice. The terms *bi-* and *bis-* are nearly equivalent, assigned with slight differences in meaning according to custom.

Diacetin (glyceryl diacetate) $CH_3OCOCH_2CHOHCH_2OCOCH_3$, (1,3-diacetin), and 1,2-diacetin, a mixture of isomers. A water-soluble plasticizer, and a solvent for cellulose nitrate and cellulose acetate.

Diacetone Alcohol (4-hydroxy-4-methyl-2-pentanone) $(CH_3)_2C(OH)$-CH_2COCH_3. A pleasant-smelling, colorless liquid, miscible with water and most organic liquids, used as a solvent for cellulosic, vinyl, and epoxy resins.

Diacetyl Peroxide See ACETYL PEROXIDE.

Diafoam A term sometimes used for SYNTACTIC FOAM, w s, that contains gas bubbles in addition to microspheres.

Diallyl Chlorendate (DAC) A reactive monomer used as a flame-resisting agent in diallyl phthalate, epoxy, and alkyd resins. It can be used in the monomeric form (a high-viscosity liquid); or in the polymeric form, alone or in conjunction with other flame retardants.

Diallyl Isophthalate $C_6H_4(COOCH_2CH{=}CH_2)_2$. A polymerizable monomer, used in laminating and molding.

Diallyl Maleate $H_2C{=}CHCH_2OOCCH{=}CHCOOCH_2CH{=}CH_2$. A monomer that polymerizes readily when exposed to light or temperatures above 50°C.

Diallyl Phthalate (diallyl *o*-phthalate, DAP) C_6H_4-1,2-$(COOCH_2CH{=}CH_2)_2$. In the monomeric form, DAP is a colorless liquid ester with a viscosity about equal to that of kerosene, widely used as a crosslinking monomer for unsaturated polyester resins, and as a polymerizable plasticizer for many resins. It polymerizes easily, either gradually or rapidly, increasing in viscosity until it finally becomes a clear, infusible solid. The name DAP is used for both the monomeric and polymeric forms. In the partly polymerized form, DAP is used in the production of thermosetting molding powders, casting resins, and laminates.

Di-Alphinyl Phthalate A plasticizer derived by the esterification of primary aliphatic alcohols in the range C_7 to C_9. It is used in place of dioctyl phthalate in some applications.

Dialysis The separation of different-sized molecules by transport in solution through a semipermeable membrane, driven by the difference in chemical potential between the liquids separated by the membrane. The membranes may be any of a wide range of materials, e g, parchment or microporous films of cellulosic polymers.

4,4'-Diaminodiphenylmethane (DDM, methylenediamine, MD) $NH_2C_6H_4CH_2C_6H_4NH_2$. A silvery, crystalline material obtained by heating formaldehyde anilide with aniline hydrochloride and aniline. It is used as a curing agent for epoxy resins and as an intermediate in making diisocyanates for urethane elastomers and foams by reaction with phosgene. Possible occupational hazards in the use of DDM are toxic hepatitis and liver damage.

4,4'-Diaminodiphenylsulfone (DDS) $H_2NC_6H_4SO_2C_6H_4NH_2$. An epoxy curing agent that gives the highest deflection temperatures among amine-cured epoxies.

Diamyl Phthalate $C_6H_4(COOC_5H_{11})_2$. A plasticizer derived by esterification of phthalic anhydride with amyl alcohol, compatible with most vinyls, polymethyl methacrylate, and cellulosics.

Diaphragm Gate (web gate) In an injection or transfer mold, a gate that forms a complete thin web across the opening of the part, used in the molding of annular and tubular objects.

Diatomite (diatomaceous earth, DE, kieselguhr, infusorial earth, siliceous earth, tripolite) The naturally occurring deposit of skeletons of small unicellular algae called *diatoms*, consisting of from 83 to 89% silica. Its many uses include fillers for plastics.

DIBA Abbreviation for DIISOBUTYL ADIPATE, w s.

Dibasic Pertaining to acids having two replaceable hydrogen atoms e g, sulfuric acid, H_2SO_4, or to acid salts in which two of the three hydrogens have been replaced by metal(s), e g, K_2HPO_4.

Dibasic Lead Phosphite $2PbO \cdot PbHPO_3 \cdot 0.5H_2O$. A white, crystalline powder long used as a heat and light stabilizer for PVC and other chlorine-containing resins. It has good electrical properties, and acts as an antioxidant and screening agent for UV light. Less used today because of concern to keep lead out of the environment.

Dibasic Lead Phthalate A heat and light stabilizer for vinyl insulation, opaque film and sheeting, and foam.

Dibasic Lead Stearate $2PbO \cdot Pb(OOCC_{17}H_{35})_2$. A good heat stabilizer with lubricating properties. See also LEAD STEARATE.

Dibenzyl Ether $(C_6H_5CH_2)_2O$. A plasticizer for cellulose nitrate.

Dibenzyl Sebaçate $(C_6H_5CH_2OOC)_2(CH_2)_8$. A nontoxic plasticizer, often used in vinyl compounds for lining container closures.

Diborane B_2H_6 (boron hydride, boro-ethane) A colorless gas that has been used as a catalyst in the polymerization of ethylene.

DIBP Abbreviation for DIISOBUTYL PHTHALATE, w s.

2,3-Dibromo-2-Butene-1,4-Diol (dibromobutenediol) $HOCH_2C(Br)=C(Br)CH_2OH$. A low-molecular-weight, chemically reactive, brominated primary glycol. It is used as a building block for condensation polymers that can be incorporated into a wide variety of polymers including esters, urethanes, and ethers. It is also used as a flame-retardant monomer for polyurethanes and thermoplastics, and as a substitute for methylene-bis-*o*-chloroaniline (MOCA) in urethane foams.

Dibromoneopentyl Glycol A high-melting solid, available in powder or flake form, used as a flame retardant for polyester resins. A more convenient liquid material is made by using dibromoneopentyl glycol to form a polyester alkyd that is dissolved in styrene, the resulting liquid being more easily used in polyester reactors. The material is also adaptable to urethane foams, polymeric plasticizers, and coating resins.

2,3-Dibromopropanol A brominated alcohol used as a component in making fire-retardant urethane foams.

Dibutoxyethoxy Ethyl Adipate $[C_4H_9OC_2H_4OC_2H_4COO(CH_2)_2]_2$. A plasticizer for cellulose nitrate, ethyl cellulose, polyvinyl butyral, and polyvinyl acetate.

Dibutoxyethyl Adipate $(C_2H_4COOC_2H_4OC_4H_9)_2$. A primary plasticizer for PVC and many other resins, imparting low-temperature flexibility and UV resistance. It is widely used as a plasticizer for polyvinyl butyral in the interlayer of safety glass.

Dibutoxyethyl Phthalate (DBEP) $C_6H_4(COOC_2H_4OC_4H_9)_2$. A primary plasticizer for vinyls, methacrylates, nitrocellulose, and ethyl cellulose, imparting low-temperature flexibility and UV resistance. Incorporation of up to 20% of DBEP into vinyl calendering compounds eliminates defects such as streaks and blisters. In plastisols, DBEP imparts low initial viscosity and low fusion temperatures.

Dibutoxyethyl Sebaçate $(CH_2)_8(COOC_2H_4OC_4H_9)_2$. A primary plasticizer for PVC and PVAc, with low-temperature resistance.

Dibutyl Adipate $(C_4H_9COO)_2(CH_2)_4$. A plasticizer for vinyl and cellulosic resins.

Dibutyl Butyl Phosphonate $C_4H_9P(O)(OC_4H_9)_2$. A plasticizer and antistatic agent.

Di-*tert*-Butyl-*p*-Cresol (DBPC, butylated hydroxytoluene, BHT) $[C(CH_3)_3]_2C_6CH_3H_2OH$. A white crystalline solid used as an antioxidant in polyethylene, vinyl monomers, and many other substances.

Di-*tert*-Butyl Peroxide $(CH_3)_3COOC(CH_3)_3$. A member of the alkyl peroxide family, used as an initiator in vinyl chloride polymerization, polyester reactions, and as a crosslinking agent.

Dibutyl Fumarate $C_2H_2(COOC_4H_9)_2$. Derived from fumaric acid and used as a plasticizer for polyvinyl acetate, polyvinyl chloride, and PVC-AC copolymers.

Dibutyl Isosebaçate $C_8H_{16}(COOC_4H_9)_2$. A plasticizer for vinyls and other thermoplastics.

Dibutyl Phthalate (DBP) $C_6H_4(COOC_4H_9)_2$. One of the most widely used plasticizers for cellulose nitrate and other cellulose-ester and -ether lacquers and coatings. It is a primary plasticizer for many other resins, but its high volatility limits its use in vinyls.

Dibutyl Sebaçate (DBS) $(CH_2)_8(COOC_4H_9)_2$. A plasticizer, one of the most effective of the sebaçate family. It has good low-temperature properties, low volatility, and is compatible with vinyl chloride polymers and copolymers, polyvinyl butyral and ethyl cellulose. It is non-toxic, suitable for use in food wrappings.

Dibutyl Succinate $(CH_2)_2(COOC_4H_9)_2$. A plasticizer for cellulosic resins.

Dibutyl Tartrate $(CHOH)_2(COOC_4H_9)_2$. A lubricant, plasticizer, and solvent for cellulosic resins.

Dibutyltin Bis(isooctylmercapto Acetate) A stabilizer for rigid PVC, used primarily during the period from 1953 to 1970. Thereafter, improved butyltin and methyltin derivatives, and synergistic mixtures of them with other stabilizers, replaced this stabilizer for most uses.

Dibutyltin Di-2-Ethylhexoate $(C_4H_9)_2Sn(OOCC_7H_{18})_2$. A white, waxy solid made by reacting dibutyltin oxide with 2-ethylhexoic acid. Used as a catalyst for silicone curing and in polyether foams.

Dibutyltin Dilaurate (DBTDL) $(C_4H_9)_2Sn(OOCC_{11}H_{23})_2$. A lubricating stabilizer for vinyl resins, a catalyst for urethane foams and for condensation polymerizations. It is used in vinyl compounds when good clarity is needed, and also imparts excellent light stability. However, it degrades at high processing temperatures.

Dibutyltin Maleate $[(C_4H_9)_2Sn(OOCCH)_2]_x$. A white amorphous powder used as a condensation-polymerization catalyst and a stabilizer for PVC. The molecular weight of the material varies, and grades of lower molecular weight tend to be volatile and produce gases. The higher-weight polymers are very effective in rigid PVC.

Dibutyltin Sulfide $[(C_4H_9)_2SnS]_3$. An antioxidant and stabilizer for vinyl resins.

1,1-Dibutylurea (*N,N*-dibutylurea) $NH_2CON(C_4H_9)_2$. A polymerizable compound. When copolymerized with simple urea and formaldehyde, permanently thermoplastic resins are obtained.

Dicapryl Adipate $C_4H_8(COOC_8H_{17})_2$. A plasticizer for cellulosic and vinyl resins, yielding good low-temperature flexibility. Also compatible with polymethyl methacrylate and polystyrene.

Dicapryl Phthalate [(DCP, di-(2-octyl) phthalate] $(C_8H_{17}COO)_2C_6H_4$. A plasticizer for cellulosic and vinyl resins. It is similar to DOP and DIOP, but has low initial viscosity and is preferred in plastisols.

Dicapryl Sebaçate $(CH_2)_8(COOC_8H_{17})_2$. A plasticizer for vinyl resins and acrylonitrile rubbers, imparting good low-temperature flexibility.

Dicarboxylic Acid Any of a large family of organic acids containing two carboxylic (–COOH) groups. Those of greatest importance in the plastics industry are the adipic, azelaic, glutaric, pimelic, sebacic, and succinic acids, esters of which are widely used as plasticizers that impart low-temperature flexibility. They are also used in the production of alkyd and polyester resins, polyurethanes, and nylons.

Dicetyl Ether (dihexadecyl ether) A mold lubricant.

2,4-Dichlorobenzoÿl Peroxide A crosslinking agent for silicone elastomers. It is sold as a 40% active paste dispersed in silicone fluid (Cadox®).

Dichlorodifluoromethane (Freon 12) CCl_2F_2. Long used as a refrigerant, aerosol propellant, and as a blowing agent for foamed plastics that was safer than inflammable hydrocarbons, this accused destroyer of stratospheric ozone is rapidly being phased out of all its former uses.

1,1-Dichloroethylene Synonym for VINYLIDENE CHLORIDE, w s.

α-Dichlorohydrin $CH_2ClCHOHCH_2Cl$. A cellulosic-resin solvent.

Dichloromethane Synonym for METHYLENE CHLORIDE, w s.

2,6-Dichlorostyrene $Cl_2C_6H_3CH{=}CH_2$. A monomer and comonomer used mainly in plastics research.

Dichlorotetrafluoroethane (Freon 114) ClF_2CCClF_2. A fluorocarbon blowing agent used when a low boiling point (3.6°C) is required.

Dichroism (1) A property possessed by many doubly refracting crystals of exhibiting different colors when viewed from different directions. (2) The exhibition of different colors by certain solutions in different degrees of concentration of the same solute.

Dicing The process of cutting thermoplastic sheets (or square strands) into cubical pellets for further processing.

Dicumyl Peroxide $[C_6H_5C(CH_3)_2O]_2$. A vulcanizing agent for elastomers, also used to crosslink polyethylene.

Dicyandiamide (cyanoguanidine) $H_2NC(=NH)NHCN$. The widely used, but incorrect name for the dimer of cyanamide. Cyanoguanidine is used mainly in the production of melamine, but also as a stabilizer for vinyl resins and curing agent for epoxy resins.

Dicyclohexyl Azelate $C_6H_{11}OOCC_7H_{14}COOC_6H_{11}$. A plasticizer for PVC.

Dicyclohexyl Phthalate (DCHP) $C_6H_4(COOC_6H_{11})_2$. A plasticizer for PVC and many other resins. It imparts good electrical properties, low volatility, low water and oil absorption, and resistance to extraction by hexane and gasoline. In vinyls DCHP is usually combined with other plasticizers. In cellulose nitrate, polystyrene, polymethyl methacrylate, and ethyl cellulose it serves as a primary plasticizer.

DIDA Abbreviation for DIISODECYL ADIPATE, w s.

Didecyl Adipate $C_4H_8(COOC_{10}H_{21})_2$. A plasticizer for PVC and cellulosics. Its most noteworthy properties are low-temperature flexibility, low volatility, and good electricals.

Didecyl Ether $(C_{10}H_{21})_2O$. A processing and mold lubricant.

Didecyl Phthalate (DDP) $C_6H_4(COOC_{10}H_{21})_2$. A primary plasticizer for vinyl resins, also compatible with polystyrene and cellulosics. It has the lowest specific gravity of the most common phthalate plasticizers, low volatility, and resistance to extraction by soapy water.

DIDG Abbreviation for DIISODECYL GLUTARATE, w s.

DIDP Abbreviation for DIISODECYL PHTHALATE, w s.

Die (1) EXTRUSION DIE, w s. (2) The recessed block into which plastic material is injected or pressed, shaping the material to the desired form. The terms *mold cavity* or *cavity* are more often used. (3) Steel-rule die. See DIE CUTTING.

Die Adapter (extrusion) See ADAPTER.

Die Blade (die lip) In extrusion, a deformable member attached to a die body that determines the slot opening and that may be adjusted to produce uniform thickness across the film or sheet being made.

Die Block That part of an extrusion die that holds the forming bushing and core.

Die Body In the US, same as ADAPTER, w s. In Great Britain, the outer body or barrel of an extrusion die.

Die Cart A sturdy, height-adjustable framework on casters designed to support a heavy extrusion die, such as a sheet die.

Die Characteristic Of an extrusion die, the relationship between rate of flow of melt through the die, the pressure drop through the die, and the viscosity of the melt. It is defined by the flow equation:

$$Q = K \cdot \Delta P / \eta$$

in which Q = the volumetric flow rate, P = the pressure drop, η = the effective viscosity at the temperature and shear rate of the melt in the die, and K = the die characteristic. K is dependent on die dimensions and has the dimensions of volume. The die characteristic is linked to the SCREW CHARACTERISTIC, w s, by ΔP, which must be the same for both, whereas η usually will not be the same for both.

Die Core The tapered element in an extrusion die for pipe or tubing that guides the material to the webs of the spider. Sometimes called the *torpedo* or *spreader*.

Die Cutting (blanking, clicking, dinking) The process of cutting shapes from sheets of plastic by pressing a shaped knife edge through one or several layers of sheeting. The dies are often called *steel-rule dies*, and pressure is applied smartly by hydraulic or mechanical means.

Die-Entry Angle At the entrance to the land of an extrusion die, the angle included between the adjacent, inside die surfaces. In many rheometer orifices and in some profile dies, the entry angle is 180°, the extreme of abruptness in actual use. Streamlined-entry dies stretch out the reduction and change of cross-sectional shape over a distance, so the included angle at the die lip may be 20 to 60°. Extreme gradualness is designed into high-speed wire-coating dies, where the approach to the lip is made in several stages of decreasing taper, the final included angle being only 6 to 10°.

Die Gap (die-lip opening) In film and sheeting dies, the perpendicular distance between the lips of the die land at any point along the the width of the die, usually measured after the die has been brought up to temperature but before extrusion has begun. During extrusion, internal pressure inside the die causes the gap to enlarge slightly, more at the center than at the ends of a center-fed die, necessitating adjustments to attain uniform thickness across the film or sheet.

Die Land In an extrusion die, the land is the portion of the die wherein the dimensions of the opening are constant from a point within the die to the discharge point, giving the extrudate its final shape.

Dielectric A material with electrical conductivity less than 10^{-6} siemens per centimeter (1 μS/cm), thus so weakly conductive that different parts of a sheet can hold different electrical charges. In radio-frequency heating, the term dielectric is used for the material being heated. The term is also used for the nonconductive material separating the conductive elements of a capacitor. Polymeric materials are widely employed as dielectrics. The two most important properties of a dielectric are its DIELECTRIC CONSTANT and DIELECTRIC STRENGTH, defined below. Values for some polymeric and other dielectrics at 60

Hz and room temperature are listed here.

Material	Dielectric constant	Dielectric strength, kV/mm
air	1.00054	0.8
alumina*	3.5	1.6–6.3
Pyrex glass	4.5	13
bond paper	3.5	14
polyethylene	2.3	18–39
polypropylene	2.1–2.7	18–26
nylon 6/6	3.6–4.0	12–16
glass-reinforced nylon 6/6	4.0–4.4	19
crystal polystyrene	2.5–2.65	20–28
acrylic resin	3.3–3.9	16
phenoxy resin	4.1	16–20
thermoplastic polyurethane	6–8	33–43
GP phenolic	5–10	12–17
epoxy/glass	5.5	14

*At 1 MHz

Dielectric Absorption An accumulation of electrical charges within the body of an imperfect dielectric material when it is placed in an electric field.

Dielectric Breakdown Voltage (breakdown voltage, disruptive voltage) The voltage at which electrical breakdown of a specimen of electrical insulating material between two electrodes occurs under prescribed conditions. The ASTM test for dielectric breakdown voltage and dielectric strength is D 149.

Dielectric Constant (permittivity constant) Between any two electrically charged bodies there is a force (attraction or repulsion) that varies according to the strength of the charges, q_1 and q_2, the distance between the bodies, r, and a characteristic of the medium separating the bodies (the dielectric) known as the dielectric constant, ε. This force is given by the equation:

$$F = q_1 \cdot q_2 / (\varepsilon \cdot r^2).$$

For a vacuum, $\varepsilon = 1.0000$; values for some other materials are listed above. In practice, the dielectric constant of a material is found by measuring the capacitance of a parallel-plate condenser using the material as the dielectric, then measuring the capacitance of the same condenser with a vacuum as the dielectric, and expressing the result as a ratio between the two capacitances. When the dielectric is a polymeric material whose atoms or molecules may change their positions in an alternating electric field (a *polar* material), frictional energy is dissipated as heat and is characterized by the DISSIPATION FACTOR, w s. The ASTM method for measuring dielectric constant (permittivity) and related properties of solid electrical insulating materials is D 150. A second method, for polyethylene, is given in D 1531 (Section 10.02).

Dielectric Heating (electronic heating, RF heating, radio-frequency heating, high-frequency heating, microwave heating) The process of heating poor (but polar) conductors of electricity (dielectrics) by means of high-frequency fields. At frequencies above 10 MHz sufficient heat for rapid sealing and welding of many plastics can be generated at low, safe voltages. The process of dielectric heating consists of placing the material to be heated between two shaped electrodes that are connected to a high-frequency current supply. These electrodes act as the plates of a capacitor and the material serves as the dielectric separating them. As the field changes polarity, charge-bearing atoms or groups of the dielectric undergo reorientation in an effort to keep their positive poles toward the electrode that is momentarily negative, thus generating molecular friction that is dissipated as heat. The theoretical rate of heating is given by the equation:

$$P = 2\pi \cdot f \cdot C \cdot V^2 \cdot tan\delta$$

where P = the power input(W), f = the electrical field frequency (Hz), C = the capacitance (F), V = the voltage difference between the electrodes (V), and $tan\delta$ = the loss factor. The actual rate of heating will be somewhat less because of nonuniformity of the field, air gaps between electrodes and material, and losses to the surroundings. Dielectric heating is most effective for materials such as PVC and phenolics that have high loss factors because of their numerous polar groups. Nonpolar plastics with low loss factors, such as polystyrene and polyethylene, are impractical to heat dielectrically. See also DIELECTRIC HEAT SEALING, MICROWAVE HEATING.

Dielectric Heat Sealing A sealing process widely used for vinyl films and other thermoplastics with sufficient dielectric loss, in which two layers of film are heated by DIELECTRIC HEATING, w s, and pressed together between two electrodes, an applicator and a platen, the films serving as the dielectric of the so-formed condenser. The applicator may be a pinpoint electrode as in "electronic sewing machines", a wheel, a moving belt, or a contoured blade. Frequencies employed range up to 200 MHz, but are usually 30 MHz or less to avoid interference problems. See also HEAT SEALING.

Dielectric Loss Energy dissipated as heat within a polar material subjected to a rapidly alternating, strong electric field. See DIELECTRIC HEATING.

Dielectric Loss Angle (dielectric phase difference) The complement of the DIELECTRIC PHASE ANGLE (w s), i e, 90°– θ.

Dielectric Phase Angle The angular difference in phase between the alternating voltage (usually sinusoidal) applied to a dielectric and the resulting current. The angle is often symbolized by θ, the cosine of which is the POWER FACTOR, w s.

Dielectric Strength (electric strength) A measure of the voltage required to puncture an insulating material, expressed in volts per mil of thickness (SI: V/mm). The voltage is the root-mean-square (rms) voltage difference between the two electrodes in contact with opposite surfaces of the specimen at which electrical breakdown occurs under prescribed test conditions, e g, as in ASTM D 149.

Dielectrometer An instrument that enables one to measure several dielectric properties of a material. They are capacitance (dielectric constant), phase angle and loss tangent (dissipation factor), usually over ranges of temperature and frequency. Determination of these properties is the subject of ASTM Methods D 149 and, in Section 10.02, D 1531.

Die Lines In extrusion and blow molding, longitudinal marks caused by damaged die surfaces or by extrudate buildup on dies.

Die-Lip Buildup In extruding film and sheet, the gradual accumulation of a bead of resin on the face of the die, parallel to the slit. Similar buildup can occur on dies for shapes.

Die Manifold See MANIFOLD.

Diene Polymer Any of a family of polymers based on unsaturated hydrocarbons having two double bonds (*diolefins*). When the double bonds are separated by only one single bond, as in 1,3-butadiene (CH_2=CH–CH=CH_2), the diene and its double bonds are said to be *conjugated*. In an *un*conjugated diene the double bonds are separated by at least two single bonds and act more independently. The family includes polymers of butadiene, isoprene, cyclopentadiene, and copolymers with ethylene and propylene.

Die Plate (1) In injection molds, a member that is attached to the fixed or to the moving head of the press; *mold plate*. (2) In extrusion, especially of pellets or shapes, the die plate is that part of the die assembly that is bolted to the outlet of the die body and contains the orifices that form the melt into continuous strands or a particular cross-sectional shape.

Die Pressure In extrusion, the pressure of the melt entering the die.

Dieseling See BURN MARK.

Die Spider In extrusion, the legs or webs supporting the die core within the head of an in-line pipe, tubing, or blown-film die. In many pipe dies, the spider legs are cored to permit application of air or water for cooling the mandrel.

Die Swell See EXTRUDATE SWELLING.

Die-Swell Ratio (extrudate-swelling ratio) In extrusion, particularly in extruding parisons for blow molding, the ratio of the outer parison diameter or parison wall thickness to, respectively, the outer diameter of the parison die or the die gap. The ratio is affected by the polymer type, its temperature, the die geometry, and the extrusion rate. Some writers have defined the die-swell ratio as the ratio of the *cross section* of the extrudate shortly after emergence to that of the die opening.

2,2-Diethoxyacetophenone (DEAP) A photo-initiator used for curing acrylate coatings, either in an inert atmosphere or in air with UV light. It absorbs impinging light.

Diethoxyethyl Adipate $(CH_2)_4(COOC_2H_4OC_2H_5)_2$. A plasticizer for cellulosic resins.

Diethoxyethyl Phthalate $C_6H_4(COOC_2H_4OC_2H_5)_2$. A plasticizer for cellulosic and vinyl resins.

Diethyl Adipate $(CH_2)_4(COOC_2H_5)_2$. A plasticizer for cellulosic resins.

Diethylaluminum Chloride $(C_2H_5)_2AlCl$. A colorless liquid that bursts into flame instantly upon contact with air (pyrophoric) and reacts violently with water. It is used as a catalyst in olefin polymerization.

3-Diethylaminopropylamine $(C_2H_5)_2NCH_2CH_2CH_2NH_2$. An epoxy curing agent.

Di(2-ethylbutyl) Azelate $(CH_2)_7(COOC_6H_{13})_2$. A plasticizer for PVC, its copolymers, and cellulose esters. It is very compatible and efficient in vinyls, finding its largest use in high-clarity film and sheeting.

Di(2-ethylbutyl) Phthalate (dihexyl phthalate, DHXP) $C_6H_4(COO\text{-}C_6H_{13})_2$. A fast-fluxing, highly solvating plasticizer used in vinyl plastisols and extrusion compounds and for cellulose esters.

Diethyl Carbonate (ethyl carbonate) $(C_2H_5)_2CO_3$. A solvent for cellulosic and many other resins.

Diethylene Glycol Bis-(Allyl Carbonate) $(CH_2{=}CHCH_2OCOOCH_2\text{-}CH_2)_2O$. The monomer from which, by addition polymerization, is prepared the transparent, water-white polymer CR-39, one of the allyl-resin family. Since it is difficult and tricky to handle, the resin has found little commercial use aside from optical applications. For these, its excellent optical properties and resistance to scratching, and its low density—half that of glass—have given it wide use in eyeglasses. Other uses include optical filters, instrument windows, welders' masks, and large windows in atomic-energy plants.

Diethylene Glycol Dibenzoate $(C_6H_5COOC_2H_4)_2O$. A plasticizer for cellulosic resins, polymethyl methacrylate, polystyrene, PVC, polyvinyl acetate, and other vinyls. It imparts good stain resistance.

Diethylene Glycol Dipelargonate $[CH_3(CH_2)_7COOC_2H_4]_2O$. A simple diester of pelargonic acid used mainly as a secondary plasticizer for vinyl resins, but also as a plasticizer for cellulosics. Within the limits of its compatibility it provides economical low-temperature flexibility.

Diethylene Glycol Dipropionate $(CH_3CH_2COOC_2H_4)_2O$. A plasticizer for cellulosic plastics.

Diethylene Glycol Distearate $(C_{17}H_{35}COOC_2H_4)_2O$. A plasticizer for cellulose nitrate and ethyl cellulose.

Diethylene Glycol Monoacetate $HO(CH_2)_2O(CH_2)_2OOCCH_3$. A solvent for cellulose nitrate and cellulose acetate.

Diethylene Glycol Monobutyl Ether $C_4H_9OCH_2CH_2OCH_2CH_2OH$. A solvent with a high boiling point, used in coatings when very slow drying rates are desired. It is also useful as a dispersant in vinyl organosols, and as an intermediate for the production of plasticizers.

Diethylene Glycol Monoethyl Ether $C_2H_5OCH_2CH_2OCH_2CH_2OH$. A solvent for cellulosics.

Diethylene Glycol Monolaurate See DIGLYCOL LAURATE.

Diethylene Glycol Monoricinoleate See DIGLYCOL RICINOLEATE.

Diethylenetriamine (DETA, 2,2'-diaminodiethylamine) $(NH_2CH_2CH_2)_2NH$. A pungent liquid providing fast cures with epoxy resins, even at room temperature.

Di(2-ethylhexyl) Adipate (dioctyl adipate, DOA) $[C_2H_4COOCH_2CH(C_2H_5)C_4H_9]_2$. A primary plasticizer for vinyls, cellulose nitrate, polystyrene, and ethyl cellulose. In vinyls, DOA is often used in combination with phthalate and other plasticizers, imparting good resilience, low-temperature flexibility, and resistance to extraction by water. It is FDA-approved for use in vinyl food-packaging films.

Di(2-ethylhexyl) Azelate (dioctyl azelate, DOZ) $(CH_2)_7[COOCH_2CH(C_2H_5)C_4H_9]_2$. A plasticizer for vinyl chloride polymers and copolymers. It is one of the most compatible of the low-temperature, monomeric plasticizers, has low volatility, and imparts low extractability by water and soapy water. This plasticizer has been approved for food-contact use.

Di(2-ethylhexyl) 2-Ethylhexyl Phosphonate $C_8H_{17}PO(OC_8H_{17})_2$. A plasticizer and stabilizer.

Di(2-ethylhexyl) Hexahydrophthalate (dioctyl hexahydrophthalate) $C_6H_{10}[COOCH_2CH(C_2H_5)C_4H_9]_2$. A light-colored liquid, a plasticizer for vinyls.

Di(2-ethylhexyl) Isophthalate (dioctyl isophthalate, DIOP) $C_6H_4[COOCH_2CH(C_2H_5)C_4H_9]_2$. A primary plasticizer for PVC, most notable for low volatility and its resistance to marring by nitrocellulose lacquers, in addition to good general-purpose properties. It is also compatible with polyvinyl butyral, vinyl chloride-acetate copolymers, cellulosic resins, polystyrene, and chlorinated rubber.

Di(2-ethylhexyl) Phthalate (dioctyl phthalate, DOP) $C_6H_4[COO-CH_2CH(C_2H_5)C_4H_9]_2$. The most widely used plasticizer for PVC, also compatible with ethyl cellulose, cellulose nitrate, and polystyrene. It is generally recognized as imparting the best all-around good properties to vinyls, and is often used as the standard against which other plasticizers are evaluated. DOP has been approved by the FDA for use in packaging films for greaseless foodstuffs of high water content.

Di(2-ethylhexyl) Sebaçate (dioctyl sebaçate) $(CH_2)_8(COOC_8H_{17})_2$. A plasticizer for vinyl chloride polymers and copolymers, cellulosic plastics, polystyrene, and polyethylene. It imparts good low-temperature properties.

Di(2-ethylhexyl) Succinate (dioctyl succinate) $(CH_2)_2(COOC_8H_{17})_2$. A plasticizer.

Di(2-ethylhexyl)-4-Thioazelate $C_8H_{17}OOC(CH_2)_2CS(CH_2)_4COOC_8-H_{17}$. A plasticizer for cellulose nitrate, ethyl cellulose, polymethyl methacrylate, polystyrene, and vinyl resins.

Diethyl Phthalate (ethyl phthalate) $C_6H_4(COOC_2H_5)_2$. A plasticizer and solvent for nearly all thermoplastics and coumarone resins. It has been approved by FDA for use in food packaging.

Diethyl Sebaçate $(CH_2)_8(COOC_2H_5)_2$. A plasticizer with good low-temperature properties, compatible with PVC and many other thermoplastics.

Diethyl Succinate $(CH_2)_2(COOC_2H_5)_2$. A plasticizer for cellulosic resins.

Diethyl Tartrate $[-CH(OH)COOC_2H_5]_2$. A solvent and plasticizer for cellulosic resins.

1,1-Diethylurea $H_2NCON(C_2H_5)_2$. A white solid polymerizable with simple urea and formaldehyde to form permanently thermoplastic resins.

Differential Refractometer An instrument used in connection with a chromatographic column in SIZE-EXCLUSION CHROMATOGRAPHY (w s) of polymers (in dilute solution). Two streams, one pure solvent, the other the eluting polymer solution, pass through the instrument, whose signal, proportional to the difference in refractive indices of solution and solvent, is interpreted as polymer concentration. The instrument is also used to determine the rate of change of refractive index with concentration (dn/dc), an important parameter in the computation of weight-average molecular weights from light-scattering measurements in dilute polymer solutions.

Differential Scanning Calorimeter (DSC) An instrument that measures the rate of heat evolution of absorption of a specimen while it is undergoing a programmed temperature rise. A recorder displays the data as a trace of increase in heat per increase in temperature (dq/dT)

versus temperature. The DSC has been used to study curing reactions and related properties of thermosetting resins, and heats of decomposition of resins.

Differential Thermal Analysis (DTA) An analytical method in which the specimen material and an inert substance are heated concurrently in separate minipans and the difference in temperature between the two is recorded, along with the temperature of the inert substance. DTA has been useful in the study of phase transitions and curing and degradation reactions in polymers.

Diffusion Bonding An experimental method of joining materials in which extremely flat and finely finished surfaces are clamped together under high vacuum for a period of hours or days, sometimes at a high temperature, allowing the atoms of each member to diffuse into the other. As of early 1992, the method had been used only with inorganic crystalline materials.

Diffusion Couple An assembly of two materials in such intimate contact that each diffuses into the other.

Diffusivity (1) (diffusion coefficient) The constant of proportionality in Fick's first law of diffusion, which states that the mass or molar rate of transport (flux) of one molecular species into another is equal to the diffusivity times the gradient of concentration. Several related units are in use, e g, cm^2/s, ft^2/h. The SI unit is m^2/s, corresponding to flux in $mol/(s \cdot m^2)$ and gradient in $(mol/m^3)/m$. See also FICK'S LAW. (2) THERMAL DIFFUSIVITY, w s.

Diglycidyl Ether of Bisphenol A (DGEBA) The main constituent of most commercial epoxy resins prior to curing, DGEBA is formed by the reaction of excess epichlorohydrin with bisphenol A in the presence of aqueous sodium hydroxide. It has the structural formula:

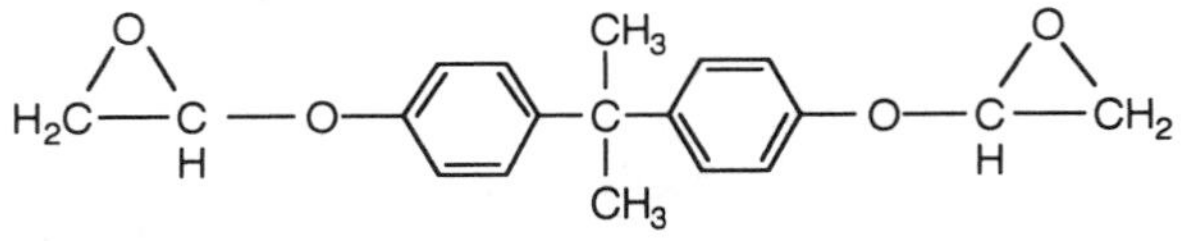

Diglycol Laurate (diethylene glycol monolaurate) $C_{11}H_{22}COOC_2$-$H_4OC_2H_4OH$. A plasticizer for ethyl cellulose, cellulose nitrate, polyvinyl butyral, and vinyl chloride-acetate copolymers.

Diglycol Ricinoleate (diethylene glycol monoricinoleate) $CH_3(CH_2)_5$-$CH(OH)CH_2CH{=}CH(CH_2)_7COOC_2H_4OC_2H_4OH$. A plasticizer for ethyl cellulose and cellulose nitrate.

Dihalide A compound containing two halogen atoms (of valence = –1) per molecule.

Di-*n*-hexyl Adipate $(CH_2)_4(COOC_6H_{13})_2$. An important low-temperature plasticizer for synthetic rubbers and several plastics, including some cellulosics, PVC, and polyvinyl acetate.

Di-*n*-hexyl Azelate (DNHZ) $(CH_2)_7(COOC_6H_{13})_2$. A plasticizer for cellulose and vinyl resins. It has been approved for food-contact use, has low volatility and good compatibility.

Dihexyl Phthalate (DHXP) See DI(2-ETHYL BUTYL) PHTHALATE.

Dihexyl Sebaçate $(CH_2)_8(COOC_6H_{13})_2$. A plasticizer for vinyls.

Dihydrodiethyl Phthalate C_6H_6-1,2-$(COOC_2H_5)_2$. A plasticizer for PVC and cellulose nitrate, with partial compatibility with other thermoplastics.

Diisobutyl Adipate (DIBA) $[-C_2H_4COOCH_2CH(CH_3)_2]_2$. A plasticizer for most synthetic resins, including cellulosics, PVC, and other vinyls, and FDA-approved for use in food-packaging films. In vinyls, it is a very active solvent and lowers processing temperatures to levels that permit elimination or lower percentages of stabilizers.

Diisobutyl Aluminum Chloride $[(CH_3)_2CHCH_2]AlCl$. A catalyst for polymerizing olefins.

Diisobutyl Azelate $(CH_3)_2CHCH_2OOC(CH_2)_7$. A plasticizer for cellulosics, polymethyl methacrylate, PVC, and polyvinyl acetate.

Diisobutyl Ketone (2,6-dimethyl-4-heptanone) $[(CH_3)_2CHCH_2]_2CO$. A high-boiling ketone with moderate solvent power for cellulose nitrate and vinyl copolymers. Having limited solvency for PVC, it is used as a viscosity modifier in organosols.

Diisobutyl Phthalate (DIBP) $C_6H_4[COOCH_2CH(CH_3)_2]$. A plasticizer for vinyls, cellulosics, and polystyrene.

Diisocyanate Any compound containing two isocyanate (–NCO) groups, used in the production of polyurethanes. Many methods have been reported for synthesizing diisocyanates, but the one most widely used is reacting phosgene with an amine in a solvent. Toluene diisocyanate (TDI), the most commonly used, is an 80-20 mixture of 2,4- and 2,6-toluene diisocyanate isomers. Also used are diphenylmethane-4,4'-diisocyanate (MDI), a modified toluene diisocyanate, and polymethylene polyphenyl isocyanate (PAPI). The diisocyanates are key ingredients in the production of urethane foams, fibers, coatings, and solid elastomers. See also ISOCYANATE and POLYURETHANE.

Diisodecyl Adipate (DIDA) $(-C_2H_4COOC_{10}H_{21})_2$. A plasticizer for PVC in lower concentrations, e g, up to 30 phr, at which it imparts low-temperature flexibility and resistance to lacquer marring. It is often used in combination with phthalate and phosphate plasticizers. DIDA is completely compatible with vinyl chloride-acetate copolymers, cellulose acetate-butyrate with high butyral content, cellulose nitrate, ethyl cellulose, and chlorinated rubber. In polystyrene, it may be used up to 25 phr.

Diisodecyl-4,5-Epoxy-Tetrahydrophthalate A plasticizer for PVC that also acts as a stabilizer and fungistat. Too, it is compatible with cellulose nitrate, ethyl cellulose, polymethyl methacrylate, polystyrene, and other vinyl polymers and copolymers.

Diisodecyl Glutarate $(CH_2)_3(COOC_{10}H_{21})_2$. A plasticizer for PVC having low-temperature properties equal to those of dioctyl adipate but with lower volatility and greater resistance to soapy water.

Diisodecyl Phthalate (DIDP) $C_6H_4(COOC_{10}H_{21})_2$. A general-purpose plasticizer for vinyl resins, imparting good water resistance and suitable for processing at high temperatures due to its low volatility. It is also compatible with most synthetic resins, e g, cellulose nitrate, ethyl cellulose, and polystyrene.

Diisononyl Phthalate (DINP) $C_6H_4(COOC_9H_{19})_2$. A plasticizer for PVC, cellulosics, and polystyrene. It has lower volatility than dioctyl phthalate with equivalent low-temperature performance and poorer efficiency.

Diisooctyl Adipate (DIOA) $(-C_2H_4COOC_8H_{17})_2$. A primary plasticizer for vinyls, cellulose nitrate, polystyrene, polymethyl methacrylate, and ethyl cellulose. Its performance is similar to that of dioctyl phthalate.

Diisooctyl Azelate (DIOZ) $C_7H_{14}(COOC_8H_{17})_2$. A plasticizer for cellulosic resins and polymers and copolymers of vinyl chloride. In vinyls it imparts good low-temperature properties and other characteristics similar to those obtained with dioctyl azelate.

Diisooctyl Fumarate $(=CHCOOC_8H_{17})_2$. A plasticizer for PVC.

Diisooctyl Isophthalate $C_6H_4(COOC_8H_{17})_2$. A plasticizer for PVC, also compatible with cellulose nitrate, ethyl cellulose, polystyrene and other vinyl resins.

Diisooctyl Monoisodecyl Trimellitate A plasticizer for cellulosic and vinyl resins.

Diisooctyl Phthalate (DIOP) $C_6H_4(COOC_8H_{17})_2$. A primary plasticizer for PVC, ethyl cellulose, cellulose nitrate, and polystyrene. In vinyls, its performance is similar to that of dioctyl phthalate except that it is slightly less volatile than DOP and produces better viscosity characteristics in plastisols. DIOP is FDA-approved for food-packaging materials and medical applications involving contact with water, but not with fats.

Diisooctyl Sebaçate (DIOS) $(CH_2)_8(COOC_8H_{17})_2$. A plasticizer for vinyl and other resins.

Diisopropylene Glycol Salicylate $HOCH_2CH(CH_3)CH(CH_3)CH_2O–OCC_6H_4OH$. An ultraviolet absorber in plastics.

Dilatancy Flow characterized by reversible, instantaneous increase in viscosity with increasing shear rate. The opposite of *pseudoplasticity*.

Dilatometer An instrument for measuring the volume changes of a liquid or solid as the sample's temperature changes. A simple liquid dilatometer consists of a small bulb holding a known mass of sample and sealed to a graduated capillary of known diameter. As the sample is heated, its expansion is indicated by its rise up the capillary. From these measurements the VOLUME COEFFICIENT OF THERMAL EXPANSION, w s, can be estimated. (The inverse principle is used in liquid-in-glass thermometers.)

Dilauryl Ether (didodecyl ether) $(C_{12}H_{23})_2O$. A lubricant for plastics processing.

Dilinoleic Acid $C_{34}H_{62}(COOH)_2$. An unsaturated, dibasic acid used as a modifier in alkyd, nylon, and polyester resins.

Diluent A substance that dilutes another substance. In an organosol, a diluent is a volatile liquid such as naphtha that has little or no solvating effect on the resin, but serves to lower the viscosity of the mix, and is evaporated during processing. In the paint industry, a diluent is any substance capable of thinning paints, varnishes, etc. The term is sometimes also used for a liquid added to a thermosetting resin to reduce its viscosity; and for an inert powdered substance added to an elastomer or resin merely to increase the volume. (Compare: FILLER.)

Dilute-Solution Viscosity (solution viscosity) (1) A catchall term that can mean any of the interrelated and quantitatively defined viscosity ratios of dilute polymer solutions or their absolute viscosities. (2) The kinematic viscosity of a solution as measured by timing the rate of efflux of a known volume of solution, by gravity flow, through a calibrated glass capillary that is immersed in a temperature-controlled bath. Two common types of viscometer are the Ostwald-Fenske and Ubbelohde. From the viscosities of the solution η and the solvent η_0, and the solution concentration c, five frequently mentioned "viscosities" (viscosity ratios, actually) can be derived, as follows:

$$\text{Relative viscosity} = \eta_r = \eta/\eta_0;$$

$$\text{Specific viscosity} = \eta_{sp} = \eta_r - 1;$$

$$\text{Reduced viscosity} = \eta_{red} = \eta_{sp}/c;$$

$$\text{Inherent viscosity} = \eta_{inh} = (\ln \eta_r)/c;$$

$$\text{Intrinsic viscosity} = [\eta] = \text{the limit as } c \rightarrow 0 \text{ of } \eta_{sp}/c,$$

$$= \text{the limit as } c \rightarrow 0 \text{ of } \eta_{inh}$$

The intrinsic viscosity, because it is extrapolated to zero concentration from a series of measurements made at different concentrations, is independent of concentration. However, *different solvents yield different intrinsic viscosities with the same polymer*, so the solvent used must be identified. Some ASTM tests for viscosities of dilute polymer and plastics solutions are D 789, D 1243, D 1601, D 2857, and D 4603, all in Section 08. D 445 and D 446, describing the proper use of the viscometers mentioned above, are in Section 05.

Dilution Ratio (1) As used in the surface-coatings industry, dilution ratio is the volume ratio of diluent to solvent in a blend of these two constituents that just fails to completely dissolve 8.00 g of nitrocellulose in 100 ml of the blend. The procedure is described in ASTM D 1720, Section 06. It is used to determine the most economical, yet adequate amount of high-cost active solvent required in a nitrocellulose-lacquer system. (2) In most other contexts, dilution ratio is the quotient of the concentration of the undiluted solute divided by that of the diluted solution, both concentrations in the same units.

Dimensional Stability The ability of a plastic part to maintain its original dimensions, within set tolerances, over its design life in the expected use environment.

Dimer (1) A molecule formed by union of two identical simpler molecules. For example, C_4H_8 is a dimer of C_2H_4, as N_2O_4 is of NO_2. (2) A substance composed of dimers.

Dimer Acid A coined, generic term for high-molecular-weight, dibasic acids that combine and polymerize with alcohols and polyols to make plasticizers, etc. A trimer acid is analogous, having three acid groups.

Dimethoxyethyl Adipate $C_4H_8(COOC_2H_4OCH_3)_2$. A plasticizer for cellulose-ester polymers.

Di(2-methoxyethyl) Phthalate $C_6H_4(COOCH_2CH_2OCH_3)_2$. A plasticizer, especially for cellulose acetate, but compatible, too, with other cellulosics, polystyrene, and vinyls.

Dimethyl Acetamide (DMAC) $CH_3CON(CH_3)_2$. A colorless liquid, a solvent for resins and plastics, a catalyst, and an intermediate.

2-Dimethylamino Ethanol (DMAE) $(CH_3)_2NCH_2CH_2OH$. A colorless liquid derived from ethylene oxide and dimethylamine, a catalyst for urethane foams. It has little odor and toxicity, and resists staining.

3-Dimethylaminopropylamine $(CH_3)_2N(CH_2)_3NH_2$. A colorless liquid used as a curing agent for epoxy resins.

Dimethylaniline (DMA) $C_6H_5N(CH_3)_2$. The term usually means the tertiary amine, *N,N*-dimethylaniline, though ring-substituted isomers are known. This amine is useful as an accelerator in polyester molding and sprayup.

Di(methylcyclohexyl) Adipate $C_4H_8(COOC_6H_{10}CH_3)_2$. A plasticizer compatible with most thermoplastics.

Di(methylcyclohexyl) Phthalate $C_6H_4(COOC_6H_{10}CH_3)_2$. A plasticizer for cellulosics, polystyrene, PVC, and other vinyl resins.

Dimethylformamide (DMF) $HCON(CH_3)_2$. A colorless and very active solvent for PVC, nylon, polyurethane, and many other resins and elastomers, with fairly low volatility. Its strong solvent power makes it useful as a solvent booster in coating, printing, and adhesive work, and in paint strippers. Because it is toxic and readily absorbed through the skin, DMF must be handled, and disposed of, with care.

Dimethyl Glutarate (DMG) $C_3H_6(COOCH_3)_2$. A liquid chemical intermediate, a source of dicarboxylic acid, used in making plasticizers, polyester resins, synthetic fibers, films, adhesives, and solvents.

Dimethyl Glycol Phthalate $C_6H_4(COOCH_2CH_2OCH_3)_2$. A solvent and plasticizer for cellulosic resins.

Dimethylisobutylcarbinyl Phthalate $C_6H_4[COOCH(CH_3)CH_2CH(CH_3)_2]_2$. A plasticizer for most common thermoplastics.

Dimethyl Ketone Synonym for ACETONE, w s.

Dimethylol Urea $CO(NHCH_2OH)_2$. A colorless crystalline material resulting from the combination of urea and formaldehyde in the presence of salts or alkaline catalysts, the first or A-stage of urea-formaldehyde resin.

Dimethyl Phthalate $C_6H_4(COOCH_3)_2$. A nontoxic plasticizer for most common thermoplastics, but with limited compatibility with PVC.

Dimethyl Polysiloxane See POLYDIMETHYLSILOXANE.

2,2-Dimethyl-1,3-Propanediol See NEOPENTYL GLYCOL.

Dimethyl Sebaçate $[-(CH_2)_4COOCH_3]_2$. A solvent and plasticizer for cellulosic and vinyl resins, also compatible with most other thermoplastics.

Dimethyl Sulfoxide (DMSO) $(CH_3)_2SO$. An active polar solvent useful for dissolving such polar polymers as polyacrylonitrile and for certain polymerization reactions.

Dimethyl Terephthalate (DMT) C_6H_4-1,4-$(COOCH_3)_2$ A white crystalline solid generally obtained by the oxidation of *p*-xylene. The carbomethoxy groups of DMT are typical of those attached to a benzene ring, and their ready participation in alcoholysis reactions is the basis for most of the uses of the material. DMT is used in making polyethylene terephthalate (PET) and polyester fibers therefrom.

Dimple Synonym for SINK MARK, w s.

DIN Abbreviation for Deutsches Institut für Normung (German Institute for Standardization).

***N,N'*-Dinitroso-*N,N'*-Dimethylterephthalamide** A blowing agent that is a weak explosive in powder form and thus unsafe to handle, but is available in desensitized form by treatment with mineral oil (Du Pont Nitrosan®). This blowing agent is unique in that its low decomposition temperature permits the expansion of vinyl plastisol prior to gelation (93°C). Subsequent fusion at 177°C produces open-cell vinyl foam. Closed-cell foam can be produced by heating to fusion in a closed mold, releasing the pressure and subsequently heating in an oven at 100°C.

Dinitrosopentamethylenetetramine (DNPT) A blowing agent widely used for foam rubber, but of limited use in the plastics industry due to its high decomposition temperature and unpleasant residual odor.

Dinitrosoterephthalamide (DNTA) A chemical blowing agent for vinyls, liquid polyamide resins, and silicone rubbers. It is especially noted for its low decomposition exotherm.

Dinking See DIE CUTTING.

Dinonyl Adipate (DNA) $(-C_2H_4COOC_9H_{19})_2$. A nonyl-alcohol ester used as a plasticizer for cellulosic, acrylic, styrene, and vinyl polymers.

Dinonyl Phthalate (DNP) $C_6H_4(COOC_9H_{19})_2$. A general-purpose plasticizer for vinyl resins, with low volatility and good electrical properties.

DINP Abbreviation for DIISONONYL PHTHALATE, w s.

DIOA Abbreviation for DIISOOCTYL ADIPATE, w s.

Dioctyl- See DI(2-ETHYLHEXYL)- for several compounds for which this shorter prefix is more commonly used.

Di-*n*-octyl, *n*-Decyl Phthalate (DNODP) A mixed plasticizer for PVC and several other thermoplastics.

Dioctyl Ether $(C_8H_{17})_2O$. A mold and processing lubricant.

Dioctyl Fumarate (DOF) $(=CHCOOC_8H_{17})_2$. An unsaturated plasticizer for vinyl resins.

Di(2-octyl) Phthalate (DOP) See DICAPRYL PHTHALATE.

Dioctyl Terephthalate (DOTP) C_6H_4-1,4-$(COOC_8H_{17})_2$. Although the physical properties of DOTP are similar to those of DOP (the ortho isomer), DOTP is less volatile, imparts slightly better low-temperature flexibility, and is more resistant to lacquer marring.

Di-*n*-octyltin-*S*,*S*'-bis(Isooctyl Mercaptoacetate) A stabilizer for PVC that has been approved for use in food-grade bottles up to 3 phr when made to certain purity specifications.

Di-*n*-octyltin Maleate Polymer. Like the preceding tin stabilizer, this one, too, has been FDA-approved for use in food-packaging compositions up to 3 phr.

Dioctyltin Stabilizer See ORGANOTIN STABILIZER.

Diol An acronym for ***dihydric alcohol***, i e, an alcohol containing two hydroxyl (–OH) groups. Ethylene glycol, $HOCH_2CH_2OH$, and 1,5-pentanediol, $HO(CH_2)_5OH$, are examples.

Diolefin Polymer See DIENE POLYMER.

DIOP Abbreviation for DIISOOCTYL PHTHALATE, w s.

DIOS Abbreviation for DIISOOCTYL SEBAÇATE, w s.

1,4-Dioxane (diethylene ether, diethylene dioxide, dioxyethylene ether) $\overline{OCH_2CH_2OCH_2CH_2}$. A solvent for cellulose esters and other plastics. Dioxane is rarely used today because of its suspected carcinogenicity.

DIOZ Abbreviation for DIISOOCTYL AZELATE, w s.

Dip Coating A coating process wherein the object to be coated, preheated or at room temperature, depending on the materials, is dipped into a tank of fluid resin, solution, or dispersion, withdrawn and subjected to further heat or drying to solidify the deposit. See FLUIDIZED-BED COATING for a similar process employing powdered resin.

Dipentene (*dl*-limonene, *p*-mentha-1,8-diene) A cyclic diene, a colorless liquid with a lemony odor, used as a solvent for coumarone and alkyd resins, rubber, and natural resins.

Dip Forming (dip molding) A process similar to dip coating, except that the fused, cured or dried deposit is stripped from the dipping mandrel. As most frequently used for making vinyl-plastisol articles, the process comprises dipping into the plastisol a preheated form shaped to the desired inside dimensions of the finished article; allowing the plastisol to gel in a layer of the desired thickness against the form surface; withdrawing the coated form; heating the deposit to fuse the layer; cooling and stripping off the deposit. Some articles may be inverted after stripping so that the textured inside surface becomes the external surface of the finished article.

Diphenyl Synonym for BIPHENYL, w s.

Diphenylamine (DPA) $(C_6H_5)_2NH$. A crystalline solid, a stabilizer for several plastics.

Diphenyl Carbonate $(C_6H_5O)_2CO$. The monomer from which polycarbonates are produced.

Diphenyl Decyl Phosphite $(C_6H_5O)_2POC_{10}H_{21}$. A nearly colorless liquid used as a stabilizer for vinyl and polyolefin resins.

Diphenyl Mono-*o*-Xenyl Phosphate $(C_6H_5O)_2(C_6H_5C_6H_4O)PO$. A plasticizer for cellulosic plastics, polystyrene, and, with limited compatibility, vinyls.

Diphenyl Octyl Phosphate (DPOP) $(C_6H_5O)_2(C_8H_{17}O)PO$. A flame-retardant plasticizer for PVC and cellulosic resins. It has been approved by FDA for use in food packaging.

Diphenyl Phthalate (DPP) $C_6H_4(COOC_6H_5)_2$. A powder that melts at 75°C. It is used as a solid plasticizer for rigid PVC, cellulosic, and other resins.

Diphenyl Xylenyl Phosphate $(C_6H_5O)_2[(CH_3)_2C_6H_3O]PO$. A plasticizer for cellulose acetate-butyrate, cellulose nitrate, polystyrene, PVC, and vinyl chloride-acetate copolymers.

Dipole (1) A combination of two electrically or magnetically charged particles of opposite sign that are separated by a small distance. (2) Any system of charges, such as a circulating electric current, having the property (a) that no net force acts upon it in a uniform field; or, (b) a torque proportional to sin θ, where θ is the angle between the dipole axis and a uniform field, *does* act on the charges. In polymers, atoms such as Cl bearing negative charge induce opposite charge in neighboring atoms, thus creating dipoles. Tiny movements of these dipoles in reaction to the changing directions of rapidly alternating fields are the basis for DIELECTRIC HEATING, w s.

Dip Molding See DIP FORMING.

Dipropylene Glycol (2,2'-dihydroxydipropyl ether) $(CH_3CHOH-CH_2)_2O$. A high-boiling glycol ether with a low order of toxicity, widely used as a solvent and chemical intermediate. As a solvent it is used with cellulose acetate and nitrate, and is one of the few known solvents for polyethylene. Thus it is used in screening tests to identify polyethylene. As an intermediate, dipropylene glycol reacts with dibasic acids to form alkyd resins, polyester plasticizers, and urethane-foam intermediates.

Dipropylene Glycol Dibenzoate $[C_6H_5COOCH(CH_3)CH_2-]_2O$. A plasticizer for PVC, also compatible with most common thermoplastics, imparting good stain resistance.

Dipropylene Glycol Monosalicylate (salicylic acid, dipropylene glycol monoester) $HOC_6H_4COOCH(CH_3)CH_2OCH_2CHOHCH_3$. A light-colored oil used in ultraviolet-screening agents and plasticizers.

Dipropyl Ketone $(CH_3CH_2CH_2)_2O$. A stable, colorless liquid, a solvent for many resins.

Dipropyl Oxalate $(CH_3CH_2CH_2OOC-)_2$. An active, high-boiling (211°C) solvent.

Dipropyl Phthalate $C_6H_5(COOC_3H_7)_2$. A plasticizer for cellulose acetate and cellulose acetate-butyrate.

Direct Gate A gate that has the same cross section as the runner, i e, the absence of a distinguishable gate.

Disbond (v) In an adhesive-bonded joint, to separate at the bond surface. (n) Such a separation.

Discharge-Inception Voltage In a dielectric-strength test, the voltage at which discharges begin in the voids within the specimen.

DISCO An epoxy-based composite prepreg containing discontinuous but oriented fibers, nearly as strong as unidirectional composites of the same fiber content, and formable into complex shapes and over corners.

Discoloration (1) Any change from an initial color possessed by a plastic. (2) A lack of uniformity in color where color should be uniform over the whole area of a plastic object. In the second sense, where they are applicable, one may use the more definite terms *mottle*, *segregation*, or *two-tone*. Discoloration can be caused by inadequate blending of ingredients, by overheating, exposure to light, irradiation, or chemical action.

Dished Showing a symmetrical concave distortion of a flat or curved surface of a plastic object, so that, as normally viewed, it appears more concave than its design calls for. See also WARP and SINK MARK.

Disk-and-Cone Agitator A mixing device comprised of disks or cones rotating at speeds between 20 and 60 rev/s or higher. The disks/cones displace fluid contacting their surfaces by centrifugal force. They are used in preparing pastes and dispersions.

Disk Extruder Any of a variety of novel devices in which pellets or powders are melted by contact with a heated, rotating disk and scraped off the disk into a screw pump or other pressure-developing machine; or devices in which friction of pellets between rotating disks and nearby stationary surfaces causes the plastic to melt, thence to be discharged to a pressure-developing machine for shaping into a product. One such device, the "Diskpack" extruder, has been developed and marketed by the Farrel Group of Emhart Corporation. See also EXTRUDER, ELASTIC-MELT.

Disk Feeder A horizontal, flat or grooved, rotating disk at the bottom of a hopper feeding a continuous extruder that controls the feed rate by varying the speed of rotation of the disk, or by varying the clearance between the disk and the scraper that directs the feed from the disk into the feed throat of the extruder.

Disk Gate See DIAPHRAGM GATE.

Dispersant In an organosol, a liquid component that has a solvating or peptizing action on the resin, thus aiding in dispersing and suspending it.

Disperse Phase In a SUSPENSION or EMULSION, w s, the "disperse phase" refers to the particles of one material individually dispersed, more or less stably, in the continuously connected domain of another material, usually a liquid, known as the *continuous phase*.

Dispersing Agent A material added, usually in relatively small percentage, to a suspending medium to promote and maintain the separation of discrete, fine particles of solids or liquids. Dispersants are used, for example, in the wet grinding of pigments and for suspending water-insoluble dyes.

Dispersion (1) A system of two or more phases comprising one or more finely divided materials distributed in another material. Types of dispersions are *emulsions* (liquids in liquids), *suspensions* (solids in liquids), *foams* (gases in liquids or solidified liquids), and *aerosols*, (liquids in gases). In the plastics industry, the term dispersion usually denotes a finely divided solid dispersed in a liquid or in another solid. Examples are fillers and pigments in molding compounds, plastisols and organosols. (2) A mixing process in which the particles or globules of the discontinuous phase are reduced in size. This may be accomplished by breaking solid particles, as in ball-milling, or by shearing viscous regions, as in kneading or extrusion. See also SEGREGATION (2).

Dispersion Resin A special type of PVC resin with very small spherical particles, usually one μm or less in diameter, permitting them to be mixed with plasticizers by simple stirring techniques. They are used in compounding plastisols and organosols.

Dispersive Mixing Any mixing process in which the principal action is reduction of the size of particles or interlayers. The progress of the process is judged by two criteria, INTENSITY OF SEGREGATION and SCALE OF SEGREGATION, w s.

Displacement Angle In filament winding, the angle whose tangent equals the quotient of the advancement distance of the winding ribbon on the equator after one complete turn, divided by ($\pi \times$ the equatorial diameter).

Dissipation Factor (electrical) The ratio of the conductance of a capacitor in which the test material is the dielectric to its susceptance; or the ratio of its parallel reactance to its parallel resistance. Most plastics have a low dissipation factor, a desirable property in electrical insulations for high-frequency applications because it minimizes the waste of electrical energy as heat. On the other hand, for those plastics with high dissipation factors, e g, PVC and phenolic, the property provides a fast and economical method of even heating using microwaves and is the basis of electronic preheating of thermosetting molding powders and of electronic sealing of flexible-vinyl films. See DIELECTRIC HEATING.

Dissipation Factor (mechanical) The ratio of the loss modulus to the modulus of elasticity in a plastic part undergoing rapid, cyclic changes (or even reversals) of stress. As in electrical dissipation, the product of mechanical dissipation is heat, heat that can raise the temperature of the part and cause it to weaken, creep rapidly, or even fail prematurely.

Distearyl Ether (dioctadecyl ether) $(C_{18}H_{37})_2O$. A mold lubricant.

Distortion (1) A change in the shape of a solid body, often associated with temperature differences or stress gradients within the body. (2) An apparent change of shape as perceived through an optically imperfect, transparent membrane or reflected from an imperfect mirror.

Distributive Mixing Any mixing operation in which the principal action is convection rather than dispersion, resulting in elements of the different phases being more intimately blended, but without much size reduction. Typical is blending of masterbatch pigmented pellets with virgin pellets prior to extrusion or molding, by tumbling.

Di-*tert*-Butyl Peroxide $(CH_3)_3COOC(CH_3)_3$. A stable liquid used as a catalyst for polymerizations at high temperatures of a variety of olefin and vinyl monomers, e g, ethylene, styrene, and styrenated alkyds.

Ditetrahydrofurfuryl Adipate $(OC_4H_7CH_2OOCCH_2CH_2-)_2$. A heterocyclic plasticizer for cellulose acetate-butyrate.

Ditridecyl Phthalate (DTDP) $C_6H_4(COOC_{13}H_{27})_2$. A primary plasticizer for PVC, also compatible with cellulosics and polystyrene. In vinyls, it imparts resistance to high temperatures and to extraction by hot soapy water, excellent flexibility, and anti-fogging properties.

Ditridecyl Thiodipropionate $(C_{13}H_{27}OOCCH_2CH_2\ )_2S$. A stabilizer, plasticizer, and softening agent.

Diundecyl Phthalate (DUP) $C_6H_4(COOC_{11}H_{23})_2$. A plasticizer characterized by low volatility and good low-temperature properties compared to other phthalates.

Divergent Die A die for hollow articles in which the internal channels leading to the orifice increase in cross section toward the lip.

Divinyl B Synonym for BUTADIENE, w s.

Divinylbenzene (DVB, vinylstyrene) $C_6H_4(CH{=}CH_2)_2$. A monomer derived from styrene, used in making ion-exchange resins, synthetic rubbers, and casting resins. The commercial product contains 55% mixed *m*- and *p*-isomers, the rest being ethylvinylbenzenes. It is often used along with styrene as a reactive monomer in the production of polyester resins, to which it imparts a higher degree of crosslinking and superior chemical resistance.

Di-*o*-xenyl Phenyl Phosphate $(C_6H_5C_6H_4O)_2(C_6H_5O)PO$. A plasticizer for cellulosics, polystyrene, and vinyls.

Di-*p*-xylylene (DPX, DPXN) $(-CH_2C_6H_4CH_2-)_2$. The stable dimer of *p*-xylylene, a white powder. When heated to 600°C in an evacuated chamber, the monomer is regenerated and instantly polymerizes to form a tough, impervious film on any cold surface between the heating pot and the vacuum source. See PARYLENE.

DMA Abbreviation for DIMETHYLANILINE, w s.

DMEP Abbreviation for DI(2-METHOXYETHYL) PHTHALATE, w s.

DMF Abbreviation for DIMETHYL FORMAMIDE, w s.

DMG Abbreviation for DIMETHYL GLUTARATE, w s.

DMP Abbreviation for DIMETHL PHTHALATE, w s.

DMSO Abbreviation for DIMETHYL SULFOXIDE, w s.

DMT Abbreviation for DIMETHYL TEREPHTHALATE, w s.

DNA Abbreviation for DINONYL ADIPATE, w s.

DNHZ Abbreviation for DI-*n*-HEXYL AZELATE, w s.

DNODA Abbreviation for di-*n*-octyl-*n*-decyl adipate. See *n*-OCTYL-*n*-DECYL ADIPATE.

DNODP Abbreviation for DI(*n*-OCTYL-*n*-DECYL) PHTHALATE, w s.

DNP Abbreviation for DINONYL PHTHALATE, w s.

DNPT Abbreviation for DINITROSOPENTAMETHYLENETETRAMINE, w s.

DNTA Abbreviation for DINITROSOTEREPHTHALAMIDE, w s.

DOA Abbreviation for dioctyl adipate or DI(2-ETHYLHEXYL) ADIPATE, w s.

Doctor (v) To spread a coating on a substrate in a layer of uniform, controlled thickness by means of a blade or roll.

Doctor Blade (doctor bar, doctor knife) A flat bar that regulates the amount of liquid material on the rollers of a coating machine, or to control the thickness of a coating after it has been applied to a substrate.

Doctor Roll A roll that operates at a different speed or in the opposite direction from the primary roll of a coating machine, thus regulating the uniformity and thickness of material on the roll before it is applied to a substrate.

4-Dodecycloxy-2-Hydroxybenzophenone An ultraviolet inhibitor for polyethylene and polypropylene, also suggested as suitable for PVC, polystyrene, polyesters, and surface coatings such as those based on cellulosic and acrylic resins.

DOF Abbreviation for DIOCTYL FUMARATE, w s.

Dogbone A slang term for the dumbbell shape of tensile-test specimens having relatively wide ends for gripping and a narrower, gage-length center within which all plastic deformation and breakage occur. See TENSILE BAR.

Doily In filament winding, the planar reinforcement that is applied to a local area between windings to provide extra strength in an area where a cut-out is to be made; for example, a port opening.

DOIP Abbreviation for dioctyl isophthalate. See DI(2-ETHYLHEXYL) ISOPHTHALATE.

DOP Abbreviation for dioctyl phthalate. See DI(2-ETHYLHEXYL) PHTHALATE.

Dope A solution of a cellulosic plastic, historically cellulose nitrate, used for treating fabrics.

DOS Abbreviation for dioctyl sebaçate. See DI(2-ETHYLHEXYL) SEBAÇATE.

Dosimeter A detector worn by workers to measure the amount of an environmental agent, usually radioactivity, but sometimes noise or noxious gases, to which the worker has been exposed during a working shift or longer period of time.

DOTP Abbreviation for DIOCTYL TEREPHTHALATE, w s.

Double Bond A type of covalent bond, common in organic chemistry, in which two pairs of electrons are shared between two elements. The double bond may be symbolized either by ":" or "=", as in ethylene, $CH_2:CH_2$, or dimethyl ketone, $(CH_3)_2C=O$.

Doubler In filament winding, a local area with extra wound reinforcement, either wound integrally with the part or wound separately and bonded to the part.

Double-Ram Press A press for injection or transfer molding in which two distinct systems of the same kind (hydraulic or mechanical), or of different kinds, create respectively the injection or transfer force and the clamping force.

Double-Screw Extruder See EXTRUDER, TWIN-SCREW.

Double-Shot Molding (two-shot molding, insert molding, two-color molding, over-molding) A process for making two-color or two-material parts by means of successive molding operations. The basic process includes the steps of injection molding one part, transferring this part to a second mold as an insert, and molding the second material against the first. Examples of parts made by double-shot molding are computer keys, pushbuttons, telephone-keypad buttons, and other such products in which indicia must resist heavy wear and remain permanently legible. In a modification of the process, cups and the like with differently colored insides and outsides are made automatically by means of a machine equipped with two injection molders and a swinging platen carrying two cup cores indexed around a central tie-bar, bringing the molds into position for each successive shot. This arrangement permits simultaneous molding of both shots. Some of these products are made today by COINJECTION, w s.

Double-Skinned Sheet A plastic sheet consisting of a relatively thick center section bonded on each face to thinner layers that differ in color or composition from the center section. Today such sheets are usually made by coextrusion of the three plastic layers and, in many cases, two additional adhesive layers between the core and the skins.

Dough (dough-molding compound) A term sometimes used for a reinforced-plastic mixture of dough-like consistency in an uncured or partly cured state. A typical dough consists of polyester resin, glass fiber, calcium carbonate, lubricants and catalysts. The compounds are formed into products either by hand layup or, more usually, by compression molding.

Dowel (dowel pin) A hardened steel pin, usually having a slight taper, used to maintain alignment between two or more parts of a mold or machine.

Dowel Bushing A hardened steel insert in the portion of a mold that receives the dowel pin.

Dowtherm® The tradename (Dow Chemical Co.) of a liquid heat-transfer medium consisting of biphenyl and phenyl ether in eutectic ratio. High-boiling and stable, it is less used today than formerly because of its suspected carcinogenicity.

DOZ Abbreviation for dioctyl azelate. See DI(2-ETHYLHEXYL) AZELATE.

DP Abbreviation for DEGREE OF POLYMERIZATION, w s.

DPA Abbreviation for DIPHENYLAMINE, w s.

DPCF (DPCP) Abbreviation for diphenyl cresyl phosphate, a plasticizer.

DPP Abbreviation for DIPHENYL PHTHALATE, w s.

Draft A slight taper in a mold wall, proceeding inward from the parting surface to the bottom of the cavity, designed to facilitate removal of the molded object from the mold. When the taper is reversed, tending to impede removal of the article, the term *back draft*, or *reverse draft* is employed.

Draft Angle In a mold, the angle in the profile plane with its vertex at the bottom of the mold or cavity, between the side of the mold and the vertical plane. In most cases, 1° is adequate for smooth ejection of parts from the mold.

Drag Flow (1) In general, the laminar flow of a viscous liquid that is contained between two surfaces, one of which is moving relative to the other. (2) In the metering section of an extruder screw, the rate of drag flow is the component of total material flow in the down-channel direction caused by the relative motion between the screw and cylinder. If the screw were discharging freely, the output rate would be equal to the drag-flow rate. (3) In a wire-coating die, the flow generated by the relative motion of the wire through the stationary die. In both extruders and wire-coating dies, pure drag flows almost never occur, but are altered in complex ways by opposing or augmenting pressure fields and the nonNewtonian nature of the melts.

Drape (1) With reference to plastics films and coated fabrics, their ability to hang without creases and to form graceful folds when used as draperies, shower curtains, and the like. (2) In sheet thermoforming, the ability of the preheated sheet to conform to the mold under the influence of gravity.

Drape-Assist Frame In sheet thermoforming, a frame made from thin wires or thick bars shaped to the periphery of the depressed areas of the mold and suspended above the sheet to be formed. During forming, the assist frame drops down, drawing the softened sheet tightly into the mold and thereby preventing webbing between high areas of the mold and permitting closer spacing in arrays of multiple molds.

Drape Forming (drape vacuum forming, drape thermoforming) Forming a thermoplastic sheet into three-dimensional articles by clamping the sheet in a movable frame, heating the sheet, then lowering it to drape over the high points of a male mold. Vacuum is then applied to complete the forming. See also SHEET THERMOFORMING.

Drawdown In extrusion, the process of pulling the extrudate away from the die at a lineal speed greater than the average velocity of the melt in the die, thus reducing extrudate's cross-sectional dimensions. The term is also used by blow molders to denote the decrease in parison diameter and wall thickness due to gravity.

Drawdown Ratio In extrusion or fiber spinning, the ratio of the cross-sectional area of the die opening to that of the finished product. In making sheet or cast film, where the sheet width is nearly equal to the width of the die opening, the ratio of the thickness of the die opening to that of the final sheet is sometimes spoken of as the drawdown ratio.

Drawing The process of stretching a thermoplastic filament, sheet, or rod to reduce its cross-sectional area and/or to improve its physical properties by ORIENTATION, w s.

Draw Ratio (1) A measure of the degree of stretching during the orientation of a fiber or filament, expressed as the ratio of the cross-sectional area of the undrawn material to that of the drawn material. (2) In monofilament manufacture, the filament is wrapped several times around a vertical, driven roll, passed through a warming oven, then wrapped again around a second roll running faster than the first one. In this way the filament is stretched and oriented and its cross section is reduced. The ratio of the surface speed of the second, faster roll to that of the first equals the draw ratio.

Draw Resonance A phenomenon occurring in film and filament extrusion in which the extrudate is drawn into a quenching bath at a certain critical speed that creates a cyclic pulsation in the cross-sectional area of the extrudate. The pulsation increases with rising drawing speed until the filament or film eventually breaks at the bath surface. Draw resonance has been observed while extruding polypropylene, polyethylene, and polystyrene.

Dribbling or Drooling A condition sometimes occurring in injection molding between shots in which melt drips from the withdrawn nozzle. Drooling was a common problem with 6/6 nylon in the 1950s because of depolymerization caused by moisture pickup in feed hoppers and overheating in the old, torpedo-type heating cylinders.

Drop-Weight Test (falling-weight test) Any test of impact resistance in which known weights are dropped once or repeatedly on the test specimen. Examples are ASTM D 4272 (plastic films), D 3029, D 4226, and D 4495 (rigid PVC sheet and parts), and F 736, Section 15, (polycarbonate sheet). Also see FREE-FALLING-DART TEST.

Drum Coloring See DRY COLORING.

Drum Extruder A plasticating machine having a rotating cylindrical element inside a concentric or eccentric housing. Pellets or powder are fed into the gap between drum and housing at the top. Shear action melts the solids and the melt exits through a slot about 270° around from the feed point, usually passing into a pressure-developing device that can form an extruded product. A wiper bar above the die prevents melted material from recirculating.

Drum Tumbler A device used to mix plastic pellets with color concentrates and/or regrind. The materials are charged into cylindrical drums that are tumbled end-over-end or rotated about an inclined axis for a time sufficient to thoroughly blend the ingredients.

Dry Blend (n) A dry, free-flowing mixture of resin powder, typically PVC, with plasticizers and other additives, prepared by blending the components in a large, closed mixing bowl with a high-speed rotor at its bottom, stopping the action at temperatures comfortably below the fluxing point. Dry blends are generally more economical feedstocks for extrusion than molding powders and pellets made by plasticating and extrusion, but in some cases have been difficult to process.

Dry-Blend (v) To combine ingredients, typically in a high-speed mixer, to produce a dry blend.

Dry Coloring The process of combining colorants to molding compounds and resin pellets by tumble-blending them with dyes, pigments, or color concentrates. This process enables custom molders and extruders to carry a large inventory of uncolored compound, preparing smaller batches of colored compounds to customers' specifications.

Drying Oil Any of several plant-derived polyunsaturated oils, such as linseed, oiticia, and tung, that, when exposed to air, form dry, tough, durable films. Linseed oil to be used in paints and varnishes today is usually boiled with cobalt or manganese salts of linoleic or naphthenic acids to shorten its "drying" time.

Dry Laminate A laminate containing insufficient resin for complete bonding of the reinforcement.

Dry Layup The construction of a laminate by layering preimpregnated, partly cured reinforcements in or on a mold, usually followed by bag molding or autoclave molding.

Dry Purge In extrusion, preparatory to shutting down operation after the feed hopper has been emptied, running the extruder until no more plastic emerges from the head end.

Dry Spinning See SPINNING.

Dry Spot An imperfection in reinforced plastics, an area of incomplete surface film where the reinforcement has not been wetted with resin (ASTM D 883).

Dry Strength The strength of an adhesive joint determined immediately after drying or curing under specified conditions or after a period of conditioning in a standard laboratory atmosphere. See also WET STRENGTH (2).

Dry Winding Filament winding with preimpregnated roving, as distinguished from *wet winding* in which unimpregnated roving is pulled through a resin just prior to winding on a mandrel.

DSC See DIFFERENTIAL SCANNING CALORIMETRY.

DSTDP Abbreviation for DISTEARYLTHIODIPROPIONATE, w s.

DTA Abbreviation for DIFFERENTIAL THERMAL ANALYSIS, w s.

DTDP Abbreviation for DITRIDECYL PHTHALATE, w s.

Dual-Sensor Control An improved system for controlling cylinder and plastic temperatures in extruders. For each heating zone, there are two temperature sensors, one in a shallow well slightly beneath the heater, the other deep, just outside the lining layer and near the plastic. An average of the two signals is used to control the electrical heat input (Eurotherm/Welex).

Ductile Fracture (ductile rupture) The breaking or tearing, most commonly in tension, of a test specimen or part after considerable unrecoverable stretching (plastic strain) has occurred. Since the mode of fracture depends on conditions as well as material, the distinction between ductile and brittle fracture, which latter occurs after relatively little, recoverable strain, is not always clear. Low temperatures, especially below the glass transition (T_g), and high rates of strain favor brittle behavior, while the opposites favor ductile behavior.

Ductility The amount of plastic strain that a material can undergo before rupture. A ductile material generally shows a YIELD POINT, w s.

Dulmadge Mixing Section Invented by F. E. Dulmadge, this mixing section, usually located near the end of an extruder or injection screw, is really three short sections separated by gaps with no flights. In each section, the regular screw flight is replaced by about 20 narrow, closely spaced flight segments with lead angles of about 70°. The melt approaching the first section is subdivided into 20 substreams that, on exiting, are swirled circumferentially before entering the second section, and again, on leaving that one, before entering the third. The main goal was to improve melt-temperature uniformity.

Dumas Nitrogen A long used method for determining the percentage of nitrogen in organic materials, invented by A. Dumas in 1830. The sample was first oxidized with red-hot copper oxide, followed by reduction of NO_x to N_2 over hot copper. Following absorption of CO_2 in KOH solution and condensation and absorption of most of the water, the volume of N_2 was measured. The method has largely been supplanted by automated analyzers that directly determine C, H, and N spectroscopically and find O by difference.

Dumbbell See DOGBONE.

Duplicate Cavity Plate A removable plate that retains cavities, used where two-plate operation is necessary for loading inserts, etc.

Durene (durol, 1,2,4,5-tetramethylbenzene) $C_6H_2(CH_3)_4$. A substance occurring in coal tar, but usually prepared from xylene and methyl chloride in the presence of $AlCl_3$. It has been patented (US 4,000,120) as an additive to make packaging films of polyolefins and polystyrene photodegradable under direct action of sunlight.

Durometer An instrument used for measuring the hardness of a material. See ASTM Test D 2240, also INDENTATION HARDNESS.

Durometer Hardness See INDENTATION HARDNESS.

DVB Abbreviation for DIVINYLBENZENE, w s.

Dwell (1) A pause in the application of pressure to a mold, made just before the mold has completely closed, to allow the escape of gas from the molding material. (2) In filament winding, the time that the traverse mechanism is stationary while the mandrel continues to rotate to the appropriate point for a new traverse to begin. (3) In heat sealing, dwell time is the time during which pressure and heat (or microwave energy) are applied to the area to be sealed.

Dye An intensely colored substance that imparts color to a substrate to which it is applied. Retention of the dye in the substrate may be by means of adsorption, solution, mechanical bonding, or by ionic or covalent chemical bonding. The dye substance is devoid of crystal structure. Dyes used for coloring plastics usually dissolve in the plastic melt, unlike PIGMENTs (w s), which remain dispersed as undissolved particles.

Dynamic Fatigue Usually the same as *fatigue*. The "dynamic" modifier is sometimes used by persons who think of creep and creep failure as "static fatigue". See entries at FATIGUE....

Dynamic Mechanical Analyzer An instrument that can test in an oscillating-flexural mode over a range of temperature and frequency to provide estimates of the "real", i e, in-phase, and "imaginary", i e, out-of-phase parts of the complex modulus. The real part is the elastic component, the imaginary part is the loss component. The square root of the sum of their squares is the complex modulus. With polymers, the components and the modulus are usually dependent on both temperature and frequency. ASTM D 4065 spells out the standard practice for reporting dynamic mechanical properties of plastics. See also MECHANICAL SPECTROMETER.

Dynamic Mechanical Spectrum The information obtained from testing with a MECHANICAL SPECTROMETER, w s. A plot or tabulation of complex modulus or its components vs frequency of oscillation or temperature or both. The mode of stress may be tensile/compressive, flexural, or torsional (shear). Because both abscissa (frequency) and ordinates can range widely, bilogarithmic plots are usual.

Dynamic Stress A stress whose magnitude and/or direction vary with time, typically cyclically and sinusoidally.

Dynamic Stress Relaxometer An instrument that measures the relaxation response of an elastomeric material to a prescribed shear deformation over a range of temperature. Basic elements of the instrument are a cone-shaped stator cavity and a conical rotor, both electrically heated. The sample is placed in the stator, which rises to a position to form a constant specimen thickness, forcing out the excess material. After heating to the desired temperature, the rotor is rotated quickly through a small angle to a known shear deformation. The subsequent drop-off in torque, which results from the relaxation of stress within the sample, is recorded over the time it takes for it to decay.

Dynamic Valve A device sometimes incorporated in an extruder head to control flow by adjusting the clearance between conical elements, one stationary, the other rotating with the screw. Dynamic valves of various designs have also been used between the stages of two-stage, vented extruders as an aid to balancing the flow rates in the stages and preventing extrusion out the vent.

Dynamic Viscosity (1) ABSOLUTE VISCOSITY, w s, as distinguished from KINEMATIC VISCOSITY, w s. See also VISCOSITY. (2) In sinusoidally varying shear, the part of the stress in phase with the rate of strain, divided by the strain rate.

Dyne In the (now deprecated) cgs system of units, the force required to accelerate a mass of one gram by one centimeter per second-squared. The dyne = 1×10^{-5} newton.

Dypnone (phenyl α-methyl styryl ketone, 1,3-diphenyl-2-butene-1-õne) $C_6H_5COCHC(CH_3)C_6H_5$. A plasticizer and ultraviolet absorber.

E e

e (1) The base of natural logarithms, 2.71828.... (2)The charge on an electron, 1.60199×10^{-18} coulomb.

E (1) SI abbreviation for prefix EXA-, w s. (2) Symbol commonly used for modulus of elasticity in tension (see MODULUS OF ELASTICITY); for activation energy in the ARRHENIUS EQUATION,w s; and for electric potential.

EAA Abbreviation for ETHYLENE-ACRYLIC ACID COPOLYMER, w s.

EC Abbreviation for ETHYL CELLULOSE, w s.

ECTFE See POLY(ETHYLENE-CHLOROTRIFLUOROETHYLENE), w s.

EDC Abbreviation for ETHYLENE DICHLORIDE, w s.

Eddy Current (Foucault current) The current induced in a mass of conducting material by a varying magnetic field.

Eddy-Current Coupling (eddy-current clutch) An economical type of speed variator used in older extruder drives. Inserted between a constant-speed AC induction motor and the speed reducer, it consists of two rotating elements, face-to-face, the space between being filled with a magnetically susceptible fluid. By varying the closeness of magnetic coupling between the two elements, any degree of connection from none to virtually solid can be obtained. The coupling delivers constant torque and is efficient at top rated speed but efficiency diminishes with output speed, with half the motor energy being dissipated as heat (a heat exchanger is usually required) at half speed. The advent of reasonably priced, solid-state DC power supplies has made the eddy-current coupling less attractive than it was in the 1960s.

Edge Bead In some cast-film and sheet-extrusion operations, the narrow border at the edge of the sheet, usually somewhat thicker, that must be trimmed off prior to winding or stacking the product.

EDM Abbreviation for ELECTRICAL-DISCHARGE MACHINING, w s.

EEA Abbreviation for ETHYLENE-ETHYL ACRYLATE COPOLYMER, w s.

Effective Modulus Synonym for CREEP MODULUS, w s.

Efficiency (1) Of a machine or process, the ratio of the energy delivered to that supplied. (2) Effectiveness of operation or production, particularly as related to some standard for same. (3)Reciprocal of the unit cost of items for sale (*cost efficiency, economy*). (4) See also EFFICIENCY OF REINFORCEMENT and PLASTICIZER EFFICIENCY.

Efficiency of Reinforcement (fiber-efficiency factor) The percentage of fiber in a reinforced-plastic structure or part contributing to the property of concern. For example, in a unidirectionally reinforced bar, the theoretical efficiency for Young's modulus and fiber-direction tensile strength is 100%. For a sheet molded from chopped-strand mat with all fibers randomly oriented in the sheet plane, the efficiency for in-plane properties is 37%. With chopped strands randomly oriented in three dimensions, efficiency falls to 20%.

E Glass A low-alkali ($Na_2O + K_2O = 0.6\%$) borosilicate glass, like Pyrex®, with good electrical properties, the most widely used glass-fiber reinforcement for plastics. Major constituents are: SiO_2, 54%; CaO, 17%; Al_2O_3, 15%; B_2O_3, 8%; MgO, 4.7%. The fibers are vulnerable to surface abrasion so are always sized before stranding. Average fiber properties are: density = 2.54 g/cm^3; tensile modulus (axial) = 72 GPa; tensile strength = 3.5 GPa.

EHMWPE Abbreviation for *extra-high-molecular-weight polyethylene*, any of a subfamily of linear PE resins having molecular weights in the range 250,000 to 1,500,000. See also POLYETHYLENE.

Eilers Equation A modification of the EINSTEIN EQUATION (w s) relating the viscosity η_f of a Newtonian liquid filled with spherical particles to the viscosity of the pure liquid η_0 and extending to concentrations above 10% of filler. It is:

$$\frac{\eta_f}{\eta_0} = \left[\frac{1 + 1.25 \cdot f}{1 - S \cdot f}\right]^2$$

where f is the volume fraction of spheres and S is an empirical coefficient usually between 1.2 and 1.3. See also MOONEY EQUATION.

Einstein Equation An equation relating the viscosity η_f of a sphere-filled, Newtonian liquid to that of the unfilled liquid η_0, for volume fractions f of spheres up to about 10%. It is:

$$\eta_{rel} = \eta_f / \eta_0 = 1 + k_E \cdot f$$

where k_E is the Einstein coefficient, = 2.5 for spheres. The Einstein equation has been extended to higher concentrations by adding terms in f^2, f^3, etc, and by correcting for the limiting volume fraction f_m that can be filled by uniform spheres. A versatile model of this type is

$$\eta_{rel} = \frac{1 + af + bf^2 + cf^3}{1 - (f/f_m)}$$

where a, b, and c are empirical constants and $f_m \approx 0.7$. The MOONEY EQUATION, w s, is a form of this model for which b and $c = 0$. See also the EILERS EQUATION and FRANKEL-ACRIVOS EQUATION.

Ejection Ram A small supplementary hydraulic ram fitted to a molding press to operate piece-ejection devices.

Ejector Pin (ejector sleeve, knockout pin, KO pin) A rod, pin, or sleeve that pushes a molding off a force or out of a cavity of a mold. Attached to an ejector bar or plate, it is actuated by the ejector rod(s) of the press or by auxiliary hydraulic or compressed-air cylinders.

Ejector Plate A plate that backs up the ejector pins and holds the ejector assembly together.

Ejector-Return Pin (return pin, surface pin, safety pin, position pushback) A projection, usually one of several that push back the ejector assembly as the mold closes.

Ejector Rod A bar that actuates the ejector assembly when a mold is opened.

Elaidamide $CH_3(CH_2)_7CH{=}CH(CH_2)_7CONH_2$. The amide of *trans*-9-octadecenoic acid, a steroisomer of OLEAMIDE, w s, used in fractional percentages as a slip agent for polyethylene to be made into film.

Elastic Constant Any of the several constants of a constitutive relaionship between stress (of any mode) and strain in a material. For an isotropic material stressed in its elastic range, there are (at any temperature) four interrelated constants: tensile modulus, E, shear modulus, G, bulk modulus, B, and Poisson's ratio, μ. Two expressions of the relations are:

$$G = E/2(1 + \mu) \text{ and } B = E/3(1 - 2\mu).$$

More constants are needed to define the behavior of nonisotropic materials, in the most general case, 36.

Elastic deformation A change in dimensions of an object under load that is fully recovered when the load is released. That part of the total strain in a stressed body that disappears upon removal of the stress. See also PLASTIC DEFORMATION.

Elastic Design Engineering design for load-bearing members based on the assumption that stress and strain are proportional and will be kept well within the elastic range, with working stresses set at half or less of the yield stress. Elastic design based on short-time test measurements may be applicable to the design of plastic products that will be loaded intermittently and for short periods. However, the universal phenomenon of creep in plastics means that, under sustained loads, *visco*elastic rather than elastic behavior is the norm. Even so, elastic-design methods can often be used by employing CREEP MODULUS and CREEP STRENGTH (w s) in place of short-time parameters.

Elasticity The ability of a material to quickly recover its original dimensions after removal of a load that has caused deformation. When the deformation is proportional to the applied load, the material is said to exhibit *Hookean elasticity* or *ideal elasticity.*

Elasticizer A term sometimes used for a compounding additive that contributes elasticity to a resin. For example, chlorinated polyethylenes and chlorinated copolymers of ethylene and propylene are blended with PVC compositions for this purpose.

Elastic Limit The greatest stress that a material can experience which, when released, will result in no permanent deformation. With some polymers this limit can be well above the PROPORTIONAL LIMIT, w s. See also YIELD POINT.

Elastic-Melt Extruder See EXTRUDER, ELASTIC-MELT.

Elastic Memory See MEMORY.

Elastic Modulus See MODULUS OF ELASTICITY.

Elastic Nylon See NYLON 6/10.

Elastic Recovery That fraction of a given deformation that behaves elastically. A perfectly elastic material has a recovery of 100% while a perfectly plastic material has no elastic recovery. Elastic recovery is an important property in films used for stretch packaging because it relates directly to the ability of a film to hold a load together. Retention of the elastic-recovery stress over a period of time is also important.

Elastodynamic Extruder See EXTRUDER, ELASTIC-MELT.

Elastomer A material that at room temperature can be stretched repeatedly to at least twice its original length and, immediately upon release of the stress, returns with force to its approximate original length. This definition is one criterion by which materials called plastics in commerce are distinguished from elastomers and rubbers. Another criterion is that, unlike thermoplastics which can be repeatedly softened and hardened by heating and cooling without substantial change in properties, most elastomers are given their final properties by mastication with fillers, processing aids, antioxidants, curing agents, etc, followed by vulcanization (curing) at elevated temperatures. However, a few elastomers are thermoplastic. Polymers usually considered to be elastomers, at least in some of their forms, are listed below.

Chemical Name	Abbreviations
Acrylonitrile-chloroprene copolymer	NCR
Acrylonitrile-isoprene copolymer	NIR
Butadiene-acrylonitrile copolymer	GR-N, NBR, PBAN
Chlorinated polyethylene	CPE
Chlorosulfonated polyethylene	CSM, CSR, CSPR
Ethylene ether polysulfide	EOT
Ethylene-ethyl acrylate copolymer	EEA, E/EA
Ethylene polysulfide	ET
Ethylene-propylene copolymer	EPM, EPR
Ethylene-propylene-diene terpolymer	EPD, EPDM. EPT, EPTR
Fluoroelastomer (any)	FPM
Fluorosilicone	FVSI
Hexafluoropropylene-vinylidene fluoride copolymer	FPM
Isobutene-isoprene copolymer	Butyl, GR-I
Organopolysiloxane	SI
Acrylic ester-butadiene copolymer	ABR, AR
Polybutadiene	BP, BR, CBR
Polychloroprene	CR
Polyepichlorohydrin	CO, CHR
Polyisobutene	PIB
Polyisoprene, natural	NR
Polyisoprene, synthetic	CI, IR, PIP
Polyurethane (polyester)	AU, PUR
Polyurethane (polyether)	EU
Polyurethane (polyether and polyester)	TPU
Styrene-butadiene copolymer	SBR, GR-S
Styrene-chloroprene copolymer	SCR
Polyethylene-butyl graft copolymer	TPO
Styrene-butadiene-styrene triblock polymer	SBS

Elastomeric Adhesive See ADHESIVE.

Electret A disk of polymeric material that has been electrically polarized so that one side has a positive charge and the other a negative charge, analogous to a permanent magnet. Electrets may be formed of poor conductors such as polymethyl methacrylate, polystyrene, nylon, and polypropylene, by heating and cooling them in the presence of a strong electric field.

Electrical-Discharge Machining (EDM, spark erosion) A method of machining molds and extrusion dies in which a conductive tool (often brass) has the inverse shape of the cavity or hole to be machined. A high-voltage DC difference is applied between the tool and the workpiece. Capacitive discharge erodes steel from the workpiece at about eight times the rate that the tool itself is eroded. Roughing and finishing tools are used. The process is accurate, produces good detail, can cut thin, deep slots, and can be used with hardened steels, thus averting the distortion sometimes caused by hardening after conventional machining. Metal removal is relatively slow.

Electrically Conductive Plastic Business-machine housings, structural components, and static-control accessories often require plastics that have some degree of electrical conductivity. Additives imparting such conductivity are metal powders, carbon black, carbon fibers, and metallized-glass fibers and spheres. See also CONDUCTING POLYMER.

Electrical Resistance (1) Ohmic resistance to the flow of electrical current. Related properties of plastics are INSULATION RESISTANCE, SURFACE RESISTIVITY, and VOLUME RESISTIVITY, w s. (2) The ability of plastics to withstand various electrical stresses. See ARC RESISTANCE, BREAKDOWN VOLTAGE, CORONA RESISTANCE, DIELECTRIC STRENGTH.

Electrochemical equivalent In an electrolytic cell, the mass of a metal deposited with the passage of one coulomb of electricity. (For many years, prior to the advent of SI, the coulomb was *defined* as the charge that would deposit 0.001118 g of silver from a silver nitrate solution. In the SI system, the coulomb is defined as 1 A·s.)

Electrode A terminal member in an electrical circuit designed to promote an electrical field between it and another electrode. In the plastics industry electrodes are used in microwave heat sealing and surface treating of films. One of the electrodes may be a press platen or a roll.

Electroformed Mold A mold made by electroplating a model which is subsequently removed from the metal deposit. The deposit is sometimes reinforced with cast or sprayed metal backings to increase its strength and rigidity. Such molds are used in slush casting of vinyl plastisols and other forming processes done at low pressures.

Electroforming A method of making molds for plastics processes, usually those employing low or moderate pressures, in which a pattern made of preplated wax or flexible material is electroplated.

Electroless Plating The deposition of metals on a catalytic surface from solution without an external source of current. The process is used as a preliminary step in preparing plastic articles for conventional electroplating. After cleaning or etching, the plastic surface is immersed in solutions that react to precipitate a catalytic metal *in situ*, for example, first in an acidic stannous chloride solution, then into a solution of palladium chloride. Palladium is reduced to its catalytic metallic state by the tin. Another way of producing a catalytic surface is to immerse the plastic article in a colloidal solution of palladium followed by immersion in an accelerator solution. The electroless plating bath is a solution of several components, including nickel or copper salts, chelating agents, stabilizers, and reducers. Metal reduced from the salt plates on the active sites on the palladium surface, the plated plastic being removed from the bath when the thickness of the electroless deposit is from 3 to 7 μm. The plastic article thus treated can now be plated with nickel or copper by the electroless method, forming a conductive surface that then can be plated with other metals by conventional electroplating.

Electroluminescence Generation of light by high-frequency electrical discharge through a gas or by applying an alternating current to a phosphor.

Electrolysis The process of producing chemical changes by passing an electric current through a conductive (i e, ionic) solution. Ions migrate to the electrodes and are converted to compounds, free metals, or gases. The process is the basis of electroplating of metals. When a current i (A) flows for a time interval t (s) and deposits a metal whose equivalent mass is e (g), the mass (g) deposited is equal to $e \cdot i \cdot t$.

Electromagnetic Adhesive An intimate blend of a material that absorbs electromagnetic energy with a thermoplastic of the same composition as the sections to be bonded. The adhesive is applied in the form of a liquid, a ribbon, a wire, or a molded gasket to one of the surfaces to be joined. The two surfaces are brought into contact, then the adhesive is rapidly heated by eddy currents induced by a high-frequency induction coil placed close to the joint. This melts the adhesive which, after cooling, bonds the surfaces together.

Electromagnetic Welding See INDUCTION WELDING.

Electromotive Force The difference in electric potential that causes current to flow in a circuit. The SI unit is the volt (V), defined as the difference of potential between two points of a conductor carrying a constant current of 1 ampere when the power dissipated between the two points equals 1 watt. Thus, in SI, $1\ V \equiv 1\ W/A$.

Electron An elementary subnuclear particle having a unit negative electrical charge, $1.60219 \cdot 10^{-19}$ coulomb, its mass (at rest) is 1/1837 that of a hydrogen nucleus, or $9.10953 \cdot 10^{-28}$ gram, and its diameter is $5.6359 \cdot 10^{-13}$ cm. Every atom consists of a nucleus of protons and neutrons, and as many orbiting electrons as there are protons in the nucleus. Cathode rays and beta rays are electrons.

Electron-Beam Radiation Magnetically accelerated electrons focused by electric fields have been used for crosslinking polyethylene in special applications such as wire coating, to improve modulus and temperature resistance. Treatment levels must be carefully controlled since overexposure will cause degradation. Electron beams are also used to cure epoxy-resin coatings, eliminating the need for photo-initiators.

Electronic Heating See DIELECTRIC HEATING.

Electron Micrograph An image formed on a photographic medium with an electron microscope, an instrument in which the subject is examined with an electron beam focused by electric fields. The beam may be transmitted through the specimen or, after precoating the surface to be examined with gold or platinum, reflected from it. Magnifications to 100,000 times are achievable with good resolution. Electron micrography has been a powerful tool in determining the structures of plastics crystals and the topography of fractured surfaces.

Electronic Treating See CORONA-DISCHARGE TREATMENT.

Electron-Volt (eV) The kinetic energy acquired by any charged particle carrying unit electronic charge when it falls through a potential difference of 1 volt. One eV is equal to $1.60219 \cdot 10^{-19}$ joule. Multiples of this unit in common use are the keV (10^3), MeV (10^6), and GeV (10^9). The GeV is also written BeV.

Electrophoretic Deposition A direct-current process analogous to electroplating, used to coat electrically-conductive articles with plastics, deposited from aqueous latices or dispersions. The cathode may be a noncorrodible metal such as stainless steel, generally serving as the container in which the process is performed. The DC potential is usually under 100 V. The deposited coatings are baked to remove residual water. Among available polymer latices suitable for the process are PVC, polyvinylidene chloride, acrylics, nylons, polyesters, polytetrafluoroethylene, and polyethylene.

Electroplating on Plastics Articles of almost any of the common plastics can be plated by conventional processes used on metals after their surfaces have been rendered conductive by precipitation of silver or other conductor (see ELECTROLESS PLATING). A layer of copper is usually applied first, followed by a final plating of gold, silver, chromium, or nickel. Acrylonitrile-butadiene-styrene resins have been most widely used for electroplated articles. Others in commercial use for the process include cellulose acetate, some grades of polypropylene, polysulfones, polycarbonate, polyphenylene oxide, nylons, and rigid PVC. See also METALLIZING.

Electrostatic Fluidized-Bed Coating A process combining elements of the fluidized-bed method of coating and electrostatic spraying. Pointed electrodes are inserted through the porous bottom of a fluidized-bed container. When the bed is aerated in the usual manner, a potential of about 100 kV is applied between the electrodes and ground. The associated charge repels the fluidized plastic particles into the space above the bed, from which they are attracted to a grounded article to be coated. The article may be at room temperature when inserted in the powder bed, the coating temporarily adhering by electrostatic charge. Subsequent heating fuses the coating.

Electrostatic Printing A printing process employing electrostatic charge to transfer powdered ink from an electrically charged stencil to a plastic film or sheet. The film to be printed is interposed between a grounded metal plate and the stencil. Areas corresponding to those *not* to be printed are masked on the stencil as in conventional screen printing. The powdered ink is brushed on the back side of the screen, where it receives a charge propelling it toward the grounded plate as an image cloud until intercepted by the film. Post-heating is usually required to fuse the ink to the substrate.

Electrostatic Spray Coating A spraying process that employs electrical charges to direct the paths of atomized particles to the work surface. Dry plastic powders are charged with static electricity as they emerge from a spray gun, the nozzle of which is attached to the negative terminal of a high-voltage DC power supply. The charged particles are attracted to the grounded object, which must be at least slightly electrically conductive. The powder coating is subsequently heated to obtain a smooth, homogeneous layer.

Ellis Model A three-constant model of pseudoplastic flow that merges Newton's law of flow, applicable at very low shear rates, with the POWER LAW (w s) at high rates and provides a smooth transition between the two. For one-dimensional flow the equation is:

$$-\frac{dv_z}{dx} = \frac{\tau_{xz}}{\eta_0}\left[1+\left|\frac{\tau_{xz}}{\tau_{1/2}}\right|^{\alpha-1}\right]$$

where v_z is the z-directed velocity perpendicular to coordinate x, τ_{xz} is the opposing shear stress, η_0 is the zero-shear (newtonian) viscosity, α is an exponent larger than 1, approximately equal to the reciprocal of the *flow-behavior index*, and $\tau_{1/2}$ is the shear stress at which the viscosity (= shear stress/shear rate) is half η_0.

Elmendorf Tear Strength The Elmendorf tear tester, originally developed to test papers and fabrics, has been adapted for plastics films in ASTM D 1922. Acting by gravity, a calibrated pendulum swings through an arc, tearing the specimen from a precut slit. The energy absorbed is indicated by a pointer and scale. Some other modes of measuring tear resistance are spelled out in ASTM D 1004, D 1938, and D 2582.

Elongation In tensile testing, the fractional increase in length of a marked test length as the test specimen is stretched and stress rises. At any point during the test, percent nominal elongation = 100 × the increase in gage length ÷ original length. *Ultimate elongation* is the elongation just prior to rupture of the sample and is the "elongation at break" reported in most properties tables. See also TRUE STRAIN.

Elongational Flow (extensional flow) Flow caused by stretching a material, usually a hot melt, as in fiber drawing, film blowing, parison drawdown, and biaxial stretching of sheet. This flow is always accompanied by a reduction in cross section. The *rate* of elongation, at any moment during one-dimensional, elongational flow, is given by

$$de/dt = (1/s)(ds/dt) = d(\ln s)/dt$$

where e = the true elongation and s = the strand length at time t.

Elongational Viscosity (extensional viscosity, Trouton viscosity) The viscosity that characterizes an element undergoing ELONGATIONAL FLOW (above). It is equal to the tensile stress divided by the rate of elongation and for polymers it depends on the rate, but may increase with rate, unlike the usual reduction of shear viscosity with rate. Tensile viscosities are apt to be many times larger than shear viscosities for the same resin, temperature, and deformation rate. Values in the range of 10^4 to 10^7 Pa·s have been reported. For Newtonian liquids, the elongation viscosity is three times the shear viscosity (at the same temperature).

Eluent In gas and liquid chromatography, the fluid that carries the solute out of the column.

Elution The removal, in chromatography, of the species adsorbed on the column matrix by a flowing liquid or gas.

Elutriation The separation of less massive particles from a sample of distributed particle sizes or densities by upward flow of a liquid or gas.

EMA Abbreviation for ETHYLENE-ACRYLIC ACID COPOLYMER, w s.

Embedding The process of encasing an article in a resinous mass, performed by placing the article in a simple mold, pouring a liquid resin into the mold to completely submerge the article, sometimes under vacuum so as to suck out hidden air bubbles, curing the resin, and removing the encased article from the mold. In the case of electrical components, the lead wires or terminals may protrude from the embedment. The main difference between embedding and potting is that in potting, the mold is a container that remains fixed to the resinous mass. The liquid resin may contain microspheres to reduce the final mass of the embedment. See also ENCAPSULATION, IMPREGNATION, and POTTING.

Embedment Decorating A technique for decorating reinforced-plastics articles in which a mat or web of fibrous material printed with a design is embedded in the surface of the article and covered with a transparent gel coat. The technique can be adapted for use in hand layup, continuous laminating, pultrusion, matched-die molding, etc.

Embossing Any of several techniques used to create depressed patterns in plastics films or sheeting. In the case of cast film, embossing can be accomplished directly by casting on an inverse-textured belt or roll. Calendered films are frequently embossed by rolls following the calender rolls. Other films or coated fabrics can be embossed subsequent to manufacture by reheating them and passing them through embossing rolls, or compressing them between plates. Extruded sheets, up to 3 mm or thicker, are commonly embossed as the sheets emerge from the extruder with an embossing roll on the takeoff.

Emery A mixture of mostly corundum and some magnetite that, because of its great hardness, is widely used for grinding and polishing, including tumble-polishing of plastics moldings. Emery is available in many grades from coarse to extremely fine, as powder and bonded to paper and cloth.

EMI Abbreviation for 2-ETHYL-4-METHYLIMIDAZOLE, w s.

Emulsifying Agent A substance used to facilitate the formation of an EMULSION, w s, from two or more immiscible liquids, and/or to promote the stability of the emulsion. As surface-active agents, emulsifiers act to reduce interfacial tensions between the several phases. They also act as protective colloids to promote stability.

Emulsion Strictly, an emulsion is a two-phase, substantially permanent, intimate mixture of two incompletely miscible liquids, one of which is dispersed as finite globules in the other. However, in plastics and other industries the term is sometimes broadened to include colloidal suspensions of solids such as waxes and resins in liquids. In an emulsion, the liquid that forms globules is known as the *dispersed, discontinuous*, or *internal phase*. The surrounding liquid is called the *continuous* or *external phase*. The dispersed phase may be held in suspension by mechanical agitation or by the addition of small amounts of emulsifying agents.

Emulsion Polymerization A polymerization process in which the monomer or mixture of monomers is emulsified in a low-viscosity aqueous medium by means of soaps or other surface-active, solubilizing and emulsifying agents. The emulsion does not require intensive stirring as in suspension polymerization, and produces polymers of higher molecular weight than those produced by bulk or suspension processes. The polymers remain in emulsion, and must be recovered from the latex by freezing or chemical precipitation. The polymerization medium usually also contains a water-soluble initiator, catalyst, or chain-transfer agent. Examples of polymers produced from emulsions are polyvinyl acetate, acrylo-nitrile-styrene terpolymer, PVC, polyethylene, acrylics, and polystyrene.

Enamel A dispersion of pigment in a liquid that forms a solid adherent film, on the surface to which it is applied, by means of oxidation, polymerization, or other chemical reaction. The liquid vehicle of an enamel usually contains a thermosetting resin and a solvent. An initial soft film is formed by evaporation of the solvent, then the film hardens or cures at room temperature or during baking.

Enantiomer (enantiomorph) An asymmetric molecule that is the mirror image of its stereoisomer. The two isomers are given the prefixes *dextro-* and *levo-*, e g, *d-* and *l*-lactic acid. The physical properties of pure enantiomers are equal within experimental error, yet mixtures of the two, called *racemic* mixtures, may have different properties. For example, 50-50 *dl*-lactic acid melts 20°C lower than its pure enantiomers.

Enantiomeric Configurational Unit Either of two stereoisomeric groups in a polymer that are mirror images at the plane containing the main-chain bonds.

Enantiotropic Denoting asymmetric crystal forms (enantiomorphs) capable of existing in reversible equilibrium with each other.

Encapsulation The process of applying a fairly thick coating that conforms to the shape of the coated object. The coating, of either thermoplastic or thermosetting resin, may be applied by brushing, dipping, spraying, or thermoforming. The process is much used for the protection and insulation of electrical components and assemblies. See also EMBEDDING, IMPREGNATION, POTTING,and MICROENCAPSULATION.

Encapsulization The enclosure of adhesive particles with a protective film that prevents them from coalescing until such time as proper pressure or solvation is applied.

End (1) A strand of roving consisting of a given number of filaments gathered together. The group of filaments is considered to be an *end* or *strand* before twisting, and a *yarn* after the twist has been applied. (2) An individual warp yarn, thread, or fiber.

End-Capping Conversion, by chemical reaction, of the end groups of polymer chains to less reactive, more stable groups, thus preventing "unzipping" of chains and rendering the polymer itself more stable to processing. An example is the conversion of –OH end groups in polyoxymethylene to acetate groups.

End Group A chemical group or radical forming the end of a polymer chain. Although end groups constitute a minute fraction of the polymer, they may vary considerably from the main-chain chemical structure and may exert effects on polymer properties that are stronger than one would expect from their numbers.

End-Group Analysis The quantitative determination, by chemical or spectrographic methods, of the number of end groups present in a sample of polymer. With linear, unbranched polymers, the mass of the sample divided by half the measured number of end groups equals the NUMBER-AVERAGE MOLECULAR WEIGHT, w s.

Endo- (1) A chemical prefix denoting an inner position, for example in a ring rather than in a side chain, or attached as a bridge within a ring. (2) When prefixing "-thermic", denoting that the reaction so labeled takes heat from the surroundings when proceeding from left to right. In both senses, the opposite of EXO-, w s.

Endothermic Pertaining to a chemical reaction or an operation that is accompanied by the absorption of heat.

Endurance Limit (fatigue limit) The stress level below which a specimen will withstand cyclic stress indefinitely without exhibiting fatigue failure. Rigid, elastic, low-damping materials such as thermosetting plastics and some crystalline thermoplastics do not exhibit endurance limits.

Endurance Ratio The ratio of the ENDURANCE LIMIT, under cyclic stress reversal, to the short-time, static strength of a material. If the mode of stress is not specified, tension/compression may be presumed.

Engineering Plastic (1) A broad term covering those plastics, with or without fillers and reinforcements, that have mechanical, chemical, electrical, and/or thermal properties suitable for industrial applications. R. B. Seymour, an outstanding authority, defined them as "...polymers...thermoplastic or thermosetting, that maintain their dimensional stability and major mechanical properties in the temperature range 0–100°C." He listed the "big five" (among neat resins) as nylons, polycarbonate, acetals, polyphenylene ether, and thermoplastic polyesters. Among many others are acrylics, fluorocarbons, phenoxy, acrylonitrile-butadiene-styrene terpolymer, polyaryl ether, polybutylene, chlorinated polyether, polyether, and many polymers reinforced with ADVANCED FIBERs, w s. At the high end of the spectrum of performance, usable temperature range, and price, unreinforced engineering plastics are dubbed ADVANCED RESINS, w s.

Engraved-Roll Coating See GRAVURE COATING.

Enhancement Ratio In a filled or reinforced plastic, the ratio of the modulus of the filled material to that of the neat resin. The ratio is likely to be higher at low strains than high strains because of slippage between filler and matrix.

Enthalpy (heat content) Thermodynamically, the enthalpy of a system is the sum of the internal energy and the pressure-volume product. We are usually concerned about *changes* in enthalpy rather than absolute values. Enthalpies of polymers are usually stated per unit mass, e g, kJ/kg, and are ordinarily referred to room temperature, 20 to 25°C, at which temperature enthalpy is arbitrarily set to zero. The rate of change of enthalpy with temperature is HEAT CAPACITY, w s.

Entrance Angle (entry angle) In an extrusion die, the total included angle, never more than 180°, of the main converging surfaces of the flow channel leading to the land area of the die.

Entropy A measure of the unavailable energy in a thermodynamic system, commonly expressed in terms of its changes on an arbitrary scale, the entropy of water at 0°C being assigned the value of zero. When a system (or sample of material) is heated or cooled from one temperature, T_1, to another, T_2, the change in its entropy is given by the equation:

$$\Delta S = \int_{T_1}^{T_2} \frac{dQ}{T}$$

Environmental Stress Cracking The formation of internal or external cracks in a plastic caused by tensile stresses well below its short-time strength, and induced by exposure to heat, solvent vapor, or chemically active solutions. ASTM Test F 1248 describes the measurement of the environmental stress-cracking resistance of polyethylene pipe in the presence of a surface-active agent. Other ASTM tests treating this subject are D 1693, D 2561, D 2951, and, in Section 15.03, F 484, F 791, and F 1164.

EOS Abbreviation for EQUATION OF STATE, w s.

EP (1) Abbreviation for EPOXY RESIN or EPOXIDE, w s. (2) (Usually E/P) Abbreviation sometimes used for copolymers of ethylene and propylene.

EPDM Abbreviation for elastomeric terpolymer from ethylene, propylene, and a conjugated diene. See ETHYLENE-PROPYLENE RUBBER.

Epi- (1) A prefix signifying a chemical compound or group differing from a parent compound or group by having a bridge connection. (2)EPI: See EPICHLOROHYDRIN.

Epichlorohydrin (chloropropylene oxide, EPI) An active solvent for cellulosic and other resins, and a key reactant for epoxy resins having the structure shown below.

O
H_2C—C—CH_2Cl
H

EPI is highly reactive with polyhydric phenols such as bisphenol A and forms glycidyls with many compounds containing active hydrogens.

Epichlorohydrin Rubber (CO, CEO) Any of several elastomers comprising polymers and copolymers of epichlorohydrin, with good high-temperature resistance, low-temperature flexibility, resistance to fuels, oils, and ozone, and low gas permeability. The homopolymer (CO) is a saturated, aliphatic polyether with a chloromethyl side chain. The ECO type is an equimolar copolymer of epichlorohydrin and ethylene oxide.

Episulfide The sulfur analog of epoxide in which the sulfur is part of a ring. The two most important members are ethylene sulfide and propylene sulfide. Episulfides are starting materials for POLYSULFIDE RUBBER, w s.

EPM Abbreviation for ETHYLENE-PROPYLENE RUBBER, w s.

Epoxidation A chemical reaction in which an oxygen atom is joined to an olefinically unsaturated molecule to form a cyclic, three-membered ether. The products of epoxidation are known as *oxiranes* or EPOXIDEs, w s.

Epoxide Any compound containing the oxirane structure, a three-membered ring containing two carbon atoms and one oxygen atom. The most important members are ethylene oxide and propylene oxide.

Epoxide Equivalent The mass of resin in grams that contains one gram-equivalent of EPOXIDE, w s.

Epoxidized Soybean Oil See EPOXY PLASTICIZER.

Epoxy (epoxy group, oxirane group) A label denoting an oxygen atom joined in a ring with two carbon atoms, as shown below.

```
     O
    / \
  —C———C—
   |   |
```

β-(3,4-Epoxycyclohexyl) Ethyltrimethoxy Silane A coupling agent for reinforced polyester, epoxy, phenolic, melamine, and many thermoplastics.

Epoxy Foam Two basic types of epoxy foams are in use, *chemical foams* and *syntactic foams.* Chemical-foam compositions contain the resin, curing agent, blowing agent, wetting agent, and a small percentage of an inert organic compound such as toluene to dissipate the exothermic heat of curing, and thus control the foaming action. Because foaming is rapid, the curing agent is withheld until all the other ingredients have been mixed, to be added just prior to casting. These systems may also contain amine-terminated polyamide resins to impart resiliency to the foam. In syntactic foams, the voids are provided by hollow phenolic microspheres and the resin does not foam but acts as a binder for the spheres. Epoxy foams are used in casting, in potting and encapsulating of electrical aasemblies, in insulating coatings for chemical-storage tanks, and in cores of laminates for aircraft and boats.

Epoxy-Novolac Resin A two-step resin made by reacting epichlorohydrin with a phenol-formaldehyde condensate. Such resins are also known as thermoplastic, B-stage phenolic resins that are in a state of partial cure. Whereas bisphenol-based epoxy resins contain up to two epoxy groups per molecule, the epoxy novolacs may have seven or more such groups, producing more tightly crosslinked structures in the cured resins. Thus they are stronger and superior in other properties.

Epoxy Number The number of gram-equivalents of epoxy groups per 100 grams of polymer, equal to 1/100 of the reciprocal of the epoxide equivalent.

Epoxy Plasticizer (epoxide plasticizer) Any of a large family of plasticizers obtained by the epoxidation of vegetable oils or fatty acids. The two main types are (a) epoxidized unsaturated triglycerides, e g, soybean oil and linseed oil; and (b) epoxidized esters of unsaturated fatty acids, e g, oleic acid, or butyl-, octyl-, or decyl- esters. Most epoxy plasticizers have a heat-stabilizing effect and they are often used for stabilization in conjunction with other stabilizers. Epoxidized oils generally have good resistance to extraction and migration and low volatility, but they cannot be used as sole plasticizers in unfilled vinyl compounds and hence are not considered to be primary plasticizers. Certain epoxidized soybean oils have been FDA-approved for food-contact use.

Epoxy Resin Any of a family of thermosetting resins containing the oxirane group. (See EPOXY for structure.) Originally made by condensing epichlorohydrin and bisphenol A, epoxy resins are now more generally formed from low-molecular-weight diglycidyl ethers of bisphenol A and modifications thereof; or, another type, by the oxida-ion of olefins with peracetic acid. Depending on molecular weight, the resins range from liquids to solids. The liquids, used for casting, potting, coating, and adhesives, are cured with amines, polyamides, anhydrides, or other catalysts. The solid resins are often modified with other resins and unsaturated fatty acids. Epoxy resins are widely used in reinforced plastics, having strong adhesion to glass fibers. Epoxies based on epoxidized heterocyclic hydantoin are useful in electrical composites because their thermal expansion coefficient can be matched to that of copper. Their low viscosities are effective in wetting the various reinforcing materials used with them. Fast-curing epoxies are based on the diglycidyl ether of 4-methylol resorcinol (DGEMR). The methylol group appears to effectively catalyze the curing reactions. This resin is curable with all types of conventional epoxy hardeners including aliphatic and aromatic amines, anhydrides, and amidoamines. DGEMR cures approximately thirty times as fast as a conventional bisphenol A epoxy and two to five times as fast as older fast-gelling epoxies, and at lower temperatures. DGEMR may be formulated with flexibilizers and fillers without prolonging gel time. These same properties make the resin well suited for adhesives, coatings, and low-temperature applications. See also EPOXY-NOVOLAC RESIN.

Epoxy Stabilizer (epoxide stabilizer) Most EPOXY PLASTICIZERs, w s, also serve as stabilizers because of the ability of the epoxide group to accept HCl, or to serve as an intermediate, in the presence of metallic salts, to convert HCl to a metallic chloride. Epoxy stabilizers are most often used in conjunction with barium-cadmium and other stabilizers, with which they have a synergistic effect.

EPR Abbreviation for ETHYLENE-PROPYLENE RUBBER, w s.

EPS Abbreviation for expanded polystyrene. See POLYSTYRENE FOAM.

Equation of State (EOS) As with gases, an equation giving the specific volume of a polymer from the known temperature and pressure and, sometimes, from its morphological form. An early example is the modified Van der Waals form, successfully tested on amorphous and molten polymers. The equation is:

$$v = b + RT/M(P + \pi)$$

where b = the "unfree" specific volume occupied by the polymer molecules, roughly 90+ percent of v, R = the universal molar-energy constant, T = the absolute temperature, M = an empirical molecular weight that for several thermoplastics has been closely equal to the mer weight, and π = an empirical internal pressure much larger than the highest injection-molding pressures. A number of more complex models have since been introduced and tested against experimental data.

Equivalent Weight (combining weight, equivalent mass) The atomic or formula weight of a given element or ion divided by its valence in a reaction under consideration. Elements entering into combination always do so in quantities proportional to their equivalent weights. In oxidation-reduction reactions the equivalent weights of the reacting entities are dependent upon the change in oxidation numbers of the particular substances.

Erosion Breakdown In an electrical-conductor insulation, deterioration caused by chemical attack of corrosive chemicals such as ozone and nitric acid that are formed by corona discharge from a high-voltage cable. This breakdown can occur even in the most chemically resistant polymers, such as fluorocarbons, after long exposure to the condition.

Erucamide $CH_3(CH_2)_7CH{=}CH(CH_2)_{11}COONH_2$. The amide of *cis*-13-docosenoic acid, used in fractional percentages as a slip agent in polyethylene film resins.

Erucyl Alcohol $CH_3(CH_2)_7CH{=}CH(CH_2)_{11}CH_2OH$. A monounsaturated, fatty alcohol used as a mold lubricant.

Erythrene Old synonym for BUTADIENE, w s.

ESC Abbreviation for ENVIRONMENTAL STRESS CRACKING, w s.

ESO Abbreviation for epoxidized soybean oil. See EPOXY PLASTICIZERS.

ESR Abbreviation for *electron-spin-resonance spectroscopy.*

Ester An organic compound analogous to an inorganic salt. Esters are formed by reacting an acid with an alcohol, or by the exchange of a replaceable hydrogen atom of an acid for an organic alkyl radical. Esters of many monofunctional alcohols and organic acids are oily, fruity-smelling liquids, forming important families of solvents and plasticizers. When the alcohol selected is polyfunctional, that is, contains two or more –OH groups, and the acid is di- or polybasic, long chains of repeating units can be formed by their reaction. These are POLYESTERS, w s.

Ester Interchange (ester exchange) A reaction between an ester and another compound in which occurs an exchange of alkoxy or acyl groups, resulting in the formation of a different ester. When an ester is reacted with an alcohol, the process is called *alcoholysis*; reaction between an ester and an acid is called *acidolysis*. Ester interchanges are used in producing plasticizers, polyvinyl alcohol, acrylics, polyesters, and polycarbonates.

ETFE Abbreviation of ETHYLENE-TETRAFLUOROETHYLENE COPOLYMER, w s.

Ethanal Synonym for ACETALDEHYDE, w s.

Ethanol Synonym for ETHYL ALCOHOL, w s.

Ethanolurea $NH_2CONHCH_2CH_2OH$. A white compound melting at 71–74°C. It condenses with formaldehyde to form permanently thermoplastic, water-soluble resins. Simple urea can be incorporated in the condensation reaction to give modified resins with any desired degree of water solubility and flexibility, both of which properties increase with urea content.

Ethene IUPAC's name for ETHYLENE, w s.

Ethenoid Plastics (1) Plastics made from monomers containing the polymerizable double-bond group C=C, for example ethylene. Thermosetting ethenoid resins are made from monomers or linear polymers capable of giving crosslinked structures as a result of double-bond polymerization. (2) A British generic term that includes acrylic, vinyl, and styrene plastics.

Ether (1) Any organic compound in which an oxygen atom is interposed between two carbon atoms or organic radicals in the molecular structure. Ethers are often derived from alcohols by elimination of one molecule of water from two molecules of alcohol. (2) Specifically, diethyl ether, $(C_2H_5)_2O$.

Ethyl Acetanilide (ethyl phenylacetamide) $C_6H_5N(C_2H_5)COCH_3$. A substitute for camphor in the manufacture of celluloid.

Ethyl Acetate (acetic ester) $CH_3COOC_2H_5$. A colorless liquid made by heating acetic acid and ethyl alcohol in the presence of sulfuric acid, then distilling. It is a powerful solvent for ethyl cellulose, polyvinyl acetate, cellulose acetate-butyrate, acrylics, polystyrene, and coumarone-indene resins. Although it is highly flammable, it is the least toxic of common industrial solvents.

Ethylacetic Acid Synonym for BUTYRIC ACID, w s.

Ethyl Acrylate $CH_2=CHCOOC_2H_5$. A polymerizable monomer, from which acrylic resins used in paints are made.

Ethyl Alcohol (alcohol, ethanol, grain alcohol) C_2H_5OH. An alcohol used, in denatured form, as a solvent for ethyl cellulose, polyvinyl acetate, and polyvinyl butyrate. The industrial grade of *un*denatured alcohol usually contains 5 wt% water. The pure compound is called ABSOLUTE ALCOHOL, w s.

Ethyl Aldehyde Synonym for ACETALDEHYDE, w s.

Ethyl Aluminum Dichloride $C_2H_5AlCl_2$. A clear, yellow, flammable liquid, a catalyst for olefin polymerization.

Ethyl Aluminum Sesquichloride $(C_2H_5)_3Al_2Cl_3$. A catalyst for olefin polymerization.

Ethyl Benzoate (benzoic ether) $C_6H_5COOC_2H_5$. A colorless liquid derived by heating ethyl alcohol and benzoic acid in the presence of sulfuric acid. It is a solvent for cellulosics.

2-Ethylbutyl Acetate $C_2H_5CH(C_2H_5)CH_2OOCCH_3$. A solvent for cellulose nitrate.

Ethyl-*n*-butyl Ketone (2-heptanone) $C_2H_5CO(CH_2)_3CH_3$. A stable, colorless liquid with medium volatility, used in solvent mixtures for cellulosic and vinyl resins. When used in vinyl organosols it imparts good long-time viscosity stability.

Ethyl Butyrate (ethyl butanoate) $C_3H_7COO(C_2H_5)$. A solvent for cellulosics.

Ethyl Carbamate Synonym of URETHANE, w s.

Ethyl Carbonate Synonym for DIETHYL CARBONATE, w s.

Ethyl Cellosolve Synonym for CARBITOL, w s.

Ethyl Cellulose (EC) An ethyl ether of cellulose formed by reacting cellulose steeped in alkali with ethyl chloride. Since the repeating units are etheric, it is chemically different from other cellulosics, which are esters, and is therefore not compatible with them. EC resin can be injection molded, extruded, cast into film, or used as a coating material. It has the lowest density of all cellulosic plastics, good toughness and impact resistance, and is dimensionally stable over a wide temperature range.

Ethyl Chloride (chloroethane) C_2H_5Cl. A colorless gas at ambient conditions, used in the production of ethyl cellulose by reaction with sodium cellulose.

Ethyl Citrate See TRIETHYL CITRATE.

Ethylene (bicarburetted hydrogen, ethene) $H_2C{=}CH_2$. A colorless, flammable gas derived by cracking petroleum and by distillation from natural gas. In addition to serving as the monomer for polyethylene, it has many uses in the plastics industry including the synthesis of ethylene oxide, ethyl alcohol, ethylene glycol (used in making alkyd and polyester resins), ethyl chloride, and other ethyl esters.

Ethylene-Acrylic Acid Copolymer (EAA) Either of the block or random copolymers of ethylene and acrylic acid whose ionic character gives strong adhesion to metals and other surfaces. Their toughness has created uses in multilayer packaging films and golf-ball covers. Two trade names are DuPont's Surlyn® (block) and Dow's EAA (random).

***N,N'*-Ethylene Bis-Stearamide** (Acrawax C®) A lubricant used in acrylonitrile-butadiene-styrene resins, PVC, and polystyrenes.

Ethylene-bis Tris-(2-cyanoethyl) Phosphonium Bromide (ECPB) A flame retardant for thermoplastics. In polymethyl methacrylate, 20% ECPB caused the resin to become opaque and reduced its burning rate to zero.

Ethyl Carbonate (glycol carbonate, 1,3-dioxolan-2-õne) $(CH_2O)_2CO$. A solvent for many polymers and resins.

Ethylene Carboxylic Acid A little used synonym for ACRYLIC ACID, w s.

Ethylene Chloride Synonym for ETHYLENE DICHLORIDE, w s.

Ethylene Chlorotrifluoroethylene Copolymer (ECTFE, E/CTFE) A fluoroplastic (Ausimont's Halar®) with good mechanical, thermal, electrical, processing and resistance properties.

Ethylene Dichloride (EDC, 1,2-dichloroethane, ethylene chloride, dutch oil) CH_2ClCH_2Cl. A colorless, oily liquid used in producing vinyl chloride monomer and as a solvent for phenolic and cellulosic resins.

Ethylene-Ethyl Acrylate Copolymer (EEA, E/EA) A family of elastomeric resins similar in appearance to polyethylene, but possessing properties like those of rubber and flexible vinyls.

Ethylene Glycol (ethanediol, ethylene alcohol, glycol) $HOCH_2CH_2$-OH. A clear, syrupy liquid used as a solvent for cellulosics, particularly cellophane, and in the production of alkyd resins and polyethylene terephthalate.

Ethylene Glycol Diacetate $(CH_3COOCH_2-)_2$. A very slowly evaporating solvent for cellulosic and acrylic resins, sometimes used as a fugitive plasticizer for vinyls and acrylics.

Ethylene Glycol Dibenzoate $(C_6H_5COOCH_2-)_2$. A plasticizer for cellulosic resins, having limited compatibility with some vinyl resins.

Ethylene Glycol Dibutyrate (glycol dibutyrate) $(C_3H_7COOCH_2-)_2$. A plasticizer for cellulosic plastics.

Ethylene Glycol Dipropionate (glycol propionate) $(C_2H_5COOCH_2-)_2$. A plasticizer for cellulosic plastics.

Ethylene Glycol Monoacetate (glycol monoacetate) $HOCH_2CH_2OOCCH_3$. A solvent for cellulose nitrate and cellulose acetate.

Ethylene Glycol Monobenzyl Ether (benzyl cellosolve) $C_6H_5CH_2OC_2H_4OH$. A solvent for cellulose acetate.

Ethylene Glycol Monobutyl Ether (2-butoxyethanol, butyl cellosolve) $C_4H_9OC_2H_4OH$. A colorless liquid used as a solvent for cellulosic, phenolic, alkyd, and epoxy resins, especially in varnish and other coating formulations.

Ethylene Glycol Monobutyl Ether Acetate $C_4H_9OC_2H_4OOCCH_3$. A colorless liquid with a fruity aroma, used as a high-boiling solvent for cellulose nitrate, epoxy resins, and as a film-coalescing aid for polyvinyl-acetate latex.

Ethylene Glycol Monobutyl Ether Laurate $C_4H_9OC_2H_4OOCC_{11}H_{23}$. A plasticizer for cellulosics, polystyrene, and vinyls.

Ethylene Glycol Monobutyl Ether Oleate $C_4H_9OC_2H_4OOC(CH_2)_7CH{=}CH(CH_2)_7CH_3$. A plasticizer for cellulose nitrate, ethyl cellulose, and PVC.

Ethylene Glycol Monobutyl Ether Stearate $C_4H_9OC_2H_4OOCC_{17}H_{35}$. A plasticizer for cellulose nitrate, ethyl cellulose, polystyrene, and polyvinyl butyral.

Ethylene Glycol Monoethyl Ether (cellosolve, ethyl cellosolve) $C_2H_5OC_2H_4OH$. A solvent for cellulose nitrate, phenolic, alkyd and epoxy resins. It is colorless, nearly odorless, has a low evaporation rate, and imparts good flow properties to coatings.

Ethylene Glycol Monoethyl Ether Acetate (cellosolve acetate) $C_2H_5OC_2H_4OOCCH_3$. A solvent for cellulose nitrate, ethyl cellulose, vinyl polymers and copolymers, polymethyl methacrylate, polystyrene, epoxy, coumarone-indene and alkyd resins.

Ethylene Glycol Monoethyl Ether Laurate $C_2H_5OC_2H_4OOCC_{11}H_{23}$. A plasticizer for cellulosic and vinyl resins, and polystyrene.

Ethylene Glycol Monoethyl Ether Ricinoleate $C_2H_5OC_2H_4OOC(CH_2)_7CH{=}CHCH_2(OH)CH(CH_2)_5CH_3$. A plasticizer.

Ethylene Glycol Monomethyl Ether (2-methoxyethanol, methyl cellosolve) $CH_3OC_2H_4OH$. A solvent for cellulose esters.

Ethylene Glycol Monomethyl Ether Myristate $CH_3OC_2H_4OOCC_{13}H_{27}$. A plasticizer for cellulosic plastics, PVC, and polyvinyl butyral.

Ethylene Glycol Monomethyl Ether Oleate $CH_3OC_2H_4OOC(CH_2)_7$-$CH=CHC_8H_{17}$. A plasticizer for cellulosic and vinyl resins.

Ethylene Glycol Monomethyl Ether Stearate $CH_3OC_2H_4OOCC_{17}$-H_{35}. A plasticizer for cellulosics and polystyrene, having limited compatibility with other thermoplastics.

Ethylene Glycol Monooctyl Ether $C_8H_{17}OC_2H_4OH$. A solvent for cellulose esters, and a plasticizer.

Ethylene Glycol Monophenyl Ether $C_6H_5OC_2H_4OH$. A solvent for cellulosics, vinyls, phenolics, and alkyd resins.

Ethylene Glycol Monoricinoleate (ethylene glycol ricinoleate) $CH_3(CH_2)_5CH(OH)CH_2CH=CH(CH_2)_7COOC_2H_4OH$. A plasticizer and an intermediate for urethane polymers.

Ethylene-Methyl Acrylate Copolymer (EEA, E/EA) An elastomer vulcanizable with peroxides or diamines. It resists attack by oils and temperatures to 175°C.

Ethylene Oxide (epoxyethane) A three-membered ring compound with the formula H_2COCH_2, it is a colorless gas at room temperature, important as a raw material for the production of ethylene glycol, cellosolves, higher alcohols, acrylonitrile, and ethanolamines.

Ethylene Plastic See POLYETHYLENE.

Ethylene-Propylene Rubber (E/P, EPDM, EPM, EPR) Any of a group of elastomers obtained by the stereospecific copolymerization of ethylene and propylene (EPM), or of these two monomers and a third monomer such as an unconjugated diene (EPDM). Their properties are similar to those of natural rubber in many respects, and they have been proposed as potential substitutes for natural rubber in tires.

Ethylene-Tetrafluoroethylene Copolymer A copolymer of ethylene and tetrafluoroethylene (DuPont Tefzel®), ETFE is readily processed by extrusion and injection molding. It has excellent resistance to heat, abrasion, chemicals, and impact, with good electrical properties.

Ethylene-Urea Resin A type of AMINO RESIN, w s.

Ethylene-Vinyl Acetate Copolymer (EVA, E/VAC) Any copolymer containing mainly ethylene with minor proportions of vinyl acetate. They retain many of the properties of polyethylene but have considerably increased flexibility, elongation and impact resistance. They resemble elastomers in many ways, but can be processed as thermoplastics.

Ethylene-Vinyl Alcohol Copolymer (EVAL®, E/VAL, EVOH) A family of copolymers made by hydrolyzing ethylene-vinyl acetate copolymers with high VA content. Those containing about 20-35% ethylene are useful as barriers to many vapors and gases, though not to water. Because of their water sensitivity, they are usually sandwiched between layers of other polymers.

Ethyl Formate C_2H_5OOCH. A solvent for cellulose acetate.

2-Ethyl-1,3-Hexanediol $C_3H_7CH(OH)CH(C_2H_5)CH_2OH$. A stable, colorless, nearly odorless, high-boiling liquid with weak solvent action. In two-part urethane systems, the material acts as a viscosity reducer at room temperature. When the urethane mixture is heated to cure it, the diol reacts into the urethane matrix to eliminate solvent emissions.

2-Ethylhexyl- An 8-carbon, branched-chain radical of the formula $C_4H_9CH(C_2H_5)CH_2$–, often called *octyl* in the plastics industry. For example, the common plasticizer di-2-ethylhexyl phthalate is commonly referred to as dioctyl phthalate and by its abbreviation, DOP.

2-Ethylhexyl Acetate (octyl acetate) $C_4H_9CH(C_2H_5CH_2OOCCH_3$. A high-boiling retarder solvent with low evaporation rate and limited water solubility, used primarily in coating formulations based on cellulose nitrate. It is also used as a dispersant in vinyl organosols.

2-Ethylhexyl Acrylate $C_4H_9CH(C_2H_5)CH_2OOCCH{=}CH_2$. One of the monomers for acrylic resins, especially for those used in water-based paints.

2-Ethylhexyl Alcohol (2-ethylhexanol, octyl alcohol) $C_4H_9CH(C_2$-$H_5)CH_2OH$. An involatile solvent with many uses in the plastics industry. As a solvent, it is used in coatings for stenciling, silk screening, and dipping. As an intermediate, the alcohol is an important raw material for the production of the 2-ethylhexyl esters of dibasic acids used as plasticizers, such as dioctyl phthalate, adipate, and azelate.

2-Ethylhexyl Decyl Phthalate A mixed diester plasticizer for cellulosics, polystyrene, PVC, and polyvinyl acetate.

2-Ethylhexyl Epoxytallate An epoxy ester used mainly as a combined plasticizer and stabilizer in vinyl compounds. At concentrations as low as 5 phr it reacts synergistically with many metallic stabilizers to provide stability comparable to similar combinations based on epoxidized soybean oils. As a partial replacement for other plasticizers, it imparts good low-temperature flexibility. It is also compatible with vinyl chloride-vinyl acetate copolymers, high-butyral cellulose acetate-butyrate resins, ethyl cellulose, polystyrene, and chlorinated rubbers.

2-Ethylhexyl Isodecyl Phthalate (octyl isodecyl phthalate) C_8H_{17}-$OOCC_6H_4COOC_{10}H_{21}$. A mixed ester compatible with PVC, vinyl chloride-acetate copolymers, cellulose acetate-butyrates with higher butyrate contents, cellulose nitrate, and, in lower concentrations, with polyvinyl butyral. In vinyls, it is somewhat less volatile than dioctyl phthalate and has equivalent low-temperature properties.

2-Ethylhexyl-*p*-Hydroxybenzoate $C_8H_{17}OOCC_6H_4OH$. A plasticizer for polyamides.

Ethyl-α-Hydroxyisobutyrate $C_2H_5OOC(OH)C(CH_3)_2$. A solvent for cellulose nitrate and cellulose acetate.

Ethylidene Acetobenzoate (ethylidene benzoacetate) $C_6H_5COO(CH_3$-$CH)COCH_3$. A solvent for cellulosics and synthetic resins.

Ethyl Lactate $C_2H_5OOCCH(OH)CH_3$. A solvent for cellulosic and other resins.

Ethyl Levulinate $C_2H_5OOC(CH_2)_2COCH_3$. A solvent for cellulose acetate.

Ethyl Methacrylate $H_2C{=}C(CH_3)COOC_2H_5$. A readily polymerizable monomer used for certain types of acrylic resins.

2-Ethyl-4-Methylimidazole (EMI) An epoxy-resin curing agent with the heterocyclic structure shown below:

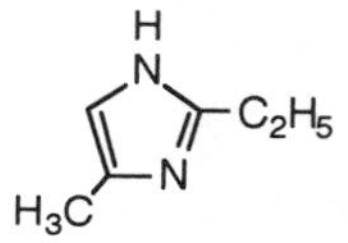

EMI is used with epoxies formed from epichlorohydrin and bisphenol A or –F, and for novolac epoxy resins. It provides ease of compounding, long pot life, low viscosity, and non-staining characteristics, and yields castings with excellent mechanical and electrical properties.

Ethyl Oleate $C_2H_5OOCC_{17}H_{33}$. This monounsaturated fatty ester is a solvent, lubricant, and plasticizer.

Ethyl Oxalate (diethyl oxalate) $(H_5C_2OOC{-})_2$. A solvent for cellulosics and many synthetic resins.

Ethyl Phthalate See DIETHYL PHTHALATE.

Ethyl Phthalyl Ethyl Glycolate $C_2H_5OOCC_6H_4COOCH_2COOC_2H_5$. A plasticizer compatible with PVC and most common thermoplastics. It has been approved by the FDA for use in food packaging.

Ethyl Propionate (propionic ester) $C_2H_5OOCC_2H_5$. A solvent for cellulose ethers and esters.

***N*-Ethyl-*p*-Toluene Sulfonamide** A solid plasticizer for rigid PVC.

EU Abbreviation for polyether type of polyurethane rubber.

Eutectic Said of a specific mixture of two or more substances that has a lower melting point than that of any of its constituents alone or any other percentage composition of the constituents.

EVA (E/VAC) See ETHYLENE-VINYL ACETATE COPOLYMER.

EVE See VINYLETHYL ETHER.

EVOH See ETHYLENE-VINYL ALCOHOL COPOLYMER.

Exa- The SI prefix meaning $\times 10^{18}$.

Excitation In ultraviolet curing, the first state of the polymerization process, in which the photo-initiator, such as benzophenone amine, is stimulated by UV into a singlet or triplet state, with subsequent formation of free radicals.

Exo- A chemical-structure prefix meaning attachment to a side chain rather than to a ring. Compare ENDO-.

Exotherm (1) The temperature/time curve of a chemical reaction giving off heat, particularly the polymerization of a casting resin. ASTM Test D 2471 delineates a procedure for measuring this curve, which in practice is strongly dependent on the amount of material present and the geometry of the casting. (2) The amount of heat given off per unit mass of the principal reactant. As yet (1992) no standard method for determining this heat in reacting plastics has been adopted.

Exothermic Denoting a chemical reaction that is accompanied by the evolution of heat. An example in the plastics industry is the curing reaction of an epoxy resin with an amine hardener.

Expandable Plastic A plastic formulated so as to be transformable into a cellular plastic by thermal, chemical, or mechanical means.

Expanded Plastic See CELLULAR PLASTIC.

Expanded Polystyrene (XPS) See POLYSTYRENE FOAM.

Expanding Agent See BLOWING AGENT.

Expansivity See COEFFICIENT OF THERMAL EXPANSION.

Extender (1) In plastics compounding, a material added to the mixture to reduce its cost per unit volume. The material may be a resin, plasticizer, or filler. See also BLENDING RESIN. (2) A substance, generally having some adhesive capacity, added to an adhesive formulation to reduce the amount of the primary (i e, more costly) binder required per unit of bonding area. See BINDER and FILLER.

Extensibility (1) The ability of a material to stretch or elongate upon application of sufficient tensile stress. It is expressed as a percentage of the original length. (2) The value of said ability just prior to rupture of the specimen; ultimate elongation.

Extensiometer A rheometer for measuring the extensional flow properties of molten polymers. In one early form, the *Cogswell rheometer*, useful at tensile viscosities over 10^5 Pa·s, unidirectional tensile force was exerted on a polymer rod by a dead weight acting through a cam and pulley. As the cam rotated, the moment arm exerted by the weight on the rod decreased in proportion to the rod cross section so as to maintain constant stress.

Extensional Strain Rate See ELONGATIONAL FLOW.

Extensional Viscosity See ELONGATIONAL VISCOSITY.

External Plasticizer A plasticizer that is added to a resin or compound, as opposed to an *internal plasticizer* that is incorporated into the resin during the polymerization process.

External Undercut Any recess or projection on the outer surface of a molded part that prevents its direct removal from its mold cavity. Parts with such undercuts may be molded by splitting the mold vertically and opening the split to withdraw the part; or by providing *side draws*, i e, mold parts that are withdrawn to the sides to relieve the undercuts.

Extractable (n) The amount of soluble material extracted from a polymer specimen when it has been exposed to a solvent under specified conditions. In ASTM TEST D 4754, disks of the polymer and glass separator beads are alternately threaded onto a stainless-steel wire and slipped into a vial containing the test solvent. The tightly closed vial is placed in an oven, and the solvent is sampled and analyzed periodically.

Extraction The transfer of a constituent of a plastic mass to a liquid with which the mass is in contact. The process is generally performed with a solvent selected to dissolve one or more specific constituents; or it may occur as a result of environmental exposure to a solvent.

Extraction Extrusion An extrusion operation in which a volatile component present in the feedstock is removed by flash vaporization through a vent connected to a vacuum pump. The volatile component is typically a small amount of water, but may be monomer or solvent. In a two-stage, single-screw extruder, the vent is located over the deep extraction section that begins the second stage of the screw. A few double-vented (three-stage) machines have been made. Some twin-screw machines have greater capacity for removal of volatiles.

Extrudate (1) The product or material delivered from an extruder, for example film, pipe, profiles, and wire coatings. (2) The extruded melt just as it emerges from the die.

Extrudate Roughness See MELT FRACTURE.

Extrudate Swelling (Barus effect) The increase in thickness or diameter, due mainly to the release of stored elastic energy, as a hot melt emerges *freely* from an extrusion die. In many commercial operaions, because the extrudate is drawn away at speeds higher than the mean flow velocity in the die, swelling is more than offset by draw down and is not actually observed. Swelling tends to increase with extrusion rate, abruptness of the approach to the die land, and inversely with land length and melt temperature.

Extruded-Bead Sealing (melt-bead sealing) A method of welding or sealing continuous lengths of thermoplastic sheeting or thicker sections by extruding a bead of the same material between two sections and immediately pressing the sections together. The sensible heat in the bead is sufficient to achieve the weld to the adjacent surfaces.

Extruded Foam Cellular plastic produced by extrusion with the aid of a blowing agent—a decomposable, gas-generating chemical—or by injection into the extruder of a gas such as nitrogen or carbon dioxide, or a highly volatile liquid, such as pentane.

Extruded Shape (profile) Any of a huge variety of cross-sectional shapes produced in continuous or cut lengths by extrusion through profile dies. Some complex shapes reach their final form with the aid of postforming jigs that alter the shape before the extrudate has cooled and become firm.

Extruder (1) A machine for producing more or less continuous lengths of plastics having constant cross sections, such as strands ("spaghetti"), rods, sheets, tubes, profiles, and cable coatings. The essential elements are a thick-walled, tubular barrel, usually heated electrically and provided with some kind of cooling means; a rotating screw, or a ram within the barrel; a hopper at one end from which the material to be extruded is fed to the screw or ram; and a die at the opposite end for shaping the extruded mass. Commercial extruders may be divided into three general types: single-screw (most common), twin- or multi-screw, and ram; each type in turn having several variations and particular applications. (2) A company or person who operates extruders.

Extruder Barrel (extruder cylinder) A thick-walled, cylindrical steel tube, lined with a special hard alloy to resist wear, that forms the housing for the extruder screw and contains, between itself and the screw, the plastic material as it is conveyed from feed hopper to die. Barrels are usually surrounded by heating and cooling media, such as electrical heater bands, cast-in-aluminum calrods and tubing for coolant, induction heaters, or, rarely, by a compartmented jacket for the circulation of hot oil or steam. Electrically heated barrels are usually furnished with some means of cooling, such as air blowers or coils through which fluid may be circulated. Small holes drilled radially into (but not through) the barrel accommodate temperature sensors whose signals are used to control the means of heating and cooling.

Extruder Breaker Plate See BREAKER PLATE.

Extruder Drive The system comprising an AC or DC motor, speed reducer (gearbox), screw-shaft bearings, coupling, and controls that supply power to the screw and regulate its speed. To facilitate startup and accommodate various operating conditions, modern extruders are always equipped with some means of varying the screw speed, with close control, over a wide range.

Extruder, Dual-Ram A modification of the original ram extruder employing two identical units, stroking alternately, and delivering to the same die. Aided by valves, the result is the conversion of a batch operation into one that makes continous extrusions from sinterable resins such as polytetrafluoroethylene and ultra-high-molecular-weight polyethylene.

Extruder, Elastic-Melt (elastodynamic extruder) A type of extruder in which the material is fed into a fixed gap between stationary and rotating, vertical disks, is melted by frictional heat, and flows in a spiral path toward the center of rotation, from which it is discharged into a secondary device that can develop the high pressure required for extrudate shaping. Only rubbery polymers with certain viscoelastic properties are suitable for the process.

Extruder, Hydrodynamic A device similar to the elastic-melt extruder (see preceding entry) in that the plastic pellets are sheared between relatively rotating disks. However, the disks in a hydrodynamic extruder are shaped to provide positive driving force, whatever the properties of the melt. It, too, can provide efficient melting while developing little pressure.

Extruder, Piggy-Back A system in which the two chief functions of a plasticating extruder—melting and pressure development—are made independently controllable by using two extruders in tandem. The first receives the cold feed, melts it, and delivers the melt to the second extruder, which is essentially a melt pump with possible mixing and/or extraction zones, and which develops the die pressure. Though the piggy-back principle is sound, the market has favored the combination of extruder with gear pump. However, see EXTRUDER, TANDEM.

Extruder, Planetary Screw A multi-screw device in which a number of satellite screws, generally six, are arranged around one longer central screw. The portion of the central screw extending beyond the the satellite screws serves as the final pumping screw as in a single-screw extruder, while the planetary screws aid in plastication and permit the discharge of volatiles toward the hopper. A few of these have served in processing fine powders such as PVC dry-blend.

Extruder, Ram An extruder in which the material is advanced through the barrel and die by means of a ram or plunger rather than by a screw. Melting is accomplished either by preheating the feedstock close to the fusing temperature and by conductive heating from the barrel wall, or both. The ram extruder was the earliest type to be used in the plastics industry, dating back to 1870 when cellulose nitrate was extruded into rods. Among plastics today, polytetrafluoroethylene, which softens to a gel but does not achieve a true melt state, is the one mainly processed with ram extruders.

Extruder Screen Pack A layered group of woven metal-wire screens placed at the end of the screw and supported by the BREAKER PLATE, w s, to prevent contaminants from obstructing or passing through the die. Screen packs have also been used to create additional resistance to flow, thus raising the head pressure and increasing the level of viscous working and mixing. Today the second function is more conveniently accomplished with an adjustable valve in the adapter between screw and die. (See VALVED EXTRUSION.) The pack usually contains several screens of different meshes, the finest one facing the incoming melt and supported by successively coarser ones with their thicker wires. Screens gradually become loaded with the foreign material they retain, and, with fewer openings remaining clear, the head pressure gradually rises, requiring increases in screw speed to maintain the rate. Eventually clean screens must be substituted for dirty ones. A device that permits making this change with minimal disturbance of the extrusion operation is a SCREEN CHANGER, w s. Frequency of changing can range from once a week in plants processing mostly virgin resin and carefully handled regrind, to every half hour in plants recovering scrap resins.

Extruder Screw A solid or cored shaft with a continuous helical channel (sometimes two channels) cut into it, usually extending from the feed throat of the extruder barrel to the die end of the barrel. In most screws, the channel varies in its volume per turn of the helix, being larger at the end receiving the feed with its low bulk density, to much less—roughly one-third—in the pumping section at the delivery end. The reduction in volume is usually accomplished by reducing the channel depth but can be done by reducing the helical lead; sometimes a combination of both has been used. The reduction in volume serves several purposes: feeding, compressing the particles and forcing the interstitial air back out the feed hopper, melting the polymer, and developing pressure to overcome resistance at the die. Extruder screws are made of tough steel—SAE 4140 is common—are usually chrome plated, and have flight tips hardened by one of several techniques. (See NITRIDING and HARD-FACING ALLOY.) Many clever designs have been, and continue to be, developed and marketed for extruder screws. Some terms used in describing extruder screws are defined below.

Barrier Screw See SOLIDS-DRAINING SCREW.

Constant-Lead Screw (uniform-pitch screw) A screw with a flight of constant helix angle over its whole length.

Constant-Taper Screw A screw of constant lead and uniformly increasing root diameter over its length (rarely made today).

Cored Screw A screw with a hole bored along its axis for circulation of a heat-transfer medium or insertion of a heater. The core may extend only through the feed section or further, even to the screw tip.

Decreasing-Lead Screw A screw in which the helix angle decreases steadily over the length of the screw. Channel depth is usually constant.

Metering-Type Screw A screw whose final section is of constant lead and relatively shallow depth.

Multiple-Flighted Screw (multi-flight screw) A screw having more than one flight, thus having two or more parallel channels.

Single-Flighted Screw A screw having just one flight—the usual case, presumed if not otherwise stated.

Two-Stage Screw Essentially two metering-type screws in series, typically used for vented operation. The first stage consists of a feed section, compression zone, and metering zone. The second stage consists of a deep, constant-depth section (decompression zone), usually running only fractionally full to permit expansion of bubbles and release of volatiles, followed by a short, steep compression zone and a metering zone of about one-third greater capacity than that of the first stage. There may be a restriction at the end of the first stage.

Vented Screw A two-stage screw with a screw vent in the decompression zone, permitting volatiles to escape through the screw core.

Water-Cooled Screw A screw cored in its feed section to permit circulation of water there.

Extruder, Single-Screw An extruder with one tubular barrel within which a solid or cored screw rotates.

Extruder Size Traditionally, the nominal inside diameter of the extruder barrel, usually stated in inches or millimeters. However, the output power rating of the drive is more directly related to output capability than is the diameter. See also L/D RATIO.

Extruder, Tandem A pair of extruders used sequentially for the production of foamed-polystyrene board. The first extruder, usually fitted with a two-stage screw operating at high speed, melts the resin and intimately mixes it with the nucleating agent and the blowing agent injected at the second stage. It feeds directly into a larger, slowly turning extruder where the foamable melt is cooled to a lower temperature, higher viscosity, and higher melt strength, permitting the extruded foam to be controllably expanded to the desired density as it emerges from the die onto a long conveyor.

Extruder, Twin-Screw (double-screw extruder) An extruder with a barrel consisting of two side-by-side intersecting cylinders internally open to each other along their intersection. There are two basic types. If the two internal cylinders are tangentially joined, the two screws are also nearly tangential and are normally of opposite "hands" and counter-rotating, turning downward at their juncture (Welding Engineers). This design permits the use of long vented sections and that feature, together with the milling action of the screws, makes it possible to remove large percentages of volatiles from the feedstock. One of the two screws is extended to become the metering section that provides die pressure. In the second basic type, the two cylinders of the barrel intersect more deeply and the screws intermesh with each other. Rotation may be co- or counter-, depending on whether the screws have the same or opposite "hands". In these machines the intermeshing of the screws traps the plastic and moves it much more positively than in the tangential type, with virtually no back flow, and shear working is less severe. The screws of both types are assembled from segments, each segment designed to perform a particular function. An important application for the second type has been extrusion of rigid PVC compounds from powder, heating gently by conduction and fluxing near the end of the screws and into the die entry and extruding at die pressures up to 100 MPa! Other applications are compounding fluffy polyolefin powders, volatiles extraction, and compounding. Twin-screw machines have also proved useful in reactive extrusion, in which chemical reactions, including polymerization, are performed within the extruder.

Extruder, Vented An extruder provided with a vent opening, most often equipped with a vacuum pump to draw off water vapor or other volatiles. In nonintermeshing twin-screw machines, because of the mutual forwarding action developed by the screws, large vent openings are feasible. In single-screw machines, a two-stage screw is required and a circular vent hole through the top center of the barrel is located above the deep-flighted decompression section following Stage 1. Some vented extruders have valved first stages to help control flow from Stage 1 into the vented zone. In others, the first metering section is designed to deliver considerably less than the capacity of the second metering section to discharge against the total resistance of the screen pack (if any), adapter and die. A resistance-altering valve may also be used at the head end to adjust the balance between stages for the most uniform rate of discharge.

Extrusion Any process by which lengths of constant cross section are formed by forcing a material, e g, a molten plastic, through a die. Typical shapes extruded are hose, tubing, flat films and sheets, wire and cable coatings, parisons for blow molding, filaments and fibers, strands cut hot or cold to make pellets for further processing, webs for coating and laminating, and many of the above in multiple layers by COEXTRUSION, w s.

Extrusion, Autothermal (autogenous extrusion, "adiabatic" extrusion) An extrusion operation in which the entire increase in enthalpy of the plastic, from feed throat to die, or very nearly all of it, is generated by the frictional action of the screw. In such an operation, which most commercial single-screw extrusions approach closely, the functions of the barrel heaters are to preheat the machine at startup and, during steady operation, to prevent heat loss from the plastic through the barrel to the surroundings.

Extrusion Blow Molding The most common process of BLOW MOLDING, w s, in which the parison is formed by extrusion.

Extrusion Casting A term sometimes employed in the industry for the process of extruding unsupported film, especially a composite of two or more integral resin layers formed by COEXTRUSION, w s. Such extrusion-cast composite films possess desired properties on each of the respective sides, e g, heat-sealability on one side and stiffness on the other, or different levels of slip, or different colors.

Extrusion Coating Coating a substrate by extruding a layer or molten resin onto the substrate with sufficient pressure to bond the two together without the use of an adhesive. A common application of the process is the coating of foil, paper, or fabric with polyethylene, by extruding a web directly into the nip of a pair of rolls through which the substrate is passing.

Extrusion Coloring The method of adding colorants to a plastic compound by dry-blending the colorant with the solid granular resin, extruding the mixture into strands, and cutting these strands into pellets for use in subsequent processing operations.

Extrusion Die The orifice-containing element, mounted at the delivery end of an extruder, that shapes the extrudate. Elements of the die assembly are (1) the die block, (2) an adapter connecting the die to the extruder, (3) a manifold within the die that distributes the melt to the orifice, (4) in the case of dies for hollow sections, a mandrel inserted in the flow channel to form the interior surface of the extrudate, (5) a spider that holds the mandrel in position, and (6) the land section, i e, the orifice that gives the extrudate its emergent shape. Extrusion dies are classified in four ways according to the relation between the screw axis and the direction of flow of the extrudate: *straight (in-line), offset, angle,* and *crosshead.* In an in-line die, the die-discharge channel is coaxial with the screw. In an offset die, those directions are parallel but not coaxial. In an angle die the axis of the die-discharge channel is at an angle, typically 45°, to that of the screw. A crosshead die is an angle die in which the two axes are perpendicular. Sheet and film dies are also classified as to the type of feed. In a *center-fed die*, the melt enters at the lateral center of the manifold and divides into two equal streams that move in opposite directions to the ends of the die. In an *end-fed die*, the melt enters one end of the manifold and flows toward the other. Theory tells that it is easier to maintain thickness uniformity over the width of the extruded film or sheet if the die is center-fed, but end-fed dies are sometimes used for logistical reasons, especially in extrusion coating.

Extrusion Laminating A process in which a plastic layer is extruded between two layers of substrate(s). See EXTRUSION COATING.

Extrusion Plastometer (melt-indexer) A simple viscometer consisting of a heated vertical cylinder with two bores, a central one that contains a close-fitting piston and a recess for an orifice block, the other, nearby, for a thermoswitch. The orifice is 2.1 mm in diameter and 8 mm long. Plastic particles are loaded into the bore, allowed to heat for 6–8 min, then the weighted piston is released, and the extrudate is collected for a measured time interval. The melt-flow index (MFI) is stated as the rate of extrusion in grams per ten minutes. The instrument and its use are described in ASTM D 1238. Originally developed in 1953 for low-density polyethylene, the melt-indexer is now used with many other polymers with specific temperatures and piston weights. It is *essential* to state the condition (A through X) at which an MFI was measured. Most of the conditions result in shear rates far below those typical of commercial processing. (MFI = 1 g/10 min corresponds to about 2 s^{-1}.) Thus, while the measurement is useful for product identification and quality control, it is a poor indicator of processability. An estimate of a resin's pseudoplasticity may be obtained by running two or more tests with substantially different piston weights.

Extrusion Pressure (1) Broadly, pressure indicated anywhere within an extruder. (2) The pressure at the delivery, or head end of the screw in a screw extruder, immediately upstream of the screen pack if one is present. (3) In a ram extruder, the pressure at the face of the ram in contact with the plastic.

Exudation The undesirable appearance on the surface of an article of one or more of its constituents that have migrated or *exuded* to the surface. In vinyls, such constituents may be residual emulsifier from the resin, stabilizer, lubricant, or plasticizer. Secondary plasticizers in particular have a tendency to exude when used in excessive percentages. Exudation may appear on a product shortly after it has been made, but more often it is delayed for periods ranging from several weeks to years. Products that do not exude for long periods under ideal storage conditions can be caused to exude by exposure to pressure, heat, high humidity, light, or other environmental agent.

Eyring Model (Prandtl-Eyring model) A rheological model, proposed for plastic melts, that contains two constants that must be evaluated from experimental flow measurements on the material and at the temperature of interest. The model calls for Newtonian behavior at very low shear rates, with a gradual transition into ever stronger pseudoplasticity as the shear rate increases. It has the form

$$\tau_{xz} = A \sinh^{-1}\left(\frac{1}{B}\frac{dv_z}{dx}\right)$$

dv_z is the velocity in the z-direction and in the x-surface, dv_z/dx is the shear rate, and A and B are the two constants. The inverse hyperbolic sine function above is shorthand for

$$\ln\left[\frac{1}{B}\frac{dv_z}{dx} + \sqrt{\left\{\left(\frac{1}{B}\frac{dv_z}{dx}\right)^2 + 1\right\}}\right]$$

Thus it is clear that this model is difficult to apply even to simple geometries. Also, the great body of data on polymer melts shows that, once clearly above the Newtonian, low-shear region, they are well represented by the POWER LAW, w s, and do not become increasingly more pseudoplastic.

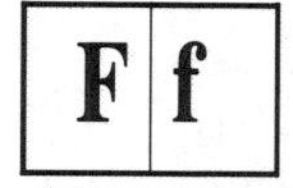

f (1) SI abbreviation for FEMTO-, w s. (2) Symbol for frequency of an oscillating system.

F (1) Chemical symbol for the element fluorine. (2) SI abbreviation for FARAD, w s.

Fabric (cloth) A flexible structure, usually thin relative to its width and length, made up of intermingled yarns, fibers, filaments, or wires. In woven fabrics, the elements are alternately crossed over and under one or more of those oriented in the other directions, typically two perpendicular directions. In nonwoven fabrics, such as felts, fibers are randomly oriented.

Fabricate In the broadest sense, this term means to manufacture, devise, or to make an assembly of parts and sections. In the plastics industry it refers to the assembly or modification of preformed plastics articles by processes such as welding, heat sealing, adhesive joining, machining, and fastening. "Fabrication" is *not* generally used to mean basic manufacturing processes such as extrusion, calendering, molding and the like.

Factor of Safety (safety factor) The ratio of the strength of a material or structure to the allowable stress in the material. The factor of safety is used to provide for uncertainties of design and expected service, such as aging and corrosion, differences between test samples and finished products, approximations in calculation of stresses, etc. It may be as low as 2 for ductile materials used in well understood loading situations to as much as 20 for brittle materials subject to occasional impact in addition to steady loads. The safety factor may be decided by the designer's engineering judgment or, for many standard structures, by established engineering or building codes.

Fading See LIGHT RESISTANCE.

Fadometer An apparatus for determining the resistance of resins and finished products to fading by subjecting the articles to high-intensity ultraviolet rays of approximately the same spectral distribution as those found in sunlight. See also LIGHT RESISTANCE.

Falling-Ball Viscometer (falling-sphere viscometer) An instrument well suited to determining polymer-melt viscosity at extremely low shear rates, i e, the limiting Newtonian viscosity. A sphere more dense than the melt is placed between two premolded slugs of the test polymer within a steel cylinder, which is then kept for a preset time in a temperature-controlled oven. From the initial and final positions of the sphere the viscosity can be calculated by Stokes' law (with corrections). By repeating the test with spheres of different densities, a range of low shear rates can be explored.

Falling-Dart Impact Test In addition to the ASTM tests mentioned at FREE-FALLING-DART TEST, w s, several similar tests exist for products such as pipe and bottles as well as sheeting. One procedure is the *staircase method*, also known as the UP-AND-DOWN METHOD, w s, which, for a given quantity of testing, provides a good estimate of the impact energy at which 50% of such samples may be expected to break. In the *probit method*, groups of samples are tested at preselected drop heights ranging from that at which most or all of the samples fail to that at which very few or none fail. This method also provides an estimate of the 50% point but, in addition, provides a better estimate of the standard deviation than does the staircase method. Combinations of the two methods have been used to optimize the amount of information per test specimen.

False Body The deceptively high apparent viscosity of a pseudoplastic fluid at low shear rates, which disappears at high shear rates. See also THIXOTROPY.

False Neck In blow molding of containers, a neck construction that is additional to the neck finish of the container and that is only intended to facilitate the blow-molding operation. Afterwards the false neck is removed from the container.

Family Mold (composite mold) A multicavity mold containing variously shaped cavities, each of which produces a component of an item that is assembled from the components. For example, a family mold for a model-airplane kit would contain a cavity for each part, and components of a complete kit would be produced in one shot.

Fan Gate A shallow gate becoming wider (and usually thinner) as it extends from the runner to the cavity.

Fantail Die (fishtail die) An extrusion die, usually one making a wide strip or sheet, in which the flow passage diverges from the adapter to the die lip.

Farad (F) The SI unit of electrical capacitance. A capacitor with a 1-V potential between its plates and holding a charge of 1 coulomb has a capacitance of 1 F. Thus 1 F ≡ 1 C/V.

Fatice Sometimes called "artificial rubber" or a "rubber substitute", fatice is made by vulcanizing with sulfur a vegetable oil such as soybean, rapeseed, or castor oil. It is used as a processing aid and extender in natural-rubber compounds and synthetic elastomers.

Fatigue Curve (S/N curve) A plot of the maximum cyclic stress applied to a fatigue specimen versus the number of cycles to failure, the abscissa being a logarithmic scale. Typically the S/N curve is linear or slightly concave upward, sloping gently downward, sometimes flattening at the low-stress (right) end, suggesting that there may be an ENDURANCE LIMIT, w s.

Fatigue Failure The cracking or rupture of a plastic article under repeated cyclic stress, at a stress well below the normal short-time breaking strength as measured in a static (0.5-cycle) test.

Fatigue Life The number of cycles of specific alternating stress required to bring about the failure of a test specimen.

Fatigue Limit Synonym for ENDURANCE LIMIT, w s.

Fatigue Notch Factor The ratio of the fatigue strength of a specimen with no site of stress concentration (notch) to that of a duplicate specimen having a notch.

Fatigue Ratio The ratio of fatigue strength at a given number of cycles of stated alternating tensile stress to the static tensile strength.

Fatigue Strength The maximum-stress level at which a material subjected to cyclic alternating stress fails after a given number of cycles. This is a number read off a FATIGUE CURVE, w s, that is derived by measuring the cycles to failure of numerous specimens subjected to various known maximum alternating stresses.

Fatty Acid Any monobasic organic acid obtained by the hydrolysis (saponification) of a natural fat or oil. Linoleic, linolenic, oleic, palmitic, and stearic acids are used in the synthesis of many plasticizers and stabilizers for plastics.

Fatty Polyamide (Versamid®, oldest of many trade names) A polymer formed by the condensation of a dibasic acid having a bulky side group and from 13- to 21-carbon chains, or the dimer acids, C–36, with di- or polyamines. The commercially important dimer acids are addition products of unsaturated C–18 fatty acids and can take several forms, giving different structures to the polyamides. They are used in hot-melt adhesives, inks, as epoxy flexibilizers and, in amine form, as curing agents for epoxies.

Fay To smooth and fit together, as with two surfaces about to be lap-joined.

FDA Abbreviation for Food and Drug Administration, the US agency within the Department of Health, Education and Welfare that is concerned with the safety of products marketed for consumer use, particularly those substances that might be ingested, applied to the skin, or used in therapy or prostheses.

Fe Chemical symbol for the element iron (Latin: ferrum).

Feedback Control A system of controlling a machine or process in which the difference between a measured output variable and its target value is amplified and, through automation, causes an appropriate adjustment of an input machine variable or process condition that will move the output nearer to its target.

Feed Block In coextrusion, a massive metal block in which the streams of the several extruders are brought together to form the layers of a single stream just before it enters the die.

Feed Bushing Synonym for SPRUE BUSHING, w s.

Feedforward Control Process control in which early process variables are monitored and their disturbances are fed to a process model that computes adjustments of the variables needed to provide the desired process outputs. Compare FEEDBACK CONTROL.

Feed Hopper An inverted conical or pyramidal vessel mounted over the feed port of an extruder or injection-molding machine that contains a supply of pellets or powder being fed. Feed hoppers typically have a slender window from bottom to top along one side to permit observation of the feedstock level.

Feed Plate In injection molds, the plate contacting the injection nozzle and containing the sprue and, usually, most of the runner system. Used with a floating cavity plate, the system provides for separation of runners and sprue from the moldings and stripping of both into separate collectors or chutes.

Feed Port An opening at the rear end of the barrel of an extruder or injection molder through which plastic powder or pellets fall into the rotating screw or, in older injection machines, in front of the withdrawn ram (rare now).

Feedscrew See EXTRUDER SCREW.

Feed Zone The first (rear) zone of an extruder screw that is fed from the hopper, usually of constant lead and greater depth than other zones, and offically terminating at the beginning of the compression zone.

Feldspar Any of several anhydrous minerals containing aluminum silicates or alkali or alkaline-earth metals (Na, K, Ca, Ba) which, when ground, make low-cost, modulus-raising, nontoxic fillers for plastics.

Felt (felting) A nonwoven, fibrous material made up of randomly oriented fibers held together by stitching, a chemical binder, or by action of heat or moisture.

Femto- (f) The SI prefix meaning $\times 10^{-15}$.

FEP Abbreviation for FLUORINATED ETHYLENE-PROPYLENE RESIN, w s.

Ferrite (hard ferrite) A compound having the general formula $MFe_{12}O_{19}$, in which M is usually a divalent ion such as barium or strontium. These materials are strongly magnetic and can be incorporated into plastics to make bonded permanent magnets, rigid or flexible, and in many forms, including strips.

Ferrocene (dicyclopentadienyl iron) $(C_5H_5)_2Fe$. A coordination compound of ferrous iron and cyclopentadiene, soluble in PVC and stable to 400°C. Its uses include smoke suppression in PVC, curing agent for silicone resins, intermediate for high-temperature polymers, and ultraviolet absorber.

Festooning Oven An oven used to dry, cure, or fuse plastic-coated fabrics with uniform heating. The substrate is carried on a series of slowly rotating shafts with long loops or "festoons" between the shafts.

FF Abbreviation for furan-formaldehyde polymer. See FURAN RESIN.

Fiber (fibre) A single homogeneous strand of material having a length of at least 5 mm, that can be spun into a yarn or roving or made into a fabric by interlacing in a variety of methods. Fibers can be made by chopping filaments (converting). *Staple fibers* may be 1.2 to 8 cm in length with lineal density from 0.1 to 0.5 mg/m. The natural fibers used by mankind from the earliest times were first supplemented by rayon and acetate, both of which are derived from cellulose. The first commercially successful, wholly synthetic fiber was nylon, introduced in 1939. Then followed acrylic fibers in 1950, polyesters in 1951, and various other polymeric fibers in subsequent years. In 1967 the wholly synthetic, "man-made" fibers surpassed the natural fibers in volume produced. See also SYNTHETIC FIBER.

Fiber Content The percent by volume of fiber in a fiber-filled molding compound, or a molding or a laminate. Fiber content is sometimes stated as weight percent.

Fiber Direction (fiber orientation) In a laminate, the direction(s) in which most of the fibers' lengths lie, relative to the length axis of the part.

Fiberfill Molding (Fiberfil™) A term used for an injection-molding process employing as a molding material pellets containing short bundles of fiber surrounded by resin.

Fiberglass (Fiberglas™) See GLASS-FIBER REINFORCEMENT.

Fiber Optics A term employed for light-transmitting fibers of glass and some plastics, such as polymethyl methacrylate. Each fiber is coated with a material with a refractive index lower than that of the fiber itself, and many fibers may be gathered in a bundle that is jacketed with polyethylene or other flexible plastic. Such bundles transmit light from one end to the other even though curved. Applications are in aircraft and automobile instrument panels, telephone lines, electronics, displays, medical techniques, and packaging.

Fiber-Reinforced Plastic (FRP) Any plastic material, part or structure that contains reinforcing fibers, such as glass, carbon, synthetic, or metal fibers generally having strength and stiffness much greater than that of the matrix resin, thereby improving those properties. Because glass fibers were used so early and widely, FRP is often used to mean *glass-fiber-reinforced plastics*. See also ADVANCED COMPOSITES, COMPOSITE LAMINATES, and REINFORCED PLASTIC.

Fiber-Resin Interface (fiber-matrix interface) The surfaces shared by the fibers and the resin in a fiber-reinforced plastic structure. This interface, and the effects of various *sizes* and chemical treatments on the interfacial bond, are subjects of many past and ongoing studies. Because of the pretreatment of fibers with sizes, the interface has a small but finite thickness.

Fiber-Resin Ratio An expression, as a ratio of fiber to resin, of the FIBER CONTENT, w s.

Fiber Show (fiber prominence) In reinforced plastics, a condition in which ends of reinforcement strands, rovings, or bundles unwetted by resin appear on or above the surface. It is believed to be caused by a deficiency in the glass, and may not appear until the part is fully cured. Remedies include measures to improve wet-out, use of resins of optimum viscosity, and reducing exotherm rates, which cause stresses within the laminate, and gel coating after the main body of the part has partly cured.

Fiber Spinning See SPINNING.

Fiber Streak (fiber whitening) A group of fibers within a translucent laminate that were incompletely wetted by resin, appearing as a whitish defect.

Fiber Stress The stress acting on the reinforcing fibers in a laminate under load (estimated).

Fibre Alternate spelling of FIBER, w s.

Fibrid A generic term for fibers made of synthetic polymers.

Fibril A single crystal having the form of a fiber.

Fibrillation The phenomenon in which a filament or fiber shows evidence of smaller-scale fibrous structure by a longitudinal raveling of the filament under rapid, excessive tensile or shearing stress. Separate fibrils can then often be seen in the main filament trunk. The whitening of polyethylene when severely strained at room temperature is a manifestation of fibrillation.

Fibrous-Glass Reinforcement See GLASS-FIBER REINFORCEMENT.

Fick's Law (Fick's first law) The basic law of diffusion of different molecular species into each other. Fick's law states that the flux of a given component will be in the direction in which the concentration of that component decreases most steeply (i e, opposite the gradient), at a rate given by the product of the mutual diffusivity and the gradient. Most published diffusivity data are for two-component systems. Fick's law, which has many equivalent forms, is the defining equation for diffusivity.

Filament A variety of fiber characterized by extreme length, which permits its use in yarn with little or no twist and usually without the spinning operation required for fibers. See also MONOFILAMENT.

Filamentary Composite A reinforced-plastic structure in which the reinforcement consists of FILAMENTs (w s), usually oriented to most efficiently withstand the stresses imposed on the structure. The filaments are not woven and in a single lamina they will all be parallel.

Filament Count The number of filaments in a yarn.

Filament Winding A method of forming reinforced-plastic articles comprising winding continuous strands of resin-coated reinforcing material onto a mandrel. Reinforcements commonly used are single strands or rovings of glass, asbestos (rare today because of carcinogenicity fright), jute, sisal, cotton, and synthetic fibers. Polyester resins are most widely used, followed by epoxies, acrylics, nylon, and various others. To be effective, the reinforcing material must form a strong adhesive bond with the resin. The mandrels may be permanent structures remaining in the finished article, or of flexible or destructible material, or able to be disassembled, i e, capable of being removed after curing. The process is performed by drawing the reinforcement from a spool or creel through a bath of resin, then winding it on the mandrel under controlled tension and in a predetermined pattern. The mandrel may be stationary, in which event the creel structure rotates about the mandrel; or it may be rotated on a lathe about one or more axes. By varying the relative amounts of resin and reinforcement, and the pattern of winding, the strength of filament-wound structures may be controlled to resist stresses in specific directions. After sufficient layers have been wound, the structure is cured at room temperature or with heat.

Filament-Wound Made by FILAMENT WINDING, w s.

Filing Manual filing is sometimes used to bevel, smooth, deburr, and fit the edges of plastic moldings and sheets. The process is limited to parts that cannot be tumbled easily, and to plastics with suitable hardness and heat resistance.

Fill (n, adj) Synonym for WEFT, w s.

Fill-and-Wipe A decorating process for articles molded with depressed designs, wherein the general area containing the designs is coated with paint by brushing, spraying, or rolling, then surplus paint is wiped from the undepressed areas surrounding the depressions.

Filled Plastic Any plastic compound containing a significant percentage of a solid, usually not fibrous or resinous, material whose main purpose may be to dilute the resin, or to provide certain enhanced properties in the compound.

Filler A relatively inert substance added to a plastic compound to reduce its cost per unit volume and/or to improve such mechanical properties as hardness, modulus, and impact strength. A filler differs from a REINFORCEMENT (w s) in two respects. Filler particles are generally small and roughly equidimensional, and they do not markedly improve the tensile strength of a product. Reinforcements, on the other hand, are fibrous, having one dimension much longer than the others, and they do markedly improve tensile strength. The most commonly used, general-purpose fillers are clays, silicates, talcs, carbonates, and paper. (Asbestos fines, once popular, are little used today because of their connection with lung cancer among workers in asbestos plants.) Some fillers also act as pigments, e g, chalk, titanium dioxide, and carbon black. Graphite, molybdenum disulfide, and polytetrafluoroethylene powder impart lubricity, while antimony oxide, alumina trihydrate, and halogen-containing compounds provide flame retardancy. Magnetic properties can be obtained by incorporating magnetic mineral fillers such as hard ferrite. Metallic fillers such as lead or its oxides are used to increase density or screen nuclear radiation. Carbon black and powdered aluminum improve thermal and electrical conductivity, as do other powdered metals. More detailed descriptions of specific fillers and their applications are given at the following entries.

ACETYLENE BLACK
ALPHA CELLULOSE
ALUMINA TRIHYDRATE
ALUMINUM SILICATE
ASBESTOS
BARIUM SULFATE
BENTONITE
BLANC FIXE
CALCIUM CARBONATE
CALCIUM SILICATE
CALCIUM SULFATE
CARBON BLACK
CELLULOSE
CERAPLAST
CLAY
CORK
DIATOMITE
FELDSPAR
FERRITE
FLAME RETARDANT
FUMED SILICA
GLASS FLAKES
GLASS SPHERES
KAOLIN
KERATIN
LIGNIN
LITHOPONE
MICA
MICROBALLOONS
MICROSPHERES
MOLYBDENUM DISULFIDE
NEPHELINE
NOVACULITE
NUTSHELL FLOUR
QUARTZ
SILICA
SOYBEAN MEAL
TALC
TERRA ALBA
VERMICULITE
WOOD FLOUR
ZINC OXIDE

Filler Rod (welding rod) A rod of plastic material used in HOT-GAS WELDING, w s, made of the same material as the plastic to be welded.

Filler Specks Visible particles of a filler, such as wood flour or asbestos, that stand out in color contrast against a background of plastic binder.

Fillet A concavely curved transition at the angle formed by the junction of two plane surfaces, i e, a rounded inside corner. Also, the material making up the transition. Where the surfaces are likely to endure bending toward or away from each other, the fillet distributes and reduces the stress that would otherwise be magnified at the corner.

Filling Yarn See WEFT.

Film (1) Customarily in the plastics industry, a web of plastic that is 0.25 mm or less in thickness. Thicker webs are called *sheet*. Films are made by extrusion, casting from solution, and calendering. (2) In convective heat transfer, the thin, supposedly stagnant layer of fluid next to a heated or cooled surface (such as a pipe wall) that contributes part (or all) of the resistance to transfer of heat from the main body of the fluid to a medium on the opposite side of the wall (or to the wall itself). A closely related concept exists in mass transfer.

Film Blowing (blown-film extrusion) The process of forming thermoplastic film wherein an extruded plastic tube is continuously inflated by internal air pressure, cooled, collapsed by rolls, and subsequently wound into rolls on thick cardboard cores. The tube is usually extruded vertically upward, and air is admitted through a passage in the center of the die as the molten tube emerges from the die. An AIR RING, w s, is always employed to speed and control the initial cooling close to the die. Air is contained within the blown bubble by a pair of pinch rolls that also serve to collapse and flatten the film. Thickness of the film is controlled not only by the die-lip opening but also by varying the internal air pressure and by the rates of extrusion and take-off. Extremely thin films (< 0.01 mm) and films with considerable biaxial orientation can be produced by this method.

Film Casting The process of making an unsupported film or sheet by casting a fluid resin, a resin solution, or a plastic compound on a temporary carrier, usually an endless belt or circular drum, followed by solidification by cooling or drying, and removal of the film from the carrier. The term *film casting* is also used for the process of extruding a molten polymer through a slot die onto a chilled roll.

Film Coefficient (1) In convective heat transfer, the rate of heat flow through a "stagnant" fluid film adjacent to a solid surface, per unit area of the film, divided by the temperature difference through the film. (2) A similarly structured definition applying to mass transfer through films at fluid interfaces.

Film Die A die for the extrusion of flat or blown film. Flat-film dies are usually of the CROSSHEAD and COATHANGER designs, w s, with one lip locally adjustable so as to achieve uniform thickness across the film (see FLEXIBLE-LIP DIE). Blown-film dies are cylindrical, end- or side-fed, with the concentricity/eccentricity of the core and body adjustable for circumferential uniformity of film thickness. That uniformity, which is critical in both types of films for winding even rolls, also depends on the performance of the air ring. Blown-film dies are often oscillated slowly about their axes to distribute the remaining nonuniformity evenly over the final roll width, in that way assuring that the unrolled film will lie flat in spite of slight thickness variations.

Film Extrusion Making plastic films by extruding molten plastic through a FILM DIE, w s, by FILM BLOWING or FILM CASTING, w s.

Film Slitting See SLITTING.

Fin See FLASH.

Fines In the classification of powdered or granular materials according to particle size, fines are the portion of the material whose particles are smaller than a stated minimum size. When the particle-size distribution is determined by SIEVE ANALYSIS, w s, the fines are those particles passing the finest sieve and found on the pan, usually designated as "minus 000 mesh", where 000 is the mesh number of that finest sieve.

Finish (1) (size) In reinforced plastics, a compound containing a coupling agent and (optionally) a lubricant and/or binder, used to pretreat glass fibers prior to using them as reinforcements. (2) In the container industry, the plastic forming the opening of a container and shaped to mate with a particular closure. (3) The surface texture of a molding, a machined or polished surface, or other article. When measured, it is usually stated as the root-mean-square roughness in nanometers or microinches.

Finishing (1) The removal of flash, gates, and defects from plastic articles. See DEFLASHING, DEGATING and TUMBLING. (2) The development of a desired texture and/or color on the surfaces of an article when such are not accomplished in compounding and forming the article. See GRIND, POLISHING, SANDING, and the processes listed under DECORATING.

Finish Insert (neck insert) In blow molding bottles, a removable part of the mold that aids in forming a specific neck finish of the bottle.

Finite Element In the mathematical analysis of physical processes, a subdivision of the region of interest sufficiently small so that numerical approximations may safely be substituted for derivatives in the solution of differential equations. Using a computer program, numerical integration is then performed repeatedly over many finite elements to obtain a detailed solution for the entire region. In this way, for example, two-dimensional flows in extrusion dies, stresses in complex structures under load, and transient processes such as thermoforming have been elucidated. The procedures are collectively known as the *finite-element method.*

Fire Resistance Of plastics used in building, the ability to fulfill for a stated time the required stability, integrity, thermal insulation and/or other expected duty, specified in a standard fire-resistance test. See FLAMMABILITY.

Fisheye A visible fault in transparent or translucent plastics, particularly films or thin sheets, appearing as a small globular mass (*gel particle*) and thought to be caused either by stray resin particles of much higher molecular weight than that of the polymer as a whole, or by inclusion of foreign particles.

Fishtail Die Synonym for FANTAIL DIE, w s.

Fissure A term used in the cellular-plastics industry to denote a separation, crack, or split in a formed cellular article.

Five Regions of Viscoelasticity As an amorphous polymer is heated from an extremely low temperature it gains more dimensions of molecular motion and its mechanical behavior changes through five qualitative regions: glassy, slowed elastic, rubbery, rubbery flow, and viscous flow.

Flake (1) A term used to signify the dry, unplasticized, basic form of cellulosic plastics. (2) GLASS FLAKES, w s.

Flame Hardening A cheap method, obsolete today, of initially hardening flight tips of extruder screws in which the surface is rapidly chilled (tempered) after being heated with a flame (a process not easy to control). The hardness imparted is gradually lost over a few hundred hours of normal operation, requiring that the process be repeated frequently, defeating the initial saving. See CASE HARDENING and HARD-FACING ALLOY.

Flame Polishing A method of finishing a plastic article, particularly a just formed extrudate, in which a carefully controlled flame or stream of hot gas is directed at the surface, melting a thin skin of resin that, when quenched, has a high gloss.

Flame-Retardant (adj) This term has long been used to indicate the ability of a plastic to resist combustion as shown by a laboratory test, or one that will not continue to burn or glow after a source of ignition has been removed. The term does *not* mean that a flame-retardant plastic will not burn in a conflagration. A few neat resins, such as PVC and the fluoro- and chlorofluorocarbons, are flame-retardant. See FLAMMABILITY.

Flame Retardant (n) A material that reduces the tendency of plastics to burn. Flame retardants are usually incorporated as additives during compounding, but sometimes applied to surfaces of finished articles. Some plasticizers, particularly the phosphate esters and chlorinated paraffins, also serve as flame retardants. *Inorganic flame retardants* include antimony trioxide, hydrated alumina, monoammonium phosphate, dicyandiamide, zinc borate, boric acid, and ammonium sulfamate. Another group, called *reactive-type flame retardants*, includes bromine-containing polyols, chlorendic acid and anhydride, tetrabromo- and tetrachlorophthalic anhydride, tetrabromo bisphenol A, diallyl chlorendate, and unsaturated phosphonated chlorophenols.

Flame-Spray Coating A coating process utilizing powdered metals or plastics, in which the powdered materials are heated to the sintering temperature in a cone of flame enroute from a spraygun orifice to the article being coated.

Flame Treating A method of rendering inert thermoplastics, particularly polyolefins, receptive to inks, lacquers, paints, and adhesives by briefly bathing the surface of the article in a highly oxidizing flame. This treatment oxidizes the surface slightly, creating carbonyl and possibly peroxide groups, thereby increasing its surface energy.

Flammability With respect to plastics, flammability is a very broad term that has been the focus of a potpourri of tests and standards generated by many organizations, predominantly in the US by Underwriters Laboratory (UL) and ASTM. The behavior of various plastics when burning, and tests designed to evaluate flammability, encompass six categories: ignitability, burning rate, heat evolution, smoke production, products of combustion, and endurance of burning. For the convenience of our readers, the most commonly used tests for these aspects of flammability and fire resistance are collected here as subtopics rather than being listed separately in strict alphabetical order throughout the dictionary.

Arc-Ignition Test (UL high current) A flammability test for plastics used in electrical applications certified by UL. High-current electrodes resting on the sample are gradually moved together until they arc, then moved apart until the arc breaks. The number of breaks required to ignite the sample is reported.

Arc-Ignition Test (UL high voltage) A flammability test for plastics used in electrical applications certified by UL. Two electrodes, 4.0 mm apart, rest on the surface of the sample. Voltage is raised until an arc forms. The time required for the plastic to ignite is reported.

ASTM Test D 568 for Flammability of Flexible Plastics A small-scale screening procedure for comparing the relative flammability of plastics in the form of flexible, thin sheets or films. The specimen, of standard length and width, is suspended vertically and exposed to a gas flame at its lower end. The time and extent of burning are measured and reported if the specimen does *not* burn 38 cm. An average burning rate is reported if the specimen burns to the 38-cm mark.

ASTM Test D 635 for Flammability of Self-Supporting Plastics (also recognized by FTM 2021) This method covers a laboratory screening procedure for comparing the relative flammability of self-supporting plastics in the form of bars, molded or cut from sheets, plates, or panels. A bar of the material is supported horizontally at one end. The free end is exposed to a specified gas flame for 30 seconds. Time and extent of burning are reported if the specimen does *not* burn 100 mm. An average burning rate is reported for a material if it burns to the 100-mm mark from the ignited end. This method is used to establish relative burning characteristics and should not be used as a fire-hazard test.

ASTM TEST D 757 for Incandescence Resistance of Rigid Plastics in a Horizontal Position (discontinued by ASTM in 1987 but possibly still being used).

ASTM Test D 1433 for Flammability of Flexible Thin Plastic Sheeting Supported on a 45° Incline (discontinued by ASTM in 1988 but possibly still being used).

ASTM Test D 1929 for Ignition Properties of Plastics Sometimes called the Setchkin Technique, this test determines the self-ignition and flash-ignition properties of plastics using a hot-air furnace. The sample is placed in a vertical refractory tube, which is inside a furnace tube that is vertically heated by electrical current passing through Nichrome wire in an asbestos sleeve wound around the tube. Air is admitted at a controlled rate, and air temperature at ignition is measured by thermocouples.

ASTM Test D 2584 for Ignition Loss of Cured Reinforced Resins A weighed specimen of about 5 g is placed in a crucible, then into an electric muffle furnace at 565±28°C and kept there until all carbonaceous material has burned away. It is then removed, cooled in a desiccator, and reweighed. The percent loss in weight is reported.

ASTM Test D 2843 for Density of Smoke from the Burning or Decomposition of Plastics Used for testing plastics for compliance with building codes. The sample is placed in a chamber and is ignited by a propane flame. Smoke density is measured by a photocell across a horizontal 30-cm light path. Most building codes permit using a material if the maximum light absorption is less than 50%. Uniform Building Code accepts up to 75%.

ASTM Test D 2863 for Flammability Using the Oxygen-Index Method A procedure for determining the minimum concentration of oxygen in a flowing mixture of oxygen and nitrogen that will just support flaming combustion of a plastic specimen. The mixture, with an initial low concentration estimated from experience with similar materials, is caused to flow upward in a test column. The vertically positioned specimen, supported by a frame if necessary, is exposed to an ignition flame at its top. If ignition does not occur at the starting concentration, the O_2 percentage is increased until ignition occurs. The volume percent O_2 is reported as the oxygen index for the material. This test is regarded by some as useful only for materials that ignite with difficulty in air.

ASTM Test D 3014 for Flammability of Rigid Cellular Plastics This test is a screening procedure for comparing relative flammability of rigid foams and should not be used as a fire-hazard classification. The specimen is mounted in a vertical chimney with a glass front and is ignited with a bunsen burner for 10 seconds. The height and duration of flame and the weight percent retained by the specimen are recorded.

ASTM Test D 3801 for Measuring the Comparative Extinguishing Characteristics of Solid Plastics in a Vertical Position In this test specimens 13 by 127 mm of any thickness up to 12.7 mm are supported vertically at their upper ends and exposed at their lower ends to a bunsen-burner flame for two 10-s intervals. Recorded are flaming times before extinguishment and time of glowing extinguishment after the second application of the flame.

ASTM Test D 3894 for Evaluation of Fire Response of Rigid Cellular Plastics Using a Small Corner Configuration The cellular-plastic specimen sheets are assembled into a simulated room corner or ceiling corner, then ignited with a gas burner. The flame spread, temperature rise, and damage are recorded.

ASTM TEST D 4804 for Determining the Flammability Characteristics of Nonrigid Solid Plastics This method contains two procedures, one in which the specimen is positioned horizontally, the other, vertically. Sheet specimens 5 × 20 cm are formed into tubes by being wrapped around a mandrel parallel to the long side and taped, then the mandrel is removed. In Method A, the hanging specimen is twice subjected to a burner flame for 3 s and flaming times and time to extinguishment of glowing are recorded. In Method B, the specimen is horizontally supported on a fixture of wire gauze and ignited at one end. The time it takes for the flame to progress from one mark to another 7.5 cm distant is recorded.

ASTM Test D 5048 for Measuring the Comparative Burning Characteristics and Resistance to Burn-Through of Solid Plastics Using a 125-mm Flame Bar or plaque specimens are subjected to a 12.5-cm flame for five 5-s intervals 5 s apart. The time of flaming plus glowing after the fifth application is recorded, also whether or not there is burn-through and dripping.

ASTM Test E 84 for Surface Burning Characteristics of Building Materials (in Sec 04.07) This is probably the most widely used test for spread of surface flame of plastics, as well as other building materials. The specimen is a 7.6-m-long, 50-cm-wide slab mounted face down so as to form the roof of a 7.6-m-long tunnel 44 cm wide. Two gas burners are located 30 cm from the fire end and 19 cm below the slab's underside. The flame is adjusted according to a formula so that a similarly dimensioned and mounted sample of red-oak flooring would spread flame 50 cm from the end of the igniting fire in 5.5 minutes. A thermocouple located 30 cm from the vent end of the tunnel is used to measure heat release, which is recorded each 30 s or oftener. The end point of the test is taken as the time for the vent-end thermocouple to reach 527°C, which in the case of red oak is 5.5 minutes. The area under the time-temperature curve for the 10-minute test is used as a measure of heat release or fuel contribution. A light source mounted on a horizontal section of the vent pipe with the light beam directed downward to a photocell is used to measure smoke density. This test is recognized in UL 723 and NFPA 225, and is also called the Steiner Tunnel Test.

ASTM Test E 119 for Fire Tests of Building Construction and Materials (in Sec 04.07; fire-endurance test, UL 263, MFPA 251) Used for testing walls, floors, ceilings, roofs, etc, as required by various building codes. A full-size wall section is used as a partition in a room-sized furnace. In ASTM E 119 one side of the partition is exposed to gas fire with temperatures reaching 538°C in 5 min, 704° in 10 min, 927° after 1 h, and 1110° after 4 h. To pass, temperature on the opposite side of the specimen should not exceed 121°C. Similar tests specify somewhat different temperatures and time limits.

ASTM Test E 162 for Flammability of Materials Using a Radiant Heat-Energy Source (in Sec 04.07; radiant-panel test) This test, developed at the former National Bureau of Standards, employs a radiant heat source consisting of a 30- × 46-cm vertically mounted porous refractory panel maintained at 670±4°C. A specimen measuring 15 × 46 cm is supported in front of the panel with the 46-cm dimension inclined 30° from the vertical. A pilot burner ignites the top of the specimen, so that the flame front propagates downward along the underside exposed to the radiant panel. The temperature rise recorded by stack thermocouples, above their base level of 180 to 230°C, is used as a measure of heat release. A smoke-sampling device that collects smoke particles on glass-fiber filter paper is used to measure smoke density.

Bureau of Mines Flame-Penetration Test Used for plastic foams in mines. The time required to burn through a 2.5-cm-thick layer of foam exposed to a continuous flame from a propane torch, the temperature of which is 1177°C, is reported.

FAA Horizontal-Flammability Test Required by the Federal Aviation Administration for components of aircraft. Conditions are the same as in their vertical test (which follows) except that the samples are horizontal during the test. Maximum acceptable burning rate is 6.4 cm/min for acrylic windows, instrument assemblies, and seat belts. Small molded parts are acceptable with burning rates less than 10 cm per minute.

FAA Vertical Flammability Test Required by Federal Aviation Administration for materials including surface finishes and decorative components of aircraft crew and passenger compartments. A bunsen-burner flame is applied to a vertical specimen for 60 seconds. Flame time, burn length, and burn time of drips are noted. To pass, burn must be less than 15 cm, flame time must not exceed 15 s, and drippings must stop burning within 3 seconds.

Factory Mutual Calorimeter Heat-Contribution Test Used by insurance underwriters to test building components. The sample is burned in a gasoline-fired furnace and a time-temperature curve is recorded. A noncombustible sample is substituted and the time-temperature curve is duplicated by burning propane. The heat of combustion of the propane used is taken as the heat-contribution of the sample.

Flame Propagation Test, UL Subject 94 A test for self-extinguishing polymers in sheet form for applications certified by UL. A vertically oriented sample is exposed to a bunsen-burner flame for 10 s. If burning ceases in less than the ensuing 30 s, the flame is applied for another 10 s. Flaming droplets are allowed to fall on cotton. If the average burning time is less than 5 s and drips don't ignite the cotton, the material is rated Self-Extinguishing, Group 0. If the time is less than 25 s and drips don't ignite the cotton, the material is rated Self-Extinguishing, Group I. If the cotton is ignited, the rating is Self-Extinguishing, Group II.

Horizontal Burn Test MVSS302 Used by the Department of Transportation for all materials used in automotive interiors. The horizontal specimen is ignited by a 15-s application of a bunsen-burner flame. When the flame has burned 3.8 cm of the sample, time is measured until the material either ceases to burn or burning has progressed 25 cm, and the rate must not exceed 10 cm per minute.

Hot-Wire-Ignition Test A flammability test used for plastics in electrical applications certified by UL. A sample bar is wrapped with resistance wire which is heated electrically to red heat. The time for the sample to ignite is measured.

Methenamine-Pill Test (ASTM D 2859, sec 07.01) A flammability test for carpets and floor coverings, for compliance with Department of Commerce standards for carpets larger than 2.2 m^2. A methenamine pill is ignited on the sample carpet and allowed to burn out. The burned area is measured.

NBS Smoke-Chamber Test The sample is placed in a completely closed cabinet, supported vertically in a frame and exposed to an electrical heat source. Smoke production is measured by a vertical photometer within the chamber, results being expressed in terms of specific optical density.

Flash (fin) The thin, surplus web of material that is forced into the parting line between mating mold surfaces during a molding operation and which remains attached to the molded article. For methods of removing flash, see DEFLASHING.

Flash Gate A long, shallow rectangular gate in an injection mold, extending from a runner that lies parallel to an edge of a molded part along the flash or parting line of the mold.

Flash Groove (spew groove) A groove in a mold force that allows the escape of excess material during a compression-molding operation.

Flash Line See PARTING LINE.

Flash Mold A mold in which the mating surfaces are perpendicular to the clamping action of the press so that, as the clamping force increases, the distance between the mating surfaces decreases, thus permitting excess molding material to escape as flash as the mold closes.

Flashover (1) A flammability term. Flashover occurs when hot, combustible gases are generated in burning sections of a building, become mixed with sufficient oxygen upon spreading to non-burning areas, and ignite to cause total surface involvement, but without a progressive flame-spreading stage. (2) In the electrical industry, an electric discharge around the edge or over the surface of insulation (ASTM D 1711, sec 10.01).

Flash Point The lowest temperature at which a combustible liquid will give off a flammable vapor that will momentarily burn when exposed to a small flame. ASTM has developed many tests of flash point for different liquids, of which perhaps the most widely used is D 93 (sec 04.09, 05.01, 06.03), employing the Pensky-Martens closed tester. D 1929 for plastics measures both the minimum flash-ignition temperature (flash point) and the minimum self-ignition temperature, using a muffle furnace. See ASTM TEST D 1929 under FLAMMABILITY.

Flash Ridge That part of a flash mold along which the excess material escapes until the mold is fully closed.

Flat-Entry Die An extrusion die in which the approach to the land has no taper, i e, one having a 180° ENTRANCE ANGLE, w s.

Flat Film Film made by extrusion from a flat die onto a polishing roll. Not to be confused with LAY-FLAT FILM, w s.

Flexibilizer A term rarely used for an additive that makes a plastic more flexible. See PLASTICIZER.

Flexible Foam See POLYETHYLENE FOAM, POLYURETHANE FOAM, and VINYL FOAM.

Flexible-Lip Die In film and sheet extrusion, a die in which a deep groove, reaching almost to the inside surface of the upper die body just behind the lip, has been machined. Adjusting bolts that can either push or pull on the lip pass through the gap from the upper body. The relative flexibility of the thin steel web from which the upper lip extends facilitates die adjustment to minimize sheet-thickness variations across the sheet. See also AUTO-FLEX DIE.

Flexible Mold A mold made of rubber, elastomer, or flexible thermoplastic, used for casting thermosetting plastics or other materials such as concrete and plaster. The mold can be stretched to permit removal of the cured casting, even one with undercuts.

Flex Life Informally, the number of bending-reversal cycles causing a part to fail in a particular service. More specifically, the number of cycles to failure of a test specimen repeatedly bent in a prescribed manner. The ASTM test for plastics is D 671. The specimen, molded or cut from sheet, is subjected to load reversal at 30 Hz at a predetermined level of outer-fiber stress until it either fails or the test is discontinued. By setting up different stresses for successive specimens, one can develop a graph of stress at failure vs number of cycles to failure (usually plotted on semilogarithmic coordinates), i e, the flex-life curve or FATIGUE CURVE, w s.

Flexographic Printing A rotary process employing flexible rubber or elastomeric printing plates adhered to a roll, inked by a screen roll which in turn is coated from a feed roll immersed in ink.

Flexural Modulus (flex modulus) The ratio, within the elastic limit, of the applied stress in the outermost fibers of a test specimen in three-point, static flexure, to the calculated strain in those outermost fibers, according to ASTM Test D 790 or D 790M. For a given material and similar specimen dimensions and manufacture, the modulus values obtained will usually be a little higher than those found in a tensile test such as D 638, and may differ, too, from the moduli found in the cantilever-beam test, D 747.

Flexural Strength (flexural modulus of rupture) The maximum calculated stress in the outermost fibers of a test bar subjected to three-point loading at the moment of cracking or breaking. ASTM Test D 790 and D 790M are widely used for measuring this property. For most plastics, flexural strength is usually substantially higher than the straight tensile strength.

Flight In an extruder screw, the helical ridge of metal remaining after machining the screw channel.

Flight Clearance In screw extruders, half the difference between the inside diameter of the barrel and the diameter of the flight surface, usually about 0.1% of the nominal diameter in a new machine. Clearances in older machines may vary along the screw, intentionally or because of differential wear.

Flight Depth (screw depth, channel depth) In screw extruders, the radial dimension, at any point along the screw, from the flight surface to the screw root. Some users of the term have taken it to mean half the difference between the internal diameter of the barrel and the screw-root diameter, but this difference is larger than the true flight depth by the amount of radial clearance between flight and barrel. In most (but not all) single-screw extruders, flight depth is much greater at the feed end, decreasing toward the delivery end of the extruder, where it is usually constant for at least several flights. See CHANNEL-DEPTH RATIO and EXTRUDER SCREW.

Flitter Synonym for GLITTER, w s.

Floating Chase A mold member, free to move vertically, that fits over a lower plug or cavity, and into which an upper plug telescopes.

Floating Platen In compression molding, a platen located between the main head and the press table in a multi-daylight press and capable of being moved independently of them.

Floating Punch A male mold member attached to the head of a press in such a manner that it is free to align itself in the female part of the mold when the mold is being closed.

Floats A term used in the past for asbestos filler in the form of very fine, short fibers with associated dust.

Flock Short fibers of cotton or synthetic fibers such as polyester, acrylic, or nylon. They are used as reinforcements in phenolic, allylic, and other thermosetting molding compounds, also for decorating plastics by the process of FLOCKING, w s.

Flocking (flock coating) A method of finishing sometimes employed for plastics articles whereby the article is coated with a tacky, slow-drying adhesive, then is dusted with a fibrous material cut into very short lengths to give a finish resembling suede, plush, etc. Fibers for flocking are available in a wide range of materials including acrylic, nylon, polyester, polyolefins, and natural fibers. Machinery for flocking films and fabrics includes gravure printing stations for applying the adhesive in desired patterns and flock heads that distribute a precalculated layer of flock to the web, and retrieve and recirculate surplus flock.

Flood Feeding The usual way of feeding an extruder or screw-injection molder, in which the feed material flows from the feed hopper by gravity and completely fills the feed section of the screw. The actual throughput is thus controlled by screw design, die resistance and temperature conditions within the screw, in contrast to what occurs in STARVE FEEDING, w s.

Flooding An undesirable separation of pigment in a wet paint film, causing a color change or mottling. The condition is caused by improper milling of pigment, an excess of solvent, or too low viscosity.

Flory Temperature Synonym for THETA TEMPERATURE, w s.

Flow-Behavior Index See POWER LAW.

Flow Birefringence (streaming birefringence) The difference, Δn, between the refractive indices of a flowing polymer solution or melt in the direction of flow and a direction perpendicular to the flow. Usually measured, using one of two techniques, by directing a light beam downward through the liquid in a rotational viscometer. The amount of birefringence is related to the degree of orientation of the polymer chains, in turn related to shear stress and first normal-stress difference in the flowing medium. The measurements are useful for testing molecular theories and rheological models, also for understanding processing problems such as extrudate roughness.

Flow Coating A painting process in which the article to be painted is drenched with a paint, either by pouring or by spraying with a mist in a closed or semi-closed chamber. The parts are sometimes rotated during and after drenching to avoid sags and runs. The process is used for coating metallized parts and other irregularly shaped articles that are difficult to paint by ordinary spraying methods.

Flow Curve A graph, usually on bilogarithmic coordinates, of shear stress vs shear rate or, sometimes, of apparent or true (corrected) viscosity vs either shear rate or shear stress. For nearly all polymer melts, log-log plots of shear stress vs shear rate of sufficient range exhibit a Newtonian region of slope = 1 at extremely low shear rates, a brief transition region of decreasing slope ("knee"), and a higher-shear region in which the slope is nearly constant or very gradually decreasing and is in the range 0.7 to 0.25. See also PSEUDOPLASTIC FLUID and VISCOSITY.

Flowers of Antimony Synonym for ANTIMONY TRIOXIDE, w s.

Flow Line Synonym for WELD LINE, w s.

Flow Mark A defect in a molded article characterized by a wavy surface appearance, caused by improper flow of the resin into the mold, which itself may have a number of causes.

Flow Molding (1) A variation of INJECTION MOLDING, w s, used for thick-walled parts. A large gate is used and pressure is maintained on the injected melt so as to force a little more melt into the mold as each part solidifies from the outside inward, thus minimizing shrinkage and improving consistency of the parts' final dimensions. (2) A process of heating materials such as a cloth-backed, plasticized PVC with a high-frequency electric field while pressing the material into a mold made of silicone rubber. In 10 to 15 s, the PVC is formed to the contours and surface texture of the mold, which can give it the look of hand-tooled leather or similar effects.

Flow Properties See MELT-FLOW INDEX, VISCOSITY, PSEUDOPLASTIC FLUID, and RHEOLOGY.

Fluid A gas or a liquid, or, in the supercritical region, a hybrid.

Fluidity (1) The ease with which a liquid flows under stress. (2) Specifically, the reciprocal of viscosity. The SI unit of fluidity is $(Pa \cdot s)^{-1}$ or $1/(Pa \cdot s)$, replacing the deprecated cgs unit, the *rhe*. 1 rhe = 10 $(Pa \cdot s)^{-1}$.

Fluidization A gas-solid or liquid-solid contacting process in which a stream of fluid is passed upwards through a bed of small solid particles, causing them to lift, expand, and behave as a boiling liquid. The process is widely used in the chemical industry for performing reactions in which the solid is either a reactant or a catalyst. In the plastics industry, the main application is in FLUIDIZED-BED COATING, w s.

Fluidized-Bed Coating The process of applying plastics coatings to objects of other, higher-melting materials, often metals, wherein a powdered resin is placed in a container provided with a porous or perforated bottom through which a gas is directed upward to keep the resin particles in a state of agitated levitation. The part to be coated is preheated above the resin's softening temperature and lowered into the fluidized bed until a deposit of the desired thickness has formed, then the part is withdrawn and allowed to cool.

Fluorescent Brightening Agent See BRIGHTENING AGENTS.

Fluorescent Pigment Any pigment that absorbs light at certain frequencies and re-emits light at lower frequencies, thus making articles containing these pigments appear to possess an actual glow of their own. A type known as *daylight fluorescent pigments* (*dayglo pigments*) responds to radiation in both the ultraviolet and visible ranges, causing the effect of glowing in normal daylight. These pigments are comprised of fluorescent dyes incorporated in a clear-resin matrix, ground to powder form. Urea and melamine resins have been used as matrices, also a modified sulfonamide resin.

Fluorinated Ethylene-Propylene Resin (FEP, PFEP) This member of the fluorocarbon family is a copolymer of tetrafluoroethylene and hexafluoropropylene, possessing most of the desirable properties of PTFE, yet truly meltable and, therefore, processable in conventional extrusion and injection-molding equipment. It is available in pellet form for those operations and as dispersions for spraying and dipping.

Fluorocarbon Blowing Agent A family of inert, noncorrosive liquid compounds containing carbon, chlorine and fluorine, originally developed as refrigerants. They are compatible with all resins and leave no residues in molds. For years they were widely used in structural-foam extrusion, in which they were incorporated with the polymer by direct injection through the barrel of the first of two tandem extruders. Fluorinated hydrocarbons are numbered by a three-digit system developed by Du Pont for use with its trade name Freon®. The first digit is the number of carbon atoms in the molecule – 1, omitted when it is 0 (the methane group), leaving just two digits for them. The second digit equals the number of hydrogens in the molecule + 1; the third digit is the number of fluorine atoms. The remaining atoms in these saturated compounds are chlorine. Thus the once-common blowing agents were Freon 11 (trichlorofluoromethane), Freon 12 (dichlorodifluoromethane), Freon 113 (trichlorotrifluoroethane), and Freon 114 (dichlorotetrafluoroethane). Because there is strong evidence that these compounds, when released to the atmosphere, migrate to the upper levels and catalyze the destruction of UV-blocking ozone by chain reactions, they are being phased out of use by international agreement and are being replaced, in the manufacture of plastics foams, by hydrocarbons, such as neopentane, and by hydrohalocarbons.

Fluorocarbon Elastomer (fluoroelastomer) Any fluorocarbon polymer of low T_g and no crystallinity, therefore rubbery. As rubbers, these materials are resistant to high temperature and most chemicals and solvents. Most of the commercial materials are copolymers. In the US, Du Pont's Viton® materials are most familiar.

Fluorocarbon Resin Any of a family of thermoplastics chemically similar to the polyolefins, with all the hydrogen atoms replaced by fluorine. They are made by addition polymerization from olefinic monomers composed only of fluorine and carbon. The main members of the family are POLYTETRAFLUOROETHYLENE, FLUORINATED ETHYLENE-PROPYLENE RESIN, and POLYHEXAFLUOROPROPYLENE, w s.

Fluoroethylene Synonym for VINYL FLUORIDE, w s.

Fluorohydrocarbon Resin Any resin polymerized from an olefinic monomer composed of carbon, fluorine, and hydrogen only. Included are POLYVINYLIDENE FLUORIDE, POLYVINYL FLUORIDE, POLYTRIFLUOROSTYRENE, w s, and copolymers of halogenated and fluorinated ethylenes.

Fluoroplastic (fluoropolymer) A plastic based on polymers made from monomers containing one or more atoms of fluorine, or copolymers of such monomers with other monomers, the fluorine-containing monomer(s) being in the greatest amount by mass (ASTM D 883). This is a broad family including POLYTETRAFLUOROETHYLENE,

FLUORINATED ETHYLENE-PROPYLENE RESIN, POLYCHLOROTRIFLUOROETHYLENE, and POLYVINYLIDENE FLUORIDE (w s, all); and a variety of copolymers of chlorofluorinated hydrocarbons. See also FLUOROCARBON RESIN, FLUOROHYDROCARBON RESIN, CHLOROFLUOROCARBON RESIN, CHLOROFLUOROHYDROCARBON RESIN, ETHYLENE-TETRAFLUOROETHYLENE COPOLYMER, ETHYLENE-CHLOROTRIFLUOROETHYLENE COPOLYMER, PERFLUOROALKOXY RESIN.

Flushed Pigment A pigment that has been transferred from an aqueous medium in which it was first manufactured to an oil medium. The flushing process comprises mixing the aqueous pigment paste with oil in such a manner that the pigment preferentially transfers to the oil phase. The free water is then poured off, and the remaining water is removed by heating under vacuum.

Fluted Core An integrally woven reinforcing material consisting of ribs between two skins, thus providing unitized sandwich construction.

Fluted Mixing Section In extrusion, a screw section in which several short barrier flights are placed so that the plastic must flow through the high-shear clearance between the flight tip and the barrel. Fluted sections may have the flights parallel to the screw axis (Maddock-Union Carbide) or at an angle to the axis (Egan).

Flux (n) (1) In chemistry and metallurgy, a substance, e g, borax or fluorspar, used to promote fusion of metals or minerals. (2) In plastics compounding, the term flux is sometimes used for an additive that improves flow properties, e g, coumarone-indene resin in the milling of vinyl compounds. (3) In heat and mass transfer, the rate of transfer per unit of cross-sectional area perpendicular to the direction of transfer. (4) LUMINOUS FLUX, w s.

Flux (v) To melt, fuse, or make liquid. In the early years of the vinyl-plastisol art this term was often used for "fuse" before the latter term came into general use.

Fluxing Temperature Synonym for FUSION TEMPERATURE, w s.

Foam See CELLULAR PLASTIC.

Foam-Backed A term describing a fabric laminated to or coated with a layer of rubber or plastic foam.

Foam Casting (foam molding) A process with many variations, depending on the polymers used. In general, a fluid resin or prepolymer containing catalyst is foamed before or during molding by mechanical frothing, or by gas dissolved in the mixture or vapor from a low-boiling liquid. See also REACTION INJECTION MOLDING.

Foamed Plastic Synonym for CELLULAR PLASTIC, w s.

Foam Extrusion See EXTRUDED FOAM.

Foam Fabrication The process of cutting large slabs, logs or "buns" of foamed plastics into sections of desired dimensions. The raw slab is conveyed through an array of saws, knives, or hot wires that first remove the uneven top, bottom and sides, then slice the remaining rectangular block into boards, finally cross-cutting the boards to standard lengths.

Foaming Agent Synonym for BLOWING AGENT, w s.

Foam-in-Place Refers to deposition of foams at the site of the work, e g, between two containing walls for insulation, as opposed to bringing the work to the foaming machine.

Fogging See BLOOM and BLUSHING.

Foil Decorating See IN-MOLD DECORATING.

Folding Machine A machine that folds sheet plastics such as cellulose acetate into shapes such as identification-card envelopes, sheets for ring binders, visible indexes, and the like. An electrically heated blade softens the plastic and folds it into a tight, 180° crease.

Footcandle A deprecated unit of surface-lighting intensity, equal to 10.76391 lux. The lux (lx) is defined as 1 lumen per square meter (lm/m^2). See also LUMINOUS FLUX.

Force (1) Either half of a compression mold (top force or bottom force), but usually the half that forms the concave surfaces of the molded part. (2) The male half of the mold, which enters the cavity and exerts pressure on the resin, causing it to flow. (3) A basic familiar quantity, familiar to everyone, customarily defined as that which changes a body's state of rest or motion, according to Newton's second law of motion:

$$F = \frac{d}{dt}(m\,v)\,/\,g_c = m\,a/g_c$$

The second form applies when mass m is constant, as in all industrial applications. Here F = the force, v = the velocity, t = time, a = the body's acceleration = dv/dt, and $1/g_c$ is the proportionality constant relating the dimensions of the three primary quantities. Historically the units of the primary quantities, in whatever system, have been devised so that g_c is exactly 1 and it is often omitted from the equation. In the SI system, 1 newton is defined as the force that will accelerate a 1-kilogram mass 1 (meter per second) per second, making g_c exactly 1.0000 $N{\cdot}s^2/(kg{\cdot}m)$. In the English system, however, where "pound" is used to mean either the pound-mass or the pound-force (lb or lb_f), the value of g_c = 32.174 $lb_f{\cdot}s^2/(lb{\cdot}ft)$! *Weight* is the downward force exerted on bodies in the earth's gravitational field and results from the as yet unexplained mutual attraction of masses as set forth in Newton's law of universal gravitation. Because the earth is nearly spherical, the distances between bodies on and close to its surface and the earth's center are almost constant and so, therefore is the earth's attractive force on a unit mass at its surface. For convenience, that force is replaced by g, the standard "acceleration due to gravity", 9.806,650 m/s^2 (32.174 ft/s^2). Weight, then, = $m \cdot g/g_c$. For precise work, corrections are made for the variation in g with latitude and altitude, but g_c is always constant, even in outer space. Force has other aspects that are unconnected with rate of change of momentum. For example, the forces developed in a tightened bolt or the force exerted by a hydraulic ram on molding compound in a closed mold. One should keep in mind that force is *not identical* to mass times acceleration.

Force Feeder See CRAMMER-FEEDER.

Force Plate The plate that carries the plunger or force plug of a mold and the guide pins or bushings. Since the force plate is usually drilled for steam or water lines, it is sometimes called the *steam plate*.

Force Plug (plunger, piston) The portion of a mold that enters the cavity block and exerts pressure on the molding compound, designated as the *top force* or *bottom force* by position in the assembly.

Ford Viscosity Cups A series of three cylindrical cups with conical bottoms, differing only in the diameters of the orifices at the apexes of the cones, each cup having a capacity of about 100 ml. From the time of efflux, the sample volume and the orifice diameter, the kinematic viscosity of a liquid may be estimated. ASTM Test D 1200 (sec 06.01) describes the Ford cup procedures to be used with paints and varnishes.

Forgeability The ability of a solid material to flow without rupture under sudden intense compression. Some plastics have been forged into useful articles.

Forging See COLD FORMING and SOLID-PHASE FORMING.

Formability The relative ease with which a plastic sheet or rod may be given another permanent shape. See THERMOFORMABILITY.

Formaldehyde (formic aldehyde, methanal, oxymethylene) HCHO. A colorless gas with a pungent, suffocating odor, obtained most commonly by the oxidation of methanol or low-boiling petroleum gases such as methane, ethane, etc. The gas is difficult to handle, so it is sold commercially in the form of aqueous solutions (formalin), solvent solutions, as its oligomer, paraformaldehyde, and as the cyclic trimer, 1,3,5-trioxane (α-trioxymethylene). High-molecular-weight, commercial polymers of formaldehyde are called *polyoxymethylene* or ACETAL RESIN, w s. Formaldehyde is also used in the production of other resins such as PHENOLIC RESIN (phenol-formaldehyde) and AMINO RESIN (urea formaldehyde), w s.

Formalin (formol) An aqueous solution of formaldehyde, usually 37–40% in strength.

Form Grinding A method of forming circularly symmetrical parts from plastic rod or tubing, employing a grinding wheel shaped to the inverse of the desired contour, a smaller hardened-steel regulating wheel that presses the plastic rod or tube against the grinding wheel, and a work-rest blade that supports the work between the two wheels. Water is usually supplied as the coolant, from which the scrap powder can be recovered and reused.

Formic Acid (methanoic acid) HCOOH. The first of the aliphatic acids, with a pungent odor and probably existing mostly in the dimer form, formic acid is produced by the reaction of carbon dioxide and dry sodium hydroxide followed by treatment with sulfuric acid. It is a solvent for phenol-formaldehyde resins, some polyesters, polyurethanes and nylons.

Formica® A trademark of the Formica Corporation for high-pressure, decorative laminates of melamine-formaldehyde, phenolic and other thermosetting resins with paper, linen, canvas, glass cloth, etc, often misused by the public in a generic manner.

Forming A general term encompassing processes in which the shapes of plastics pieces such as sheets, rods, or tubes are changed to a desired configuration, usually with the aid of heat. The term is not usually applied to operations such as molding, casting, or extrusion in which shapes or articles are made from molding materials and liquids. See also FABRICATE and THERMOFORMING.

Forming Box See VACUUM SIZING.

Forming Cake In filament winding, the collection (*package*) of glass-fiber strands on a mandrel during the operation.

Formol Synonym for FORMALDEHYDE, w s.

Formula Weight (molecular weight) Of a chemical compound, the number obtained by summing the atomic weights of all the compound's atoms. The term is not used with polymers, which are always mixtures of compounds of different molecular weights.

Fossil Resin A natural resin obtained from fossilized remnants of plant or animal life. An example is AMBER (w s), a fossilized resin derived from an extinct species of pine.

Foundry Resin A thermosetting resin used as a binder for sand in metals founding. The types most commonly used are water-soluble phenol-formaldehyde resins that become insoluble when cured, and cold-setting furfuryl alcohol resins that cure in the presence of an acid catalyst.

Four-Dimensional Braid A type of braided reinforcement used to achieve specially directed strength and resistance to interlaminar shear, often involving mixed fibers of different materials.

Fourier Number (N_F) A dimensionless group important in analysis of unsteady heat transfer in solids, such as sheets being heated or cooled in thermoforming, or cooling of extrudates and moldings. $N_F = \alpha \cdot t/x^2$, where α = the material's thermal diffusivity, t = the heating or cooling time, and x is a thickness or half-thickness in the direction of heat flow.

Fourier's Law of Heat Conduction The fundamental equation for steady heat flow through solids. It is

$$q = -\mathrm{k} A \frac{dT}{dx}$$

where q = the rate of heat flow, k = the thermal conductivity of the material at temperature T, and A = the area through which heat flow is occurring, normal to coordinate x within the material in the direction of heat flow. This is the defining equation for THERMAL CONDUCTIVITY, w s. The quotient q/A is known as the *heat flux* or *thermal flux*. If the thermal conductivity is constant over the range of temperatures involved or is linearly dependent on temperature, an integrated, simpler version of the law, with an average conductivity, can be used to find the heat flow through a layer of thickness Δx:

$$q = \mathrm{k}_{\mathrm{avg}} A\left(T_2 - T_1\right)$$

where T_2 and T_1 are the temperatures on the hot and cold sides of the layer.

Fourier-Transform Infrared Spectroscopy (FTIR) Infrared spectroscopy (IR) is most commonly used for the identification of unknown pure organic compounds. In FTIR, infrared radiation of a broad range of wavelengths is passed through an interferometer and a pathlength difference is introduced into one part of the light beam. This IR beam is then passed through the sample, which absorbs light at energies corresponding to various bond-vibration and -rotation frequencies. The beam is then focused on a detector, and a computer calculates the absorption of the IR frequencies by the sample, identifying compounds present and their concentrations in the sample.

Fractionation A method of determining the molecular-weight distribution of polymers based on the fact that polymers of higher molecular weight are less soluble than those of lower molecular weight. Two basic methods in use are (1) precipitation fractionation, in which phases are separated from a solution of the polymer by incremental addition of nonsolvents, stepwise lowering of the solution temperature, or volatilization of the solvent; and (2) extraction fractionation, in which fractions of increasing molecular weight are preferentially extracted from a layer of polymer that has been deposited on a substrate. In either method, a series of fractions is obtained which must be recovered and characterized with respect to molecular weight. A third method that has had some success is ultracentrifugation. See ULTRACENTRIFUGE. Fractionation is also used to prepare polymer specimens having narrow molecular-weight distributions.

Fracture The separation of a body, usually characterized as either brittle or ductile. In brittle fracture, the crack propagates rapidly with little accompanying plastic deformation. In ductile fracture, the crack propagates slowly, usually following a zigzag path along planes on which a maximum resolved shear stress occur, and there is substantial plastic deformation. Slower loading and higher temperature favor ductile behavior.

Fracture Mechanics The study, both theoretical and experimental, of the behavior of cracks in stressed bodies. A basic principle is that fracture is driven by the energy released by the growth of the crack, which begins at a small imperfection such as is found in all bodies.

Fracture Toughness (K_C) The critical value of the STRESS-INTENSITY FACTOR, w s, in a material beyond which a crack will start to grow.

Frankel-Acrivos Equation An equation, derived wholly from theoretical considerations, giving the relative viscosity of suspensions of monodisperse spheres in Newtonian liquids. It is

$$\eta_r = \frac{\eta}{\eta_0} = \frac{9}{8}\left[\frac{(f / f_m)^{1/3}}{1-(f / f_m)^{1/3}}\right]$$

where η_0 = the viscosity of the pure liquid, η = the viscosity of the liquid containing a volume fraction f of spheres whose density is close to that of the liquid. f_m represents the maximum attainable volume fraction, which depends on the assumed geometry of packing, about 0.7 to 0.74. The equation well represents the best available data to loadings approaching $f / f_m = 1$. Compare EILERS EQUATION and MOONEY EQUATION.

Free-Falling-Dart Test A method of measuring the impact resistance of thermoplastic films by dropping a dart with a hemispherical head onto a film specimen held in a clamping frame. As described in ASTM D 1709, the dart is dropped from a fixed height onto each of ten specimens and the percent failure is noted. Another increment of weight is added and ten more specimens are tested. This process is repeated until 50% of the specimens fail. The weight of the dart at this point, times the drop height, is a measure of the film's impact strength. In ASTM D 4272, another falling-dart test for films, the dart is instrumented and the energy consumed in penetrating the film is computed. ASTM D 3029 describes a similar test for rigid sheets. See also FALLING-DART IMPACT TEST.

Free Forming A variant of SHEET THERMOFORMING, w s, in which a bubble is blown into the clamped, heat-softened sheet, either by applying a vacuum to the side that will be convex or pressure to the underside. The method has been used most with cast-acrylic sheet for applications where the best possible optical properties are foremost, such as airplane canopies.

Free Phenol The uncombined phenol existing in a phenolic resin after curing, the amount of which is indicative of the degree of cure. The presence of such free phenol can be detected by the GIBBS INDOPHENOL TEST, w s.

Free Radical An atom or group of atoms having at least one unpaired electron. Most free radicals are short-lived intermediates with high reactivity and high energy, difficult to isolate. They are important agents in many polymerization processes and have been detected on corona-discharge-treated films by electron-spin resonance (ESR).

Free-Radical Polymerization A reaction initiated by a FREE RADICAL (w s) derived from a polymerization catalyst. Polymerization proceeds by the chain-reaction addition of monomer molecules to the free-radical ends of growing chain molecules. Major polymerization methods such as bulk, suspension, emulsion, and solution polymerization involve free radicals. The free-radical mechanism is also useful in copolymerization, in which alternating monomeric units are promoted by the presence of free radicals.

Free Volume In a liquid or amorphous solid, the specific volume minus the volume occupied by the molecules themselves. Free volume increases with rising temperature, causing viscosity to diminish. At T_g it may be a few percent in polymers. Free volume is mainly responsible for the compressibility of liquids and solids.

Freeze Grinding See CRYOGENIC GRINDING.

Freeze Line Synonym for FROST LINE, w s.

French Mold A two-piece mold for special shapes: tall, top-heavy, leaning to one side, or with extremely fine detail.

Freon® See FLUOROCARBON BLOWING AGENT.

Frequency The cyclic rate of vibration in any vibrating system—electrical, sonic, mechanical, or radiant-energetic. The SI unit is the hertz (Hz), equal to one complete cycle per second. For wave motions such as sound and electromagnetic radiation, frequency equals the velocity of propagation divided by the wavelength.

Friction (1) The force resisting the sliding or rolling of one body relative to another with which it is in contact. The *coefficient of friction*, μ, is the quotient of the force required to cause or maintain motion divided by the normal force N exerted by the bodies upon each other, i e, $\mu = F/N$. With most materials, the force required to initiate motion is somewhat greater than that required to maintain it, so a distinction is drawn between *static* and *dynamic* coefficients of friction. Coefficients of *rolling* friction are generally much lower than coefficients of sliding friction. Dynamic sliding friction is the most important one in plastics processing, but all three have roles in product design and operation. (2) In flowing fluids, particularly liquids, interlaminar friction is believed to be the basis of viscosity.

Frictional Heating (1) Heat evolved when two surfaces rub together. For a frictional force F and relative velocity V, the rate of heat evolution is $F \cdot V$ (SI: J/s). (2) Viscous dissipation within liquids undergoing shear flow. At any point in a laminar-flowing liquid, the rate of viscous dissipation, per unit volume, is equal to the product of the shear rate times the shear stress, which is also equal to the product of the local viscosity and the square of the shear rate. Both these types of frictional heating are important mechanisms of plastication in extruders and injection molders.

Friction Calendering A process in which an elastomeric compound is forced into the interstices of woven or cord fabrics while passing through the rolls of a calender. The heated compound is fed into the top opening of three adjacent rolls, so that it will cling to the middle roll. The fabric to be impregnated is fed into the lower opening between the rolls. The distance between the rolls is regulated so as to squeeze the fabric without crushing it, and the rolls are operated at slightly different speeds so that the compound is wiped by friction into the meshes of the fabric.

Friction Welding (angular welding) A term encompassing SPIN WELDING, w s, and the newer process of applying rapid angular oscillations to heat the plastic parts to be joined. This variation of the spin-welding process is used for parts that are not symmetrical about an axis of rotation. The equipment must be programmed to stop when the parts are properly positioned for joining

Friedel-Crafts Catalysts Strongly acidic metal halides such as aluminum chloride, aluminum bromide, boron trifluoride, ferric chloride, and zinc chloride, used in the polymerization of unsaturated hydrocarbons, e g, olefins. (Friedel-Crafts reactions using such catalysts are named for Charles Friedel and James Crafts, who first used them in 1877.) These acidic halides are also known as *Lewis acids*.

Frosting A light-scattering surface resembling fine crystals. See also CHALKING, BLOOM, and HAZE.

Frost Line (freeze line) In the extrusion of blown film, a ring-shaped transition zone of frosty appearance located at the level at which the film reaches its final diameter and is changing from melt to solid.

Frothing A technique for applying urethane foam in which blowing agents or small air bubbles are introduced under pressure into the liquid mixture of foam ingredients.

Frozen-in Strain (residual strain) Strain that remains in an article after it has been shaped and cooled to its final form, due to a non-equilibrium configuration of the polymer molecules. Such strains occur when cooling is carried below a certain temperature before stresses of a molding or forming operation have been allowed to relax. Frozen-in strains/stresses can cause warping of large flat members and can lead to crazing at low applied stress levels. Aside from measures that can be taken during processing, post-annealing, sometimes done with the aid of shape-holding jigs, is used to complete the relaxation of stresses.

FRP Abbreviation for FIBER-REINFORCED PLASTIC, w s.

Full-Flighted Screw An extruder screw in which the flights extend over the entire length of the screw.

Fumed Silica (pyrogenic silica) An exceptionally pure form of silicon dioxide made by reacting silicon tetrachloride in an oxy-hydrogen flame. Individual particles of fumed silica, ranging in size from 7 to 40 nm, tend to link together by a combination of fusion and secondary bonding to form chain-like aggregates with high surface areas that retard the flow of liquids in which they are dispersed. Thus fumed silica is a useful thickening agent, imparting thixotropy to liquid resins that are normally Newtonian, e g, certain polyesters. Fumed silica is also used in dry molding powders to make them free-flowing and easier to disperse with colorants. Improved electrical properties, prevention of blocking, and reduction of plasticizer migration are other benefits attributed to fumed silica in vinyl compounds.

Functionality The ability of a molecule or group to form covalent bonds with another molecule or group in a chemical reaction. Compounds may be mono-, di-, tri-, or polyfunctional, depending on the number of functional groups capable of participating in a reaction.

Fungicide An agent incorporated in a plastic compound to control fungus growth, usually by killing the organisms. Most plastics with a few exceptions, notably some of the cellulosics, are inherently resistant to fungus attack. However, many plasticizers are highly susceptible to attack. Examples of fungicides used in plastics are copper-8-quinolinate and tributyltin oxide. Agents that retard fungal growth without killing the organisms are called *fungistats*. See also BIOCIDE.

Funginertness (fungus resistance) Not susceptible to the formation of fungus growths.

Fungistat An agent incorporated in plastics compounds to control fungus growth without killing the fungi. See also BIOCIDE.

Furan Prepreg Latent catalysts announced in 1972 made it possible to produce furan-resin prepregs of acceptable shelf life, which avoided the difficulties experienced with the wet-layup process. These composites possess good heat and chemical resistance, excellent surface hardness and fire resistance.

Furan Resin (furfuryl resin) A dark-colored, thermosetting resin obtained primarily by the condensation polymerization of furfuryl alcohol in the presence of a strong acid, sometimes in combination with formaldehyde or furfural (2-furaldehyde). The term also includes resins made by condensing phenol with furfuryl alcohol or furfural, and furfuryl-ketone polymers. The resins are available as liquids in a wide range of viscosities that cure to highly crosslinked, brittle substances. They are used for impregnating cured plaster structures, as binders for foundry-sand cores, for binding high explosives, and as wood adhesives. The cured resins exhibit good resistance to chemicals and solvents. Improved resin/catalyst systems have made the older furan systems obsolete and enabled the use of fire-retardant furans in hand layup, sprayup, and filament-winding, competing with polyesters.

Furfural (2-furaldehyde, ant oil, furfuraldehyde) A liquid obtained by distilling acid-digested corn cobs or the hulls of oats, rice, or cottonseed, and having the structure shown below.

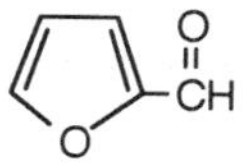

Furfural is colorless when first distilled, but darkens on exposure to air. It is used as a solvent, and in the production of furans and tetrahydrofurans. See also FURAN RESIN.

Furnace Black A type of CARBON BLACK, w s, made in a refractory-lined furnace.

Fusion With respect to vinyl plastisols and organosols, fusion is the state attained by heating when the resin particles have completely dissolved in the plasticizers and solvents present, so that upon cooling a homogeneous solid solution results. Should not be confused with CURE, w s. In the early years of these materials, the terms *gelation, gelatinization,* and *fluxing* were widely used for the state now known as fusion. *Curing* was also used, though improperly.

Fusion Bonding Any of several methods that create a thin layer of melt on the plastic surfaces to be joined, which are then pressed together. Fusion bonding is limited to identical or melt-compatible polymers.

Fusion Temperature In vinyl dispersions, the temperature at which FUSION, w s, occurs. The *optimum fusion temperature* is that at which thermal degradation has not occurred and maximum physical properties are obtained in the final product. The term *minimum fusion temperature* is sometimes employed for the temperature at which a substantial percentage, usually about 75%, of maximum physical properties are attained. Fusion temperature is also called *fluxing temperature* and, in some European literature, *gelling temperature*.

Fuzz An accumulation of short, broken filaments collected from passing glass strands, yarns, or rovings over a contact point. The fuzz may be collected, weighed and used as an inverse measure of abrasion resistance.

G g

g (1) SI abbreviation for gram. (2) Symbol for the acceleration due to gravity at the earth's surface, and, in particular, for the standard value, 9.806,650 m/s^2. It is sometimes used, erroneously, for the proportionality constant (g_c) in Newton's law of momentum change. See FORCE (3).

G (1) SI abbreviation for GIGA-, w s. (2) Symbol for shear modulus. In dynamic testing, G' symbolizes the "real" or in-phase part and G'' the "imaginary" or out-of-phase component.

Gage Alternate spelling of GAUGE, w s.

Gamma- A prefix usually abbreviated as the Greek letter γ- and usually ignored in alphabetizing compound names, signifying that the so-labeled substitution is on the third carbon away from the main functional group of the molecule, and generally synonymous with the label "4-".

Gamma Ray A quantum of electromagnetic energy emitted from some radioactive materials as they decay. Gamma rays are similar to, but of much shorter wavelength (typically $\approx$ 0.03 nm) and much higher energy than ordinary X rays. They are highly penetrating, capable of passing through several centimeters of lead.

Gamma Transition Synonym for GLASS TRANSITION, w s.

Gap In filament winding, an unintentional space between two windings that should lie next to each other.

Gardner-Holt Bubble Viscometer See AIR-BUBBLE VISCOMETER.

Gardner Impact Test (falling-weight test) A test for the impact resistance of rigid plastic sheets or parts. ASTM D 3029 describes two methods, F and G. In F, a weight with a hemispherical nose falls through a tube to strike the specimen, and G (for which the Gardner Impact Tester is approved), in which the falling weight hits a round-nose striker resting on the specimen. Either the weight or the height of drop may be varied, the former being recommended because of the sensitivity of most plastics to the velocity of impact (determined by height). Several tests on new specimens may be made at each impact-energy level (height × weight), and the fraction of breaks at each level is noted. This procedure efficiently provides information on both the mean energy to break and the standard deviation. Where the mean is of greatest interest, the UP-AND-DOWN METHOD, (w s) provides a good estimate with less testing.

Gas Black See CARBON BLACK.

Gas Chromatography (GC) A method of chemical analysis in which the specimen is vaporized and introduced into a stream of carrier gas (usually helium), the stream then passed through a column packed with adsorbent particles that separate the stream into its constituent molecules. These fractions pass through the column—first adsorbed from, then desorbed into the ongoing stream of carrier gas—at characteristic rates. Their concentrations in the effluent are measured by a detector such as a thermal-conductivity cell. The recorded concentrations are displayed on a strip chart with time as the abscissa. From the positions and areas of the peaks the identities and relative concentrations of the constituents in the sample may be determined. See also CHROMATOGRAPHY.

Gas-Injection Molding A specialized technique for molding low-density structures in which a mixture containing cork particles, or glass or phenolic microspheres, glass fibers, and a thermosetting resin is injected into a mold by fluidizing it in a gas stream.

Gas-Liquid Chromatography A variation of gas chromatography in which the chromatographic column is packed with a finely divided solid impregnated with a nonvolatile organic liquid. The sample to be analyzed is injected into the inlet of the column where it is quickly and completely vaporized. The carrier-gas stream carries it into the packed section, where the vapors contact the impregnated solids. They are absorbed by the nonvolatile liquid phase, then later desorbed into the carrier gas. The vapor of each compound spends a characteristic fraction of time in the condensed phase and the remainder in the mobile gas phase. Each chemical species will tend to migrate at its own rate and will be separated from other species in the time of emergence from the column. The detector senses their concentrations in the effluent and its signal strength is recorded and displayed vs time on a strip chart.

Gas Permeability See PERMEABILITY.

Gas-Phase Polymerization A polymerization process developed by Union Carbide for high-density polyethylene, particularly for a grade for making paper-like films. Purified ethylene and a highly active chromium-containing catalyst in dry-powder form are fed continuously into a fluidized-bed reactor. The resin forms as a powder, thereby avoiding the gel, discoloration, and contamination problems often associated with conventional polymerization processes.

Gas-Transmission Rate (GTR) The quantity of a given gas passing through a unit area of the parallel surfaces of a plastic film in unit time under the conditions of the test. Those conditions, including temperature and partial pressure of the gas on both sides of the film, must be stated. The SI unit of GTR is mol/($m^2 \cdot s$) but others, some of them involving mixed metric and English units, are still in common use. ASTM tests for gas-transmission rate through plastic film and sheeting are D 1434 and D 3985, both in section 15.09. See also PERMEANCE and PERMEABILITY.

Gas Welding See HOT-GAS WELDING.

Gate In injection and transfer molding, the channel through which the molten resin flows from the runner into the cavity. It may be of the same cross section as the runner, but most often is restricted to 3 mm or even much less. A gate whose diameter is less than 0.5 mm is known as a *pinpoint gate*. A *submarine gate* is shaped to conduct the melt below the parting line of the mold and into the cavity at a point just below its edge. Other types of gates are the *fan gate* (a diverging, thin gate), and the *tab gate* (one that extends the runner into the molded part). The term *gate* is also used for the portion of the plastic molding formed by the gate orifice.

Gate Blush (gate splay) A blemish or disturbance in the gate area of an injection-molded article. It occurs when the melt fractures as it emerges from the gate due to sudden release of large elastic stresses.

Gauge (gage) (1) Any instrument that measures and indicates such quantities as thickness, pressure, temperature, or liquid level. (2) The thickness of a plastic sheet or film, usually given in mils or mm. (3) Any of the standard wire and sheet-metal scales in which the gauge numbers are inversely related to wire diameter or sheet thickness. See STRAIN GAUGE and WIRE GAUGE.

Gauge Band A term used in the packaging-film industry for a thickness irregularity found in rolls of film. A thick area at some locality over the width of a flat film will produce a raised ring in a finished roll. Similarly, a thin area will cause a depressed ring. Such films, when unwound, tend not to lie perfectly flat. With shrinkable films, thin areas are troublesome because it is more likely that they will cause tearing or burn-through during the shrink cycle.

Gauge Length On a dogbone-shaped, tensile-test specimen before stress is applied, the distance between two marks on the narrow part ("waist") of the specimen, perpendicular to the direction of pull, that will be used to measure elongation. If the specimen is of uniform thickness and width, gauge length may be taken as the nip-to-nip distance between the clamping fixtures.

Gauss A now deprecated measure of magnetic-flux density equal to 10^{-4} TESLA, w s.

g_c The proportionality constant in Newton's law of momentum change. g_c is often needed to convert viscosities between force units and mass units: $\mu_f = g_c \cdot \mu_m$. In SI, g_c = 1.000,000 $N \cdot s^2/(kg \cdot m)$, so a viscosity of, say, 100 Pa·s is also equal to 100 kg/(m·s). See FORCE.

Gay-Lussac's Law Synonym for CHARLES' LAW, w s.

Gear Box (gear reducer) Synonyms for SPEED REDUCER, w s.

Gear Pump A pump consisting of a sturdy housing within which two intermeshing toothed wheels rotate, and inlet and outlet ports. The gears typically have widths about equal to their diameters and may be spur, helical, or, rarely because of their cost, herringbone gears. The inlet port is attached at the waist of the **8**-shaped casing where the gears are moving apart; the outlet is at the opposite side. Liquid trapped between the tight-fitting teeth and the casing is borne around to the discharge side where it is displaced and expelled by the meshing of the teeth. Small gear pumps have long been used in producing staple fiber, requiring extremely high melt pressures (up to 100 MPa). Since about the mid-1970s, larger ones have come into increasing use for general extrusion operations where die resistances or filtration requirements generate high pressures, or where close dimensional tolerances on extrudates are desired. Gear pumps have been successfully retrofitted to extruders, providing net higher output, better product quality, and quick return on the additional investment.

Gel (1) A semisolid system consisting of a network of solid aggregates within which a liquid is trapped. (2) The initial, jelly-like solid phase that develops during the formation of a resin from a liquid. (3) With respect to vinyl plastisols, gel is a state between liquid and solid that occurs in the initial stages of heating, or upon prolonged storage. Note: all three types of gel have very low strength and do not flow like liquids. They are soft, flexible, and may rupture under their own weight unless supported externally (ASTM D 883). See also GELATION. (4) A defect in plastic film such as polyethylene or PVC characterized by tiny, hard, glassy particles that appear in an otherwise clear film. These *gel particles* are believed to be bits of resin of much higher-than-average molecular weight, perhaps crosslinked. Some supposed "gels" have been identified as dirt particles picked up during inproper handling of regrind. They can also be introduced into a compound via an additive such as an impact-strength modifier or a processing aid.

Gelation The formation of a GEL, w s. With regard to vinyl plastisols and organosols, gelation is the change of state from the liquid suspension to the solid condition that occurs in the course of heating and/or aging, when the plasticizer has been mostly absorbed by the resin, resulting in a dry but weak and crumbly mass. Within normal proportions of resin and plasticizer, this state is attained when the resin particles have soaked up so much plasticizer that they touch each other. As heating progresses, the swollen particles begin to fuse together, resulting in some cohesive strength. "Gelation" is considered to continue until useful levels of mechanical properties are attained, such as have developed at the CLEAR POINT, w s. Much confusion has existed regarding the meaning of gelation because in the early years of the art, especially in Europe, *gelation* was used in place of the term *fusion*, now employed in the US.

Gelation Time With reference to thermosetting resins, the interval of time between the mixing in of a catalyst and the formation of a gel.

Gel Coat In reinforced plastics, a thin outer layer of resin, sometimes containing pigment, applied to give the structure its surface gloss and finish. It also serves as a barrier to liquids and ultraviolet radiation. The gel coat is the first to be applied to the mold in the layup process (after any mold-release agent), becoming permanently bonded to the succeeding layers of reinforcement and resin. In SPRAYUP, w s, the gel coat is applied last, sometimes after partly curing the main structure.

Gel Effect See AUTOACCELERATION.

Gelling Agent See THICKENING AGENT.

Gel Particle See GEL (4).

Gel-Permeation Chromatography Original name for, but now fading synonym of SIZE-EXCLUSION CHROMATOGRAPHY, w s.

Gel Point The stage at which a liquid begins to exhibit pseudo-elastic properties. Note: this point may be detected as the inflection point on a visscosity-time plot. See GEL (2) (ASTM D 883).

Gel Time The time required for a liquid to form a gel at a specified temperature.

Geodesic Pertaining to the shortest distance, lying on a surface, between two points on that surface, e g, a straight line on a plane or great circle on a sphere.

Geodesic Isotensoid A filamentary structure in which there exists a constant stress in any given filament at all points in its path.

Geodesic-Isotensoid Contour In filament-wound, reinforced-plastic pressure vessels, a dome contour in which the filaments are placed on geodesic paths so that the filaments will experience uniform tensions throughout their lengths under pressure loading.

Geodesic Ovaloid A contour for end domes, the fibers forming geodesic lines on the surface of revolution. The forces exerted by the filaments are proportioned to meet hoop and meridional stresses at any point.

Geometric Mean (logarithmic mean, log mean) Of a sample of n measurements of some variable, the nth root of the product of all the measurements, i e, $(x_1 \cdot x_2 \cdot \ldots x_i \cdot \ldots x_n)^{1/n}$. This is identical with the antilogarithm of the arithmetic mean of the logarithms of the x_i, i e, antilog$[(\log x_1 + \log x_2 + \ldots \log x_i + \ldots \log x_n)/n]$.

Geometric Metamerism See METAMERISM.

Germicidal-Lamp Test A quick screening method for evaluating the relative resistance of vinyl compounds to discoloration and degradation upon exposure to light and weather. The specimens are placed approximately three inches below a germicidal lamp with a principal UV wavelength of 253.7 nm. The materials are rated according to the degree of discoloration and plasticizer spewing after a specified interval of time, e g, 24 or 48 h.

Gibbs Indophenol Test A test that detects the presence of free phenol in molded-phenolic parts after curing, as an indication of completeness of cure. A few drops of dibromoquinone chloroimide reagent are added to an aqueous extract of the resin that has been rendered slightly alkaline. A bright blue color indicates the presence of phenol.

Giga- (G) (pronounced: `jiga) SI prefix meaning $\times 10^9$.

g-Index A measure of polydispersity of molecular-weight distribution in polymers given by

$$g = (M_z/M_w - 1)^{0.5}$$

where M_z and M_w are the VISCOSITY-AVERAGE and WEIGHT-AVERAGE MOLECULAR WEIGHTS, w s.

Glass Any of a wide variety of rigid, amorphous solid materials, but in particular, any of the transparent vitreous glasses formed by cooling molten mixtures of silica, sodium oxide, other alkali-metal and alkaline-earth oxides, lead oxide, alumina, and, for special purposes, other oxides. (Many organic polymers experience "glassy states" at low temperatures but they are seldom referred to as "glasses".) Glass compositions used in fibrous form as plastics reinforcements include vitrified quartz (pure silica), C GLASS, E GLASS, and S GLASS, w s. The glasses as a family are characterized by high moduli, low tensile strength (except as fibers), very low ultimate tensile elongation, brittleness, and good electrical properties and chemical resistance.

Glass-Fiber Reinforcement Any of a family of reinforcing materials for plastics based on single, drawn filaments of glass ranging in diameter from 3 to 20 μm. Single filaments are produced by mechanically drawing molten-glass streams, then these filaments are usually coated with a SIZE or COUPLING AGENT (w s) which protects them from abrasion (to which the uncoated filaments are very vulnerable) and improves bonding with resins. The filaments are then gathered into bundles called *strands* or larger bundles called *rovings*. The strands may be used in continuous form for filament winding or pultrusion; or chopped into short lengths for incorporation into molding compounds or during sprayup; or formed into fabrics and mats of various types for use in hand layup, matched-die molding, and other laminating processes. Glass-fiber reinforcements are classified according to their special properties. E glass is electrical-grade glass, the most common general-purpose type. C glass is the chemically resistant grade, and S glass denotes high tensile strength. Glass fibers coated with nickel by electron-beam deposition are used making molding compounds for electrically conductive articles. Another form of glass reinforcement is GLASS FLAKES, w s.

Glass Finish (size) See COUPLING AGENT.

Glass Flakes A reinforcing filler produced by blowing molten Type E glass into very thin tubes, then smashing the tubes into small fragments. The flakes pack closely in thermosetting-resin systems, producing strong products with good moisture resistance.

Glassine Thin transparent paper treated with urea-formaldehyde resin, used for packaging.

Glass Mat (fibrous-glass mat) A thin web of nonwoven glass fibers that may or may not contain a small percentage of resin binder.

Glass Microspheres See MICROSPHERES.

Glass-Reinforced Plastics (GRP) See REINFORCED PLASTIC.

Glass-Rubber Transition Synonym for GLASS TRANSITION, w s.

Glass Spheres Solid glass spheres of diameters ranging from 5 to 5000 μm are used as fillers and/or reinforcements in both thermosetting and thermoplastic compounds. The size used most frequently passes a 325-mesh sieve, with an average sphere diameter of 30 μm. The spheres are available coated with various silane coupling agents to improve bonding between polymer and glass. The spheres improve physical properties and reduce costs of materials and end products. Some small glass spheres are considered to be MICROSPHERES, w s.

Glass Stress In a filament-wound part, typically a pressure vessel, the stress calculated from the load and the cross-sectional area of the reinforcement only.

Glass Transition (gamma transition, second-order transition, glass-rubber transition) A reversible change that occurs in an amorphous polymer or in amorphous regions of a partly crystalline polymer when it is heated from a very low temperature into a certain range, peculiar to each polymer, characterized by a rather sudden change from a hard, glassy, or brittle condition to a flexible or elastomeric condition. Physical properties such as coefficient of thermal expansion, specific heat, and density, usually undergo changes in their temperature derivatives at the same time. During the transition, the molecular chains, normally coiled, tangled, and motionless at the lower temperatures, become free to rotate and slip past each other. This temperature varies widely among polymers; for example, the glass-transition temperature (T_g) for polystyrene is about 100°C while that of a 75/25 copolymer of butadiene and styrene is near –50°C.

Glass-Transition Temperature (T_g) The approximate midpoint of the temperature range over which the GLASS TRANSITION (w s) occurs. T_g is not obvious (like a melting point), and is detected by changes, with rising temperature, in secondary properties such as the rate of change with temperature of specific volume or electrical or mechanical properties. Moreover, the observed T_g can vary significantly with the specific property chosen for observation and on experimental details such as the rate of heating or electrical frequency. A reported T_g should therefore be viewed as an estimate. The most reliable estimates are normally obtained from the loss peak in dynamic-mechanical tests or from dilatometric data.

Glitter (flitter, spangles) A family of decorative pigments comprising light-reflective flakes of sizes large enough so that each flake produces an individually seen sparkle or reflection. They are incorporated into plastics during compounding.

Gloss The relative luminance-reflectance factor of a specimen in the direction normal to the surface; the degree to which a surface approaches perfect optical smoothness in its capacity to reflect light. Gloss of plastics is measured with a glossmeter described in ASTM Test D 523. The same meter is used in measuring the resistance of shiny plastic surfaces to abrasion (D 673).

Glow Discharge See CORONA DISCHARGE.

Glowing Combustion Burning of a solid material without flame but with emission of light from the zone of burning.

Glycerol (1,2,3-propanetriol, glycerin, glycerine, glycyl alcohol) HO-$CH_2CHOHCH_2OH$. The term *glycerol* applies to the pure product;

glycerin (or *glycerine*) applies to commercial products containing at least 95% glycerol. Glycerol is a colorless, viscous liquid with a sweet taste, long produced as a by-product of soap manufacture (animal fats and vegetable oils are triglycerides of long-chain "fatty" acids). More recently, glycerol has been synthesized from propylene or sucrose (sugar). Its uses in the plastics industry include the manufacture of alkyd resins (esters of glycerol and phthalic anhydride), the plasticizing of cellophane, and the production of urethane polymers. Several of its esters are plasticizers for various resins.

Glycerol Diacetate See DIACETIN.

Glycerol Ether Acetate $C_3H_5(OCH_2CH_2OOCCH_3)_3$. A plasticizer for cellulosics and polyvinyl acetate.

Glycerol Monoacetate $CH_3COOCH_2CHOHCH_2OH$. A plasticizer for cellulose acetate, cellulose nitrate, and vinyl resins.

Glycerol Monolactate Diacetate $CH_3CHOHCOOC_3H_5(OOCCH_3)_2$. A plasticizer for cellulose acetate, imparting resistance to gasoline.

Glycerol Monolaurate (glyceryl monolaurate, lauryl glycerin) C_{11}-$H_{23}COOCH_2CHOHCH_2OH$. A plasticizer for cellulosic and vinyl resins and polystyrene.

Glycerol Monoöleate $C_{17}H_{33}COOCH_2CHOHCH_2OH$. A yellow oil, approved by FDA as a plasticizer for food packaging.

Glycerol Monoricinoleate $C_6H_{13}CHOHC_{10}H_{18}COOCH_2CHOHCH_2$-$OH$. A plasticizer for cellulose nitrate, ethyl cellulose, and polyvinyl butyral.

Glycerol-Phthalic Anhydride Resin An alkyd resin made by modifying glycerol phthalate with an equal portion of oil, fatty acid, and natural or synthetic resin. It is used in varnishes, lacquers, and enamels.

Glycerol Triacetate (triacetin) $C_3H_5(OOCCH_3)_3$. A plasticizer for cellulosic resins, polymethyl methacrylate, and polyvinyl acetate. It has been approved by FDA for food-contact use.

Glycerol Tribenzoate (tribenzoin) $C_3H_5(OOCC_6H_5)_3$. A colorless, crystalline solid, a solid plasticizer for PVC.

Glycerol Tributyrate (tributyrin) $C_3H_5(OOCC_3H_7)_3$. A plasticizer for cellulose esters.

Glycerol Tripropionate $C_3H_5(OOCC_2H_5)_3$. A plasticizer for cellulosic resins and polyvinyl acetate.

Glyceryl Tri-(12-acetoxystearate) $C_3H_5(OOCC_{17}H_{34}OCOCH_3)_3$. A plasticizer for cellulosic and vinyl resins.

Glyceryl Tri-(Acetoxyricinoleate) $C_3H_5(OOCC_{10}H_{18}CHOHC_6H_{13}O$-$COCH_3)_3$. A plasticizer for cellulose nitrate, ethyl cellulose, and PVC.

Glycidol (2,3-epoxy-1-propanol) $\overline{OCH_2C}HCH_2OH$. A stabilizer for vinyl resins.

Glycidyl- The terminal epoxy group, $\overline{OCH_2C}HCH_2$–.

γ-Glycidoxypropyltrimethoxysilane $\overline{OCH_2C}HCH_2O(CH_2)_3Si(OCH_3)_3$. A coupling agent for fibrous glass used in reinforced thermosetting and thermoplastic resins.

Glycidyl-Ester Resin Any of a family of epoxide resins derived from the condensation of epichlorohydrin with polycarboxylic acids, first available in commercial quantities in 1968. The preferred curing agents for these resins are anhydrides, polycarboxylic acids, aromatic amines, and phenolics. They are resistant to electrical tracking and weather, have high strength and high modulus, yet are tough, have low viscosity, long pot life, and high reactivity at moderately elevated temperatures. Limitations are poorer properties above 100°C, higher shrinkage, poor alkali resistance, and inability to cure at room temperature.

Glycidyl-Ether Resin See EPOXY RESIN.

Glycol (1) An organic compound having hydroxyl groups on adjacent carbon atoms, for example, propylene glycol, $CH_3CHOHCH_2OH$. The name is also used for nonadjacent dihydric alcohols such as the 1,3-isomer of the above compound, $HOCH_2CH_2CH_2OH$, known as *trimethylene glycol.* (2) Specifically, ethylene glycol.

Glycol Phthalate Resin A type of thermoplastic polyester used mainly for fibers and oriented films. See POLYESTER, SATURATED.

Glyptal An alkyd resin, the reaction product of glycerol and phthalic anhydride (hence: gly p t al). An early product was the basis of a household and laboratory cement, tradenamed Glyptal.

Godet Stand A device used in pairs downstream from the extruder when making monofilament and split-film fibers. One stand is placed before, and another following, a conditioning oven, each consisting of a vertical motor housing driving two or more horizontal drums. The filaments are wrapped repeatedly around the drums to prevent slipping. The post-oven drums are driven at speeds several times those of the first pre-oven set, causing the filaments to be stretched, oriented, and to become much stronger after cooling.

Goniophotometry A procedure for evaluating the manner in which materials geometrically redistribute light, described in ASTM E 167 (section 14.02).

GP (1) Abbreviation for *general-purpose*, sometimes used to denote types of resins and molding compounds suitable for a wide range of applications. (2) Abbreviation for GUTTA-PERCHA, w s.

GPC Abbreviation for *gel-permeation chromatography*. See SIZE-EXCLUSION CHROMATOGRAPHY.

GR-1 Former symbol for BUTYL RUBBER, w s.

Gradient-Tube Density Method (density-gradient method) A convenient method, described in ASTM D 1505, for routinely measuring densities of small resin samples, e g, a single pellet of molding compound. A graduated, vertical glass tube (the gradient tube) is carefully filled with graded solutions of two liquids, the densest at the bottom, the next densest above that, etc, to the least dense at the top. A specimen particle is introduced in the tube and falls to a position of equilibrium that indicates its density by comparison with the positions of calibrated hollow-glass floats. Over weeks or months, the gradient is gradually diminished by diffusion and the column must be reconstituted.

Graft Copolymer A polymer comprised of molecules in which the main backbone chain of atoms has attached to it at various points side chains containing different atoms or groups from those in the main chain. The side chains are "grafted" onto the pre-existing main chains by chemical reaction. The main chain may be a copolymer or may be derived from a single monomer.

Grafting Ratio In a graft copolymer, the weight of grafted side chains divided by the weight of the original polymer.

Grain Alcohol Synonym for ETHYL ALCOHOL, w s.

Gram-Atom (gram-atomic weight) The mass of an element in grams numerically equal to the element's atomic weight.

Gram Equivalent The weight in grams of a substance displacing or reacting with 1.00797 g of hydrogen or combining with 15.9994 g of oxygen.

Granular Polymerization Synonym for SUSPENSION POLYMERIZATION, w s.

Granular Structure Nonuniform appearance of finished plastic material due to retention of, or incomplete fusion of, particles of compound either within the mass or on the surface; or to the presence of coarse filler particles.

Granulate (1, v) To reduce plastic sheet, chunks, or scrap to particles about 2 to 5 mm in size. (2, n) A molding compound in the form of small spheres or pellets.

Granulator (scrap grinder) A machine for cutting waste material such as sprues, runners, excess parison material, trim scrap from extrusion and thermoforming, and rejected parts into particles that can be reprocessed. The most common type of granulator is comprised of several thick knives bolted to a heavy cylindrical core, parallel to the core's axis, which is also its axis of rapid rotation. The rotating knives graze stationary knives mounted in the machine's housing, giving an action that combines impact with shearing. A screen placed in the discharge opening controls the final particle size.

Granule A small particle produced in various sizes and shapes by hot or cold cutting, of extruded strands or scrap, or by certain polymerization methods. Also see PELLETS.

Graphite (black lead, plumbago) A crystalline form of carbon with atoms arranged hexagonally, characterized by a soft, greasy feel. It occurs naturally, but is produced by heating petroleum coke or other organic materials under controlled atmospheric conditions. In powder form, graphite is used as a lubricating filler for nylon and fluorocarbon resins. Too, it is added to compounds to make them electrically conductive. Pyrolytic graphite fibers, made by decomposing organic filaments at high temperatures in controlled atmospheres, are used as reinforcements for high-performance applications. The best graphite fibers are among the strongest and stiffest of all fibrous reinforcements, with strengths to 2.5 GPa and moduli to 500 GPa (360 kpsi and 72 Mpsi).

Graphitize The pyrolysis of an organic material such as pitch or polyacrylonitrile (PAN) in an inert atmosphere at temperatures between 2500 and 3000°C to produce graphite. Pyrolysis of PAN fibers is a principal source of graphite fibers.

Grasshopper A stiff bunch of parallel strands in a fibrous mat.

Gravimetric Feeder (weigh feeder) A device for feeding extruders that continuously meters, by weighing, the rate at which feedstock enters the feed port of the extruder.

Gravure Coating (engraved-roll coating) A roller-coating process in which the amount of coating applied to the substrate web is metered by the depth of an all-over engraved pattern in the application roll. This process is frequently modified by interposing a resilient offset roll between the engraved roll and the web.

Gravure Printing The depressions in an engraved printing cylinder or plate are filled with ink, the excess on raised portions being wiped off by a doctor blade. Ink remaining in the depressions is deposited on the plastic film as it passes between the gravure roll and a resilient backup roll.

Gray (Gy) The SI unit of absorbed radiation dose, defined as the absorption of one joule per kilogram of body mass, i e, 1 Gy = 1 J/kg. The former unit, the *rad*, now deprecated but still being used widely in the US, equals 0.01 Gy.

Grease Forming See MECHANICAL GREASE FORMING.

Green Strength Of a preform or incompletely cured molding, the ability to withstand handling without distorting.

Grex Number Like *cut* and *denier*, a deprecated measure of lineal density of yarns and fibers, the weight in grams of 10 km of the yarn. A grex number of 1 corresponds to an SI lineal density of 10^{-7} kg/m or 0.1 tex. See also CUT, DENIER, and TEX.

Grid Channel-shaped, mold-supporting members.

Griffith Theory The idea that, in brittle fracture, crack growth occurs, causing fracture, when the rate of decrease of stored elastic energy equals or exceeds the rate of creation of new fracture-surface energy. The Griffith equation states that, in thin sheets, tensile strength is given by: $\sigma = (2 \cdot \gamma \cdot E/(\pi \cdot c)^{0.5}$, where γ = the fracture-surface energy per unit area, E = the elastic modulus, and c = half the crack length.

Grind (v) (1) To reduce the size of breakable particles by impact or by squeezing them between relatively moving surfaces. (2) To remove material from a part by contact with an abrasive, to change either the part's shape or its finish.

Grinding-Type Resin A vinyl resin that must be ground to effect dispersion in plastisols or organosols.

Grit Blasting A mold-finishing process in which abrasive particles are blown against mold surfaces in order to produce a controlled degree of roughness. The process is often used on molds for blow molding to assist air escape near the end of the blow, and on other types of molds to produce a desired texture in the product.

Grooved Barrel In an extruder or injection molder, a barrel that has shallow grooves over the first few screw diameters from the feed opening. The grooves may be parallel to the machine axis or helical with direction opposite that of the screw. Grooves have been shown to improve feeding of difficult-to-feed materials and to reduce energy consumption per unit of throughput, but can cause overfeeding and attendant problems with screws and materials for which the grooves and screw were not designed.

GRP Abbreviation for "glass-reinforced plastic". See REINFORCED PLASTIC.

GR-S Abbreviation, now obsolete, for "Government Rubber, Styrene," a copolymer of 75 parts of butadiene with 25 parts of styrene, brought into production during World War II to replace interdicted imports of Malayan natural rubber, and also known as Buna S and SBR. See STYRENE-BUTADIENE RUBBER.

Guide Eye In filament winding, a moving metal or ceramic loop (eye) through which the fiber passes as it flows from creel to mandrel.

Guide Pin (dowel pin) In compression, transfer, and injection molding, usually one of two or more hardened steel pins that maintain proper alignment of the mold halves as they open and close.

Guide-Pin Bushing (dowel-pin bushing) In molding, a hardened bushing that receives and guides the leader pin, controlling alignment of the mold halves as the mold closes.

Guignet's Green See CHROME-OXIDE GREEN.

Gum Rubber (pure gum) Raw, unvulcanized rubber recovered from the latex of the *hevea* tree or from polymerization of synthetic rubber. Gum rubber has almost no useful properties prior to being vulcanized.

Gums Although rarely used in the plastics industry, this term is used in kindred industries to include materials that can be dissolved in water to produce viscous or mucilaginous solutions. Thus, water-soluble polymers such as polyvinyl pyrrolidone, polyvinyl alcohol, ethylene oxide resins, polyacrylic acid, and polyacrylamide are regarded as gums.

Guncotton A highly inflammable and explosive form of CELLULOSE NITRATE, w s, made by digesting clean cotton in a mixture of one part nitric acid and three parts sulfuric acid, the latter acting as a catalyst and scavenger of the water generated in the reaction. As it ages in contact with air, dry guncotton becomes gradually more unstable and dangerous to handle.

Gunk A slang term for PREMIX, w s.

Gunk Molding See PREMIX MOLDING.

Gusset A tuck in the side of a bag, usually made in symmetrical pairs in both paper and plastic-film bags, that permits the bag to assume a nearly rectangular-boxy form when opened. Gussets may be formed in the tubular BLOWN FILM (w s) from which bags are made just before the film enters the pinch rolls.

Gutta-Percha (GP, PI, *trans*-1,4-polyisoprene) A rubber-related, polymeric substance extracted from the milky sap of leaves and bark of certain trees belonging to the family *Sapotaceae*, genera *Palaquium* and *Payena*, plants native to Malaysia. Its mer has the same empirical formula as natural rubber, its *cis* isomer. Gutta-percha is a tough, horny substance at room temperature, but becomes soft and tacky when warmed to 100°C. In the past it was used in compounds for golf-ball covers, electrical insulation, cutlery handles, and machinery belting, and as a stiffening agent in natural rubber, but has been replaced in many of these applications by synthetics. See also BALATA.

Gutta-Percha, Synthetic See POLYISOPRENE.

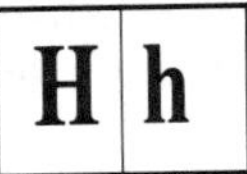

h (1) SI abbreviation for hour and for the prefix HECTO-, w s. (2) Also H, the symbol for CHANNEL DEPTH of an extruder screw, w s.

H (1)The chemical symbol for the element hydrogen. (2) Abbreviation for the SI unit of magnetic inductance, the HENRY, w s. (3) Symbol for ENTHALPY, w s.

Hagen-Poiseuille Equation (Poiseuille equation) The equation of steady, laminar, Newtonian flow through circular tubes:

$$Q = \pi \cdot R^4 \cdot \Delta P/(8 \cdot \eta \cdot L)$$

where Q = the volumetric flow rate, R and L are the tube radius and length, ΔP = the pressure drop (including any gravity head) in the direction of flow, and η = the fluid viscosity. With the roles of Q and η interchanged, this is the basic equation of capillary viscometry. Any *consistent* system of units may be used. This important equation was first derived theoretically in 1839 by G. Hagen and, a year later, inferred from experimental measurements by J. L. Poiseuille. In a laminar flow through a circular tube, a simple force balance shows that the shear stress at the wall, τ_w, = $\Delta P \cdot R/(2 \cdot L)$. By Newton's law of viscosity (see VISCOSITY), the shear rate at the wall, γ_w, must equal the shear stress divided by the viscosity. Solving the above equation for $\Delta P \cdot R/(2 \cdot \eta \cdot L)$, one obtains $\gamma_w = 4 \cdot Q/(\pi \cdot R^3)$. By applying the RABINOWITSCH CORRECTION (w s) to this expression for Newtonian shear rate, one can get the true shear rate at the wall for a nonNewtonian liquid. An often seen, equivalent, but slightly different form for the Newtonian shear rate at a tube wall is $8 \cdot V/D$, where V = the average fluid velocity = $Q/(\pi \cdot D^2/4)$ and $D = 2R$.

Halide Any compound including one (or more) of the halogen elements (F, Cl, Br, I) in its –1 valence state. The term halide is often used to indicate that any of the four principal halogens may be interchangeably present.

Halocarbon Plastic A term listed by ASTM (D 883) to mean a polymer containing only carbon and one or more halogens. The primary members of the family are the CHLOROFLUOROCARBON and FLUOROCARBON RESINs, w s.

Halogen The elements of group 7a of the periodic table; fluorine, chlorine, bromine, and iodine (F, Cl, Br, I). The fifth member, astatine, is rare, radioactive, and unstable, with a half-life less than 9 h, so is never seen in commerce.

Halogen/Phosphorus Flame Retardant Chlorine- and bromine-substituted esters of phosphoric acid or phosphonic acid, used mostly in polyurethane foams to impart fire resistance.

Hand The feel of film, fabric, or coated fabric—its flexibility, smoothness, and softness—as judged by the touch of a person.

Hand Layup See LAYUP MOLDING.

Hand Mold A mold that is removed from the press after each shot for extraction of the molded article; generally used only for short runs and experimental moldings.

Hangup (1) Stray bits of extrudate that cling to the face of the die and eventually have to be cleaned off. (2) Failure of feed material in a hopper to fall out of the hopper, caused sometimes by arching when the feed opening is only a few times larger than the pellet, but more often by clumping together of sticky particles.

Hardener See CURING AGENT.

Hard-Facing Alloy Any of several metals applied by welding to the contact surfaces (screw-flight tips) of extruder screws to reduce the rate of wear. Colmonoys® are nickel-based, containing boron, chromium, iron and silicon. Stellites® are cobalt-based, with chromium and tungsten. They are usually applied into a machined helical groove prior to cutting the main channel, then finish-ground afterward.

Hard Ferrite See FERRITE.

Hard Fiber Leaf fiber with high-lignin cell walls that is hard and stiff and is used in making twine and cordage.

Hardness The resistance of a material to abrasion, local compression, indentation, and scratching. See BARCOL HARDNESS, BRINELL HARDNESS, DUROMETER, INDENTATION HARDNESS, KNOOP MICROHARDNESS, MOHS HARDNESS, ROCKWELL HARDNESS, SCRATCH HARDNESS, VICKERS HARDNESS.

Hard Rubber The material obtained by heating a highly unsaturated diene rubber with a high percentage of sulfur. The first hard rubber was Ebonite, made from natural rubber, and black, as its name implies. Similar products have been made from several of the synthetics. The sulfur is mostly contained in 3- and 4-carbon rings with a few –C–S–C– crosslinks.

Hard-Surfacing See HARD-FACING ALLOY and CASE HARDENING.

Harmonic Mean Of a sample of n measurements of a given quantity, the inverse of the average of the inverses. Denoting the harmonic mean by $\tilde{x}$, it is given by

$$\tilde{x} = [\sum_{i=1}^{n}(1/x_i)]^{-1}.$$

Haul-Off In sheet extrusion, the three-roll stand and cooling-conveyor assembly that polishes and cools the molten sheet emerging from the die, often extended in scope to include the rubber pull rolls and winder, if any. The term has analogous meaning for other extruded products, such as pipe. See, too, SHEET LINE.

Haze The cloudy or turbid appearance of an otherwise transparent specimen caused by light scattered from within the specimen or from its surfaces (ASTM D 883). In ASTM D 1003, Test for Haze and Luminous Transmittance of Transparent Plastics, the specimen is mounted at one axial opening of an integration sphere. Total transmission of an incident beam through the specimen and the light diffused more than 2.5° from the normal direction are both measured. Percent haze is reported as 100 × the diffuse transmittance/total transmittance.

HDPE Abbreviation for *high-density polyethylene.* See POLYETHYLENE.

Head (1) In any extrusion operation, the delivery end of the extruder, usually fitted with a hinged gate that may contain a breaker plate and screen pack, to which the adapter and die are attached. (2) In blow molding, the entire apparatus by which the molten plastic is shaped into a tubular parison. This may include an adapter, a parison die, and a melt accumulator.

Head Pressure In extrusion, the pressure of the melt at the delivery, or head end of the screw, typically as signaled by a PRESSURE TRANSDUCER (w s) mounted in the barrel opposite the tip of the screw. Head pressure is one of the most important indicators of the state of any extrusion operation.

Head-to-Head Polymer A polymer in which the monomeric units are alternately reversed as in the structure shown below produced from the monomer, $CH_2{=}CHR$.

$$\left[-CH_2-\underset{R}{\underset{|}{CH}}-\underset{R}{\underset{|}{CH}}-CH_2CH_2-\underset{R}{\underset{|}{CH}}-\underset{R}{\underset{|}{CH}}- \right]_n$$

Head-to-Tail Polymer A polymer in which the monomeric units regularly repeat as in the structure shown next produced from the monomer, $CH_2{=}CHR$.

$$\left[-CH_2-\underset{R}{\underset{|}{C}H}-CH_2-\underset{R}{\underset{|}{C}H}- \right]_n$$

Heat Capacity (specific heat) The amount of heat required to raise the temperature of a unit mass of a substance one degree. In the SI system, the unit of heat capacity is J/(kg·K), but kJ/(kg·K) or J/g·K are often more convenient. Conversions from older units are: 1 cal/(g·°C) = 1 Btu/(lb·°F) = 4.186 J/(g·K). Most neat resins have heat capacities (averaged from room temperature to about 100°C) between 0.92 J/(g·K) for polychlorotrifluoroethylene and 2.9 for polyolefins. (The heat capacity of water, one of the highest of all materials, is 4.18 J/(g·K) at room temperature.) A term loosely used as synonymous with heat capacity but not truly so is SPECIFIC HEAT, w s.

Heat Cleaning (1) A batch or continuous process in which sizing on glass fibers is vaporized off. (2) Cleaning residual polymer from small extruder screws, dies or other small parts by immersing them in fused salts such as sodium nitrate.

Heat Content Synonym for ENTHALPY, w s.

Heat-Distortion Point The former name, now deprecated and fading from use, of DEFLECTION TEMPERATURE, w s.

Heated-Tool Welding See HOT-PLATE WELDING.

Heater Band An electrical heating unit shaped to fit extruder barrels, adapters, die surfaces, etc, for maintaining high temperatures of the banded items and furnishing heat to the plastics within them. Nichrome wires within the bands are embedded in insulation such as magnesium oxide and their ends are welded to screw-type terminals at the end of the band. The ends of cylindrical bands are brought together by Monel bolts to tighten them snugly against the surface to be heated. Service life is greatly increased by operating them at 80% or less of rated wattage, the usual practice. See also CAST-IN HEATER.

Heat Forming See THERMOFORMING.

Heating Cylinder (heating chamber) In elderly injection molders, that part of the machine in which the cold plastic pellets were heated to the molten condition before injection. Until the late 1950s, heating cylinders for injection machines were static devices equipped internally with *torpedoes* or *spreaders*, sometimes heated separately, that caused the charge to be distributed in a thin annulus within the cylinder. This hastened the heating but was still slow and could degrade the plastic. Today, all new commercial machines are screw-injection types, accomplishing the heating mainly by the frictional action of the screw. This heating and melting process is not only faster but produces a melt of more uniform temperature, less likely to have been scorched.

Heat Mark An extremely shallow depression or groove in the surface of a plastic article, visible because of a sharply defined rim or a roughened surface. See also SINK MARK.

Heat of Combustion The amount of heat evolved by the combustion of a unit mass, usually one gram-molecular weight, of the substance (kJ/mol). The former unit of heat of combustion, deprecated now but still widely used, was the kcal/mol, equal to 4.186 kJ/mol. Heats of combustion, which are tabulated for many pure compounds, are an important segment of the database of physical chemistry and chemical engineering, for it is from sums and differences in heats of combustion that most heats of reaction are estimated.

Heat of Fusion (enthalpy of fusion) The quantity of heat needed to melt a unit mass of a solid at constant temperature (J/kg or J/mol). Because all meltable plastics consist of broad mixtures of homologous molecules having different molecular weights, and because melting points of homologs increase with molecular weight, it isn't possible to melt a plastic at a constant temperature. Instead, melting occurs over a range of temperature. With crystalline resins, such as polyethylene, the range may be relatively small, about 20°C, and it is really only for such resins that "heat of fusion" is meaningful and measurable. To estimate the heat of fusion, enthalpy (J/g) is measured beginning at a temperature well below that at which melting begins and carried on to temperatures well above that at which the plastic has completely melted. Plots (or equations) of enthalpy vs temperature for both ranges are extrapolated to the center of the melting range, the "melting point". The difference between the higher and lower enthalpies at that point is taken as the heat of fusion of the plastic.

Heat of Polymerization (enthalpy of polymerization) The difference between the enthalpy of one mole of monomer and the enthalpy of the products of the polymerization reaction. Addition polymerizations are exothermic, values ranging from about 35 to 100 kJ/mol, and removal of this heat is an important aspect of reactor design. The reported value of an enthalpy of polymerization may be referred either to the temperature at which the polymerization is usually carried out or to a standard temperature, such as 25 or 18°C.

Heat Sealing The process of joining two or more thermoplastic films or sheets by heating areas in contact with each other to the temperature at which fusion occurs, usually aided by pressure. When the heat is applied by dies or rotating wheels maintained at a constant temperature, the process is called *thermal sealing*. In *melt-bead sealing*, a narrow strand of molten polymer is extruded along one surface, trailed by a wheel that presses the two surfaces together. In *impulse sealing*, heat is applied by resistance elements that are applied to the work when relatively cool, then are rapidly heated. Simultaneous sealing and cutting can be performed in this way. *Dielectric sealing* is accomplished with polar materials by inducing heat within the films by means of radio-frequency waves. When heating is performed with ultrasonic vibrations, the process is called *ultrasonic sealing*. See also WELDING.

Heat-Seal Strength With heat-sealed flexible films, the force required to pull apart a heat-sealed joint divided by the joined area tested. The strength of a heat seal is sometimes expressed as a percentage of the film's tear strength or tensile strength.

Heat Sensitivity The tendency of a plastic to undergo changes in properties, color, or even to degrade at elevated temperatures. Severity of change is always a matter of both temperature and time. ASTM D 794 describes the procedures to be used in determining the permanent effects on plastics of elevated-temperature exposure.

Heat-Shrinkable Film A film that is stretched and oriented while it is being cooled so that later, when used in packaging, it will, upon being rewarmed, shrink tightly around the package contents. Blown film made from plasticized PVC is the largest-volume shrink film. Heat-shrinkable tubing of several polymers is widely used in the electronics industry to protect bundles of wiring. ASTM D 2671 (section 10.01) describes a method for testing such tubing.

Heat Sink A device for the absorption or transfer of heat away from a critical part or assembly.

Heat Stability The resistance to change in color or other properties as a result of heat encountered by a plastic compound or article either during processing or in service. Such resistance may be enhanced by the incorporation of a stabilizer.

Heat Transfer The movement of energy as heat from a hotter body to a cooler body. The three basic mechanisms of heat transfer are radiation, conduction, and convection. Radiation heating occurs when heat passes from the emitting body to the receiving body through a medium, such as air, that is not warmed. Conduction heating is the flow of heat from a hot region to a cooler one in either single homogeneous substances or two substances in close contact with each other. Convection is the transfer of heat by flow of a fluid, either a gas or a liquid, and either by natural currents caused by differences in density or by forced movement caused by a fan, pump, or stirrer. All three modes of transfer are important in plastics processing.

Heat-Transfer Medium See THERMAL FLUID.

Heavy Spar See BARIUM SULFATE.

Hecto- (h) The SI prefix meaning × 100.

Helical Screw Feeder See CRAMMER-FEEDER, SCREW CONVEYOR.

Helical Transition In the transition zone (compression zone) of an extruder screw, the root surface of the screw describing an advancing spiral surface of increasing radius. See also CONICAL TRANSITION.

Helical Winding A winding in which the filament or band advances along a spiral path, not necessarily at a constant angle except in the case of a cylindrical article.

Helix Angle (1) Of an extruder screw, the lead angle of the flight with respect to the screw axis, the angle whose tangent = $t/(\pi \cdot D)$, where t = the axial distance the flight advances per turn (lead) and D = the major (or nominal) screw diameter. (2) In filament winding, the angle between the axis of rotation and the filament at its point of contact with the winding.

Henry (H) (for J. Henry) The SI unit of electric inductance, the inductance of a closed circuit in which an electromotive force of one volt is produced when the electric current in the circuit varies uniformly at a rate of one ampere per second. The magnetic flux in this situation will be just one weber (Wb). Thus, 1 henry = 1 V·s/A = 1 Wb/A.

Henry's Law (for W. Henry) The mass of a slightly soluble gas that dissolves in a definite mass of a liquid at a given temperature is directly proportional to the partial pressure of that gas. This law holds only for gases that do not chemically react with the solvent.

***n*-Heptyl *n*-Decyl Phthalate** $C_7H_{15}OOCC_6H_4COOC_{10}H_{21}$. A general-purpose plasticizer for PVC and several other thermoplastics, with lower volatility and better low-temperature performance than DIOCTYL PHTHALATE, w s. It is excellent for use in vinyl plastisols.

***n*-Heptyl *n*-Nonyl Adipate** $C_7H_{15}OOCC_4H_8COOC_9H_{19}$. A plasticizer for PVC similar to DIOCTYL ADIPATE, w s, but with better low-temperature performance and lower volatility. It is also approved for use in contact with foods.

***n*-Heptyl *n*-Dinonyl Trimellitate** An aromatic plasticizer similar to TRIOCTYL TRIMELLITATE, w s, with better low-temperature performance and lower volatility.

Hertz (Hz) Cycles per second of any periodic phenomenon. The term and its multiple, megahertz (MHz), were in the past used most frequently for waves in the radio-frequency range, but the hertz is now the SI unit of frequency.

Heterocyclic Compound A compound whose molecule includes at least one ring that contains one or more elements other than carbon and hydrogen. A simple example is *pyridine*, C_5H_5N, a benzene molecule in which one carbon is replaced by nitrogen.

Heteropolymer See HETEROPOLYMERIZATION.

Heteropolymerization A special case of addition polymerization involving the combination of two dissimilar unsaturated monomers, the product being a *heteropolymer*.

Hevea Rubber See RUBBER, NATURAL.

Hexabromobiphenyl $[2,4,6\text{-}(Br)_3C_6H_2\text{–}]_2$. A flame retardant suitable for use in thermosetting resins and thermoplastics such as acrylonitrile-butadiene-styrene resin, nylons, polycarbonate, polyolefins, PVC, polyphenylene oxide, and polystyrene-acrylonitrile. It is insoluble in water, heat-stable, and furnishes a high bromine content in the end product.

Hexachloroethane (carbon hexachloride, perchloroethane) Cl_3CCCl_3. A substitute for camphor in celluloid manufacture.

Hexachlorophene $(C_6HCl_3OH)_2CH_2$. A white, essentially odorless, free-flowing powder widely used as a bacteriostat in many thermoplastics including vinyls, polyolefins, acrylics and polystyrene.

Hexafluoroacetone Trihydrate $F_3CCOCF_3{\cdot}3H_2O$. A solvent cement, active at room temperatures, for bonding acetal resin articles to themselves and to other polymers such as nylon, acrylonitrile-butadiene-styrene, styrene-acrylonitrile, polyester, cellulosics, and natural or synthetic rubber. It is also a toxic irritant, so it must be handled with care.

Hexagonal (1) Of a closed plane figure, having six sides. (2) Having the properties of a *regular* hexagon, i e, all six sides and all six angles equal. (3) One of the six crystal systems, in which there are four principal axes, three in one plane at 120° to each other, the fourth perpendicular to the others. Atomic spacing is equal along the planar axes, different along the fourth.

Hexahydrophenol Synonym for CYCLOHEXANOL, w s.

Hexahydrophthalic Anhydride (HHPPA) $C_6H_{10}(CO)_2O$. A curing agent for epoxy resins and an intermediate for the production of alkyd resins.

Hexamethylene Synonym for CYCLOHEXANE, w s.

Hexamethylene Adipamide (nylon 6/6) A nylon made by condensing hexamethylene diamine with adipic acid. See NYLON 6/6.

Hexamethylenediamine (1,6-diaminohexane) $H_2N(CH_2)_6NH_2$. A colorless solid in leaflet form, which, when condensed with adipic acid, forms nylon 6/6. It has also been used to cure epoxy resins.

Hexamethylene-1,6-Diisocyanate (HDI) $OCN(CH_2)_6NCO$. A colorless liquid, the first aliphatic diisocyanate to be used commercially in the production of urethanes. When used with certain metal catalysts, it produces urethane polymers with good resistance to discoloration, hydrolysis, and thermal degradation.

Hexamethylenetetramine (HMT, methenamine, urotropine) $(CH_2)_6$-N_4. A bicyclic compound, the reaction product of ammonia and formaldehyde, with the structure shown below:

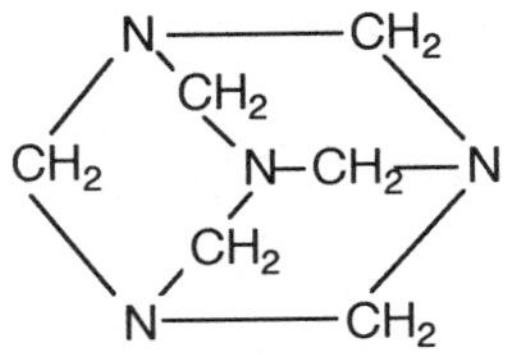

```
      N———————CH2
     / \CH2       \
    /     \        \
 CH2       N-CH2———N
    \     /        /
     \ /CH2       /
      N———————CH2
```

It is used as a basic catalyst and accelerator for phenolic and urea resins, and a solid, catalytic-type curing agent for epoxies.

Hexamethylphosphoric Triamide $[N(CH_3)_2]_3PO$. A pale, water-soluble liquid used as an ultraviolet absorber in PVC compounds.

Hexane C_6H_{14}. A straight-chain hydrocarbon, extracted from petroleum or natural gas. Commercial grades contain other hydrocarbons such as cyclohexane, methyl cyclopentane, and benzene. Hexanes are used as catalyst-carrying solvents in the polymerization of olefins and elastomers.

Hexanedioic Acid Synonym for ADIPIC ACID, w s.

Hexyl- The straight-chain radical of hexane, C_6H_{13}–.

Hexyl Acetate...$C_6H_{13}OOCCH_3$. A solvent for cellulose nitrate, cellulose acetate-butyrate, polyvinyl acetate, polystyrene, phenolics, alkyds, and coumarone-indene resins.

Hexyl Methacrylate $C_6H_{13}OOCC(CH_3)=CH_2$. A monomer used in making acrylic resins.

***n*-Hexylethyl *n*-Decyl Phthalate** (NODP) See *n*-OCTYL *n*-DECYL PHTHALATE.

HF Abbreviation for HIGH-FREQUENCY, w s.

HF Preheating See DIELECTRIC HEATING.

High-Density Polyethylene (HDPE) This term is generally considered to include polyethylenes ranging in density from about 0.94 to 0.96 g/cm^3 and higher. Whereas the molecules in low-density polyethylenes are branched and linked in random fashion, those in the higher-density polyethylenes are linked in longer chains with fewer side branches, resulting in higher-modulus materials with greater strength, hardness,and chemical resistance, and higher softening temperatures. See also POLYETHYLENE.

High-Frequency (adj) Pertaining to the part of the electromagnetic spectrum between 3 MHz and 200 MHz, employed in plastics welding, sealing and preheating operations. Frequencies of 30 MHz and below are the most used.

High-Frequency Heating Synonym for DIELECTRIC HEATING, w s.

High-Frequency Welding A method of welding thermoplastic articles in which the surfaces to be joined are heated by contact with electrodes of a high-frequency electrical generator.

High-Impact Polystyrene (HIPS) Polystyrene whose impact strength has been elevated by the incorporation of rubber particles. The best grades, produced by polymerization of styrene containing dissolved rubber, have better impact resistance than those made by milling rubber and polystyrene together. Whereas crystal PS has notched-Izod impact between 0.13 and 0.24 J/cm, HIPS resins range from 0.37 to 2.1 J/cm.

High-Intensity Mixer A type of mixer consisting of a large bowl with a high-speed rotor in its bottom, used for producing dry blends of PVC with liquid and powder additives, also for producing other powder blends. Though there is considerable heating of the charge during mixing, the main constituent is not melted and the mix is discharged in powder form. Compare with INTERNAL MIXER.

High-Load Melt Index (1) The rate of flow of a molten resin through an orifice 2.096 mm in diameter and 8.000 mm long at 190°C when subjected to a pressure difference of 2.982 MPa. This combination of temperature and pressure is now known as Condition F of ASTM D 1238, and the load is exactly ten times that of Condition E, at one time the single condition for the test and still the one most used for determining the MELT-FLOW INDEX (w s) of polyethylene and several other resins. (2) More generally in application to the wide range of thermoplastics now characterized at one of the 24 approved conditions of D 1238, any condition in which the load is ten times that of the one usually used to determine the melt-flow index of the specimen resin. The high load causes roughly 100 times the flow rate of the low load, so one flow rate or the other is likely to be difficult to measure accurately by the older, manual procedure described in D 1238. However, the two measurements can provide a rough estimate of the *flow-behavior index* (see POWER LAW) and, together, provide much better insight to the processing behavior of the resin than does the single measurement at the low load.

High-Molecular-Weight High-Density Polyethylene (HMWHDPE) See POLYETHYLENE.

High-Performance Composite See ADVANCED COMPOSITE.

High-Performance Plastic (advanced plastic) Any neat, filled, or reinforced resin, thermoplastic or thermoset, that maintains stable dimensions and mechanical properties above 100°C. Some writers consider many of the ENGINEERING PLASTICs (w s) to be high-performance plastics and the distinction is blurred. Others have limited the term to specialty thermoplastics (generally high-priced) such as polyimide, polytetrafluoroethylene, polyphenylene oxide and sulfide, liquid-crystal polymers, ultra-high-molecular-weight polyethylene, etc. See also ADVANCED RESIN.

High-Performance Polyethylene Fiber A strong, stiff fiber made from ultra-high-molecular-weight polyethylene by gel-spinning and drawing to a highly oriented state. Produced in the US by Allied-Signal as Spectra® fiber.

High Polymer A polymer with molecules of high molecular weight, sometimes arbitrarily designated as greater than 10,000. All materials commonly regarded as plastics are high polymers, but not all high polymers are plastics. See POLYMER.

High-Pressure Laminate A laminate molded and cured at pressures not lower than 6.895 MPa (1000 psi), and more commonly in the range of 8 to 14 MPa. See also DECORATIVE BOARD and LAMINATE.

High-Pressure Molding A method of molding or laminating in which the pressure used is greater than 1.4 MPa (200 psi) (ASTM D 883).

High-Pressure Powder Molding Some polymers in powder form can be molded by high-pressure compaction at room temperature followed by heating to complete sintering, curing or polymerization reactions. The process is limited to fairly simple shapes, and to polymers that do not release vapors when heated. It has been most successful with semi-crystalline polymers that can be post-heated for a sufficient time at a temperature within the crystalline endotherm of the polymer. Examples of such polymers are polyphenylene oxide, polytetrafluoroethylene, and ultra-high-molecular-weight polyethylene.

High-Pressure Spot A defect in reinforced plastics: an area containing very little resin, usually due to an excess of reinforcing material.

High-Temperature Plasticizer Any plasticizer that imparts higher than the normal resistance to high temperatures to plastics compounds in which it is incorporated. An example is di-tridecyl phthalate, which permits vinyl compounds to be used at temperatures up to 136°C.

Hindered Isocyanate See ISOCYANATE GENERATOR.

HIPS Abbreviation for HIGH-IMPACT POLYSTYRENE, w s.

Histogram (frequency plot) A bar chart compiled from a sample of 40 or more measurements of a quality characteristic or dimension on production parts or finished products. The measurements are first sorted into 8 to 20 classes, usually of uniform width, and a count is made of the number in each class. The counts may be converted to percentages of the total number measured. The abscissa is a scale covering the range of values found and the ordinate is the number or percentage found in each class.

HMT Abbreviation for HEXAMETHYLENETETRAMINE, w s.

HMWPE Abbreviation for *high-molecular-weight polyethylene*. See POLYETHYLENE.

Hob (1, n) A master model of hardened steel that is pressed into a block of softer metal to form a number of identical mold cavities. The hobbed cavities are inserted into recesses in a steel mold base and are connected by a runner system. (2, v) To form a mold cavity by forcing a hardened steel *hob* having the inverse shape of the cavity into a soft metal block (that may subsequently be hardened).

Hog A machine for reducing large pieces of plastic such as rejected moldings or extrusions to refeedable particles, similar to a GRANULATOR, w s, but more sturdily constructed, equipped with more knives, and using forced air to urge the particles through the screen.

Hold-Down Groove A small groove cut into the side wall of a mold to assist in retaining the molding in the mold as it opens.

Holomicroscopy A three-dimensional photomicrographic process utilizing laser beams and time-differential interferometry. Part of the laser beam is split off and directed through the sample under observation, while the other part of the beam takes another precise path and eventually is merged with the first half, interfering with it. The beam that passes through the sample is retarded and the amount of slowing can be estimated from the appearance of the hologram. The process has been used with polymers to clarify crystal structure.

Homogenizer A machine used to break up agglomerates and disperse elemental particles in fluids, consisting of a positive-displacement pump capable of attaining very high pressure, an orifice through which the material is forced at high velocity, and an impact ring on which the stream impinges. Homogenizers are used in the preparation of monomer emulsions and coating compounds, and for dispersing pigments and plasticizers into resins. See also COLLOID MILL.

Homologous Series A family of organic compounds that have identical functionality with each succeeding member having one more $-CH_2-$ group in its molecule than the preceding member. An example is the series *methanol* (CH_3OH), *ethanol* (CH_3CH_2OH), *propanol* (C_3H_7-OH), *butanol* (C_4H_9OH), etc.

Homologous Temperature The ratio of the absolute temperature of a material sample to the absolute temperature at which the material melts.

Homopolymer A polymer resulting from the polymerization of a single monomer and consisting of a single type of repeating unit.

Honeycomb Manufactured product consisting of sheet metal or resin-impregnated sheet material (paper, fibrous glass, etc) that has been formed into a network of open-ended, hexagonal cells, each cell's walls being shared with its immediate neighbors. Honeycombs are used as cores for sandwich constructions.

Hookean Elasticity (ideal elasticity) Stress-strain behavior in which stress and strain are directly proportional, in accordance with HOOKE'S LAW, w s.

Hookean Spring A concept visualized as a coil spring whose extension is proportional to the applied load, useful by analogy in modeling the viscoelastic behavior of polymers.

Hooke's Law The observation, by Robert Hooke, that, in a body placed in tension, the fractional increase in length is proportional to the applied load divided by the body's cross-sectional area perpendicular to the load. It was later extended to shear and compressive loading. Thomas Young later contributed the idea of the modulus and formulated Hooke's law as $e = \sigma/E$ *where* σ = the applied stress, E = the modulus of elasticity of the material, and e = the relative elongation, i e, the change in length divided by the original length. Few plastics conform to Hooke's law beyond deformations greater than one or two percent.

Hooke Model See HOOKEAN SPRING.

Hoop Stress The circumferential stress in a cylindrical body, such as a pipe, that is subjected to internal (or external) pressure. It is given by $\sigma = P \cdot D/(2 \cdot t)$ where P = the pressure, D = the mean diameter of the cylinder, $(D_o + D_i)/2$, and t = the wall thickness = $(D_o - D_i)/2$. In pipes under internal pressure, the hoop stress is twice the lengthwise stress, so dominates the processes of creep and failure.

Hopper In extrusion or injection molding, the bin mounted over the feed opening that holds a supply of molding material. The hopper may be intermittently filled or continuously fed. (See HOPPER LOADER.) Feeding from the hoppper is ordinarily by gravity, but it may be aided by vibrators, stirrers, or screw feeders. In some setups the hopper feeds into a metering device such as a GRAVIMETRIC FEEDER or VIBRATORY FEEDER, w s, that meters the rate at which the feedstock enters the extruder or molder.

Hopper Dryer A hopper through which hot air flows upward, drying and heating the feedstock. To improve extraction of moisture from the plastic the air may be passed through a desiccant prior to entering the hopper. The attendant preheating also reduces the amount of power the extruder must furnish per unit mass of material fed and can significantly increase the extruder output.

Hopper Loader (hopper filler) A device for automatically feeding molding powders to hoppers of extruders and injection molders, and maintaining the level of feedstock in the hopper. The functions of drying and blending color concentrates with the feedstock are also sometimes accomplished by loaders. There are two general types: mechanical and pneumatic. The mechanical systems use a rotating screw in a tube, operating only partly filled, or a conveyor belt or chain on which are fastened small containers that dump their contents into the hopper. Pneumatic systems employ positive pressure from a blower or pump, or negative pressure from a vacuum source, to "airvey" the material through a tube from the primary source, such as a drum or railroad car, to the machine hopper. Both types require some kind of signal of feedstock level in the hopper to control the replenishment rate.

Hotbench Test A method of determining gelation properties of plastisols, employing a bar or plate whose temperature rises from one side to the other ("temperature-gradient plate"). The sample is spread on the bar and from the positions at which various changes occur in its state, the temperatures of those changes may be estimated.

Hot-Gas Welding A welding process for plastics analogous to that used for metals, except that a stream of hot gas is used instead of a flame. Welding guns for plastics consist of a blower behind a heating element, much like a hair drier, and a nozzle that focuses the stream at the weld zone. Either air, or, better, dry nitrogen is used. The heated gas is directed at the joint that has been prepared for welding, while a rod of the plastic being welded is applied to the heated zone and melted into the groove.

Hot-Leaf Stamping See HOT STAMPING.

Hot-Manifold Mold An injection mold equipped with an internal heater located in the center of the melt stream in the manifold and nozzle system. This type of mold was developed for thermally sensitive resins to provide gentler heating and avoid the decomposition problems experienced with external heating techniques because of excessive temperature differences.

Hot-Melt Adhesive A thermoplastic adhesive applied to the surfaces to be joined in the molten state, then allowed to cool, usually under pressure. They are convenient, require no drying, and leave no voids in the joint. But they can creep and generally have lower strength than standard adhesives. ASTM has developed a number of tests for hot-melt adhesives, most of them in section 15.06 of the Annual Book of Standards.

Hot-Plate Welding (hot-tool welding) Two plastic surfaces to be joined are first held lightly against a heated metal surface, which may be coated with polytetrafluoroethylene to prevent sticking, until the surface layers have melted. The surfaces are then quickly joined and held under light pressure until the joint has cooled. See also THERMOBAND WELDING and WELDING.

Hot-Runner Mold (insulated-runner mold) An injection mold for thermoplastics in which the runners and sometimes the secondary sprues are insulated from the chilled cavity plate so that they remain hot during the entire cycle. The plastic in the runners remains molten and is not ejected with the molded part, thus avoiding the normal handling, grinding, and reprocessing of sprues and runners.

Hot-Short (adj) Inelastic, not stretchable, and easily broken in tension when hot.

Hot Stage A microscope stage equipped with heating and cooling at controllable rates, enabling one to observe changes in morphology with temperature, such as spherulite growth in polymers as they are cooled from the melt.

Hot Stamping (roll-leaf stamping, gold-leaf stamping) A method of marking plastics in which a special pigmented, dyed, or metallized foil is pressed against the plastic article by a heated die, welding selected areas of the foil to the article. The term also includes the process of impressing inked type into the material when the type is heated.

Hot-Tack Strength In heat sealing, the strength of the seal just at the end of the dwell when the die halves part (clearly more a concept than a measurable property).

Hot-Tip-Gate Molding An injection-molding technique used for thin, large-area articles molded in a single cavity. In conventional molding, a sprue connects the machine nozzle with runners leading to each gated cavity. In hot-tip-gate molding, the sprue is eliminated and material is injected directly from the nozzle through a heated bushing that serves as the sprue and gate for the individual cavity. Advantages are faster molding cycles, less material waste and reprocessing, fewer post-molding operations, and reduced sink marks and flow lines.

Hot-Wire Cutter A device used to split polystyrene-foam slabs into standard-size boards. Thin, electrically heated wires slowly melt through the slab as it is pushed against the array of wires by a conveyor belt or, for vertical cuts, by gravity.

HT-1 A type of nylon made from phenylenediamine and isophthalic or terephthalic acid, with good high-temperature properties.

Huggins Constant (k') The slope coefficient of the HUGGINS EQUATION, w s, found to be constant for a series of homologous polymers of different molecular weights dissolved in a particular solvent.

Huggins Equation In dilute polymer solutions, $\eta_{sp}/c = [\eta] + k'[\eta]^2c$, where η_{sp} and $[\eta]$ are the specific and intrinsic viscosities, c = the concentration in g/dL, and k' = the Huggins constant. By making viscosity measurements at several concentrations, plotting η_{sp}/c vs c, and extrapolating the line obtained to $c = 0$, the intrinsic viscosity can be evaluated. Then k' = the slope of the line divided by $[\eta]^2$. See, too, DILUTE-SOLUTION VISCOSITY.

Humectant A substance, e g, sorbitol, that promotes retention of moisture. Humectants have been used in antistatic coatings for plastics.

Humidity (1) The amount of a volatile compound, normally a liquid at the prevailing ambient conditions, present as its vapor in a gas. (2) Specifically, the amount of water vapor present in air, with absolute humidity expressed as the mass of water per unit mass of dry air. See RELATIVE HUMIDITY.

Humidity Blush See BLUSHING.

Hydantoin Epoxy Resin See EPOXY RESIN.

Hydrated Alumina Synonym for ALUMINA TRIHYDRATE, w s.

Hydration The addition of water to another substance. The water may react with the other material, as in the hydration of lime; may be taken up in a mixed chemical/physical affinity (absorption), as in 6/6 nylon and numerous other plastics, in time reaching an equilibrium value for the prevailing conditions; or may simply be drawn into pores of the material by capillary action (wetting). See, too, SOLVATION.

Hydraulic Drive A power system used on some extruders and molding machines in which the screw is turned by a motor driven by liquid under pressure, a kind of inverse pump. Hydraulic drives are more compact than other types for given power output, and have excellent torque characteristics, but are usually economical for extruders only when several can be driven from a single hydraulic generator. With injection molders, hydraulic power is needed for injection stroking and usually, too, mold clamping, so the hydraulic motor to turn the intermittently rotating screw at a fixed speed is simple and cost-effective.

Hydroabietyl Alcohol $C_{19}H_{31}CH_2OH$. Obtained by hydrogenation of abietic acid (a triple-ring, aromatic acid extracted from rosin) and used as a plasticizer for PVC and some cellulosic resins..

Hydrocarbon Plastics (1) Plastics based on polymers made from monomers containing only carbon and hydrogen. (2) In the plastics industry, hydrocarbon resins are considered to be those thermoplastic resins of low molecular weight made from relatively impure monomers that are derived from coal-tar fractions, cracked-petroleum distillates, and turpentine. The family includes COUMARONE-INDENE RESINs, w s; cyclopentadiene resins; petroleum resins; terpene resins; and many others of little commercial importance. Having little strength, most hydrocarbon resins are rarely used alone. Their primary applications are as binders in asphalt flooring, processing aids in elastomers and polyolefins, and coating additives.

Hydrogel A three-dimensional network of a hydrophilic polymer, generally covalently or ionically crosslinked. The most widely used one is poly(hydroxyethyl methacrylate), especially in medical applications such as implants, blood bags, and syringes.

Hydrogenated Methyl Abietate $C_{19}H_{31}COOCH_3$. A derivative of abietic acid, which is extracted from pine rosin, used as a plasticizer for cellulose nitrate, ethyl cellulose, acrylic and vinyl resins, and polystyrene.

Hydrogenation The addition of hydrogen to another substance, often an unsaturated organic compound. The process usually requires elevated temperatures, high pressures, and catalysts.

Hydrogen Equivalent The number of replaceable hydrogen atoms in one molecule of a substance, or the number of atoms of hydrogen with which one molecule could react.

Hydrolysis A double-decomposition reaction between water and another substance in which water is split into its ions and reacts to form a weak acid or base.

Hydrolytic Degradation Any breakdown of a plastic in which reaction with water or water vapor (steam) plays a role.

Hydrometer An instrument that senses and indicates the density of a liquid. A simple type consists of a glass tube with a bulb at its bottom, fine lead shot in the bulb, and a graduated scale within the tube. The amount of shot is adjusted before the top of the tube is sealed so that, when the hydrometer is floated in a liquid of density within its calibrated range, the scale reads the density of the liquid at the meniscus.

Hydrophilic "Water-loving", i e, having an affinity for, and readily wetted by water.

Hydrophobic "Water-fearing", i e, not capable of reacting with water and difficult to wet with water.

Hydroquinone (*p*-dihydroxybenzene, hydroquinol, *p*-hydroxyphenol, quinol,) $C_6H_4(OH)_2$. A white crystalline material derived from aniline, used, as are many of its derivatives, as an inhibitor of free-radical polymerization in unsaturated polyester resins and in monomers such as vinyl acetate. Hydroquinone is almost colorless and can retain its inhibitory action even in the presence of oxygen.

Hydroquinone Di(β-Hydroxyethyl) Ether (HQEE) A white solid material used as a reactant in the preparation of polyesters, polyolefins, and polyurethanes. As a chain extender in urethane prepolymers, HQEE increases to 150°C the high-temperature resistance of parts molded from the prepolymer.

Hydrosol (1) In physical chemistry, a colloidal suspension in water. (2) In the plastics industry, a suspension of resin such as PVC or nylon in water, not necessarily of colloidal nature. See also LATEX.

Hydrostatic Design Basis (HDB) One of a series of 10^5-h or 50-yr strength values spelled out in ASTM D 2837 and forming the basis for choosing the wall thickness of pipe for various diameters and service pressures and temperatures.

Hydrostatic Design Stress The estimated sustained hoop stress that can exist in a pipe at expected service conditions and over the life of the pipe with a high degree of certainty that failure will not occur. See preceding entry.

Hydrostatic Strength The hoop, stress calculated from the following equation, at which a pipe fails because of rising internal pressure, usually in about one minute. The equation is: $\sigma = P \cdot D_m/(2 \cdot t)$, in which P = the internal (gauge) pressure at rupture, D_m = the initial mean diameter (= outer diameter – t), and t = the initial wall thickness. This strength is usually less than the tensile strength of the pipe material as determined in a standard tensile test because of the presence in the pipe of longitudinal stress equal to half the hoop stress. The pressure at which the failure occurs is known as the *quickburst pressure*.

Hydrous Alumina Synonym for ALUMINA TRIHYDRATE, w s.

Hydrox-, Hydroxy- A chemical prefix denoting the presence of the –OH group in a compound.

2(2'-Hydroxy-3,5-Ditertiarybutylphenyl)-7-Chlorobenzotriazole An off-white, nontoxic, crystalline powder with high thermal stability, used as an ultraviolet absorber for polyolefins, PVC, polyurethanes, polyamides, and polyesters.

Hydroxyethyl Acetamide See *N*-ACETYL ETHANOLAMINE.

Hydroxyethyl Cellulose Any of a family of polymeric ethers formed by reacting alkali cellulose with ethylene oxide. Water solubility and applications depend on the degree of substitution of hydroxyethyl groups for –OH groups, and include textile sizes, adhesives, thickeners, and stabilizers for vinyl polymers.

Hydroxyethylmethyl Methacrylate A monomer that polymerizes to a hydrophilic polymer that is rigid when dry but when saturated with water becomes a soft, clear material (Hydron®). Applications include masonry coatings, soft contact lenses, and other biomedical devices.

Hydroxyl Value A measure of hydroxyl groups (univalent –OH) in an organic material. In plasticizers, the hydroxyl value includes –OH groups present in any free unesterified alcohol as well as those of the plasticizer molecule itself. In some plasticizers, large hydroxy values signal that the plasticizer may become incompatible on aging. In urethane technology, hydroxyl number is an important factor in the selection of polyols to achieve desired characteristics in elastomers and foams.

2-Hydroxy-4-Methoxybenzophenone An ultraviolet absorber for numerous thermoplastics.

2(2'-Hydroxy-4'-Methylphenyl) Benzotriazole ("Tinuvin P") A nontoxic crystalline powder with high thermal stability, an ultraviolet absorber for polystyrene, acrylics, PVC, polyesters, and polycarbonates.

2-Hydroxy-4-Methoxy-5-Sulfobenzophenone An ultraviolet absorber for thermoplastics.

2-Hydroxy-4-*n*-Octoxybenzophenone $C_{21}H_{26}O_3$. A pale yellow powder, an ultraviolet absorber for PVC and several other plastics. It is compatible with highly plasticized vinyls and has a very low order of toxicity.

Hydroxypropylglycerin A pale-straw-colored liquid used as a plasticizer for cellulosic resins and as an intermediate for alkyd and polyester resins.

Hydroxypropyl Methacrylate (HPMA) $CH_3CH(OH)CH_2OOC(CH_3)=CH_2$. A reactive monomer copolymerizable with a wide variety of acrylic and vinyl monomers, used for thermosetting resins and surface coatings.

Hygrometer An instrument that senses and indicates the relative humidity (of moisture) in air.

Hygroscopic Having a strong tendency to absorb moisture from the air. Some resins are hygroscopic, and therefore usually require drying before being extruded or molded.

Hysteresis (1) The lagging of the physical effect on a body behind its cause, particularly during repeated cycling. In iron-core electromagnets, for example, magnetic induction in the iron lags the magnetic intensity. In fatigue testing of plastics, the elongation can lag the stress (or vice versa), forming a *hysteresis loop*. Hysteresis is accompanied by dissipation of energy as heat. If a fatigue specimen is cycled at too high a frequency and/or stress, the resulting heat buildup can seriously bias the test and even soften the specimen. (2) The influence of prior history that causes the repetition of a process or its reversal to take a different path from the original. In tensile loading of plastics past the proportional limit but short of failure, the path (stress vs strain) followed upon unloading will generally be at lower stresses than were recorded during loading at the same strains. Open or closed loops may be observed.

Hysteresis Loop The closed area between two curves on a graph plotting results of a changing stress on a test specimen, first with ascending values, then with descending values. Such hysteresis loops are present in flow curves of thixotropic liquids and in stress-strain plots of most plastics.

Hytrel® Du Pont's tradename for a family of copolyester elastomers. Typical reactants from which the elastomers are derived are terephthalic acid, polytetramethylene glycol, and 1,4-butanediol. Powders and pellets are available for extrusion and molding. The products are highly resilient, have good flex-fatigue life at low and high temperatures, and are resistant to oils and chemicals. Some grades termed *segmented copolyesters* are excellent modifiers for PVC, improving processability and imparting resistance to abrasion, impact and fungi.

Hz SI abbreviation for HERTZ, w s.

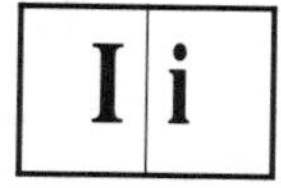

i (1) (Also I) Symbol for electric current. (2) Symbol for $\sqrt{-1}$, the coefficient of "imaginary", i e, out-of-phase, components in complex quantities.

I (1) Chemical symbol for the element iodine. (2) Symbol for electric current. (3) Symbol for moment of inertia.

ID Abbreviation for *internal diameter.*

Ideal-Gas Law The combination of Charles' and Boyle's laws, usually stated in the form $P \cdot V = n \cdot \mathrm{R} \cdot T$, where P = the absolute pressure, V = the gas volume, n = the number of moles of gas present, T = the absolute temperature, and R = the universal molar gas (or energy) constant. R has many numerical values, to be consistent with the units chosen for the variables. In SI, R = 8.31439 J/(mol·K). In some statements of the ideal-gas law, n stands for the *mass*, rather than moles, of gas. In this form, R has a different value for each gas considered.

Ideal Liquid This term has several meanings in physical chemistry and hydrodynamics. The one relevant to plastics is NEWTONIAN LIQUID, w s.

Ideal Solid A material that obeys HOOKE'S LAW, w s.

Ignition The beginning of burning.

Ignition Loss See ASTM D 2584 at FLAMMABILITY.

Ignition Temperature See FLAMMABILITY and FLASH POINT.

Ignition Time See FLAMMABILITY.

IM Abbreviation for INJECTION MOLDING, w s.

Immediate Set The deformation found by measurement immediately after the removal of the load causing the deformation in a short-time test.

Immiscible Of liquids, incapable of forming a solution, as kerosene and water. Compare INCOMPATIBLE.

Impact Adhesive See CONTACT ADHESIVE.

Impact Modifier A general term for any additive, usually an elastomer or plastic of different type, incorporated in a plastic compound to improve the IMPACT RESISTANCE, w s, of finished articles. The improvement is customarily assessed by performance of test specimens in standard tests.

Impact Polystyrene See HIGH-IMPACT POLYSTYRENE.

Impact Resistance The relative durability of plastics articles to fracture under stresses applied at high speeds. A widely used impact test, ASTM D 256, employs the Izod pendulum striker swung from a fixed height to strike a specimen in the form of a notched bar mounted vertically as a cantilever beam. The Charpy tester, an alternative in D 256, uses a specimen in the form of a horizontal beam supported at both ends. ASTM lists some fourteen different impact tests for plastics and plastics products. However, application of test results is limited by the simple geometry and notching of the specimens, so they are often augmented by subjecting actual manufactured products to conditions simulating end-use conditions, for example, dropping filled plastic bottles on a hard floor. See also BRITTLENESS TEMPERATURE, DROP-WEIGHT TEST, FREE-FALLING-DART TEST, and TENSILE-IMPACT TEST.

Impact Strength The quantitative measure of the ability of a material to withstand shock loading in a standard test. For plastics the test is usually either the Izod or Charpy test described in ASTM D 256, and the result is calculated as the energy expended (work done) in breaking a specimen, divided by its width or thickness. Specimens for both tests are usually notched on the side opposite that where they are struck, though the notch position may be reversed and unnotched specimens may be tested. In SI the convenient reporting unit is J/cm of notch width, or, for unnotched specimens, J/cm. To provide a much simpler stress distribution free of notch effects, researchers developed the TENSILE-IMPACT TEST, w s. See also the preceding entry and IZOD IMPACT STRENGTH.

Impingement Mixing Very intense, rapid and thorough mixing accomplished by causing two fast-moving liquid streams or sprays to meet within a confined space. Typically the streams are resin and curing agent or resin streams containing catalyst in one and accelerator in the other. The technique is used in reaction-injection molding, in sprayup of reinforced plastics, and in foam-in-place molding.

Impregnation The process of thoroughly soaking a material of an open or porous nature with a resin. When webs or shapes of reinforcing fibers are impregnated with a thermosetting resin in the B stage or with a thermoplastic, and such webs are intended for subsequent shaping or laminating, the masses are called SHEET-MOLDING COMPOUND or PREPREG, w s. The main difference between impregnation and ENCAPSULATION, w s, is that in encapsulation an outer protective coating is formed with little or no penetration of the resin into the article, whereas in impregnation there is little or no protective coating.

Impulse Sealing (thermal-impulse sealing) The process of joining thermoplastic sheets or films by pressing them between elements equipped to provide a pulse of intense heat to the sealing area for a very short time, followed immediately by cooling. The heating element may be a length of thin resistance wire such as nichrome, or a bar heated by a high-frequency electric field and cored for water cooling. See also HEAT SEALING.

Incandescent Luminous in the yellow-to-white range because of being at a high temperature (1250 to 1550°C).

Inching A very low rate of mold closing used in many molding operations and often automatically included in the cycle, for the last millimeter or so before the mold faces meet.

Inclusion A foreign body or impurity phase in a solid.

Inclusion Complex See ADDUCT.

Incompatible Said of two materials, such as two resins, or a resin and plasticizer, that are incapable of forming a solution or even a stable two-phase blend and that tend to separate after being mixed.

Indene A bicyclic, aromatic hydrocarbon derived from the fraction of coal tar boiling between 176 and 182°C, having the structure:

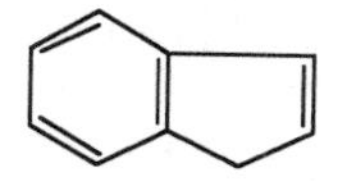

It is used in the production of COUMARONE-INDENE RESIN, w s.

Indentation Hardness The hardness of a material as determined by either the size of an indentation made by an indenting tool under a fixed load, or the load needed to produce penetration of an indenter to a predetermined depth. The instruments commonly used with plastics are the Shore Durometer (indenter A for soft resins and elastomers, D for hard materials) described in ASTM D 2240, and the Barcol Impressor, ASTM D 2583. In D 2240, the authors say, "No simple relationship exists between indentation hardness determined by this method and any fundamental property of the material tested. For specification purposes it is recommended that Test Method D 1415 (section 09.01) be used for soft materials and method A of D 530 or Test D 785 be used for hard materials." D 530 and D 785 use the Rockwell test in which a ball of suitable diameter is pressed into the sample with known force, and the indentation depth is noted. A similar test format is used in D 1415.

Index of Refraction (refractive index) The ratio of the velocity of light in a vacuum to its velocity in the transparent material of interest. It generally varies with the wavelength of the light, being higher at the shorter wavelengths, also with temperature. For precise work, therefore, measurements are corrected to the wavelength of the D line of the sodium spectrum and to 20°C. When a light beam passes from a less dense medium to a more dense one at an acute angle θ_1 with the interfacial normal, it will be bent closer to the normal in the more dense medium, defining the angle of refraction, θ_2. The index of refraction is given by $\sin \theta_1 / \sin \theta_2$. This relationship is the basis for measurement of refractive index with an instrument such as the Abbé refractometer (ASTM D 542 for transparent plastics). Index of refraction is useful for identification of plastics, as well as minerals and liquids, and is involved in many quantitative analytical methods.

India Rubber See RUBBER, NATURAL.

Indicia Any markings such as symbols, lettering, small pictures, etc, applied to a plastic article to identify it.

Indophenol Test See GIBBS INDOPHENOL TEST.

Inductance The change in a magnetic field due to the variation of current in a conducting circuit causing an induced counter-electromotive force in the circuit itself. The henry (H) is the inductance of a closed circuit in which an electromotive force of one volt is produced when the current in the circuit varies at the rate of one ampere per second. Thus, $1\ H = 1\ V/(A/s) = 1\ V \cdot s/A = 1\ Wb/A$.

Induction Heating A method of heating electrically-conductive materials, usually metallic parts, by placing the part or material in an alternating magnetic field generated by passing an alternating current through a primary coil. The alternating magnetic field induces eddy currents in the piece that generate hysteretic heat. Plastics, being poor conductors, cannot be heated directly by induction, but the process is used indirectly in welding of plastics and has been used to heat extruders and dies.

Induction Period A usually brief time after reactants are brought together during which no measurable reaction occurs, and thereafter followed by reaction. An induction period is ordinarily due to the presence of a stabilizer or inhibitor.

Induction Welding A method of welding thermoplastic materials by placing a conductive metal insert between two plastic surfaces to be joined, applying pressure to hold the surfaces together, heating the metallic insert in an alternating magnetic field (see INDUCTION HEATING) until the adjacent plastic is softened and welded, then killing the field and cooling the joint.

Industrial Talc A crude mineral filler that may contain, besides TALC (w s), some other minerals such as asbestos.

Inert Additive A material added to a plastic compound to improve properties, but which does not react chemically with any other constituents of the composition.

Infrared Heating A heating process used mostly in sheet thermoforming and for drying coatings, employing lamps or heating elements that emit invisible radiation at wavelengths of about 2 μm.

Infrared Polymerization Index (IRPI) A number representing the degree of cure of phenolic resins, defined as the ratio of absorbances by the sample, corrected for background absorbance, at wavelengths of 12.2 and 9.8 μm (far infrared). The test is often used in conjunction with the MARQUARDT INDEX, w s.

Infrared Pyrometer A narrow- or broad-band instrument that senses the peak wavelength, λ (μm) of IR radiation emanating from a warm surface. By Wien's displacement law, the absolute temperature (K) is given by $2884/\lambda$.

Infrared Spectrophotometry A technique to identify and quantitatively determine many organic substances such as plastics. All chemical compounds have characteristic intramolecular vibratory motions and can absorb incident radiant energy if such energy is sufficient to increase the vibrational motions of the atoms. With most organic molecules, vibrational motions of the substituent groups within the molecules coincide with the electromagnetic frequencies of the infrared region. An infrared spectrophotometer directs IR radiation through a film or layer or solution of the sample and measures and records the relative amount of energy absorbed by the sample as a function of the wavelength or frequency of the radiation. The chart produced is compared with charts of known substances to identify the sample. Peak area is a measure of the percentage of the various compounds present.

Infrared Thermography (thermographic NDT) A type of nondestructive testing in which damage development during fatigue testing of composites may be detected without interrupting the test by measuring locally generated infrared radiation. The damage is associated with heat developed through two mechanisms: hysteresis and frictional heating from rubbing at the surfaces of developing cracks.

Infusible Not capable of melting when heated, as are all cured thermosetting resins and a few special thermoplastics such as ultrahigh-molecular-weight polyethylene, polybenzimidazole, and aramid resins.

Inherent Viscosity (logarithmic viscosity number) In measurement of dilute-solution viscosities, the inherent viscosity is the ratio of the natural logarithm of the relative viscosity to the concentration of the polymer in g/dL of solvent. See DILUTE-SOLUTION VISCOSITY.

Inhibitor A substance capable of retarding or stopping an undesired chemical reaction. Inhibitors are used in certain monomers and resins to prolong storage life. When used to retard degradation of plastics by heat and/or light, an inhibitor functions as a STABILIZER, w s.

Initial Modulus The MODULUS OF ELASTICITY (w s) extrapolated to zero strain from measurements in the low-strain region.

Initial Viscosity A term used in the vinyl plastisol industry to denote the viscosity measured immediately after the plastisol has been mixed. The viscosity normally rises at a declining rate after mixing.

Initiator An agent that causes a chemical reaction to commence and that enters into the reaction to become part of the resultant compound. Initiators differ from catalysts in that catalysts do not combine chemically with the reactants. Initiators are used in many polymerization reactions, especially in emulsion polymerizations. Initiators most commonly used in polymerizing monomers and resins having ethenic unsaturation (–C=C–) are the organic peroxides.

Injection Blow Molding A blow-molding process in which parisons are formed over mandrel by injection molding, after which the mandrels and parisons are transferred to blow molds where the final shape is blown. While the parts are being blown, cooled and ejected, another set of parisons is being injection-molded. Advantages of the process are that it delivers a completely finished part requiring no neck trimming, etc, closer tolerances are possible, and parison-wall thicknesses can be locally varied as desired.

Injection Mold A mold used in the process of INJECTION MOLDING, w s. The mold usually comprises two main sections held together by a hydraulic or mechanical clamping press, that have sufficient strength and rigidity to contain the high pressure of molten plastic when it is injected. It is provided with channels for mold venting and for circulation of temperature-control media, and with means for product ejection.

Injection Molder (1) A person or company that operates injection-molding machines. (2) An injection-molding machine.

Injection Molding The method of forming objects from granular or powdered plastics, most often thermoplastics, in which the material is fed from a hopper into a screw-type plasticator (or heating cylinder in elderly machines), after which the screw or a ram forces the molten compound into a chilled mold. Pressure is maintained until the mass has hardened sufficiently for removal from the mold. In a variation called *flow molding*, a small additional amount of melt is forced into the mold during cooling of the initial charge to offset shrinkage. Machines employing screws for plastication (as most do nowadays) are either single- or double-stage. In single-stage machines, plastication and injection are done in the same cylinder, injection pressure being generated by forward motion of the screw while rotation is stopped. This process is called *reciprocating-screw injection molding*. In double-stage machines, the material is plasticated by a constantly rotating screw that delivers the melt through a check valve to an accumulator cylinder, from which it is injected into the mold by a piston. This process is called *screw-and-piston injection molding*. See also REACTION INJECTION MOLDING, HOT-TIP-GATE MOLDING, and TWO-SHOT INJECTION MOLDING.

Injection-Molding Pressure (1) The pressure applied to the plastic by the injection screw (or ram) during the injection stroke. (2) The pressure of the melt within the mold cavity just prior to the freezing of the gate.

Injection Nozzle A short, thick-walled, hardened-steel tube that conducts the molten plastic emerging from an injection cylinder into the mold. It usually terminates in a hemispherical tip that closely mates a recess in the mold called a *sprue bushing* and is enclosed by band heaters that are controlled by a temperature sensor.

Injection Ram In two-stage injection molders and elderly machines having no screws but using contact heating, the piston that applies pressure to the plastic during injection. In single-stage screw-injection machines, the screw doubles as the ram during a brief period of stopped rotation.

Injection Stamping A little used modification of injection molding wherein first the plastic melt is injected under relatively low pressure into a mold that is vented during this stage, then, after the cavity is filled, additional clamping pressure is applied to completely close the mold and compress or "stamp" the molded shape. Molds are designed so that even in the venting position no material exudes onto the land areas. Advantages claimed for the process are lower injection time and pressure, and shorter cycle because the injection screw can resume plastication as soon as the mold has been charged.

Ink Adhesion The tendency of printing ink to stick to the printed surface and not smear or rub off. An informal test of ink adhesion consists of pressing pressure-sensitive tape over the printed area, waiting a few minutes, then stripping it off and observing the areal fraction of ink coming away on the tape.

Inlay Printing See VALLEY PRINTING.

In-Mold Decorating The process of applying labels or decorations to plastic articles during the molding operation by which they are formed. Two basic methods, each with many variations, are in use. The first employs a preprinted label of plastic film, paper, or cloth that is positioned in the mold prior to molding. During the molding cycle, the label or its printed image fuses to and becomes an integral part of the article. In the second method, the image is printed directly onto the mold surface with wet or dry ink, or applied to the mold by an offset process. In-mold decorating is done with blow molding, injection molding, and casting.

Inorganic Pigment Any pigment derived from naturally occurring minerals or synthesized from inorganic substances. They are always opaque, unlike organic pigments and dyes, and are usually resistant to heat and light. Examples of those used in plastics are titanium dioxide, iron oxides, ultramarines, lead chromates, and cadmium compounds.

Inorganic Polymer Within the scope of the plastics industry, an inorganic polymer may be defined as a polymer without carbon in its main chain and of a degree of polymerization sufficient for the polymer to exhibit considerable mechanical strength, plastic or elastomeric properties, and the ability to be formed by processes used with plastics. Organic-group side chains may be present. The inorganic polymers of greatest commercial importance are the SILICONEs, w s.

Insert (n, first syllable accented) An article of metal or other material that is incorporated in a plastic molding either by pressing the insert into a recess in the finished molding or by placing the insert into the cavity so that it becomes a part of the molding. A common example is a metal bushing, knurled on the outside, threaded on the inside, for attaching the molded article to another article with a screw or bolt.

Insert Molding See DOUBLE-SHOT MOLDING.

In-Situ Foaming The technique of depositing a foamable plastic (prior to foaming) into the volume where it is intended that foaming shall occur. An example is the placing of foamable plastics into cavity brickwork to provide insulation. Shortly after being poured, the liquid mix foams and fills the cavity. See also CELLULAR PLASTIC.

Instant-Set Polymer (ISP) A modified polyurethane material that contains an organic modifier that is soluble in the liquid phase but insoluble in the solid phase. During the reaction the modifier precipitates out of solution and is trapped in spherical droplets about 0.5 μm in diameter. These cells act as a heat sink to retard heat build-up and thermal degradation, and also contribute to good machinability and resistance to fatigue and stress cracking. ISP can be used to cast thick parts in short molding cycles, with mechanical properties comparable to those of engineering plastics.

Instron A tensile or other testing machine made by the Instron Corp of Canton, MA. Their machines have been so omnipresent in plastics testing that "Instron" has taken on a near-generic character.

Insulated-Runner Mold See HOT-RUNNER MOLD.

Insulation Resistance (1) The electrical resistance between two conductors or systems of conductors separated only by an insulating material. The resistance of a particular insulation may be measured by dividing the voltage difference applied to two electrodes in contact with, or embedded in a unit area of the specimen by the current flowing between the electrodes. Tests for thermoplastics include ASTM D 2633, section 10.02. See also RESISTIVITY. (2) See THERMAL RESISTANCE.

Integral-Skin Molding A method of producing urethane-foam articles with substantially nonporous integral skins in one operation. Whereas normal urethane foams are inflated by carbon dioxide generated by reaction of isocyanate with excess water in the mixture (see URETHANE FOAM), integral-skin foam is expanded by the vapor of a volatilized solvent such as hexane. The mold must be heat conductive so that a layer next to the mold surfaces can be chilled to prevent foaming. The reaction mixture, typically consisting of a polyol, an isocyanate, and the blowing solvent, is introduced rapidly into a closed mold located near the geometric center of the mold. The mold is usually preheated to between 40 and 65°C, but the main heat needed to foam and cure the mass is provided by the reaction exotherm. After the mass has gelled, the skin should be pierced to equalize pressure and prevent shrinkage. The solvent blowing agent can be removed from the finished article by allowing it to stand at ambient conditions for about a day, or by oven drying for half an hour at 140°C. In some instances, integral-skin moldings have replaced composites of vinyl-covered urethane foam articles and cast vinyl skins filled with urethane foam, such as automotive arm rests, crash pads, and instrument-panel covers.

Intensity of Segregation In mixing, the average measure of the deviation in concentration of a component at any point in a mixture from the mean concentration. The STANDARD DEVIATION (w s) as determined by sampling is usually used.

Intensive Mixer See INTERNAL MIXER.

Interface The common surface separating two different phases, e g, the resin-glass interface in glass-fiber-reinforced plastics.

Interfacial Polymerization A polymerization reaction that occurs at or near the interfacial surfaces of two immiscible solutions. A simple example is the often-performed demonstration of making nylon thread from a beaker containing a lower layer of a solution of sebacyl chloride in carbon tetrachloride and an upper layer of hexamethylene diamine solution in water. A pair of tweezers is gently lowered through the upper layer, closed on the interfacial layer of polymer, then drawn upward to pull with it a continuous strand of nylon 6/10.

Interfacial Tension (interfacial surface energy) The tension in, or energy of, the interfacial surface between two immiscible liquids. Measurable, like SURFACE TENSION, w s, by capillary rise or with a du Noüy tensiometer.

Interferometry Any system of measurement based on wave interference between split rays of a light beam, one of which takes a longer or more optically dense path than the other, thus delaying it and generating a phase difference between the two rays. Interferometry has been used to measure a great variety of quantities, e g, the speed of light, film thicknesses, and chemical concentrations in solutions.

Interfusion (infusion) A novel method for emplacing hard-surfacing alloys within extruder barrels and, in future, on other parts. Few details have been released (as of late 1992) but it appears the hard-facing material in powder form is distributed around the barrel surface by centrifugal action while induction heating, and perhaps high pressure, are applied to speed the diffusion of that material into the barrel near its inner surface. The process marketers, Inductametals of Chicago, claim much slower barrel wear than is experienced by cast-in liners. See also BIMETALLIC CYLINDER.

Interlaminar Shear Stress Shear stress between layers of a laminate, an important cause of laminate failure and delamination.

Interlayer An intermediate sheet in a laminate.

Interlock Twiner A machine for making three-dimensional braided preforms in which yarn bundles are entwined by braiders to form the desired shape.

Intermediate A compound produced from raw materials that is to be used to synthesize end products. For example, benzene, originally distilled from coal tar and now made from petroleum constituents, and its derivatives, cyclohexane, cyclohexanol, and adipic acid are all intermediates in the manufacture of nylon 6/6.

Internal Bubble Cooling In blown-film production, the circulation of chilled air or carbon dioxide inside the film bubble to substantially reduce the cooling time and thereby increase the production rate. The same practices have been used in blow molding.

Internal Lubricant A lubricant that is incorporated into the compound or resin prior to processing, as opposed to one that is applied to the mold or die. Examples of internal lubricant are waxes, fatty acids and their amines, and metallic stearates such as calcium, lead, lithium, magnesium, and zinc stearates. The lubricants reduce friction and adhesion between polymers and metal surfaces, improve flow characteristics, and enhance knitting at weld surfaces and wetting properties of compounds. They are used primarily in rigid and flexible PVC, high-molecular-weight polyethylene, polystyrene, and acrylonitrile-butadiene-styrene, melamine, and phenolic resins.

Internal Mandrel Cooling In extrusion of tubing and pipe, the mandrel, which is an extension of the die core, is usually cooled internally so that the hot extrudate, cooled from both surfaces, cools faster than if cooled only on the outside, and with better control of dimensions.

Internal Mixer A heavy-duty machine in which the materials to be mixed are strenuously worked and fused by one or more rotors designed so that all parts of the charge pass repeatedly through zones of high shear. The shell and rotors may be cored to permit circulation of heat-transfer liquids to control batch temperature. One type, the BANBURY MIXER, w s, has long been used in the compounding plastics and rubbers, doing a good job of both *dispersive* and *distributive* mixing, Internal mixers have the inherent advantage of minimizing dust and fume hazards. The disadvantage of most types is that processing is batchwise rather than continuous. See BANBURY MIXER.

Internal Plasticizer An agent incorporated in, or copolymerized with, a resin during its polymerization to make it softer and more flexible, as opposed to a plasticizer added to the resin during compounding.

Internal Stabilizer An agent incorporated in a resin during its polymerization to make it more resistant to high temperatures and other processing and environmental conditions, as opposed to a stabilizer added to the resin during compounding.

Internal Undercut Any restriction that prevents a molded part from being directly removed from its core (force).

Interpenetrating Polymer Network (IPN) A kind of blend formed by swelling a crosslinked polymer with a monomer, then inducing polymerization of the monomer. Another route is to infuse a crosslinking monomer into a crosslinkable polymer.

Interpolymer A type of COPOLYMER, w s, in which the two monomer units are so intimately distributed in the polymer that the substance is essentially homogeneous in chemical composition. An interpolymer is sometimes called a *true copolymer*.

Intrinsic Viscosity (limiting viscosity number) In measurements of dilute-solution viscosity, intrinsic viscosity is the limit of the reduced and inherent viscosities as the concentration of polymer solute approaches zero. It represents the capacity of the polymer to increase viscosity. Interactions between solvent and polymer molecules give rise to different intrinsic viscosities for a given polymer in different solvents. Intrinsic viscosity is related to polymer molecular weight by the equation $[\eta] = K' \cdot M^a$, where the exponent a lies between 0.5 and 1.0, and, for many systems, between 0.6 and 0.8. See also DILUTE-SOLUTION VISCOSITY, HUGGINS EQUATION, and VISCOSITY-AVERAGE MOLECULAR WEIGHT.

Introfaction The change in fluidity and wetting capability of an impregnating material, produced by addition of an INTROFIER, w s.

Introfier A chemical that will convert a colloidal solution into a molecular one, by improving the solubility of the colloidal material.

Intumescence The foaming and swelling of a plastic when exposed to high surface temperatures or flames. It has particular reference to ablative urethanes used on rocket nose cones, and to INTUMESCENT COATING, w s.

Intumescent Coating A coating that, when exposed to flame or intense heat, decomposes and bubbles into a foam that protects the substrate and prevents the flame from spreading. Such coatings are used, for example, on reinforced-plastics building panels. Examples of such coating materials are magnesium oxychloride cement used on urethane foams, and certain epoxy coatings used on polyester panels.

Inventory In injection molding or extrusion, the amount of plastic contained in the heating cylinder or barrel.

Investment Casting (lost-wax process) A metals-casting method in which patterns made from wax or other expendable material are mounted on sprues, then "invested", i e, covered with a ceramic slurry that sets at room temperature. The set slurry is then heated to melt away the pattern, leaving a mold into which metal is poured. In one variation of the process, the set slurry is fired at 900°C to silica-bond it into a sturdy mold into which casting steel is poured to make the desired mold force or cavity for subsequent plastics molding. The finish, dimensions, and fine detail of the original pattern are faithfully reproduced in the final molded product.

Iodine Value (iodine number) The number of grams of iodine that 100 grams of an unsaturated compound will absorb in a given time under arbitrary conditions. It is used to indicate the residual unsaturation in epoxy hardeners (and many oils); a high value implies a high degree of unsaturation. Relevant ASTM tests (all in sec 06.03) are D 29, D 460, D 1541, D 1959, and D 2075. See also OXIRANE VALUE.

Ion An atom, molecule or radical that has become electrically charged by having either gained or lost an electron. When an electron is gained the negatively charged ion is called an *anion.* A positively charged ion is called a *cation.*

Ion Exchange A reversible interchange of ions between a solid phase and a liquid phase in which there is no permanent change in the structure of the solid phase. In the leading application, water softening, an ION-EXCHANGE RESIN (w s) extracts "hard", soap-precipitating calcium, magnesium, and iron ions from the water, replacing them with equivalent amounts of soluble sodium ions. Subsequently, the resin loaded with hard ions may be treated with salt solution (regenerated) to bring it back to the original sodium form, ready for reuse.

Ion-Exchange Resin Any of several small granular or bead-like resins consisting of two principal parts: a resinous matrix serving as a structural portion, and an ion-active group serving as the functional portion. The functional group may be acidic or basic. The resin matrix most often used is a copolymer of styrene and divinylbenzene. Acidic ion exchangers are made by sulfonating the resin beads with, for example, sulfuric acid, chlorosulfonic acid, or sulfur trioxide. The basic materials often contain quaternary ammonium groups. Acidic ion-exchange resins are used for softening water (see preceding entry). Complete deionization of water is accomplished by use of both acidic and basic resins, in sequence or in mixed beds. Ion-exchange resins are also used for other chemical processes such as electrodialysis.

Ionic Pertaining to an atom, radical, or molecule that is capable of being electrically charged, either negatively (anionic) or positively (cationic) or both (amphoteric), or to a material whose atoms already exist in charged state.

Ionic Initiator A substance providing either carbonium ions (cationic)or carbanions (anionic) that attack the reactive double bonds of vinyl monomers and add on, regenerating the ion species on the propagating chain.

Ionic Polymerization (cationic polymerization, anionic polymerization) A polymerization conducted in the presence of electrically charged ions that become attached to carboxylic groups on carbon atoms in the polymer chain. The carboxylic groups are produced along the polymer chain by copolymerization, providing the anionic portion of the ionic crosslinks. The cationic portion is provided by metallic ions added to the polymerization mixture. The electrostatic forces binding the chains together are much stronger than the covalent bonds between the molecules in conventional polymers. Some polymers produced by cationic polymerization are polyisobutylene, butyl rubber, polyvinyl ethers, and coumarone-indene resins. A typical product of anionic polymerization is polybutadiene, prepared with an alkali-metal catalyst.

Ionic Polyurethane A urethane resin containing electrical charges in its backbone or side chain. Cationic urethanes (those containing positive charges) can be formed by reacting diisocyanates with diols containing tertiary nitrogen to yield urethanes that are then treated by a quaternization reaction that forms positive charges in the macromolecular backbone or in side chains. Applications of such ionic urethanes are aqueous dispersions with outstanding film-forming ability, polyelectrolyte complexes that can be cast from solutions to form coatings on films or tubings used in medical practice, and electrically conducting elastomers.

Ionitriding® See NITRIDING.

Ionization Foaming The process of foaming polyethylene by exposing it to ionizing radiation which evolves hydrogen from the molten polymer, causing it to foam.

Ionization Potential The work required to remove an electron from its atomic orbit and place it at rest at an infinite distance from the atom. The customary unit of ionization potential is the electron-volt (ev), one ev being equivalent to $1.60219 \cdot 10^{-19}$ J, or 96.49 kJ/mol.

Ionomer (1) The product of an IONIC POLYMERIZATION, w s. (2) See IONOMER RESIN.

Ionomer Resin A polymer containing interchain ionic bonding. In particular, commercial thermoplastics based on metal salts of copolymers of ethylene and methacrylic acid, produced in the US by the Du Pont Co under the tradename Surlyn®. Ionomers are tough and flexible, the many grades ranging in modulus from 14 to 590 MPa. They have been approved for food-contact applications by the FDA, have outstanding resistance to puncture and impact. Applications include sporting goods (most golf-ball covers), footwear, packaging, automotive, and foams.

Ion Plating A process for deposition of metals for dielectric films onto plastic substrates with a highly adherent bond. The process is performed in a tank similar to a vacuum-metallizing tank. A negative charge is developed on a metal bias plate located behind the plastic substrate. Next, the plating material is converted to a plasma of positive ions by filament or radio-frequency heating. At this point a phenomenon known as Crooke's dark space appears, enveloping the entire surface of the substrate, and establishing a large potential difference between the ions and the charged plastic surface. This causes the ions of the plating material to strike the plastic with high kinetic energy and to form strong bonds.

IPA Abbreviation for ISOPHTHALIC ACID, w s.

IPN Abbreviation for INTERPENETRATING POLYMER NETWORK, w s.

IR (1) Abbreviation for *infrared*. (2) Abbreviation for *isoprene* *rubber* (British Standards Institution), the *cis*-1,4-type of POLYISOPRENE, w s.

Iron Oxide Pigment Heat-stable pigments ranging from yellows through blacks are obtained from various iron oxides. Reds are formed from ferric oxide, Fe_2O_3, yellows from hydrated ferric oxide, and blacks from ferroferric oxide, Fe_3O_4.

Irradiation The subjection of a material to high-energy particle radiation for the purpose of producing a desired change in properties or of determining the effects of radiation on the material. Thermosetting resins such as unsaturated polyesters, acrylic-modified polyesters, and acrylic-modified epoxies can be cured rapidly at room temperature and without catalysts by exposure to ionizing radiation. Radiation of most thermoplastics forms free radicals that react to link the molecular chains together in a three-dimensional network. This crosslinking in thermoplastics imparts higher density, increased softening temperatures, lower dielectric loss and improved chemical resistance. However, if the polymer is overdosed, serious degradation can occur. Radiation sources most widely used in the plastics industry are electron accelerators and radioisotopes such as cobalt-60.

Irregular Block A block (in a polymer structure) that cannot be described by only one species of constitutional repeating unit in a single sequential arrangement (IUPAC).

Irregular Polymer A polymer whose molecules cannot be described by only one species of constitutional unit in a single sequential arrangement (IUPAC).

Isano Oil A fatty oil extracted from an African tree of the same name, used as a flame retardant for acrylic resins. When heated to 200°C it polymerizes and may explode.

Iso- (1) The strict meaning of this prefix according to chemical nomenclature is "one methyl group on the next-to-last carbon atom, and no other branches". In the plasticizer field, the prefix is used to denote an isomer of a compound, specifically an isomer having a single, simple branching, not limited to methyl, at the end of a straight chain. (2) Equal, same, or constant, as in *isothermal*, at constant temperature.

ISO Abbreviation for *International Organization for Standardization*, headquartered in Geneva, Switzerland. ISO publishes standards in many fields, including hundreds on plastics in Field 170. ISO standards are available from AMERICAN NATIONAL STANDARDS INSTITUTE, w s.

Isoamyl Acetate Rectified AMYL ACETATE, w s.

Isoamyl Butyrate $C_5H_{11}OOCC_3H_7$. A colorless liquid derived by treating isoamyl alcohol with butyric acid, used as a solvent and as a plasticizer for cellulose acetate.

Isoamyl Salicylate (amyl salicylate, orchidae) $C_5H_{11}OOCC_6H_4OH$. A colorless liquid used as a plasticizer.

Isobaric Taking place without change in pressure.

Isobutene (isobutylene, 2-methylpropene) $(CH_3)_2C{=}CH_2$. A gas derived from petroleum, easily polymerized to form polybutene.

Isobutyl Acetate $(CH_3)_2CHCH_2OOCCH_3$. A colorless liquid with a fruity odor, used as a solvent for cellulosic plastics and lacquers. Its properties are similar to those of butyl acetate, except that its evaporation rate is higher.

Isobutyl Isobutyrate $(CH_3)_2CHCOOCH_2CH(CH_3)_2$. A colorless liquid with a fruity odor and slow evaporation rate, giving resin solutions with good flow and leveling characteristics. It is used as a solvent for nitrocellulose and vinyl resins.

Isocyanate A compound containing the isocyanate group, –N=C=O, attached to an organic radical or hydrogen. Isocyanates containing just one –N=C=O group (monoisocyanates) have limited uses in the plastics industry. The term is often used to mean a compound containing two –N=C=O groups (diisocyanate) or several such groups (polyisocyanate). However, in the case of a trimer compound containing three –N=C=O groups in a six-membered ring, the term *isocyanurate* is used. See also DIISOCYANATE.

Isocyanate Foam See URETHANE FOAM.

Isocyanate Generator (hindered isocyanate) A mixture of an isocyanate, a phenol, and a polyester that remains stable at room temperature. When heated to 70°C, the phenol and isocyanate components dissociate and react with the polyester to form a polyurethane resin.

Isocyanate Plastic A plastic based on polymers made by the polycondensation of organic isocyanates with other compounds. Reaction of isocyanates with hydroxyl-containing compounds produces polyurethanes having the urethane group –NHC(=O)O–. Reaction of isocyanates with amine-containing compounds produces polyurea having the urea group –NHCONH– (ISO). See also URETHANE and POLYURETHANE FOAM.

Isocyanurate A trimer of an isocyanate, formed by the catalytic cyclization of three isocyanate molecular groups into a six-membered ring.

Isocyanurate Foam A foam prepared from an isocyanurate. The unmodified foams have excellent flame resistance but are brittle and of little commercial value. However, isocyanurate foams modified with epoxides, polyimides, or (most commonly) urethane groups and polyols possess flame resistance far superior to that of conventional urethane foams and can be processed into a variety of foam products sutiable for insulation.

Isocyanurate Plastic A plastic based on isocyanate polymers in which trimerization of the isocyanates incorporates six-membered isocyanurate-ring groups in a chain. Note: In commercial polyisocyanurate cellular plastics, 10 to 30% of the available isocyanate groups are reacted with polyols to introduce urethane groups into the chain (ISO).

Isodecyl Octyl Phthalate $C_{10}H_{21}OOCC_6H_4COOC_8H_{17}$. A primary plasticizer for vinyls, cellulose nitrate, polystyrene, and ethyl cellulose. In vinyls, it performs better than diethylhexyl phthalate (DOP).

Isodecyl Octyl Adipate $C_{10}H_{21}OOCC_4H_8COOC_8H_{17}$. A light-colored, oily liquid used as a plasticizer for vinyls.

Isoindolinone Any of a small family of organic pigments, available in bright yellows and reds. They have good lightfastness, heat stability and bleed resistance.

Isomer From the Greek *isos* (the same, equal, alike) and *meros* (part or portion), isomers are substances comprising molecules that contain the same number and kinds of atoms, and have the same chemical formula, but that differ in structure, so that they form materials whose properties can differ widely. For example, the gas dimethyl ether, CH_3OCH_3, and the liquid ethyl alcohol, CH_3CH_2OH differ in both their chemical and physical properties, yet both have the empirical formula C_2H_6O. Isomeric polymers are formed by polymerizing isomonomers that link together in different ways.

Isomorphism A state of crystallization characterized by geometrically similar structural units.

Isooctyl Adipate See DIISOOCTYL ADIPATE.

Isooctyl Palmitate $C_8H_{17}OOCC_{15}H_{31}$. A plasticizer for polystyrene and cellulosic plastics.

Isophorone A cyclic, unsaturated ketone with the structure:

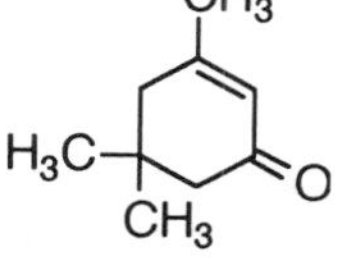

CH_3

H_3C O

CH_3

It is a powerful solvent for vinyl and cellulosic resins, with moderate power to dissolve nearly all common thermoplastic and (uncured) thermosetting resins.

Isophorone Diisocyanate (IPDI) An isocyanate used in the production of urethane elastomers and foams. It is less volatile than toluene diisocyanate, therefore easier to maintain at low levels in workers' airspace and safer to work with. Its structure is modified from that of ISOPHORONE, above, in that the oxygen has been replaced by an –N=C=O group, the double bond is gone, and the top carbon atom in the ring has an additional $-CH_2N=C=O$ linked to it.

Isophthalic Acid (benzene-1,3,-dicarboxylic acid, IPA) $C_6H_4(COOH)_2$ Used instead of phthalic anhydride in making unsaturated polyester resins that, when cured, have good stiffness and resistance to heat and chemicals.

Isoprene (3-methyl-1,3-butadiene, 2-methyl-1,3-butadiene) $CH_2=C(CH_3)CH=CH_2$. A colorless, volatile liquid derived from propylene or from coal gases or tars, chemically similar to the mer unit of natural rubber. Its polymer of the *cis*-1,4-type of polyisoprene is chemistry's nearest approach to synthesizing the natural product and it has sometimes been called "synthetic natural rubber".

Isoprene Rubber The *cis*-1,4-type of POLYISOPRENE, w s.

Isopropyl Acetate $(CH_3)_2CHOOCCH_3$. A colorless, fragrant liquid used as a solvent for cellulose nitrate, ethyl cellulose, polyvinyl acetate, polymethyl methacrylate, polystyrene, and certain phenolic and alkyd resins.

Isopropyl Alcohol (2-propanol, dimethyl carbinol) $(CH_3)_2CHOH$. A colorless solvent boiling at 82.4°C, moderately polar, which, because of its low toxicity, is enjoying greater use today than formerly in the plastics industry.

Isopropylbenzene (isopropylbenzol) Synonym for CUMENE, w s.

***p,p'*-Isopropylidenediphenol** See BISPHENOL A.

Isopropyl Myristate $(CH_3)_2CHOOCC_{13}H_{27}$. A plasticizer for cellulosic resins.

Isopropyl Oleate $(CH_3)_2CHOOCC_{17}H_{33}$. A plasticizer for cellulose nitrate, ethyl cellulose, and polystyrene, and, with partial compatibility, vinyl resins.

Isopropyl Palmitate $(CH_3)_2CHOOCC_{15}H_{31}$. A plasticizer for cellulose nitrate and ethyl cellulose.

Isostatic Pressing Forming solid articles from fine powders by enclosing the powder in a closed rubber mold, then immersing the mold in a hydraulic fluid and holding it at a controlled temperature and high pressure for 5 to 30 minutes. Because the pressure is equal in all directions, the method, though more costly than force-in-mold pressing and much more limited in the shapes producible, provides pieces of more uniform density and strength.

Isotactic (1) Denoting a polymer structure in which monomer units attached to a polymer backbone are identical on one side and/or the other side of the backbone. See also SYNDIOTACTIC. (2) Pertaining to a type of polymeric molecular structure containing a sequence of regularly spaced asymmetric atoms arranged in like configuration in a polymer chain (ASTM D 883). Materials containing isotactic molecules may exist in highly crystalline form because of the high degree of order that may be imparted to such structures. See also STEREOSPECIFIC.

Isotactic Polypropylene Polypropylene in which each mer unit has the pendant $-CH_3$ group on the same side of the chain backbone. Commercial PPs are about 90% isotactic, conferring high crystallinity and softening range. In contrast, atactic PP is rubbery and weak.

Isothermal Of processes and operations, carried out at constant temperature of the working substance.

Isotope In chemistry, one of two or more forms of an element ("nuclides") having the same number of protons in the nucleus but differing in mass number because of different numbers of neutrons. Natural elements are usually mixtures of isotopes; thus the observed atomic weights are average values weighted by isotopic relative abundance.

Isotropic Laminate A laminate in which the strength properties are equal, or approximately equal, in all directions.

Isotropy A material state in which the material properties are the same in all directions.

Itaconic Acid (methylenesuccinic acid) $HOOCC(=CH_2)CH_2COOH$. A white crystalline powder usually obtained by the oxidative fermentation of sucrose or glucose with *Aspergillus terreus*. It is capable of polymerization alone, or as a comonomer with acrylic acid, acrylonitrile, styrene, methyl methacrylate, and vinylidene chloride. It is used as an additive in acrylic resins to increase their adhesion to cellulose. By polycondensation of itaconic acid with diols, polyesters are obtained that contain methylene side groups.

IUPAC Abbreviation for International Union of Pure and Applied Chemistry, headquartered in Oxford, England.

IVE Abbreviation for VINYLISOBUTYL ETHER, w s.

Izod Impact Strength A widely used measure of impact strength, described in ASTM D 256, determined by the difference in energy of a swinging pendulum before and after it breaks a notched specimen clamped vertically as a cantilever beam. The pendulum is released from a vertical height of 0.61 m, and the vertical height to which it rises after breaking the specimen is used to calculate the energy expended in that breakage. The notch across the width of the specimen is usually on the side opposite the side impacted, but the reverse setup is provided for and specimens may also be tested unnotched. The convenient SI unit for the result is J/cm of (notched) width. ASTM D 3419 describes the preparation of screw-injection-molded test specimens of thermosetting materials for this and other tests. See also IMPACT STRENGTH and DROP-WEIGHT TEST.

J j

j Used by electrical engineers rather than *i* for $\sqrt{-1}$, to avoid confusion with *i*, current, in AC-circuit studies.

J (1) The SI abbreviation for JOULE, w s. (2) In most references to technical publications whose title contains the word "Journal", the abbreviation of that word. Sometimes "J."

Jaquet's Indicator A simple, compact tachometer based on a 6-second stopwatch movement and useful for measuring rotational speeds of shafts (with accessible ends) in the range from 50 to 5000 rpm and speeds of moving surfaces between 0.13 and 13 m/s.

Jar Mill A small BALL MILL, w s, utilizing a portable jar of porcelain or metal rather than a fixed cylinder for containing the material to be ground and the grinding media. After being charged and tightly closed, the jar is placed on a pair of rubber rollers—one driven, the other idling—and rotated for the desired time, typically overnight.

Jet-Abrasion Test A test for the abrasion resistance of coatings in which the time required for an air blast of fine abrasive particles to wear through the coating is measured (ASTM D 658, section 06.01).

Jet Molding (offset molding) A modification of injection molding designed for molding thermosets. An elongated nozzle or "jet" is attached to the front of the molding cylinder and is provided with a high-watt-density heating element and means for rapid cooling. It is also necessary to control cylinder temperatures carefully to prevent premature hardening of the resin.

Jet Spinning For most purposes, similar to melt spinning of staple fiber. Hot-gas-jet spinning uses a directed jet of hot gas to "pull" molten polymer from a die lip and instantly draw it into fine fibers.

Jetting In injection molding, a wriggly flow of resin from a small gate into the mold cavity, mistakenly referred to as turbulence and probably related to MELT FRACTURE, w s. Jetting is the antithesis of the desired laminar flow forming a smooth flow front across the mold.

It can cause strength problems in molded parts because of incomplete welding of the wormlike surfaces of the jet.

Jeweler's Rouge Very fine and pure ferric oxide (Fe_2O_3) powder used for polishing metals and plastics. *Rouge paper* contains the same abrasive glued to paper. It also comes in cloth form, called *crocus cloth.*

Jig (1) A device for positioning component parts while they are being assembled or otherwise worked on, or for holding tools. (2) A clamping device used to secure a bonded assembly until the adhesive has set. (3) A restraining frame into which freshly molded parts are placed to prevent their warping during annealing or final cooling.

Jig Welding The welding of thermoplastic components between suitably shaped jigs. Heat may be applied to the material by heating the jigs or by any other appropriate means.

Joggles (keys) A term sometimes employed for matching inserts that exactly position the parts of a multi-piece mold.

Joining The process of assembling plastic parts by means of mechanical fastening devices such as rivets, screws, clamps, etc. See also FABRICATE.

Joint The location where two separately made parts are joined with each other by adhesive bonding, welding, or fastening. See also BUTT JOINT, LAP JOINT, and SCARF JOINT.

Joint Efficiency The ratio of the strength of a completed joint to that of the weaker of the substrates that it joins.

Joule (J) The SI unit of work and energy, equal to 1 m·N, that replaces a variety of not-quite-equal older joules as well as numerous calories, all of which equal 4.18 to 4.19 J. Six different BRITISH THERMAL UNITs, w s, are all about equal to 1055 J. The joule should *not* be substituted for the SI unit of torque, the newton-meter (N·m). For precise conversion of older energy quantities to the joule, one should consult ASTM E 380 (section 14.02) or the ANSI standard with the same number.

Jute A fiber obtained from the stems of several species of the plant *Corchorus* grown mainly in India and Pakistan. It is used in the form of fiber, yarn, and fabric for reinforcing phenolic and polyester resins.

K k

k (1) Abbreviation for SI prefix, KILO-, w s. (2) Symbol for THERMAL CONDUCTIVITY, w s.

K (1) Abbreviation for KELVIN, w s. (2) Chemical symbol for potassium (Latin: kalium). (3) Symbol for BULK MODULUS, w s.

k' Symbol for HUGGINS CONSTANT, w s.

Kalrez® Du Pont's tradename for fluoroelastomers made from tetrafluoroethylene, perfluorovinylmethyl ether, and a small percentage of crosslinkable monomer. These elastomers combine the rubbery properties of VITON (w s) with the thermal stability, chemical resistance, and electrical characteristics of tetrafluoroethylene resin.

Kaolin (china clay, bolus alba) A variety of CLAY, w s, consisting essentially of the minerals *kaolinite*, *dickite*, and *nacrite* (all are $Al_2O_3 \cdot 2SiO_2 \cdot 2H_2O$). The name kaolin comes from the Chinese *kaoling*, meaning high hill, the name of the mountain in China which yielded the first kaolin sent to Europe. Deposits exist also in England, Jamaica, and the southeastern states of the US. Kaolins and china clays are used as inexpensive fillers in many plastics. In liquid resins, viscosity increases sharply with kaolin content above ten percent.

Karl Fischer Reagent A colored solution of iodine, sulfur dioxide, and pyridine in methanol. It reacts quantitatively with water, becoming colorless. It is used to determine small amounts of water in a wide range of materials, including many polymerics.

Kelvin (K) The SI unit of both temperature and difference between temperatures, equal to 1/273.16 of the thermodynamic triple point of water, i e, the temperature and pressure at which all three phases of water—ice, liquid, and vapor—are in equilibrium. A change or difference of 1 K is exactly equal to 1° difference on the Celsius (formerly centigrade) scale, and the temperature 0°C corresponds to 273.15 K.

Keratin The protein derived from feathers, hair, hoofs, horns, etc of animals by calcination. It is sometimes used as a filler in plastics, particularly urea-formaldehyde molding compounds, in which it reduces brittleness and permits drilling and tapping.

Ketohexamethylene Synonym for CYCLOHEXANONE, w s.

Ketone An organic compound containing a carbonyl group (C=O) bound to two carbon atoms. The simplest one is acetone, $(CH_3)_2C{=}O$. It and the other lower ketones are widely used as solvents for vinyl and cellulosic resins, and as intermediates in the production of resins. The most important members of the family are diacetone alcohol, diisobutyl ketone, 1,2-, 1,3-, and 1,4-diketones, ethyl *n*-butyl ketone, mesityl oxide, methyl *n*-amyl ketone, and methyl isobutyl ketone.

Kevlar® Du Pont's tradename for poly-(*p*-phenylene terephthalamide) fibers. See ARAMID.

Keys Synonym for JOGGLES, w s.

K-Factor A term sometimes used (incorrectly) for insulation value or used for THERMAL CONDUCTIVITY, w s.

Kg SI abbreviation for KILOGRAM, w s.

Kilo- (k) The SI prefix meaning $\times 10^3$.

Kilogram (kg) One of the basic units of SI, the mass of a particular platinum cylinder kept at the International Bureau of Weights and Measures in Paris. One avoirdupois pound = 0.4535924 kg.

Kieselguhr Alternate name for DIATOMITE, w s.

Kinematic Viscosity (kinetic viscosity) The absolute (dynamic) viscosity of a fluid divided by the density of the fluid. The SI unit is m^2/s, but the cgs unit, the *stoke*, which equals 10^{-4} m^2/s, is still in wide use, as is its submultiple, the centistoke.

Kinetic-Energy Correction (1) In the Izod, Charpy and tensile-impact tests, a subtraction of the kinetic energy imparted to the broken-off part of the specimen. (2) In measurement of DILUTE-SOLUTION VISCOSITY, w s, a correction for the energy required to accelerate the liquid in the reservoir to its higher velocity in the capillary. A similar correction is theoretically needed in melt rheometry, but has so far been found to be much smaller than the errors of measurement.

Kirksite An alloy of aluminum and zinc, easily castable at relatively low temperatures, often used for molds for blow molding. Its high thermal conductivity hastens cooling.

Kiss-Roll Coating A coating process by which very thin plastics coatings can be applied to substrates such as paper. A roll immersed in the coating fluid transfers a layer of coating to a second roll from which a portion of the layer is transferred to the substrate. See also ROLLER COATING.

Kling Test A method for determining the relative degree of fusion of flexible vinyl sheets, coated fabrics, and thin sections of cast or molded parts, by immersing the folded specimen in a solvent and noting the elapsed time at which disintegration commences. Typical solvent systems are based on methyl ethyl ketone, tetrahydrofuran, ethyl acetate, and carbon tetrachloride. The preferred solvent system is one that will initiate degradation within 5 to 10 min in a fully fused specimen.

Kneader A mixer with a pair of intermeshing blades, often S-shaped, used for working plastic masses of semi-dry or rubbery consistency.

Knife Coating See SPREAD COATING and AIR-KNIFE COATING.

Knit Line Another name for WELD LINE, w s.

Knockout Any part or mechanism of a mold whose function is to eject the molded article.

Knockout Bar A bar or plate in a knockout frame used to back up a row or rows of knockout pins.

Knockout Pin (KO pin) See EJECTOR PIN.

Knockout-Pin Plate Synonym for EJECTOR PLATE, w s.

Knoop Microhardness A test employing a diamond indenter whose point is an obtuse pyramid that makes an indentation of length seven times its width. This tester makes smaller, shallower indentations than Brinell or Vickers, suiting it to testing hardnesses of surfaces, as in case-hardened steel, or coatings. The Knoop hardness number (KHN) is equal to $14.2 \cdot P/L^2$, where P = the applied load in grams and L = the length in millimeters of the long axis of the indentation. ASTM lists a number of standard tests using this instrument for metals and ceramics, none for plastics, and one, D 1474, section 06.01, for organic coatings.

Knot Tenacity (knot strength) The strength of a yarn specimen containing an overhand knot to measure, by comparison with the strength of the unknotted yarn, its sensitivity to compression or shearing.

Knuckle Area In reinforced plastics, the area of transition between sections of different geometry in a filament-wound part.

Kohinoor Test A test for SCRATCH HARDNESS, w s, employing a series of pencils of different hardnesses.

Kohlrausch-Williams-Watts Equation This empirical equation aids designers of load-bearing products made of plastics and reinforced plastics. It is

$$G(t) = G_0 \cdot e^{-(t/t_0)^m}$$

where $G(t)$ is the stress-relaxed modulus of the test piece at time t from application of the load, G_0 is the modulus at the reference time t_0, usually in the range of a set of measurements on the logarithmic time scale, and m is a material-specific constant between 0.33 and 0.5 for many polymers and composites.

Ko-Kneader® Trade name of a family of unique compounding machines that combine some of the actions of internal mixers with those of extruders. The moving element is a helical screw with three rows of deep slots along its length. The screw itself advances with each rotation, then quickly falls back to start the next rotation. As it falls back, teeth attached to the cylinder pass through the slots in the screw. Because the output is pulsating with little pressure development, the Ko-Kneader usually discharges to a short melt extruder that develops die pressure and forms the final extrudate, typically strands that are cut into pellets.

Krebs Unit An obsolete unit once used in reporting viscosity measurements made with the weight-driven STORMER VISCOMETER, w s. A Krebs unit was the weight in grams that would turn a paddle-type rotor, submerged in the sample, 100 revolutions in 30 s.

Kraemer Equation For dilute polymer solutions, an equation relating the inherent viscosity to intrinsic viscosity and concentration. It is:

$$\eta_{inh} = (\ln \eta_r)/c = [\eta] + k'' [\eta]^2 \cdot c$$

where η_r = the reduced viscosity, $[\eta]$ = the intrinsic viscosity and c = the concentration in g/dL. See also HUGGINS EQUATION and DILUTE-SOLUTION VISCOSITY. Huggins' constant, k', and Kraemer's, k'', are related by: $k' - k'' = 0.5$. Thus, since k' is often between 0.6 and 0.8, k'' will often lie between 0.1 and 0.3.

K-Value An alternate name for THERMAL CONDUCTIVITY, w s.

L l

l (also L) Symbol for length.

L (1) Abbreviation for SI-permitted (but discouraged) volume unit, the liter, redefined in 1964 to be exactly 1 dm^3. (2) Symbol for length or magnetic inductance.

Lacquer A solution of a film-forming natural or synthetic resin in a volatile solvent, with or without color pigment, which when applied to a surface forms an adherent film that hardens solely by evaporation of the solvent. The dried film has the properties of the resin used in making the lacquer. The word derives from the *lac* insect, which secreted the resinous substance from which *shellac* solutions were (and still are) made. Today most lacquers are made with cellulosic, alkyd, acrylic, and vinyl resins.

Lactam A cyclic amide obtained by removing one molecule of water from an amino acid. An example is CAPROLACTAM, w s.

Lactic Acid (milk acid, α-hydroxypropionic acid) $CH_3CHOHCOOH$. A colorless or yellowish liquid with several applications in plastics. Reacted with glycerine, it forms an alkyd resin. It is a catalyst for vinyl polymerizations, and an additive for phenolic casting resins.

Ladder Polymer (double-stranded polymer) A polymer comprising chains made up of fused rings. Examples are cyclized (acid-treated) rubber and POLYIMIDAZOPYRROLONE, w s.

Lake A type of organic pigment prepared from water-soluble acid dyes, precipitated on an inert substrate by means of a metallic salt, tannin, or other reagent. Lakes were used in plastics at one time, but have been replaced by more permanent pigments.

LALLS See LOW-ANGLE LASER-LIGHT SCATTERING, w s.

Lambert A deprecated unit of illumination equal to 1 lumen/cm^2. The SI unit is the *lux* (lx), equal to 1 lm/m^2, i e, 10^{-4} lambert. See also LUMINOUS FLUX.

Lamellar Structure Platelike single crystals that exist in most crystalline polymers.

Lamina A single layer or ply within a laminate.

Laminar Flow Flow without turbulence, i e, the movement of one layer of fluid past another layer with no eddying between them. See REYNOLDS NUMBER. Most melt flow, even at high velocities, is laminar. With sudden changes in melt velocity, such as may occur at some extrusion-die entries and at injection-mold gates, laminar flow may be disrupted by MELT FRACTURE, w s. Also see JETTING.

Laminate (1, n) A product made by bonding together two or more layers of material or materials. The term most usually applies to preformed layers joined by adhesives or by heat and pressure. However, some authors apply the term to composites of plastic films with other films, foils, and papers, even though they have been made by spread coating or by extrusion coating. In the reinforced-plastics industry, the term refers mainly to superimposed layers of resin-impregnated or resin-coated fabrics or fibrous reinforcements that have been bonded together, usually by heat and pressure, to form a single piece. When the bonding pressure is at least 1000 psi (6.9 MPa), the product is called a *high-pressure laminate*. Products pressed at pressures under 200 psi are called *low-pressure laminates*; and those made with little or no pressure, such as hand layups, are sometimes called *contact-pressure laminates*. The term *parallel laminate* refers to one in which all layers are oriented approximately parallel with respect to the grain or strongest dimension in tension. In a *cross laminate*, one or some of the plies are perpendicular to the grain. The term laminate is also used to include composites of resins and fibers that are not in distinct layers, such as filament-wound structures and sprayups. Resins most widely used in laminates are the epoxies, phenolics, polyesters, melamine-formaldehydes, and silicones. The reinforcing materials include fabrics and mats of cotton, glass, nylon, ADVANCED FIBERs (w s), and metal wires. See too REINFORCED PLASTIC, COMPOSITE LAMINATE, and DECORATIVE BOARD. (2, v) To make a laminate or bond sheets together.

Laminated Glass A structure consisting of two or more parallel sheets or shells of glass interleaved with, and bonded to, layers of tough, sticky plastic, typically polyvinyl butyral or polycarbonate. The former resin is used in the "safety-glass" windshields of US-made cars. The glass will splinter under a heavy blow, but it is very resistant to penetration and the shards stick to the interlayer.

Lampblack A bulky, black soot obtained from the incomplete combustion of creosote or fuel oils, of duller and less intense blackness than CHANNEL BLACK and other CARBON BLACKs, and having a blue undertone and a small oil content.

Land (1) The horizontal bearing surface of a semipositive or flash mold by which excess material escapes. (See CUT-OFF.) (2) The bearing surface along the top of the flights of an extruder screw. (3) The final shaping surface of an extrusion die, usually parallel to the direction of melt flow. (4, plural) The mating surfaces of any mold, adjacent to the cavity depressions, that, when in contact, prevent the escape of material.

Land Area The area of those surfaces of a mold that contact each other when the mold is closed, measured in a plane perpendicular to the direction of application of the closing pressure.

Landed Force A force with a shoulder that seats on the land in a landed positive mold.

Land Length In an extrusion die, the distance across the land in the direction of melt flow between the lands.

Land Width (flight width) Of an extruder screw, the distance across the tip of the flight, perpendicular to the flight faces.

Lap (1) In filament winding, the amount of overlay between successive windings, usually provided to eliminate the formations of gaps. (2, v) To finely polish a surface with a relatively soft tool containing an embedded fine abrasive, or used with an oil suspension of same (*lapping compound*). (n) The tool used in lapping.

Lap Joint A joint made by placing one surface to be joined partly over another surface and bonding or fastening the overlapping portions. Compare BUTT JOINT and SCARF JOINT.

Lap Winding A variant of FILAMENT WINDING, w s, consisting of convolutely winding a resin-impregnated tape onto a mandrel of the desired configuration. The process has been used for making large chemical- and heat-resistant, conical or hemispherical parts such as heat shields for atmospheric-reentry vehicles.

Laser An acronym coined from the bold-face letters in **l**ight **a**pplication by **s**timulated **e**mission of **r**adiation. Early lasers were made of synthetic-ruby rod, silvered at one end face, semi-silvered at the other, and surrounded by a toroidal flashlamp. Today the NdYAG (for neodymium-yttrium aluminum-garnet) laser, with higher productivity than the ruby laser, has taken over many of its jobs. Gas lasers appear to be the direction of the future, and many types are now in use. A laser beam is made up of very concentrated, coherent light. Lasers have been useful in drilling, perforating, cutting, and welding operations with plastics and other materials. Low-power lasers have been used in highly accurate gauging and inspection systems. Medium-power CO_2 lasers are preferred for machining plastics because they produce light at a wavelength of 10.6 μm, which is completely absorbed by plastics. In 1960, dye lasers were discovered, rhodamine G, found in 1967, being the most widely used. These are colored organic compounds which, in solution, are tunable in the 250- to 1800-nm spectral domain. Excimer (from **exci**ted di**mer**) lasers can be set up in many polymers containing dye-like structures. Lasers have been useful in the analysis of polymers by laser mass spectrometry, in initiating polymerizations, and in curing polymers.

Laser-Ionization Mass Spectrometry A technique for chemical analysis of transparent and opaque plastics, capable of detecting all the elements but hydrogen and helium.

Latch Plate A plate used for retaining a removable mold core of relatively large diameter, or for holding insert-carrying pins on the upper part of a mold. Release of the pins or core is effected by moving the latch plate.

Latent Heat of Fusion See HEAT OF FUSION.

Latent Heat of Vaporization The quantity of heat required to change a unit mass (or sometimes a mole) of a liquid to its vapor, the two phases remaining in equilibrium at constant temperature, most commonly the *normal boiling point*, i e, the temperature of boiling at a pressure of 101.325 kPa. The convenient SI units are J/g, kJ/kg or kJ/mol. Because high polymers decompose without boiling, their heats of vaporization cannot be directly measured.

Latent Solvent An organic liquid that has little or no solvent effect on a particular resin until it is activated by either heat or admixture with a true solvent.

Latex (pl: latices or latexes) (1) A stable emulsion of a polymer in water, mostly used in paints and coatings. (2) The sap of the *hevea* (rubber) tree and other plants, or emulsions prepared from the same. Latices of interest to the plastics industry are based mainly on styrene-butadiene copolymers, polystyrene, acrylics, and vinyl polymers and copolymers. See also HYDROSOL.

Lattice Pattern In filament winding, a pattern with a fixed arrangement of open voids producing a basket-weave effect.

Lauroÿl Peroxide $[CH_3(CH_2)_{10}C(O)O-]_2$. A peroxide used as an initiator in free-radical polymerizations of styrene, vinyl chloride, and acrylic monomers.

Lauryl Methacrylate $H_2C{=}C(CH_3)COO(CH_2)_{11}CH_3$. A monomer used in the production of acrylic resins.

Law of Mixtures (rule of mixtures) Properties of binary mixtures lie between the corresponding properties of the pure components and are proportional to the volume fractions (sometimes weight or mole fractions) of the components. For example, the density of a blend is given by:

$$\rho_m = \rho_1 \cdot v_1 + \rho_2 \cdot v_2 = \rho_2 + (\rho_1 - \rho_2)$$

The "law" works well for properties of unidirectional composites, such as modulus, but often fails for melt viscosities of blends, where maxima and minima above or below the viscosities of the neat resins are common. Another rule suggested by Arrhenius for mixture viscosities has the form:

$$\log \mu_m = x_1 \cdot \log \mu_1 + x_2 \cdot \log \mu_2$$

where the x_i are mole fractions. While this rule has a different form than the preceding one and fits the mixture data of some systems, it, too, cannot provide maxima or minima.

Lay (1) The length of twist produced by stranding singly or in groups, such as fibers or rovings; or the angle that such filaments make with the axis of the strand during a stranding operation. The length of twist of a filament is usually measured as the distance parallel to the axis of the strand between corresponding points on successive turns of the filament (ASTM D 883). (2) The term is also used in the packaging of glass fibers for the spacing of the roving bands in the package expressed as the number of bands per inch.

Lay-Flat Film Film that has been extruded as a wide, thin-walled, circular tube, usually blown, cooled, then gathered by converging sets of rollers and wound up in flattened form.

Lay-Flat Width In blown-film manufacture, half the circumference of the inflated film tube.

Layup The tailoring and placing of reinforcing material—mat or cloth—in a mold prior to impregnating it with resin. The reinforcement is usually cut and fitted to the mold contours. See also SHEET-MOLDING COMPOUND.

Layup Molding (hand-layup molding) A method of forming reinforced plastics articles comprising the steps of placing a web of the reinforcement, which may or may not be preimpregnated with a resin, in a mold or over a form and applying fluid resin to impregnate and/or coat the reinforcement, followed by curing of the resin and extraction of the cured article from the mold. When little or no pressure is used in the curing process, it is sometimes called *contact-pressure molding*. When pressure is applied during curing, the process is often named for the method of applying pressure, e g, *vacuum-bag molding* or *autoclave molding*. A related process is SPRAYUP, w s.

LC Polymer Abbreviation for LIQUID-CRYSTAL POLYMER, w s.

LDPE Abbreviation for LOW-DENSITY POLYETHYLENE, w s. Also see POLYETHYLENE.

L/D Ratio In an extruder, the ratio of the flighted length of the screw to its nominal diameter. The Extrusion Division of SPE has established two somewhat different definitions of L/D. *Total L/D ratio* is the distance from the rear surface of the feed opening to the front end of the flight, expressed in bore diameters, e g, 20:1. The *effective (enclosed) L/D ratio* is the distance from the forward surface of the feed opening expressed in the same way. Thus, if the feed opening were rectangular and two diameters long, and the total L/D measured 24:1, the effective L/D would be 22:1.

Leaching The process of extraction of a component from a comminuted solid material by treating the material with a solvent that dissolves the component of interest but not the remaining components.

Lead (Pb, from Latin *plumbum*) (pronounced like "led") Powders of this metal have been employed as fillers in plastics products used for radiation shielding.

Lead (pronounced like "seed") Of an extruder screw, the distance parallel to the screw axis from any point on the screw thread to the corresponding point on the next turn of that thread. Compare PITCH.

Lead Angle See HELIX ANGLE.

Lead Carbonate See BASIC LEAD CARBONATE.

Lead Chrome Pigment Any of a series of inorganic pigments including yellows, oranges, and greens, used in PVC, polyolefins, cellulosics, acrylics, and polyesters.

Leader Pin See GUIDE PIN.

Leader-Pin Bushing See GUIDE-PIN BUSHING.

Leading Flight Face (leading flight) The forward or front side of the flight of an extruder screw, the rear side being the *trailing flight face*.

Lead Oxide Either yellow lead oxide, PbO (litharge) or red lead oxide, Pb_3O_4. Both are used as pigments, though much less today in the US than formerly because of concerns about lead's toxicity and the need to keep it out of the environment. The oxides are sometimes used as fillers in radiation-shielding applications.

Lead Phosphite, Dibasic See DIBASIC LEAD PHOSPHITE.

Lead Salicylate (dibasic lead salicylate) $Pb(C_6H_4OHCOO)_2{\cdot}H_2O$. A white crystalline material formerly used as a heat stabilizer.

Lead Stabilizer Any of a large family of highly effective heat stabilizers that are limited to use in applications where toxicity, sulfur staining, and lack of clarity are not objectionable. Examples are: BASIC LEAD CARBONATE, basic lead sulfate complexes, basic silicate of white lead, coprecipitated lead silicate and silica gel, dibasic lead maleate, DIBASIC LEAD PHOSPHITE, DIBASIC LEAD PHTHALATE, dibasic lead salicylate, dibasic lead silicate-sulfate, DIBASIC LEAD STEARATE, lead chlorosilicate complexes, LEAD STEARATE, and monohydrous dibasic lead sulfate.

Lead Stearate $Pb(C_{17}H_{35}COO)_2$. A white powder used as a vinyl-resin stabilizer and lubricant in extrusion compounds, and, earlier, in phonograph-record compounds.

Leakage Flow In the metering section of an extruder screw, leakage flow is the backward flow of material through the clearance between the screw-flight lands and the barrel wall. It is usually a very small negative component of the throughput. See also DRAG FLOW, NET FLOW, and PRESSURE FLOW.

Least-Squares Curve Fitting See REGRESSION ANALYSIS.

Leathercloth A term sometimes used, especially in Europe, for plastic-coated fabric with a leather-like texture.

Lecyar Model A versatile model for viscosities of polymer blends, most of which do not obey simple mixture laws. The Lecyar model can accommodate a viscosity minimum, maximum, or both in the composition range from 0 to 100% of one polymer intimately blended with another. For a given shear rate and temperature it has the form:

$$\ln \eta_B = Am_1^3 + Bm_2^3 + Cm_1^2m_2 + Dm_1m_2^2$$

in which A and B are the natural logarithms of the neat resins 1 and 2, C and D are the natural logarithms of interaction viscosities that must be evaluated from measurements on blends, and m_1 and m_2 are the mass fractions of the two components. ($m_1 + m_2 = 1$.)

LEFM Abbreviation for LINEAR ELASTIC FRACTURE MECHANICS, w s.

Lemon Yellow Any of two families of yellow pigments, one being mixtures of barium chromate with zinc carbonate, the other mixtures of lead chromate and lead carbonate.

Let-Go An area in laminated glass over which an initial adhesion between interlayer and glass has been lost (ASTM D 883).

Let-Off A device used in coating by calendering or extrusion to suspend a coil or reel from which the material to be coated is fed to the coating machine.

Letterpress Printing The process used for paper is adapted to plastics by the use of special inks and transfer rolls, and possibly a modification of press speed. Flexible printing plates are usually employed, made of vinyl or rubber.

Lewis-Acid Catalyst See FRIEDEL-CRAFTS CATALYSTS.

Lewis-Nielsen Equation An equation derived from the theory of mixtures that provides an estimate of the modulus of a short-fiber, thermoplastic composite. It is given below.

$$E_c = \frac{E_r(1 + 2AB\phi_f)}{1 - B\psi\phi_f} \quad \text{where} \quad B = \frac{(E_f / E_r) - 1}{(E_f / E_r) + 2A} \quad \text{and} \quad \psi = 1 + \frac{(1 - V_r)\,\phi_f}{2V_r}$$

In these expressions, E_c, E_f, and E_r are the moduli of the composite, fiber, and resin, ϕ_f is the volume fraction of fiber, A is the average aspect ratio of the fiber, and ψ is the maximum packing fraction. Application of the equation is limited to small strains.

Lexan® General Electric's tradename for their polycarbonate resins produced by reacting bisphenol A and phosgene, the first commercial PC resins. See also POLYCARBONATE RESIN.

Li Chemical symbol for the element LITHIUM, w s.

Lift (1) The complete set of moldings produced in one cycle of a molding press.

Light Absorbance (absorbance, absorptivity) The percentage of the total luminous flux incident upon a test specimen that is neither reflected from, nor transmitted through the specimen. Compare LIGHT REFLECTANCE and LIGHT TRANSMITTANCE.

Light Reflectance (reflectance, reflectivity) The fraction of the total luminous flux incident upon a surface that is reflected, generally a function of the color (wavelength) of the light. See also LIGHT ABSORBANCE and LIGHT TRANSMITTANCE.

Light Resistance (light fastness, color fastness) The ability of a plastic material to resist fading, darkening, or degradation upon exposure to sunlight or ultraviolet light. Nearly all plastics tend to change color under outdoor conditions, due to characteristics of the polymeric material and/or pigments incorporated therein. Tests for light resistance are made by exposing specimens to natural sunlight or to artificial light sources such as the carbon arc, mercury lamp, germicidal lamp, or xenon arc lamp. ASTM D 4459 and D 4674 describe procedures for testing light fastness of plastics for indoor applications. See also ARTIFICIAL WEATHERING.

Light Scattering In a dilute polymer solution, light rays are scattered and diminished in intensity of transmission by a number of factors including fluctuations in molecular orientation of the polymer solute. Observations of the intensity of light scattered at various angles provide the basis for an important method of measuring weight-average molecular weights of high polymers. See also LOW-ANGLE LASER-LIGHT SCATTERING.

Light Stabilizer An agent added to a plastic compound to improve its resistance to light-induced changes. See also STABILIZER and ULTRAVIOLET STABILIZER.

Light Transmittance (luminous transmittance, light transmissivity) The ability of a material to pass incident light through it, whether specular or diffuse. ASTM prescribes several tests of this property in plastics: D 1003, Haze and Luminous Transmittance of Transparent Plastics; D 1494, Diffuse Light-Transmission Factor of Reinforced-Plastics Panels; D 1746, Transparency of Plastic Sheeting; and D 3349, [Light] Absorption Coefficient of Carbon-Black-Pigmented Ethylene Plastic. *Transmissivity* is the ratio of the intensity of the transmitted light to that of the unreflected incident light. See also OPACITY.

Lignin The major non-carbohydrate constituent of wood and woody plants, functioning in nature as a binder to hold the matrix of cellulose fibers together. Lignins are obtained commercially from by-products of coniferous woods, for example by treating wood flour with a derivative of lignosulfonic acid. They are used as extenders in phenolic resins, and sometimes as reactants in the production of phenol-formaldehyde resins.

Lignin Plastic A plastic based on lignin resins (ISO, ASTM D 883).

Lignin Resin A resin made by heating lignin or by reaction of lignin with chemicals or resins, the lignin being in greatest amount by mass. (ISO, ASTM D 883).

Ligroin (ligroine, benzine) Any of several saturated petroleum-naphtha fractions boiling in the range 60 to 100°C, used as solvents. The term *benzine* is deprecated due to confusion with *benzene*, and should not be used.

Lime Synonym for CALCIUM OXIDE, w s.

Limestone See CALCIUM CARBONATE.

Limiting Viscosity Number The IUPAC term for INTRINSIC VISCOSITY, w s.

Lineal Density The meaning of DENIER, w s.

Linear Elastic Fracture Mechanics (LEFM) A theory of fracture, applicable to brittle plastics (and other brittle materials), based on the assumption that the material is Hookean up to the point of fracture, with yielding restricted to a small volume near crack tips in the stressed material.

Linear Expansion See COEFFICIENT OF THERMAL EXPANSION.

Linear Low-Density Polyethylene (LLDPE) The original low-density polyethylene (LDPE), produced at high pressure, has a highly branched structure. Using Ziegler-Natta catalysts and low pressure, with a small percentage of 1-butene or other comonomer, one can produce a more linear PE with density between 0.919 and 0.925 g/cm^3. LLDPE films have the gloss and clarity of LDPE films, but are stronger, so can be blown thinner to carry design loads. Because of higher melt viscosity, screw modifications are usually necessary for processing LLDPE in extruders designed for LDPE.

Linear Polymer A polymer in which the molecules form long chains without branches or crosslinking. The molecular chains of a linear polymer may be intertwined, but the forces tending to hold the molecules together are physical rather than chemical and therefore can be weakened by heating. Linear polymers are thermoplastic.

Linear Unsaturated Polyesters See POLYESTER, UNSATURATED.

Linear Viscoelasticity VISCOELASTICITY, w s, characterized by a linear relationship between stress, strain, and strain rate.

Liner (1) A continuous, usually flexible coating on the inside surface of a filament-wound pressure vessel, used to protect the laminate from chemical attack or to prevent leakage under stress. (2) In extruders and injection molders, the hard-alloy interior surface of the cylinder. Decades ago, some of these were separately fabricated and pressed into the steel cylinders. Today they are centrifugally cast into cylinders, or formed by INTERFUSION, w s, then ground and polished to the final internal diameter. Thickness is about 2-3 mm. See also BIMETALLIC CYLINDER.

Linoleic Acid An 18-carbon, straight-chain fatty acid with two double bonds that may be in the 9 and 12 or 9 and 11 positions. It is found in nature as its glyceryl triester in many vegetable oils and is a starting material for some plasticizers for plastics.

Linolenic Acid An 18-carbon, straight-chain fatty acid that contains three double bonds in the 9, 12, and 15 positions. Its triglyceryl ester is an important constituent of linseed and perilla oils, with excellent "drying" behavior.

Linseed Oil An oil expressed from flax seeds, a mixture of glyceryl esters of linolenic (25%), oleic (5%), linoleic (62%), stearic (3%), and palmitic (5%) acids, a DRYING OIL (w s) long used in the paint and varnish trade, now also in alkyd resins.

Linters Short fibers that adhere to cotton seeds after ginning. Used in rayon manufacture, as fillers for plastics, and as a base for the manufacture of cellulosic plastics.

Liquid Colorant In the compounding of plastisols and organosols, it has always been common practice to use liquid or pasty dispersions of colorants, which are easily stirred into the compounds. More recently, similar liquid dispersions and metering systems have been developed for adding the colorants directly onto the screw of an extruder at the base of the feed opening. The term *liquid colorant* is now used for such dispersions to be added to dry molding or extrusion compounds. The advantages of liquid colorants are better pigment dispersion, higher letdown ratios, savings in handling and pollution control, and less buildup on the screw and screen pack.

Liquid Chromatography See CHROMATOGRAPHY.

Liquid-Crystal Polymer (LC polymer, liquid-crystalline polymer, mesomorphic polymer) A polymer capable of forming regions of highly ordered structure (*mesophase*) while in the liquid (melt or solution) phase. The degree of order is somewhat less than that of a regular solid crystal. Four types have been identified: rodlike, including aromatic polyamides, esters, azomethines, and benzobisoxazoles; helical, mostly natural materials such as polypeptides; side-chain (*comb polymers*); and block copolymers with alternating rigid and flexible units. These polymers are described as *nematic*, in which the mesogens (ordered regions) show no positional order, only long-range order; *cholesteric* or *chiral*, a modified nematic phase in which the orientation direction changes from layer to layer in a helical pattern; and *smectic*, in which the mesogens have both long-range order and 1- or 2-dimensional positional order. Liquid-crystal polymers are difficult to get into the molten condition because the solid crystals generally decompose before melting. The most commercially successful ones to date are those processed in solution, e g, poly(*p*-phenylene terephthalamide) (Kevlar). LC polymers are also classified as *lyotropic* and *thermotropic*. Lyotropic ones show their liquid-crystalline character only in solution, while thermotropic ones can show it in the melt without the presence of a solvent.

Liquid Injection Molding (LIM) A process of injection-molding thermosetting resins in which the uncured resin components are metered, mixed, and injected at relatively low pressures through nozzles into mold cavities, the curing or polymerization taking place in the mold cavities. The process is most widely used with resins that cure by addition polymerization such as polyesters, epoxies, silicones, alkyds, diallyl phthalate, and (occasionally) urethanes. However, the term REACTION INJECTION MOLDING (RIM) is most often used with urethane reactants. The term *liquid-resin molding* has also been used for LIM or RIM.

Liquid Reaction Molding (LRM) Older synonym for REACTION INJECTION MOLDING, w s.

Lithium (Li) Element number 3, the least dense of all the metals (density = 0.534 g/cm^3), with valence of +1, and highly reactive. Lithium aluminum hydride ($LiAlH_4$) is an important catalyst in organic reductions and lithium is a component of many greases, e g, the high-temperature lubricant, LITHIUM STEARATE, w s.

Lithium Stearate $LiOOCC_{17}H_{35}$. A white crystalline material used as a lubricant in plastics.

Lithopone (Charlton white, Orr's white, zinc baryta) A mixed pigment obtained by the interaction (*metathesis*) of equimolar solutions of barium sulfide and zinc sulfate, from which precipitate barium sulfate and zinc sulfide, both white. It was long used as a pigment and filler in plastics, but has been largely supplanted by titanium white.

Live-Feed Molding See MULTI-LIVE-FEED MOLDING.

LLDPE Abbreviation for LINEAR LOW-DENSITY POLYETHYLENE, w s.

Ln (1) Abbreviation for LUMEN, w s. (2) Abbreviation for natural logarithm, i e, logarithm to the base e (= 2.71828...).

Load Cell An instrument, most often part of a machine for testing mechanical properties and some rheometers, that senses the force applied to the specimen (or piston).

Loading Board Alternate term for LOADING TRAY, w s.

Loading Space Space provided in a compression mold or in the pot used with a transfer mold to accommodate the molding material before it is compressed.

Loading Tray (charging tray, loading board) A device for charging measured amounts of molding compound simultaneously into all the cavities of a multi-cavity mold, comprising a compartmented tray with a slide-out bottom.

Locating Ring A ring that aligns the nozzle of an injection-molding cylinder with the entrance of the sprue bushing, and aligns the mold to the machine platen.

Locking Pressure The pressure applied to an injection or transfer mold to keep it closed during molding.

Locking Ring A slotted plate in an injection or transfer mold that locks the parts of the mold together and prevents the mold from opening while the plastic is being injected.

Logarithmic Decrement (Δ) In a damped, vibrating system, the natural logarithm of the ratio of the amplitude of any oscillation to the amplitude of the succeeding oscillation. Where damping is mild, the ratio of amplitudes several vibrations apart will usually give a more accurate estimate. The equations are:

$$\Delta \equiv \ln(A_i/A_{i+1}) \quad \text{and} \quad \Delta = (1/n)\cdot\ln(A_i/A_{i+n})$$

Logarithmic Viscosity Number The IUPAC term for INHERENT VISCOSITY, w s.

Lognormal Distribution (logarithmic normal distribution) A statistical probability-density function, characterized by two parameters, that can sometimes provide a faithful representation of a polymer's molecular-weight distribution or the distribution of particle sizes in ground, brittle materials. It is a variant of the familiar normal or Gaussian distribution in which the logarithm of the measured quantity replaces the quantity itself. Its mathematical form is

$$f(x) = \frac{1}{\sqrt{2\pi}\beta} x^{-1} e^{-(\ln x - \alpha)/2\beta^2} dx \quad \text{or}$$

$$f(\ln x) = \frac{1}{\sqrt{2\pi}\beta} e^{-(\ln x - \alpha)/2\beta^2} d\ln x$$

α and β are the mean and standard deviation of ln x. Anti-ln α is called the *log mean* of x. To test the suitability of this distribution, one plots the cumulative percent of members having weights or sizes below x, vs x, on lognormal probability paper and looks for linearity in the plot.

Long-Chain Branching In a polymer's structure, the presence of arms (branches) off the main chain that are about as long as the main chain. In making low-density polyethylene, a typical molecule may contain 50 short branches and only one or zero long branch, yet the presence of long branches greatly broadens the molecular-weight distribution. Polymers containing long branches tend to be less crystalline than the corresponding polymers without long branches.

Long-Fiber-Reinforced Thermoplastic A pelletized thermoplastic resin for injection molding, usually nylon 6/6 or polypropylene, produced by pultrusion from continuous-filament glass yarn, and cut to lengths of 9 to 13 mm (about three times the length of short-fiber pellets). Nylon containing 50 weight percent longer-fiber glass is about

15% stronger and stiffer than its short-fiber mate, with double the notched-Izod impact strength.

Longo A colloquialism used in the filament-winding industry, designating an article that is wound longitudinally or with a low-angle helix.

Loop Test A simple test (ASTM 3291) for evaluating the compatibility of vinyl resin plasticizers based on the fact that a material under compressive stress will exude plasticizer more rapidly. A specimen in sheet form is folded double, forming a loop with internal radius equal to the sheet thickness. At intervals, the bend of the loop is reversed 360° and the former inside surface of the loop is examined for evidence of plasticizer spewing.

Loose Punch A male portion of a mold constructed so that it remains attached to the molding when the press is opened, to be removed from the part after demolding.

Loss Angle The inverse tangent of the electrical dissipation factor. See DIELECTRIC LOSS ANGLE.

Loss Compliance The "imaginary" part of the complex compliance. See COMPLIANCE and COMPLEX MODULUS.

Loss Factor The product of the power factor and dielectric constant of a dielectric material.

Loss Modulus The "imaginary" component of the COMPLEX MODULUS, w s. The product of the storage modulus and the tangent of the loss angle. Loss modulus is an indicator of the conversion of mechanical energy into heat when a material is deformed.

Lost-Wax Process See INVESTMENT CASTING.

Low-Angle Laser-Light Scattering A technique for determining weight-average molecular weights of polymers in solution. The low angle—2 to 10°—reduces the number of measurements needed and simplifies their interpretation, as compared with conventional, wide-angle light scattering.

Low-Density Polyethylene (LDPE) This term is generally considered to include polyethylenes having densities between 0.915 and 0.925 g/cm^3. In LDPE, the ethylene molecules are linked in random fashion, with many side branches, mostly short ones. This branching prevents the formation of a closely knit pattern, resulting in material that is relatively soft, flexible, and tough, and which will withstand moderate heat. See also HIGH-DENSITY POLYETHYLENE and POLYETHYLENE.

Low-Pressure Injection Molding A term sometimes used for the process of injecting a fluid material such as a vinyl plastisol into a closed mold, using a grease gun or similar low-pressure equipment.

Low-Pressure Laminate Various definitions place the upper limit of pressure for this term at from 6.9 MPa down to pressures obtained by mere contact of the plies. According to ASTM D 883, the upper limit is 1.4 MPa (200 psi). The Decorative Board Section of the National Electrical Manufacturers' Association (NEMA) has recommended abandonment of the term "low-pressure laminate" in favor of *decorative board* in the case of "...a product resulting from the impregnation or coating of a decorative web of cloth, paper, or other carrying media with a thermosetting resin and consolidation of one or more of these webs with a cellulosic substrate under heat and pressure of less than 500 pounds per square inch." This includes all boards that were formerly called low-pressure melamine and polyester laminates, but not vinyls. See also CONTACT-PRESSURE MOLDING and LAMINATE.

Low-Pressure Molding Molding or laminating in which the pressure is 1.4 MPa (200 psi) or less (ASTM D 883).

Low-Pressure Resin See CONTACT-PRESSURE RESIN.

Low-Temperature Flexibility All plastics that are flexible at room temperature become less so as they are chilled, finally becoming brittle at some low temperature. This property is often measured by torsional tests over wide ranges of temperature, from which apparent moduli of elasticity are calculated. See also BRITTLENESS TEMPERATURE and CLASH-BERG POINT. Some relevant ASTM tests are D 1043, D 3295, D 3296, D 3374 (sec 07.02), and D 1055 (sec 09.01).

L-Sealer A heat-sealing device, used in packaging, that seals a length of flat, folded film on the edge opposite the fold and simultaneously seals a strip across the width at 90° from the edge seals. The article to be packaged may be inserted between the two layers of folded film prior to sealing. When it is desired to sever the continuous length of sealed compartments into individual packages, a heated wire or knife is incorporated between two sealing bars that form the bottom of the **L**. These bars then make the top seal of the filled bag and the bottom seal of the next bag to be filled.

Lubricant A substance that tends to make surfaces slippery, reduce friction, and prevent adhesion. Lubricants are added to plastics to (1) ease flow in calendering, molding, and extrusion by reducing friction between the metal and plastic surfaces; (2) assist in knitting and wetting of the resin in mixing and milling operations; and (3) impart lubricity to finished products. Among the lubricants commonly used are fatty-acid soaps, metallic soaps, i e, salts of fatty acids such as calcium or barium stearate, paraffin waxes, hydrocarbon oils, fatty alcohols, low-molecular-weight polyethylenes, synthetic waxes of the fatty-amide and -ester types, and certain silicones. Graphite, molybdenum disul fide, and fluorocarbon polymers are used to impart lubricity to fin ished articles made of acetals, nylon, and polycarbonate. Lubricity is also contributed by certain plasticizers, stabilizers, and pigment dispersions.

Lubricant Bloom See BLOOM. The term *lubricant bloom* should only be used when the exudation is known to be caused by a lubricant contained in the plastic compound or applied to it during processing.

Lucite® Du Pont's trade name for methacrylate-ester monomers and polymers, including PMMA and several other resins, and for certain products made from such resins.

Lumen (lm) The SI unit of LUMINOUS FLUX, w s.

Luminescent Pigment A pigment that produces striking effects in darkness or light. See FLUORESCENT PIGMENT and PHOSPHORESCENT PIGMENT.

Luminous Flux The total visible energy emitted by a source per unit time. The SI unit is the *lumen* (lm), defined as the luminous flux emitted in a solid angle of one steradian (sr, the solid central angle that cuts out of a spherical surface a square whose side is equal to the radius) by a point source having a uniform intensity of one candela. Therefore, 1 lm = 1 cd·sr.

Luminous Transmittance Synonym for LIGHT TRANSMITTANCE, w s.

Lux (lx) The SI unit of illuminance, defined as the illuminance produced by a luminous flux of one lumen uniformly distributed over a surface of one square meter. That is, 1 lx = 1 lm/m^2.

Lyophilic Describing a substance that easily forms colloidal suspensions. Such ability when the suspending medium is water is called *hydrophilic*. A PVC plastisol is an example of a lyophilic suspension.

Lyotropic See LIQUID-CRYSTAL POLYMER.

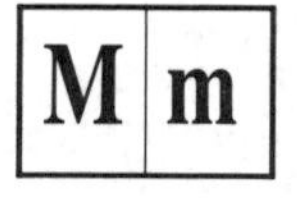

m (1) Abbreviation for METER (1), w s. (2) Abbreviation for the SI prefix, MILLI-, w s. (3) (usually italicized) abbreviation for chemical positional prefix META-, w s.

M (1) Abbreviation for prefix MEGA-, w s. (2) Symbol for molecular weight. (3) Symbol for bending moment.

mA Abbreviation for *milliampere*.

Machinability (1) In fabricating materials by such operations as drilling, lathe-turning, and milling, the ease with which the material is removed. ASTM offers tests of machinability for ferrous metals, lumber, and particle board, but none for plastics. Most plastics compounds are easily machined, but cutting tools must be kept sharp and should have larger rake angles than are required for metals. Unless these practices are observed, thermoplastic workpieces may overheat, even to the point of softening and melting. Tungsten-carbide- or diamond-tipped tools are recommended for compounds or laminates containing glass or other ceramic-fiber reinforcements. (2) In packaging operations, the ease with which a particular packaging material or medium passes through the machinery.

Machine Direction In extrusion, calendering, and pultrusion, the principal overall direction in which the plastic material moves through the equipment, and to which the TRANSVERSE DIRECTION (w s) is perpendicular.

Machine Shot Capacity See SHOT CAPACITY.

Machining of Plastics Many of the machining operations used for metals are applicable to rigid plastics, with appropriate variations in tooling and speeds [see MACHINABILITY (1), above]. Among such operations are BLANKING, boring, drilling, grinding, milling, planing, PUNCHING, routing, SANDING, sawing, shaping, tapping, threading, and turning.

Mach Number (N_{Ma}) The ratio of a fluid velocity or the relative velocity of an object moving through a fluid to the velocity of sound in the fluid.

Macromolecular Pertaining to a substance consisting of very large molecules.

Macromolecule The large ("giant") molecules that make up high polymers, both natural and synthetic. Each macromolecule may contain hundreds of thousands of atoms.

Macroscopic Visible to the naked eye, as opposed to *microscopic.*

Maddock Mixing Section (Union Carbide mixing section) Named for its inventor, Bruce Maddock, a short special zone of an extruder screw in which the regular helical channel is interrupted by a cylindrical section about two to three screw diameters long, usually located just to the rear of the metering section. Shallow slots are machined axially along the cylinder, half of them opening to the screw channel just to the rear of the section, the other half, alternating between pairs of the first group, opening to the forward channel. Material entering from the rear can only move forward by passing through the close clearances between lands and barrel, thus holding back temporarily any unmelted remnants of the feed pellets. While some mixing (both dispersive and distributive) does occur in the Maddock section, its main effect is preventing unmelted material from reaching the screen pack or die.

Magnesia Synonym of MAGNESIUM OXIDE, w s.

Magnesium Carbonate (magnesia alba, precipitated magnesium carbonate) $MgCO_3$. A white powder of low density, prepared by metathesis, used as a filler or modifier in phenolic resins. This carbonate also occurs naturally as *magnesite*.

Magnesium Glycerophosphate $MgPO_4C_3H_5(OH)_2$. A colorless powder, derived by the action of glycerophosphoric acid on magnesium hydroxide, used as a stabilizer for plastics.

Magnesium Hydrogen Phosphate Trihydrate (dibasic magnesium phosphate, magnesium monohydrogen *ortho*phosphate) $MgHPO_4 \cdot 3H_2O$. A white, crystalline powder derived by reacting *ortho*phosphoric acid with magnesium oxide, used as a nontoxic stabilizer for plastics.

Magnesium Hydroxide $Mg(OH)_2$. Used as a thickening agent for polyester resins. Its action is slower than that of magnesium oxide.

Magnesium Hydroxychloride Cement (Sorel cement, magnesium oxychloride cement) A mixture of magnesium chloride and magnesium oxide that reacts with water to form a solid mass, presumed to be magnesium hydroxychloride, $Mg(OH)Cl$. It has been useful as an intumescent coating for urethane foams and other materials such as polystyrenes, nylons, acetals, polyesters, and silicones.

Magnesium Oxide (magnesia, periclase) A white powder used as a filler and as a thickening agent in polyester resins. It occurs naturally as the mineral *periclase*, but is usually made in purer form by calcining magnesium hydroxide or carbonate.

Magnesium Phosphate, Dibasic See MAGNESIUM HYDROGEN PHOSPHATE TRIHYDRATE.

Magnesium Phosphate, Monobasic (magnesium dihydrogen phosphate) $Mg(H_2PO_4)_2 \cdot 2H_2O$. A white, hygroscopic, crystalline powder derived by reacting phosphoric acid with magnesium hydroxide. It is used as a flame retardant and stabilizer for plastics.

Magnesium Phosphate, Tribasic $Mg_3(PO_4)_2 \cdot 8H_2O$ or $\cdot 4H_2O$. A fine, soft white powder derived by reacting magnesium oxide and phosphoric acid at a high temperature, used as a nontoxic stabilizer.

Magnesium Soap A magnesium salt of a fatty acid, e g, MAGNESIUM STEARATE, w s, precipitated by an inorganic magnesium salt from a solution of sodium or potassium soaps. See also SOAP, METALLIC.

Magnesium Stearate $Mg(OOCC_{17}H_{35})_2$. A white, soft powder used as a lubricant and stabilizer.

Magnetic Filler Any permanently magnetizable material in powder form that may be incorporated into plastics to produce molded or extruded-strip magnets. Major ones in use are Alnico, rare earths, and, most used in plastics, hard FERRITE, w s.

Magnetic Separator A device that removes tramp iron and steel from a stream of mainly nonmagnetic material, such as reground plastic or mixed wastes, by passing the stream close to strong magnets. Some design parameters for magnetic separators are given in Section 21 of *Chemical Engineers' Handbook*, Sixth Edn, R. H. Perry and D. W. Green, Eds, McGraw-Hill, and the two preceding editions.

Maillefer Screw The first of several barrier-flight or solids-draining screw designs, patented in 1959 by C. Maillefer in Switzerland. The design is a three-zone metering screw modified with a secondary flight, narrower and with more clearance than the main flight. The secondary flight begins shortly beyond the feed opening of the extruder, starting from the leading face of the main flight and having a slightly faster lead. In successive turns, the fraction of the channel width to the rear of the auxiliary flight, into which the melted resin drains from the forward portion, becomes wider while the fraction in which the unmelted solids are confined becomes narrower. The secondary flight merges with the trailing face of the main flight just prior to the metering (pumping) section, where it is hoped that no unmelted solids persist. See also SOLIDS-DRAINING SCREW.

Maleic Anhydride (2,5-furandione) A compound crystallizing as colorless needles, obtained by passing a mixture of benzene and air over a heated vanadium pentoxide catalyst, and having the structure shown below.

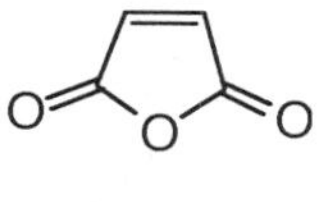

It has many applications in plastics, including the production of alkyd, polyester, and vinyl-copolymer resins, and as a curing agent for thermosetting resins such as phenolics and ureas. About half the maleic anhydride produced in the US is used in the manufacture of unsaturated polyester resins, to which it imparts fast curing and high strength.

Maltese Cross A dark shadow, having the shape of a maltese cross, seen in polymer (e g, polyethylene) spherulites when viewed under a polarizing microscope.

MAN Abbreviation for METHACRYLONITRILE, w s.

Mandrel (1) The core around which paper, fabric, or resin-impregnated fibrous glass is wound to form pipes or tubes. (2) In extrusion, an extension of the core of a pipe or tubing die, internally cooled by circulating water or other fluid, that guides and cools the internal surface of the tube as it emerges from the die proper. The mandrel is an important determiner of the final internal diameter of the tube.

Manifold A pipe or channel with several inlets or outlets. With reference to blow molding, extrusion, and injection molding, a manifold is a piping or distribution system that receives the outflow of the extruder or molder and divides or distributes it to feed several blow-molding heads or injection nozzles. Manifolds are also incorporated in cooling systems. Manifolds as components of extrusion dies are named according to shape, as follows: (a) *tear drop*—the cross-section is streamlined in tear-drop shape, narrowing as it leads to the die land; (b) *fishtail*—the flow channel leading to the die land area is roughly trapezoidal in shape in its top view and resembles a fishtail; (c) ***T**-shape*—a manifold that is fed at the center of its width and is circular in cross section (little used today); (d) *coat-hanger*—a type of manifold used in sheet and film dies, center-fed, in which the top view of the flow channel has the obtuse-isosceles-triangular shape of a coat hanger.

Man-Made Fiber Synonym for SYNTHETIC FIBER, w s.

Mannich Reaction The condensation of ammonia or a primary or secondary amine with formaldehyde and a compound containing at least one hydrogen atom of pronounced activity. The active hydrogen is replaced by an aminomethyl or substituted aminomethyl group. This reaction has been employed in producing "Mannich polyols" for use in making urethane foams.

Marble (1) Limestone that has crystallized to varying extent, often with veined inclusions, and occurring in many colors. Its preponderant constituent is CALCIUM CARBONATE, w s. (2) A smooth round sphere of any hard nonmetal in the size range from about 0.7 to 2.5 cm.

Marquardt Index In an infrared-absorption study of the cure advancement of a phenolic resin, the Marquardt index is the numerical difference in percent transmission between the absorption peaks at 12.2 and 13.3 μm. As resin cure progresses, the intensity of the 13.3-μm absorption increases more rapidly than that of the initially stronger 12.2-μm peak; thus the Marquardt index decreases as the cure advances. See also INFRARED POLYMERIZATION INDEX.

Mar Resistance The resistance of a glossy plastic surface to abrasive action. It is measured (ASTM D 673) by abrading a specimen to a series of degrees, then measuring the gloss of the abraded spots with a glossmeter and comparing the results to that of the unabraded area of the specimen. See also GLOSS.

Martens Heat-Deflection Temperature The temperature at which, under four-point loading, a bar of polymer deflects by a specified amount. For amorphous polymers, the Martens temperature is about

20°C below the glass-transition temperature. Compare DEFLECTION TEMPERATURE.

Mask A stencil used for spray-painting plastics, consisting of a relatively thin sheet shaped to fit the part to be painted with openings for areas to be painted. Masks for irregularly shaped articles are often made by electroforming a thin shell over a part, then cutting openings in the desired areas. Masks for spherical articles such as play balls can be made from spun-metal hemispheres, and those for flat articles from metal sheets.

Mass Quantity of matter, whose unit, the kilogram, is one of seven base units of the SI system. The term is often confused with *weight* in everyday use, probably because, when weighed on an equal-arm balance, the mass being determined is compared with standard masses, ordinarily referred to as "weights". Although the kilogram-force (*kilopond*) has long been used and is still, alas! being used, it has no place in the SI system. See also WEIGHT and FORCE (3).

Mass-Action Law For a homogeneous reacting system, the rate of chemical reaction is proportional to the active masses of the reacting substances, the molecular concentration of a substance in a gas or liquid being taken as its active mass.

Mass Dyeing See SPIN DYEING.

Mass Polymerization See BULK POLYMERIZATION.

Mass Spectrometry (MS) As applied to polymers, mass spectrometry is a sophisticated analytical technique in which the polymer is pyrolyzed, the fragment molecules are injected into a vacuum chamber where they are ionized with an electron gun, accelerated in an electric field, and shot through a magnetic field, the paths of the more massive molecules curving less than the lighter ones. A detector registers the mass number and ion count at each mass number and from this information develops a spectrum. A skilled polymer analyst can tell much about the original polymer from his interpretation of the spectrum of the fragments. The MS method may be supplemented by GAS CHROMATOGRAPHY, w s, which can identify the types of chemical structures in the fragments.

Mass (Fiber) Strength The force per unit of lineal density required to break a fiber. The SI measure is newtons per (kilogram per meter), or N·m/kg. Long used in the staple-fiber industry has been the unit gram-force per denier. 1 g_f/denier = 88,259 N·m/kg.

Masterbatch A term used in the rubber industry for rubber compounds containing high percentages of pigments and/or other additives, to be added in small amounts to batches during compounding. The term is often used in the plastics industry for COLOR CONCENTRATE, w s.

Master Curve The curve one gets by applying the principle of TIME-TEMPERATURE EQUIVALENCE (w s) to viscoelastic data on, say, relaxation modulus or creep.

Mastic (1) A solid resinous material obtained from the mastic tree (*Pistacia lentiscus*) and used in adhesives and lacquers. (2) *Asphalt mastic*, a composition of mineral matter with resin and solvent. (3) Any pasty material used as a waterproof coating or as a cement for setting tile.

Mastication Intense shearing of unvulcanized rubber by working in a roll mill or internal mixer to reduce its molecular weight preparatory to compounding and molding.

Mat A fabric or felt of glass or other reinforcing fibrous material cut to the contour of a mold, for use in reinforced-plastics processes such as matched-die molding, hand layup, or contact-pressure molding. The mat is usually impregnated with resin just before or during the molding process.

Matched-Metal-Die Molding (matched-die molding, matched-metal molding) The process of forming shaped articles of reinforced plastics by pressing resin-soaked mats or preforms between matching male and female mold halves. For simple shapes without compound curves or deep draws, mats cut from rolls or sheets of compacted glass fiber or other reinforcement may be used. For the more intricate shapes, preforms are made by depositing cut fibers mixed with a small amount of resin binder on a screen shaped approximately to the contours of the finished article. Fibers are deposited on the screen by spraying while air is drawn through the screen, or by applying a water slurry of the fibers to the screen and sucking the water away, or by a suction process from a rotating plenum chamber. The mat or preform is placed on one half of the mold, then a measured quantity of resin is poured on and spread in a controlled pattern. The mold is then closed and subjected to heat and pressure in the range of 1 to 3 MPa until the resin has cured. The process is used for making boat hulls, automotive parts, furniture seats, and a wide variety of shaped panels for other applications. In a variation called PREPREG MOLDING (w s), the fibrous mat has been preimpregnated with resin, fillers, pigments and other additives so that it is ready for molding without further treatment.

Matched-Mold Thermoforming A sheet-thermoforming process in which the heated plastic sheet is shaped between male and female halves of a matched mold. The molds may be of metal or inexpensive materials such as plaster, wood, epoxy resin, etc, and must be vented to permit the escape of air as the mold closes. See SHEET THERMOFORMING.

Material Well Space provided in a compression or transfer mold to allow for bulk factor, that is, to provide for the difference in volume between the loose molding compound and the final molding.

Matte Finish (mat finish) A surface or finish (as on plastics, mold surfaces, or coatings) that is without luster or gloss; a dull finish.

Maximum Allowable Concentration See THRESHOLD LIMIT VALUE.

Maximum Permissible Stress See ALLOWABLE STRESS and FACTOR OF SAFETY.

Maxwell Model (Maxwell element) A concept useful in modeling the deformation behavior of viscoelastic materials. It consists of an elastic spring in series with a viscous dashpot. When the ends are pulled apart with a definite force, the spring deflects instantaneously to its stretched position then motion is steady as the dashpot opens. See also VOIGT MODEL. A simple combination of these two types provides a fair analogic representation of real viscoelastic behavior under stress.

MBK Abbreviation for METHYL BUTYL KETONE, w s.

MBS Abbreviation for *methacrylate-butadiene-styrene* resin. These are mixtures of PMMA and butadiene-styrene copolymers, formulated in a variety of types with markedly different characteristics according to their composition and molecular weight. MBS resins can be processed by all the usual thermoplastics processes.

Mc Abbreviation for *megacycle*, one million cycles, loosely used to mean 1 MHz, one million cycles *per second*.

MD (1) Abbreviation for MACHINE DIRECTION, w s. (2) Abbreviation for *methylene dianiline*, little used because of its carcinogenicity.

MDI Abbreviation for *diphenylmethane-4,4'-diisocyanate*. See DIISOCYANATE.

MDPE Abbreviation for *medium-density polyethylene*. See POLYETHYLENE.

Mean See ARITHMETIC MEAN, GEOMETRIC MEAN, HARMONIC MEAN.

Measling The appearance of spots or stars under the surface of the resin portion of an epoxy/glass-fiber laminate (from *measles*).

Mechanical Equivalent of Heat A conversion factor that transforms work or kinetic energy into heat. Probably the best known one is 788 foot-pounds per British thermal unit; others are 2545 Btu per horsepower-hour, $4.186 \cdot 10^7$ ergs per calorie, and 3413 Btu/(kW·h). In SI there is no need for such factors because work, heat and electrical energy are all measured in joules. 1 joule = 1 meter-newton = 1 watt-second!

Mechanical Grease Forming A method of SHEET THERMOFORMING, w s, used with acrylic sheet when excellent opticals are imperative and the shape desired cannot be produced by FREE FORMING, w s. The mold surface is covered with a 1- to 2-mm-thick layer of felt soaked with melted grease that must be cleaned off the sheet after forming.

Mechanically Foamed Plastic A cellular plastic in which the cells have been produced by gases introduced by physical means. See also CELLULAR PLASTIC.

Mechanical Property Any property of a material that defines its response to a particular mode of stress or strain. Such properties include elastic moduli, strength and ultimate strain in several modes, impact strength, abrasion resistance, creep, ductility, coefficient of friction, hardness, cyclic fatigue strength, tear strength, and machinability. Many ASTM tests in Section 08 are devoted to the mechanical properties of plastics.

Mechanical Spectrometer An instrument (Rheometrics, Inc) capable of applying an alternating tensile/compressive (or flexural or torsional) deformation of constant amplitude to a plastic specimen in the frequency range from 0.002 to 80 Hz and measuring the variation of force so caused and the phase angle between the deformation and the force. From this information one can calculate the "real" and "imaginary" parts of the various moduli. Testing is described in ASTM D 4065.

Median The value in an arrayed set of repeated measurements that divides the set into two equal-numbered groups. If the sample size is odd, the median is the middle value. The median is a useful measure of the center when the distribution is strongly skewed toward low or high values. Compare ARITHMETIC MEAN.

Medium Yellow A pigment based on pure, monoclinic lead chromate.

Mega- (M) The SI prefix meaning $\times 10^6$.

Megahertz A unit of vibrational frequency equal to 10^6 cycles per second, i e, 10^6 Hz.

MEK Abbreviation for METHYL ETHYL KETONE, w s.

MEKP Abbreviation for METHYL ETHYL KETONE PEROXIDE, w s.

Melamine (2,4,6-triamino-1,3,5-triazine) A cyclic unsaturated compound, derived from cyanuric acid, with the structure shown below.

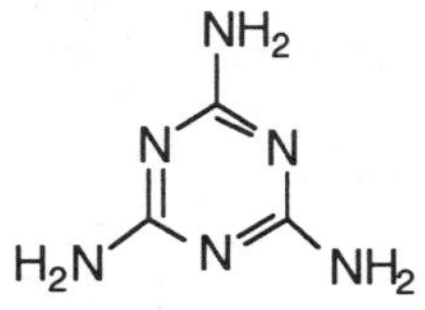

NH_2

N N

H_2N N NH_2

Melamine's main use is for MELAMINE-FORMALDEHYDE RESINs, w s.

Melamine-Formaldehyde Resin (melamine resin) Any of a group of thermosetting resins of the amino-resin family, made by reacting melamine with formaldehyde. The lower-molecular-weight, uncured melamine resins are water-soluble syrups, used for impregnating paper, laminating, etc. High-molecular-weight resins, usually cellulose-filled, are powders widely used, from 1950-1970, for plastic tableware.

Melamine/Phenolic Resin A mixture of melamine- and phenol-formaldehyde resins that combines the dimensional stability and ease of molding of phenolics with the wider range of colorability of the melamine resins.

Melt (n) A material, solid at room temperature, that has been heated to a molten condition.

Melt-Bead Sealing See EXTRUDED-BEAD SEALING.

Melt Coating See EXTRUSION COATING.

Melt-Draining Screw See SOLIDS-DRAINING SCREW.

Melt Extruder A short extruder, typically of constant channel depth and lead throughout, designed to receive a molten feed and raise its pressure for extrusion through a die, such as a pelletizing die.

Melt-Flow Index (MFI, melt index) The rate of flow, in grams per ten minutes, of a molten resin through an orifice 2.096 mm in diameter and 8.000 mm long at a specified temperature and weight of piston pressing on the melt. Numerous combinations of temperatures and weights are listed in ASTM D 1238, for various thermoplastics. This single-point flow measurement is useful in controlling production quality and resin purchasing, but most of the MFI conditions are at much lower shear than those prevailing in commercial processing, so MFI is not a reliable guide to processing behavior. MFI is inversely related to viscosity and decreases rapidly as the molecular weight in a resin family increases.

Melt Fracture In extrusion, distortion of the extrudate as it emerges from a die. The effect ranges from minor, regular ridges and valleys at 45° or 90° to the axis of the extrudate to violent wriggling and curling and, at its most extreme, breaking up of the extrudate into fragments. It usually develops when the shear stress at the die-land wall exceeds 0.1 to 0.4 MPa, and is believed to be caused by release of lateral stresses developed in the die entrance or (rarely) by alternate sticking and slipping of the stream at the die surface. Corrective measures include raising the melt temperature, reducing the extrusion rate, changing to a less viscous resin, reducing the die-entry angle, and lengthening the die land. Melt fracture probably occurs, too, in injection molds being rapidly filled through small gates and may be responsible for some types of defects occurring in molded parts near such gates. See also SHARKSKIN.

Melt Index The original name, revised around 1970, for MELT-FLOW INDEX, w s.

Melting Point (melting range) In pure, simple compounds, the temperature at which the transition from solid to liquid occurs, requiring heat input. Polymers, being broad mixtures of homologs, melt over a substantial range of temperature, the shorter chains melting first with rising temperature, the longer ones later. Crystalline polymers have narrower, more distinct melting ranges than amorphous polymers. See also HEAT OF FUSION.

Melt Instability (melt-flow instability) A term applied to the early manifestations of MELT FRACTURE, w s.

Melt Pressure The gauge pressure exerted at any point in a processing apparatus that develops pressure. In extruders, melt pressure in the head is usually monitored. In injection machines the location is analogous but melt pressures have also been measured in mold cavities. Not to be confused with (though related to) INJECTION-MOLDING PRESSURE, w s.

Melt Spinning See SPINNING.

Melt Strength The strength of a plastic while in the molten state. This property is pertinent to extrusion of parisons for blow molding, to drawing extrudates from dies, as in making monofilaments and cast film, and to sheet thermoforming. It is also important when a plastic film is reheated for shrink-packaging. This property is very difficult to measure because of the ease with which a filament stretches in elongational flow at the temperatures of interest.

Melt Temperature The temperature of molten or softened plastic at any point within the material being processed. In extrusion and injection molding, melt temperature is an important indicator of the state of the material and the process. Many types of instruments, most of them based on thermocouples or resistance thermometers, have been employed in extruders, where melt temperature is usually measured in the head and sometimes in the die. In thermoforming, temperatures of softened sheets are measured with INFRARED PYROMETERs, w s.

Melt Viscosity The resistance to shear in a molten resin, quantified as the quotient of shear stress divided by shear rate at any point in the flowing material. Elongational viscosity, which comes into play in the drawing of extrudates, is analogously defined. In polymers, the viscosity depends not only on temperature and, less strongly, on pressure, but also on the level of shear stress (or shear rate). More at VISCOSITY, POWER LAW, and PSEUDOPLASTIC FLUID.

Melting Zone In a well-designed extruder screw, the section, intended to be coincident with the TRANSITION SECTION, w s, in which most, if not all, of the melting of the feedstock occurs. The pumping section, in which the plastic is presumed to be fully melted, is sometimes called the *melt zone*.

Memory (elastic memory, plastic memory) The tendency of a plastic article to revert in dimensions to a size previously existing at some stage in its manufacture. For example, a film that has been oriented by hot stretching and chilled while under tension, will, upon reheating, tend to revert to its original prestretched size due to its "memory". See also ORIENTATION.

Mer Derived from the Greek *meros*, meaning a part or unit, the mer is the repeating structural unit of a polymer. In addition polymers such as polyethylene the mer weight is the same as the monomer's molecular weight. Saving a small correction for end groups, the molecular weight of a polymer chain equals the mer weight times the DEGREE OF POLYMERIZATION, w s. Dimers, trimers, tetramers, oligomers, and polymers contain two, three, four, several, and many mer units, respectively.

Mercuric Chloride (corrosive sublimate, mercury bichloride) $HgCl_2$. White crystals, used as a polymerization catalyst for PVC. Mercuric chloride is highly toxic, so must be handled with care and requires special disposal procedures.

Mesh Number The deprecated, but still widely used (in the US) nomenclature for screen sizes, meaning the number of wires per inch of screen width. In standard square-mesh screens used in SIEVE ANALYSIS, w s, the count and wire diameter are the same in both directions. Thus, the widths of the standard-screen openings (inches) are in approximate inverse proportion to the mesh numbers, $\approx 0.6 \cdot (\text{mesh number})^{-1}$. Modern nomenclature, in accordance with SI, designates open-mesh screens by the minimum width of the openings in millimeters.

Mesityl Oxide (4-methyl-3-pentene-2-one) $CH_3COCH{=}C(CH_3)_2$. An oily, colorless liquid used as a powerful solvent for cellulosic and vinyl resins, and as an intermediate in the production of plasticizers.

Meta- (*m-*) A prefix used in naming aromatic organic compounds, ignored in alphabetization, that designates the 3- and 5-positions relative to the substituted 1-position in a benzene ring. Compare *ORTHO-* and *PARA-* (3).

Metallic Fiber Generic term for a manufactured fiber composed of metal, plastic-coated metal, or metal-coated plastic (Federal Trade Commission). Examples are aluminum fiber covered with cellulose acetate butyrate and nickel-coated glass fiber.

Metallic-Flake Pigment Flat, thin particles of either aluminum, copper or copper alloy that reflect light specularly when incorporated into a plastic substance or coating vehicle with their reflecting surfaces approximately parallel. The aluminum pigments reflect very strongly throughout the visible spectrum, producing brilliant blue-white highlights. The copper-based pigments, called gold bronzes but actually brasses, range from the characteristic red of copper to progressively more yellow with rising zinc content.

Metallic Soap See SOAP, METALLIC.

Metallized Glass Glass spheres, flakes, or fibers that have been coated with silver or aluminum and, as fillers, provide increased electrical conductivity and light-reflecting pigmentation.

Metallizing A term covering all processes by which plastics (and some other base materials) are coated with metal. The most commonly used processes are described under ELECTROLESS PLATING, SILVER-SPRAY PROCESS, and VACUUM METALLIZING. Other methods include spraying with metallic pigments, chemical reduction, gas plating and vapor pyrolysis.

Metamer From the Greek *meta* (change, transposition, transfer) and *meros* (part or portion), the term metamer was formerly used in chemistry for a specific kind of isomer having to do with group-positional differences in molecules of the same composition and functionality. The term ISOMER, w s, is now used in this limited sense (as well as in broader ones).

Metamerism A term sometimes used in the color industry for the phenomenon exhibited by two surfaces that appear to be of the same color when viewed under one light source (e g, daylight), but that appear different when viewed under a different light source (e g, incandescent lamp). The term *geometric metamerism* refers to a change in perceived color of a surface with a change in viewing angle.

Metastable A temporary state of structure in a plastic, such as a crystalline plastic in which the final crystallinity is attained after passage of hours or days following molding. No physical or mechanical tests should be made while the test material is in a metastable condition (unless data regarding that condition are desired).

Meter (1) (m, metre) The SI unit of length, one of the seven basic units of the system, defined as 1,680,763.73 wavelengths of the radiation in vacuum corresponding to the transition between the levels $2p_{10}$ and $5d_5$ of the krypton-86 atom (an orange spectral line). One foot equals (exactly) 0.3048 m. (2) Any device for measuring a physical or chemical quantity in which the measurement is indicated digitally, or analogically on a scale. In this sense, -meter is often used as a suffix, as in *thermometer*.

Metering Screw An extruder screw whose final section, from four to ten flights, has a shallow channel of constant depth and lead. As its name suggests, the metering section of such a screw is intended to regulate the amount delivered per rotation of the screw. It also provides time for the equalization of melt temperature and helps to control the steadiness of the extrusion rate.

Metering Zone (metering section) The final portion of a METERING SCREW, w s, that builds pressure to force the melt through the screens and die. The metering section usually has a constant lead and a shallower channel than the preceding sections of the screw.

Methacrylate Ester Any of the esters of methacrylic acid (w s,) having the general formula $CH_2{=}C(CH_3)COOR$, wherein R is usually methyl, ethyl, isobutyl, or *n*-butyl to *n*-octyl. These esters are polymerizable to acrylic resins.

Methacrylate Plastic See ACRYLIC RESIN.

Methacrylic Acid (α-methacrylic acid, 2-methyl-2-propenoic acid) $CH_2{=}C(CH_3)COOH$. A colorless liquid prepared by the acid hydrolysis of acetone, from which are derived all of the methacrylate compounds. Most important of these are the esters, especially methyl methacrylate.

Methacrylonitrile (MAN, α-methyl acrylonitrile) A vinyl monomer containing the nitrile group whose homopolymers are true thermoplastics with good mechanical strength and high resistance to solvents, acids and alkalis. Modified properties can be obtained through blending, grafting, or copolymerization with other monomers such as styrene and methyl methacrylate. MAN is also used as a replacement for acrylonitrile in preparing nitrile elastomers.

γ-Methacryloxypropyltrimethoxy Silane $CH_2{=}CHCOO(CH_2)_3Si(O-CH_3)_3$. A silane coupling agent used in reinforced polyesters, epoxies, and many thermoplastics to achieve improved adhesion between resin and glass fibers.

Methanal Synonym for FORMALDEHYDE, w s.

Methanol (carbinol, methyl alcohol, wood alcohol) CH_3OH. A colorless, toxic liquid usually obtained by synthesis from hydrogen and carbon monoxide. It is sometimes called *wood alcohol*, but the methanol obtained from the destructive distillation of wood also contains additional, contaminating compounds. Methanol is used as an intermediate in producing formaldehyde, phenolic, urea, melamine, and acetal resins, and as a solvent for cellulose nitrate, ethyl cellulose, polyvinyl acetate, and polyvinyl butyral.

Method of Least Squares See REGRESSION ANALYSIS.

Methoxyethylacetoxy Stearate $C_{17}H_{34}(OCOCH_3)COOCH_2CH_2OCH_3$. A plasticizer for vinyl and cellulosic resins.

Methoxyethylacetyl Ricinoleate $C_{17}H_{32}(OCOCH_3)COOCH_2CH_2O$-

CH_3. A plasticizer for cellulosic and vinyl resins.

Methoxyethyl Ricinoleate $C_{17}H_{32}(OH)COOCH_2CH_2OCH_3$. A plasticizer for cellulosic and vinyl resins.

Methoxyethyl Stearate (1,2-propylene glycol monostearate) $C_{17}H_{35}$-$COOCH_2CH_2OCH_3$. A solvent and plasticizer for cellulosic plastics.

Methyl Abietate $C_{19}H_{29}COOCH_3$. A derivative of abietic acid (from rosin) used as a plasticizer for cellulosic, acrylic, and vinyl resins, polystyrene, and urea-formaldehyde resins.

***N*-Methyl Acetamide** (NMA) $CH_3CONHCH_3$. A solvent useful in making aromatic-mer polymers, such as polyimides.

Methyl Acetate CH_3COOCH_3. A colorless, volatile liquid with a fragrant odor, a solvent for acetyl cellulose and cellulose esters.

Methylacetyl Ricinoleate $C_{17}H_{32}(OCOCH_3)COOCH_3$. A plasticizer for some vinyl resins and polystyrene.

Methyl Acrylate $CH_2{=}CHCOOCH_3$. A colorless, volatile liquid, a monomer for acrylic resins.

Methyl Alcohol Synonym for METHANOL, w s.

Methyl *n*-Amyl Ketone (2-heptanone) $CH_3CO(CH_2)_4CH_3$. A high-boiling ketone solvent with a fruity odor, used in synthetic-resin finishes. It is especially useful in lacquers and finishes for roll coating where improved blush resistance is required.

Methyl Benzene (methyl benzol) Synonym for TOLUENE, w s.

Methyl Butadiene Synonym for ISOPRENE, w s.

Methyl Butyl Ketone (MBK, propylacetone) $CH_3COC_4H_9$. A solvent for vinyl and many other resins, often used in conjunction with methyl ethyl ketone to control the drying rate of lacquers. A higher content of MBK slows the rate.

Methyl Butynol $HC{\equiv}CCOH(CH_3)_2$. A viscosity stabilizer and solvent for some nylons.

Methyl Butyrate $CH_3(CH_2)_2COOCH_3$. A solvent for ethyl cellulose and cellulose nitrate.

Methyl Cellosolve See ETHYLENE GLYCOL MONOETHYL ETHER.

Methyl Cellulose A cellulose ether in which some of the cellulosic –OH groups have been replaced by $-OCH_3$. The degree of substitution determines properties and uses as thickeners and emulsifiers.

Methyl-2-Cyanoacrylate A fast setting adhesive used for bonding cellulosics, nylon, polyesters, acrylics, polystyrene, and polyurethanes to each other and to other materials such as woods, metals, and glass. Catalyzed by atmospheric moisture or lightly applied methanol, the adhesive polymerizes without loss of solvent. For best results, the surfaces to be bonded should mate closely.

2,2'-Methylene-bis-(6-*tert*-Butyl-4-Ether Phenol) An antioxidant for acrylonitrile-butadiene-styrene packaging, appliances, pipe, and automotive items.

2,2'-Methylene-bis-(6-*tert*-Butyl-4-Methyl Phenol) A phenolic-type antioxidant for polyolefins and acrylonitrile-butadiene-styrene resins.

4,4'-Methylene-bis-(Cyclohexyl Isocyanate) ($H_{12}MDI$) A diisocyanate used in making urethane elastomers and foams.

Methylene Chloride (dichloromethane, methylene dichloride) CH_2Cl_2. A colorless, fairly dense, nonflammable liquid used as a solvent for cellulose triacetate and vinyl resins, a solvent in the polymerization of polycarbonate resins, and as a reactant for certain phenolic resins. It was widely used as a paint stripper and solvent for cured epoxy resins, but is less used now in the effort to keep chlorinated solvents out of the atmosphere.

Methylene Group The radical $-CH_2-$ or $=CH_2$, existing only in combination.

Methyl Ethyl Ketone (MEK, 2-butanone) $CH_3COC_2H_5$. A colorless, flammable liquid with an acetone-like odor. One of the most widely used solvents for several thermoplastics including cellulosics, acrylics, polystyrene, and vinyl copolymers.

Methyl Ethyl Ketone Peroxide (MEKP, MEK peroxide) A complex peroxide mixture made by reacting hydrogen peroxide with MEK, with the approximate formula $(CH_3COOC_2H_5)_3$. MEKP is an initiator for free-radical polymerization and a curing agent for polyester resins. In combination with an accelerator such as cobalt naphthenate, MEKP can bring about cure at room temperature. Because it is unstable, it is often handled in solution. MEKP should be kept only in small quantities and stored in a freezer when not in use.

Methyl Glucoside $CH_2OHCH(CHOH)_3CHOOCH_3$. A plasticizer for alkyd, amino and phenolic resins. It is also used as a polyol for urethane-foam production.

Methyl Group The radical $-CH_3$, existing only in combination.

Methyl Hexyl Ketone (2-octanone) $CH_3COC_6H_{13}$. A colorless, high-boiling liquid with a pleasant odor, used as a solvent for epoxy coatings.

Methyl Isoamyl Ketone (5-methyl-2-hexanone) $CH_3COC_2H_4CH<(CH_3)_2$. A colorless liquid with a pleasant odor, used as a solvent for cellulose esters, acrylic resins, and certain vinyl polymers. It has a high solvent power and low evaporation rate, making it useful as a retarder solvent that promotes flow-out of coatings and reduces blushing.

Methyl Isobutyl Ketone (MIBK, hexanone, 4-methyl-2-pentanone) $(CH_3)CHCH_2COCH_3$. A solvent with a moderate evaporation rate, used with cellulosic, vinyl, alkyd, acrylic, phenolic, and coumarone-indene resins, and polystyrene.

Methyl Isopropenyl Ketone $CH_2{=}C(CH_3)COCH_3$. A flammable liquid used as a copolymerizable monomer.

Methyl Lactate $CH_3CHOHCOOCH_3$. A solvent for cellulosic plastics.

Methyl Methacrylate (MMA) $CH_2{=}CCH_3COOCH_3$. A colorless, volatile liquid derived from acetone cyanohydrin, methanol, and dilute sulfuric acid (catalyst), the most important monomer in the production of ACRYLIC RESINs, w s.

4-Methyl Morpholine $\overline{OCH_2CH_2N(CH_3)CH_2CH_2}$. A colorless liquid used as a catalyst in urethane-foam making.

Methyl Myristate (methyl tetradecanoate) $CH_3(CH_2)_{12}COOCH_3$. The methyl ester of myristic acid, with applications in stabilizers and plasticizers.

Methyl Nadic Anhydride (methyl-5-norbornene-2,3-dicarboxylic anhydride, nadic methyl anhydride, NMA) (Other isomers may be present.) A curing agent for epoxies, yielding cured resins ranging from tough, at low levels of NMA, to heat-resistant.

Methyl Oleate $C_{17}H_{33}COOCH_3$. A plasticizer for ethyl cellulose, polystyrene, and, with limited compatibility, vinyl resins.

Methylol Phenol A phenol having one or more $-CH_2OH$ groups in its ring, a first stage in the formation of phenolic resin by reaction of phenol with formaldehyde.

Methylol Urea $H_2NCONHCH_2OH$. Colorless crystals derived from combination of urea with formaldehyde, the first stage in the production of urea-formaldehyde resins.

Methyl Palmitate $CH_3(CH_2)_{14}COOCH_3$. The methyl ester of palmitic acid, with applications in stabilizers and plasticizers.

Methyl Pentachlorostearate $C_{17}H_{30}Cl_5COOCH_3$. A plasticizer for polystyrene, polymethyl methacrylate, cellulosics, and vinyl resins.

Methylpentene Resin Synonym for POLY(4-METHYLPENTENE-1), w s.

Methyl Phthalyl Ethyl Glycolate $CH_3OOCC_6H_4COOCH_2COOC_2H_5$. A plasticizer for several thermoplastics including PVC, polystyrene, and cellulosics.

Methyl Propionate $CH_3CH_2COOCH_3$. A solvent for cellulose nitrate.

1-Methyl-2-Pyrrolidinone (NMP,) $CH_3\overline{NCH_2CH_2CH_2C}=O$. A solvent with a low order of inhalation toxicity, good thermal and chemical stability, and a high flash point. It is capable of dissolving resistant resins such as polyamide-imides, epoxies, urethanes, nylon, and PVC. It is a solvent of choice for spinning PVC fibers from solution. Previously known as *N*-methyl-2-pyrrolidone.

Methyl Ricinoleate $CH_3(CH_2)_5CH(OH)CH_2CH=CH(CH_2)_7COOCH_3$. A plasticizer for cellulosic resins, polyvinyl acetate, and polystyrene.

α-Methylstyrene $C_6H_5C(CH_3)=CH_2$. A colorless liquid, easily polymerizable by heat or with catalysts, and typically copolymerized with methyl methacrylate or styrene.

Metre SI spelling of METER (1), w s.

Mev Abbreviation for *million electron volts*, a measure of kinetic energy for subatomic particles. 1 Mev = 1.60219×10^{-13} J.

MF See MELAMINE-FORMALDEHYDE RESIN.

MFC Abbreviation for *multifunctional concentrate*. See COLOR CONCENTRATE.

MFI Abbreviation for MELT-FLOW INDEX, w s.

Mg Chemical symbol for the element magnesium.

M-Glass A high-modulus glass whose fibers are sometimes used for reinforcing plastics when high modulus at moderate cost is desired. Major constituents are SiO_2 54%, CaO 13%, MgO 9%, BeO and TiO_2 8% each, Li_2O and CeO_2 3% each, and ZrO_2 2%. Fiber density is 2.89 g/cm^3, modulus (E) is 110 GPa, and tensile strength is 3.5 GPa.

MHz Abbreviation for MEGAHERTZ, w s.

MI Abbreviation for *melt index*, a term replaced by MELT-FLOW INDEX, w s.

MIBK Abbreviation for METHYL ISOBUTYL KETONE, w s.

Mica Any of a family of crystalline silicate minerals characterized physically by a perfect basal cleavage, consisting essentially of orthosilicates of aluminum and potassium. They occur naturally, mainly as the minerals *muscovite* (white mica), *phlogopite* (amber mica), and *biotite;* and are also synthesized from potassium fluosilicate and alumina.

Micas are used as fillers in thermosetting resins, imparting good electrical properties and heat resistance. A grade having high aspect ratios (HAR) with flakes 3 to 5 μm thick and aspect ratios as high as 200 can be processed, although the optimum aspect ratio appears to be about 70. The larger flakes increase flexural modulus and strength, have lower moisture content, and raise the deflection temperatures of compounds containing them.

Micelle A colloidal particle formed by the reversible aggregation of dissolved molecules. Micelles may be in the shape of spheres, cylinders, or platelets. Soaps, detergents, and other emulsifying agents used in emulsion polymerization contain micelles generally composed of from 50 to 100 molecules of emulsifier, within which the polymerization reaction may be initiated.

Micro- (μ) The SI prefix meaning $\times 10^{-6}$.

Microballoons (1) Tiny, hollow plastic spheres used to reduce evaporation of liquids such as oils by floating a layer of spheres on the surfaces of stored liquids. (2) Synonym for MICROSPHERES, w s.

Microbial Degradation See BIODEGRADATION and PINK STAINING.

Microcrystalline Pertaining to crystallinity that is visible only under a microscope, sometimes taken to mean that the crystals referred to are no larger than 1 μm.

Microcrystalline Silicate A derivative of chrysotile asbestos, consisting of tiny rod-shaped particles of hydrated magnesium silicate. The particles have hydroxyl groups on their surfaces that bond with hydrogen-bonding sites on the molecules of a fluid in which they are incorporated. The material has also been used as a viscosity-building agent in unsaturated polyester and other resins.

Microcrystalline Wax Any of a group of petroleum-derived waxes that differ from paraffin waxes in having finer crystal structure, higher melting points—between 60 and 93°C, higher liquid viscosities, and greater ductility. They are used in fiberboard coatings, paper-container linings, and polishes.

Microencapsulation The process of encasing a small solid particle or a discrete amount of liquid or gas in a capsule. The term applies to capsules ranging in diameter from a few micrometers to about 500 μm. The capsule is usually made of a synthetic plastic, although waxes, glass, and metals are also used. Methods used for forming polymeric microcapsules fall into three broad classes: phase separation, interfacial reaction, and physical methods. Phase separation methods include COACERVATION (w s), applying meltable dispersions, and spray-drying of a suspension of the material in a vaporizable solvent. Interfacial-reaction methods include interfacial polymerization, *in-situ* polymerization, and chemical-vapor deposition. The physical methods include fluidized-bed coating processes, spray coating, electrostatic coating methods, and extrusion. Typical examples of microencapsulation are "carbonless" carbon paper, timed-release drugs and fertilizers, and battery separators.

Microgel A small particle of cross-linked polymer of very high molecular weight and containing closed loops. Microgels may be present in trace amounts due to impurities in monomers, and can influence polymer properties and molecular-weight studies.

Micron This long deprecated but still used length unit and its abbreviation, the Greek letter μ, were dropped by action of the General Conference on Weights and Measures on October 13, 1967. The symbol "μ" is to be used solely as the abbreviation for the prefix MICRO-, w s. The old micron should now be spoken as micrometer (μm).

Microporous Having pores of microscopic dimensions. Some plastic films and fabric coatings are rendered microporous in order to permit the passage of water vapor ("breathing") while preventing the penetration of raindrops.

Microspheres Tiny, hollow spheres of glass or plastic used as fillers to impart low density to plastics, such plastics being known as *syntactic foams*. Plastics used to make microspheres include phenolic, epoxy and a copolymer of vinylidene chloride and acrylonitrile. The last contains a heat-activated blowing agent that expands the spheres either before their incorporation into a matrix polymer or afterward. The copolymer spheres impart better mechanical properties to the matrix than do the glass or epoxy microspheres. See also GLASS SPHERES.

Microstructure The detailed structure of plastics as seen through light and electron microscopes, approximately the magnification range of 100× to 100,000×, including such features as crystalline form, spherulites, voids, distribution of filler and pigment particles, discontinuous-phase particles in blends, and, in reinforced plastics, configuration, length distribution, and cross-section distribution of yarns and, within the yarns, the filament ends, etc.

Microwaveable Said of plastics for kitchen use, and of the utensils made from them, that are heated little or not at all by the direct action of the high-frequency waves generated by microwave ovens, and that withstand many repeated heatings by the foods contained in them without warping, shrinking, or staining.

Microwave Heating A heating process similar to dielectric heating, but using frequencies in the 10^9- to 10^{10}-Hz (radar) range. The Federal Communications Commission has allocated the specific frequencies 915, 2450, and 5850 MHz for industrial use. Microwave ovens similar to those used in restaurants and households for rapidly cooking foods have been used experimentally for preheating molding powders, vacuum-bag curing, autoclave molding, and curing of nylon overwraps. Plastic films coated with water-containing materials such as polyvinylidene chloride can be dried rapidly and economically by microwave energy. Line speeds about 5 m/s have been attained with polyethylene film, by means of a microwave cabinet only 2.4 m long.

Migration The transfer of a constituent of a plastic compound to another contacting substance.

Migration of Plasticizer In plasticized thermoplastics or elastomers, the movement of molecules of plasticizer from their interior locations when the article was originally formed to the surface layer of the article, where the plasticizer appears as a greasy or oily layer and may be rubbed off or dissolved away. The phenomenon occurs most often in vinyl compounds containing incompatible plasticizers.

Mil A unit of thickness equal to 0.001 inch, often used for specifying diameters of wires and glass fibers, and thicknesses of films. It is gradually being replaced by the SI units, the millimeter and micrometer. 1 mil = 0.0254 mm = 25.4 μm.

Mill (1, n) In the plastics industry, the term mill is generally taken to refer to a roll mill such as a two-roll mill used in compounding. More broadly, it includes all mechanical devices for converting raw materials into a condition ready for use, as well as machine tools that cut materials with rotating bits and many types of size-reduction machines.. (2, v) To process components of a plastic mixture in a two-roll mill.

Milled Fibers Small lengths of glass filaments produced by hammermilling continuous glass strands. They are useful as anticrazing and reinforcing fillers for adhesives.

Milli- (m) The SI prefix meaning $\times 10^{-3}$.

Mineral Black Black pigment made by grinding and/or heating slate, shale, or coal.

Mineral Spirits (naphtha) An aliphatic-hydrocarbon fraction of petroleum evolved in the distillation range of about 150° to 200°C. An example is "VM&P naphtha", used as a diluent in organosols.

Minimum Detectable Amount (MDA) In chemical analysis, the least amount of a substance being sought that balances two risks, Type I, the risk of falsely finding the substance to be present when in fact it is not, and Type II, the risk of not detecting that least amount. Typically, the two risks are made equal and, if both are 5%, the MDA is very nearly four times the standard deviation of the method. Lowering the risk increases MDA.

Miscibility (solubility) The greatest percentage of one liquid or polymer that forms a true, homogeneous solution, i e, a single phase, in another liquid or polymer. Few binary polymer systems are miscible over the entire range of composition, but many have limited miscibility at either end of the range. Miscibility usually increases with rising temperature. See also COMPATIBILITY.

Mixer Any of a wide variety of devices used to intermingle two or more materials to some defined state of uniformity. Some equipment intended mainly to provide size reduction may also accomplish mixing. Types used in the plastics industry are:

BALL MILL
BANBURY MIXER
CENTRIFUGAL IMPACT MIXER
CHANGE-CAN MIXER
COLLOID MILL
CONICAL DRY-BLENDER
DISK-AND-CONE AGITATOR
DRUM TUMBLER
HIGH-INTENSITY MIXER
HOMOGENIZER
INTENSIVE MIXER
INTERNAL MIXER
KNEADER
MILL
PROPELLER MIXER
RIBBON BLENDER
ROD MILL
ROLL MILL
SAND MILL
STATIC MIXER
TUMBLING AGITATOR
VIBRATORY MILL

Mixing Screw Any extruder screw that incorporates some modification (from standard designs) intended to improve mixing, mainly DISTRIBUTIVE MIXING, w s, but sometimes improving dispersion, too. One simple method is to insert one or more rings of closely spaced pegs arranged circumferentially in the screw channel and having nearly the same height as the flight. The pegs divide and redivide the melt streaming in a complex but regular path down the channel, accomplishing a kind of braiding of substreams. See also DULMADGE MIXING SECTION, MADDOCK MIXING SECTION, and CAVITY-TRANSFER MIXER.

Mixture A combination of two or more substances intermingled with varying percentage composition (unlike a true solution), in which each component retains its chemical identity.

MMA Abbreviation for METHYL METHACRYLATE, w s.

M_n (1) Abbreviation for NUMBER-AVERAGE MOLECULAR WEIGHT, w s. (2) Chemical symbol for the element manganese.

Mo Chemical symbol for the element molybdenum.

MOCA® Du Pont's tradename for methylene-bis-*o*-chloroaniline, much used until about 1980 as a curing agent for urethane rubbers and epoxy resins, prior to its being declared to be a carcinogen by OSHA.

Modacrylic Fiber A manufactured fiber in which the fiber-forming substance is any long-chain synthetic polymer composed of less than 85% but at least 35% by weight of acrylonitrile units (Federal Trade Commission).

Modified Resin Any synthetic resin into which has been incorporated a natural resin, an elastomer, or an oil that alters the processing characteristics or physical properties of the material.

Modulus (1) Derived from the Latin word meaning "small measure", a modulus is a measure of a mechanical property of a material, most frequently a stiffness property. See COMPRESSIVE MODULUS, FLEXURAL MODULUS, SHEAR MODULUS, MODULUS OF ELASTICITY, and MODULUS OF RESILIENCE. (2) The absolute value of a complex number or quantity, equal to the square root of the sum of the squares of the "real" and "imaginary" parts.

Modulus at 300% The tensile stress required to elongate a specimen to 3 times its original length (200% elongation) divided by 2. Although other elongations are used, 300% is the one most often employed for rubbers and flexible plastics.

Modulus in Compression See COMPRESSIVE MODULUS.

Modulus in Flexure See FLEXURAL MODULUS.

Modulus in Shear See SHEAR MODULUS.

Modulus of Elasticity (1, elastic modulus, tensile modulus, Young's modulus) The ratio of nominal tensile stress to the corresponding elongation below the proportional limit of a material. Since elongation is dimensionless, modulus has the units of stress. The relevant ASTM test is D 638. In contrast to structural metals such as mild steel, the stress-strain graphs for many plastics exhibit some curvature, even at very low strains. Since there is then no significant linear region whose slope would give the modulus, a SECANT MODULUS, w s, at 1 to 3% elongation may be reported for stiff materials. (2) More generally, any of the several elastic moduli characterizing behavior in shear (torsion), flexure, or change in volume under pressure (see BULK MODULUS). In SI, all types of elastic moduli are reported in pascals, usually megapascals (MPa). 1000 psi = 6.894,757 MPa.

Modulus of Resilience The energy that can be absorbed per unit volume of a stressed specimen without creating a permanent deformation. It is equal to the area under the stress-strain graph from zero to the elastic limit divided by the volume of specimen undergoing deformation.

Modulus of Rigidity See SHEAR MODULUS.

Mohs Hardness (after German mineralogist, Friedrich Mohs, 1839) A system of ranking materials according to their ability to scratch, and resist being scratched by, lower-ranking materials, diamond being the hardest material known and having the highest rank. Mohs' original scale ranked diamond as 10, corundum as 9, etc, and talc as 1. The scale has been modified to recognize some newer hard materials ranking in the large gap between corundum and diamond. The modified scale is listed below, in order of decreasing scratch hardness.

Modified Mohs Number	Material
15	diamond
14	boron carbide
13	silicon carbide
12	fused alumina
11	fused zirconia
10	garnet
9	topaz
8	quartz or Stellite®
7	vitreous silica
6	orthoclase
5	apatite
4	fluorite
3	calcite
2	gypsum
1	talc

There is a strong positive correlation between rank on the Mohs scale and KNOOP MICROHARDNESS, w s. See also SCRATCH HARDNESS.

Moiety An indefinite amount of a constituent present in a material or compound.

Moil A rarely used synonym for molding FLASH, w s.

Moisture Absorption The pickup of water vapor by a material upon exposure for a definite time interval to an atmosphere of specified humidity and temperature. No ASTM test exists for this property. Moisture absorption should not be confused with WATER ABSORPTION, w s, for which there *is* an ASTM test.

Moisture Regain The loss of weight on drying, expressed as percent of dry weight, of a predried material exposed for a specified time to a specified humidity and temperature, then oven-dried at a temperature above 100°C. ASTM D 885 (sec 07.01) describes a procedure recommended for rayon yarns and tire cords.

Moisture Sensitivity (1) The degree to which the performance of a plastic part or product is affected by changes in its moisture content or, for some persons, by changes in the relative humidity of the environment in which the product is situated. (2) The degree to which processing performance is affected by moisture pickup prior to processing.

Moisture-Vapor Transmission See WATER-VAPOR-TRANSMISSION RATE.

Molal Solution A solution that contains one mole of the solute per kilogram of the solvent.

Molar Solution (1-M) A solution that contains one mole of solute per liter of solution.

Mold (1, n) A hollow form or matrix into which a liquid or molten plastic material is placed and which imparts to the material, upon cooling or curing, its final shape as a finished article. (2, v) To impart shape to a plastic mass by means of a confining cavity or matrix, by a process usually involving high pressure and changes in temperature. The term *molding* is usually employed for processes using dry thermoplastic or thermosetting compounds, as in injection or transfer molding. The term *casting* is preferred for processes employing liquids—solutions or suspensions—that are sufficiently fluid to be poured into a mold and to fill it by gravity flow.

Mold Base An assembly of ground-flat steel plates, usually containing dowel pins, bushings and other components of injection or compression molds excepting the cavities and cores.

Mold-Clamping Force See CLAMPING FORCE.

Mold Efficiency In a multimold blow-molding system, the percentage of the total turn-around time of the mold actually required for forming, cooling, and ejecting the part.

Molding In plastics processing, any process at some stage of which the plastic is softened or melted, usually by heating, and forced to flow into a shaped cavity (mold) that essentially determines all the final dimensions of the product. Such processes are described at the following entries.

BAG MOLDING
BLOW MOLDING
CLAMSHELL MOLDING
COINING
COLD FORMING
COLD HEADING
COLD MOLDING
COLD PRESSING
COLD-RUNNER INJECTION MOLDING
COMPOSITE MOLDING
COMPRESSION MOLDING
CONTACT-PRESSURE MOLDING
DIP FORMING
DOUBLE-SHOT MOLDING
FLOW MOLDING
HIGH-PRESSURE MOLDING
INJECTION BLOW MOLDING
INJECTION MOLDING
INJECTION STAMPING
INTEGRAL-SKIN MOLDING
JET MOLDING
LIQUID INJECTION MOLDING
LOW-PRESSURE INJECTION MOLDING
LOW-PRESSURE MOLDING
MATCHED-METAL-DIE MOLDING
ONE-SHOT MOLDING
OUTSERT MOLDING
PLATFORM BLOWING
PLUNGER MOLDING
POWDER MOLDING

PULP MOLDING
REACTION INJECTION MOLDING
RECIPROCATING-SCREW INJECTION MOLDING
ROTARY MOLDING
ROTATIONAL INJECTION MOLDING
ROTATIONAL MOLDING
RUBBER-PLUNGER MOLDING
SINTER MOLDING
SLUG MOLDING
SLUSH MOLDING
SOLID-PHASE FORMING
STEAM MOLDING
STRUCTURAL FOAM
TRANSFER MOLDING
WARM FORGING

Molding Compound Granules or pellets of a resin containing all desired additives such as plasticizers, stabilizers, colorants, processing aids, and fillers, prepared by blending these ingredients with the neat resin, then reducing the hot mix to pellets by extrusion or milling, cutting, and chilling, ready for further processing into finished products. See also MOLDING POWDER.

Molding Cycle (1) The sequence of operations necessary on a molding press to produce a set of moldings. (2) The period of time occupied by the complete sequence of operations required for the production of one set of moldings.

Molding Index A practical measure of the difficulty of molding of thermosetting compounds, described in ASTM Test D 731. A calculated weight of the candidate molding powder is placed into a flash-type cup mold that has been preheated to the temperature prescribed for the material. The mold is closed and the total minimum force required to close it is reported as the molding index of the compound.

Molding Powder This term usually denotes pellets or granules of a neat resin or a MOLDING COMPOUND, w s. Also see DRY BLEND.

Molding Pressure The pressure applied to the ram of an injection machine or press to force the softened plastic to completely fill the mold cavities. It is expressed in force per unit of cross-sectional area of the ram surface acting upon the material (Pa or psi). See also INJECTION-MOLDING PRESSURE.

Molding Shrinkage (mold shrinkage, shrinkage, contraction) The fractional difference in corresponding dimensions between a mold cavity and the molding made in the cavity, both the cavity and the molding being at normal room temperature when measured. Shrinkage is often found to be different in different directions. It may be expressed as a percent, in mils/inch, or mm/m.

Mold Lubricant See PARTING AGENT.

Mold-Mat A PREPREG (w s) containing a chemical thickening agent. Mold-mats may be heated until formable (about 50°C), then compression molded or stamped to shape by dies. By means of high-energy radiation, cure can be effected very quickly.

Mold Pressure The pressure measured inside a mold cavity, usually by a flush-mounted pressure transducer, at any time during a molding cycle, but in particular the highest pressure recorded during the cycle. Compare with MELT PRESSURE.

Mold Release See PARTING AGENT.

Mold Seam A visible line on a molded or laminated piece, often very slightly raised above the general surface, impressed by the parting line of the mold.

Mold Wiper In injection molding, a device that enters between the opened mold halves during the ejection cycle, engages the molded piece, and lifts or shoves it from the mold. The wiper movement is interlocked with the mold-closing mechanism to prevent closing of the mold until the wiper is fully retracted.

Mole (mol) (1) In SI, the mole is defined as the amount of a substance of a system that contains as many elementary entities as there are atoms in 0.012 kilogram of carbon-12. The elementary entities must be specified and may be atoms, molecules, ions, electrons, other particles, or specified groups of such particles. (2) In practical usage, a mass of a substance equal to its molecular weight. To agree with the SI definition, the mass must be in grams. However, engineers often find it convenient to work with pound-moles, kilogram-moles, and even ton-moles of materials. If the mass unit isn't prefixed, the mole is always the gram-mole.

Molecular Orientation See ORIENTATION.

Molecular Sieve A porous mineral or synthetic inorganic material, such as a zeolite (hydrous silicate), usually in the form of porous pellets or fine granules, having the ability to strongly absorb molecules of other (fluid) materials. Molecular sieves have been used in plastics as carriers for blowing agents which, when heated, release expanding gases at the desired rate. Silica sieves are useful for drying air and liquids and are easily regenerated by heating in a dry environment.

Molecular Volume (molar volume) The volume occupied by one mole, numerically equal to the gram-molecular weight divided by the density at the prevailing pressure and temperature.

Molecular Weight (formula weight, molecular mass) The sum of the atomic weights of all atoms in a molecule. In most nonpolymeric compounds the molecular weight is a known constant value. In high polymers, the individual molecules range widely in the number of atoms they contain and, therefore, in molecular weight. Hence an average must be used to characterize a particular sample of polymer. The two averages most commonly used are NUMBER-AVERAGE MOLECULAR WEIGHT (M_n) and WEIGHT-AVERAGE MOLECULAR WEIGHT (M_w), w s. Methods for determining these averages include measurements of light-scattering and osmotic pressure in solutions, sedimentation in an ultracentrifuge, depression of freezing points and vapor pressures of solutions, dilute-solution viscosity, end-group titration, and spectroscopy.

Molecular-Weight Distribution The percentages by number (or weight) of molecules of various molecular weights that comprise a given specimen of a polymer. Two samples of a given polymer having the same number-average molecular weight may perform quite differently in processing because one has a broader distribution of molecular weights than the other. Two basic groups of methods are used for measuring molecular-weight distribution. Fractionation methods, which actually divide the specimen into portions having relatively narrow ranges of molecular weight, include: fraction precipitation and fraction solution (the two most widely used), chromatography, liquid-liquid partition, ultracentrifugation, zone refining, and thermogravimetric diffusion. After fractionation by any of these methods, the weight percent of each fraction is plotted vs the average molecular weight for that fraction to obtain a histogram of the distribution, which may be smoothed into a curve. Non-fractionation methods include light-scattering studies, electron microscopy, dilute-solution viscosity, size-exclusion chromatography, ultracentrifugation, and diffusion. A popular measure of the breadth of a distribution is the ratio of weight-average to number-average molecular weight, M_w/M_n. See also POLYDISPERSITY.

Molecule The smallest unit quantity of a compound that can exist by itself and still retain the chemical identity of the substance as a whole.

Molybdate Orange Pigment Any of a range of solid solutions of lead chromate, lead molybdate, and lead sulfate, used as dark-orange to light-red pigments for plastics. Their advantages are high opacity, bright color, light-fastness, good heat stability, and freedom from bleeding. Their main disadvantage is lead's toxicity.

Molybdenum Disulfide (molybdic sulfide, molybdenum sulfide,) MoS_2. A black, shiny, flaky-crystalline material used as a filler in nylons, fluorocarbons, and polystyrene to improve stiffness and strength, and, principally, to provide lubricity. The compound occurs naturally as the ore *molybdenite*.

Molybdenum FR Any of several molybdenum compounds, such as the oxide or ammonium dimolybdate, $(NH_4)_2Mo_2O_7$, added to plastics to improve their fire retardancy (FR) and smoke suppression.

Momentum Flux In hydrodynamics, an interpretation of shear stress τ in which each of the six shear components of the stress tensor is viewed as the rate of flow, per unit of shear area, and perpendicular to that area, of momentum directed along a principal axis in the surface of shear. In laminar flow through a circular orifice, with radial coordinate r and axial coordinate z, the only nonzero shear component is τ_{rz}, the flux of z-directed momentum in the r-direction. Newton's law of viscosity for this situation becomes

$$\tau_{rz} = -\mu \, (dv_z/dr)$$

where v_z is the fluid velocity at any radius r and the viscosity μ has the dimensions: (momentum/area·time)/shear rate, which reduce to M/Lt, and for which the SI unit is 1 $[(kg \cdot m/s)/(m^2 \cdot s)]/s^{-1}$ = 1 kg/m·s = 1 Pa·s.

Mono- A prefix designating the entity that follows it as the only one or as containing only one of that kind, e g, monomer, or monohydric alcohol.

Monobasic (1) Pertaining to acids having one active hydrogen per molecule, e g, hydrochloric acid, HCl. (2) Designating an acid salt in which one hydrogen (of two or three) has been replaced by a metal, e g, potassium monophosphate, KH_2PO_4.

Monocarboxylic Acid Any organic acid containing a single –COOH group in the molecule. Many of the larger acids of this type are derived from natural fats and oils and are used in the production of alkyd resins and polyesters. Esters of oleic, stearic, pelargonic, and ricinoleic acid, all monocarboxylic, are widely used as plasticizers.

Monochloroethylene Synonym for VINYL CHLORIDE, w s.

Monoclinic Of a crystal (or crystal system), having two axes that are mutually perpendicular to the third one, but not to each other. An example is the β form of elemental sulfur, which is stable between 112 and 119°C, but slowly converts to the rhombic form below 112°C.

Monodisperse Of a polymer, all the molecules having the same molecular weight. It has long been possible to make (nearly) monodisperse polystyrene of various molecular weights and a few other monodisperse polymers are now available as laboratory chemicals. Compare POLYDISPERSE.

Monofilament A single filament of indefinite length, strong enough to function as a yarn in textile operations or as an entity in other applications. Monofilaments are generally produced by extrusion. (Even spiders make theirs that way.) Their outstanding uses are in the fabrication of brush bristles, surgical sutures, fishing lines, racquet strings, screen materials, ropes, and nets. The finer monofilaments are woven and knitted on textile machinery.

Monomer A relatively simple compound that can react with itself or other compounds to form long-chain compounds by either (1) utilizing its C=C bonds for addition or (2) by having two or more functional groups that can react with receptive groups in other molecules. Monomers are the basic building blocks of polymers.

Monomeric Pertaining to a MONOMER, w s.

Monomeric Cement See ADHESIVE.

Monotropic Of a material or element, having two forms, one of which is metastable toward the other. The metastable form tends to change spontaneously to the stable form but the reverse change does not occur.

Montan Wax (lignite wax) A hard, white wax derived from *lignite*, (a lower-grade hydrocarbon fossil mineral between peat and bituminous coal). The wax is used as a mold lubricant.

Mooney Equation (after M. Mooney, an early, prolific contributor to rheological theory and practical rheometry of elastomers and polymer solutions) An empirical modification of the EINSTEIN EQUATION (w s), applicable to higher solids concentrations, and relating the viscosity of a suspension of monodisperse spheres η_f to that of the pure liquid η_0. It is

$$\ln (\eta_f / \eta_0) = 2.5 f /(1 - S \cdot f)$$

where f = the volume fraction of solids and $S \cong 1.4$ for spheres. See also EILERS EQUATION and FRANKEL-ACRIVOS EQUATION.

Mooney Scorch Time For a rubber specimen tested in a MOONEY VISCOMETER (w s), the time elapsing after the minimum torque has been reached for torque to increase by five "Mooney units".

Mooney Viscometer An instrument invented by M. Mooney in 1924, used to measure the effects of time of shearing and temperature on the comparative viscosities of rubber compounds. It consists of a motor-driven disk, tooth-surfaced on both sides, enclosed within a die cavity formed by two halves maintained at controlled temperature and closing force. The specimen is a double disk, joined at the edges and trapped between the die halves and the rotor. Its use is described in ASTM D 1646 (sec 09.01). Rotor torque in "Mooney units" (1 MU = 0.833 N·m) is recorded over a 4- or 8-min period, typically passing through a broad minimum, and the minimum is reported as the *Mooney viscosity*. Because of the complex specimen geometry, Mooney viscosity is not simply convertible to VISCOSITY, w s. The Mooney viscometer is also used in ASTM tests D 1417 and D 3346, both in sec 09.01.

Morphology The study of the physical form and structure of a material. This includes a wide range of characteristics, extending from the external size and shape of large articles to dimensions of crystal lattices, but, with polymers, it most often refers to microstructure.

Motionless Mixer See STATIC MIXER.

Mottle (n) An irregular distribution or mixture of colorants or colored materials giving a more or less distinct appearance of specks, spots, or streaks of color. Note: mottling is often deliberate although it may occur accidentally due to inadequate mixing. (adj) *Mottled*.

Mounting Plate In blow molding, the plate to which the mold is attached. See also CLAMPING PLATE.

Movable Platen The large back plate of an injection molding machine to which the back half of the mold is secured during operation. This platen is moved either by a hydraulic ram or a toggle mechanism.

MS Abbreviation for MASS SPECTROMETRY, w s.

Multicavity Mold (multiple-cavity mold, multiple-impression mold) A mold having several to hundreds of cavities so that many parts may be molded with each shot. In many cases, the parts are identical, but that need not be so. In a type of multicavity mold known as a FAMILY MOLD, w s, some of the cavities may be identical while others are different, or they may all be different.

Multifilament Yarn A manufactured yarn composed of many fine continuous filaments or strands.

Multifunctional Concentrate A plastic compound that contains high percentages of at least two of such additives as colorants, stabilizers, flame retardants, lubricants, antistatic agents, antiblocking agents, blowing agents, fillers, etc, that will be diluted in base resin to provide a tailored compound with the desired final concentrations of the additives in the extruded or molded product.

Multigated Of an injection-mold cavity, having two or more gates. Multigating is common in molding large or complex parts that would be difficult to fill through one gate. Also, shrewd placement of the several gates can direct the weld lines to areas of the part where the stresses expected in service will be low.

Multilayer Fabric A fabric for reinforced-plastics structures formed by braiding to and fro or overlapping in one direction. Layers may be biaxial or triaxial, fibers mixed, and braid angles varied.

Multilayer Film A film comprised of layers of two or more different materials, all polymeric. The goal is to make a film that has a combination of properties not achievable with a single polymer. Years ago these were not common and were produced by heat- or adhesive-laminating of separately produced films. Today, most multilayer films are made by COEXTRUSION, w s. Even in these, the principal layers are usually joined by an adhesive resin compatible with both the adjacent principals and coextruded with them. Films of nine and even more layers are routinely produced today.

Multi-Live-Feed Molding (MLFM, Scorim process) A method of filling and packing injection molds that requires that a processing head containing two or more independently controlled pistons be installed between the injection unit and the mold. The flow from the injection unit is divided into multiple streams, each going to one of the piston chambers, thence into the mold through separate gates. A programmable electronic control unit then operates the rams to move the mold contents to and fro or to apply compression, with melt being added to make up cooling shrinkage. The technique is said to eliminate part defects, particularly with fiber-reinforced compounds, by enhancing favorable orientation and breaking up potentially weak weld lines. See also PUSH-PULL MOLDING.

Multiple-Regression Analysis See REGRESSION ANALYSIS.

MVTR Abbreviation for *moisture-vapor-transmission rate*. See WATER-VAPOR-TRANSMISSION RATE.

M_w Symbol for WEIGHT-AVERAGE MOLECULAR WEIGHT, w s.

MW Abbreviation for MOLECULAR WEIGHT, w s.

Mylar DuPont's registered trade name for biaxially oriented film composed of polyethylene glycol terephthalate.

Myristoÿl Peroxide $(C_{13}H_{27}CO)O_2$. A soft, granular powder, used as a polymerization catalyst for vinyl monomers.

M_z Symbol for VISCOSITY-AVERAGE MOLECULAR WEIGHT, w s.

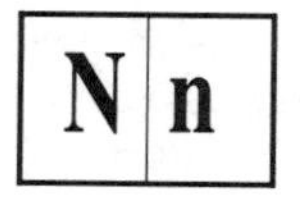

n (1) SI abbreviation for NANO-, w s. (2) *n*-: In organic chemistry, abbreviation for normal, signifying a straight (unbranched) aliphatic chain. (3) A subscript denoting the last of a series of *n* ordered numbers or data.

N (1) Chemical symbol for the element nitrogen. (2) SI abbreviation for NEWTON, w s. (3)-N: In solution chemistry, following a numeral, abbreviation for Normal or Normality (see NORMAL SOLUTION).

Na Chemical symbol for the element sodium.

NACE Acronym for NATIONAL ASSOCIATION OF CORROSION ENGINEERS, w s.

Nacreous Pertaining to, or having the appearance of, mother-of-pearl. See PEARLESCENT PIGMENT.

Nacreous Pigment See PEARLESCENT PIGMENT.

Nano- The SI prefix meaning $\times 10^{-9}$ (one billionth).

Nanometer An SI unit of length equal to 10^{-9} meter, convenient for stating light wavelengths and superseding the older angstrom unit and millimicron, both now deprecated.

Naphtha (solvent naphtha) Any of a family of petroleum and coal-tar distillates with 30°C boiling ranges within the interval 125 to 200°C. They are useful as solvents for natural resins and rubber, and as paint thinners.

Naphthalene (napththalin, tar camphor) An aromatic hydrocarbon, $C_{10}H_8$, derived from coal-tar oils or petroleum fractions and having the structure shown below.

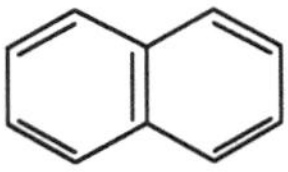

Once used as a moth repellent, it is now important as a reactant in the production of phthalic anhydride, which in turn is used for making plasticizers, alkyd resins, and polyester resins.

1,5-Naphthalene Diisocyanate (NDI) $OCNC_{10}H_6NCO$. An isocyanate used in the production of urethane elastomers and foams.

Naphthenate See NAPHTHENIC ACID.

Naphthenic Acid A carboxylic acid derived from petroleum refining and usually one of a mixture of similar compounds. The mixed acids and some of their soaps, e g, cobalt naphthenate and calcium naphthenate, are useful as catalysts or accelerators in curing polyester resins and as drying agents in paints and varnishes.

National Association of Corrosion Engineers (NACE) Address: 1440 S Creek Dr, Houston, TX 77084. A leading publisher of technical information on protection and performance of materials in corrosion environments. NACE also sponsors educational seminars on anticorrosion properties and applications of plastics in the chemical-process industries.

Natta Catalyst Any of several catalysts used in the stereospecific polymerization of olefins, e g, ethylene and propylene, particularly a catalyst made from titanium chloride and aluminum alkyl or similar materials by a special process including grinding the materials together to produce an active catalytic surface.

Natural Fiber A fiber of plant or animal origin such as cotton (nearly pure cellulose), flax, sisal, abaca, hemp, jute, etc, the wool of sheep and other animals, horsehair, and swine bristle. Cellulose from cotton linters and wood pulp is the starting material for cellulosic plastics and is by far the most important natural fiber for the plastics industry.

Natural Resin A resin produced by nature, mostly by exudation from certain trees from cuts or tears in the bark. Lac resin is secreted by the lac insect and is refined to make shellac. Some of the tree resins are copal, rosin, and sandarac, at one time widely used in wood finishes.

Natural Rubber See RUBBER, NATURAL.

NBR See ACRYLONITRILE-BUTADIENE COPOLYMER.

NC (1) Abbreviation for *numerical control.* (2) See CELLULOSE NITRATE.

NCNS See TRIAZINE RESIN.

NDOP Abbreviation for *n*-decyl-*n*-octyl phthalate. See *n*-OCTYL-*n*-DECYL PHTHALATE.

Near-Net-Shape Configuration In reinforced plastics molding, designating a fibrous-preform shape very close to the shape the preform will take after resin impregnation and curing in the mold, such that no layup and little or no trimming after molding are required.

Neat Resin Strictly, a resin containing nothing but the main identified polymer(s). Usually, the term means that, while there may be fractional percentages of stabilizers and other additives present, there are no fillers, reinforcements or pigments. Sometimes called a "pure" resin, though, since all commercial polymers are mixtures of homologs of various molecular weights, "pure" has a looser meaning here. Some writers have referred to neat resins as "barefoot" resins.

Neck-in In extrusion of film, sheet and coatings, the difference between the width of the extruded web as it leaves the die and the final width of the chilled film, etc (before any edge trimming is done).

Necking The localized reduction in cross section that may occur in a ductile material under tensile stress (ASTM D 883, slightly modified). A similar phenomenon can occur during extrusion under certain conditions as the extrudate leaves the die. Necking can also occur during drawing of fibers at temperatures below their melting ranges. Fibers of crystalline and some noncrystalline thermoplastics, e g, polyethylene, exhibit necking at a critical stress near the yield point.

Neck Insert (finish insert) In blow molding of bottles, a removable part of the mold that forms a specific neck finish of the bottle.

Needle Blow A specific blow-molding technique wherein the blowing air is injected into the hollow article through a sharpened hollow needle that pierces the parison above the finish.

Negative Catalyst (inhibitor, retarder) An agent that slows a chemical reaction.

NEMA Acronymical abbreviation for National Electrical Manufacturers Association, an organization that has strongly influenced electrical applications of plastics.

Nematic See LIQUID-CRYSTAL POLYMER.

Neo- (1) Prefix meaning *new* and denoting a compound isomerically related to an older one whose name follows the prefix. (2) A prefix denoting a hydrocarbon in which at least one carbon atom is connected directly to four other carbon atoms, e g, neopentane, $C(CH_3)_4$.

Neopentyl Glycol (NPG, 2,2'-dimethyl-1,3-propanediol) $HOCH_2C(CH_3)_2CH_2OH$. An important intermediate for the production of alkyd and polyester resins, urethane foam and elastomers, and polyester plasticizers. Gel coats base on NPG for reinforced polyesters have improved flexibility, hardness, and resistance to abrasion and weathering.

Neopentyl Glycol Diacrylate (NPGDA) $(CH_3)_2C(CH_2OOCCH{=}CH_2)_2$. A highly reactive crosslinking monomer used in photocurable coatings. It provides solvent and strain resistance as well as improved response to light.

Neopentyl Glycol Dibenzoate $(CH_3)_2C(CH_2OOCC_6H_5)_2$. A solid plasticizer for rigid PVC.

Neoprene (polychloroprene, poly-2-chloro-1-butadiene) Elastomers made from the monomer $CH_2{=}CHCCl{=}CH_2$. They are available as dry solids and latices, and are vulcanizable to tough products with excellent resistance to oils, gasoline, solvents, heat, and weathering. The original neoprene, produced by Du Pont under the tradename "Duprene", was America's first successful synthetic rubber.

Nepheline A naturally occurring mineral composed mainly of feldspar and nephelite. As a filler in PVC compounds, it has the unique property of contributing almost no opacity, so that it can be used in nearly transparent compounds. It is also used as a filler in epoxy and polyester resins.

Nesting In reinforced plastics, the placing of plies of fabric so that the yarns of one ply lie in the valleys between the yarns of the adjacent ply.

Nest Plate A retainer plate with a depressed area for cavity blocks used in injection molding.

Net Flow The output of an extruder's metering section, being, to a first approximation, the algebraic sum of the DRAG FLOW, PRESSURE FLOW, and LEAKAGE FLOW. In most plastics extruders with solid feeds and screws having the conventional three sections—feed, transition, and metering—the net flow may be conservatively estimated from the equation

$$\dot{m} = 2.0\ D^2\ N\ h\ \rho_0$$

where $\dot{m}$ is in lb/h, D is the nominal screw diameter, in., N is the screw speed in rpm, h is the channel depth in the metering section, and ρ_0 is the plastic's room-temperature density, g/cm^3. If the rate is given in kg/h and the diameter and channel depth in cm, the equation becomes

$$\dot{m} = 0.055\ D^2\ N\ h\ \rho_0.$$

Netting A crossing-strand sheet or tubular plastic structure, fused at the crossing points, produced by extrusion through (patented) oscillating dies. Plastic netting, in small pieces and baskets, is also made by injection molding.

Netting Analysis The stress analysis of filament-wound structures that neglects the strength of the resin and assumes that the filaments carry only axial tensile loads and possess no bending or shearing stiffness.

Network Polymer A polymer obtained by the polymerization of a monomer having two or more functional groups that become interconnected with sufficient interchain bonds to form a large three-dimensional network. The network can be formed during polymerization, or may be created by crosslinking the polymers after they have been formed. The vulcanization of rubber is an example of the formation of a network polymer from a preformed polymer. Copolymers of ethylene and propylene can be made into network polymers by cross-linking with ionizing radiation after reactive sites have been prepared by treating with heat or peroxides.

Network Structure An atomic or molecular arrangement in which primary bonds form a three-dimensional network. See also INTERPENETRATING POLYMER NETWORK.

Neutral Axis In a beam or column subject to bending moments, the surface near the center of the beam and perpendicular to the applied loads upon which neither tensile nor compressive stress is acting. In homogeneous beams with depth-symmetrical cross sections, the neutral axis is exactly at the center.

Neutralization A chemical reaction in which the hydrogen ion of an acid and the hydroxyl ion of a base unite to form water and a salt.

Newton (for Sir Isaac Newton) The SI unit of force that, when applied to a body having a mass of one kilogram and free to move, gives it an acceleration of one meter per second per second ($1\ m/s^2$). See also FORCE (3).

Newtonian Flow An isothermal, laminar flow characterized by a viscosity that is independent of the level of shear, so that the shear rate at all points in the flowing liquid is directly proportional to the shear stress and vice versa. Simple liquids such as water and mineral oil usually exhibit Newtonian flow, whereas polymer melts and solutions usually do not, but are *pseudoplastic*. See PSEUDOPLASTIC FLUID.

Newtonian Liquid (Newtonian fluid) A fluid for which the ratio of shear stress to shear rate is constant over the range of shear rate at a given temperature and pressure, and having zero normal-stress differences. See VISCOSITY.

Newtonian Viscosity (1) Another name for VISCOSITY, w s. (2) Of polymer melts, the viscosity at very low shear rates (< 0.01 s^{-1}) in low-density polyethylene, for example) where viscosity is independent of shear rate and the melt is essentially Newtonian. (3) In some models of pseudoplastic flow, a second limiting viscosity (μ_∞) approached as shear rate rises to extreme values and observed in some polymer solutions at high (but not infinite!) shear rates.

Newton's Second Law of Motion (Newton's law of momentum change) See FORCE.

Nextel® 3M Corporation's trade name for their high-performance fiber containing 62% Al_2O_3, 24% SiO_2, and 14% B_2O_3. Grades range in properties: density 2.71 to 3.10 g/cm^3; modulus, 150 to 240 GPa; and tensile strength, 1.3 to 2.0 GPa.

NHDP Abbreviation for *n*-hexylethyl-*n*-decyl phthalate. See *n*-OCTYL-*n*-DECYL PHTHALATE.

Ni Chemical symbol for the element nickel.

Nickel-Azo Yellow A pigment based on a nickel complex of an azo dyestuff.

NIOSH Acronymical abbreviation for National Institute for Occupational Safety and Health, a division of the Center for Disease Control (Public Health Service, under the Department of Health, Education, and Welfare). This agency conducts investigations and research projects on industrial safety and makes recommendations for the guidance of OSHA. However, it does not enforce its own findings or OSHA regulations.

Nip (1) The curved, **V**-shaped gap between a pair of counter-rotating calender rolls, chill rolls, or rubber pull rolls where incoming material is "nipped" and drawn between the rolls. (2) In safety-management parlance, "nip" is broadened to include any convergent approach of two machine elements, such as meshing spur gears or a V-belt approaching its pulley.

Nip Rolls (pinch rolls) In film blowing, a pair of rolls situated at the top of the tower that pinch shut the blown-film tube, seal air inside it, and regulate the rate at which the film is pulled away from the extrusion die. One roll is usually covered with a resilient material, the other being of metal and internally cooled.

Nitriding A type of CASE HARDENING (w s) in which a steel surface is reacted for hours or days with gaseous ammonia at temperatures from

480 and 540°C to produce a surface hardness of about Rockwell C70 to a depth from 0.2 to 2 mm. The process is used with injection-mold cavities and extrusion screws, being less expensive (also less durable) with the latter than inlaid hard facing. Special nitriding steels containing about 1% aluminum provide the best results. Years ago, nitriding was used to case harden the interiors of extruder cylinders. It was supplanted for three decades by cast-in liners, but is now making a comeback in a new variation, glow-discharge nitriding (Ionitriding®). In this process, the piece to be treated is placed in a furnace that is evacuated and becomes the cathode for high-velocity impingement of nitrogen (+) ions. An advantage is that the usual slight growth of the part due to nitrogen inclusion is offset by loss of metal atoms. Nitrided steels remain as hard at 400°C as at room temperature.

Nitrile Barrier Resin (high-nitrile polymer) One of a family of polymers generally containing greater than 60% acrylonitrile, along with comonomers such as acrylates, methacrylates, butadiene, and styrene. Both straight copolymers and copolymers grafted onto elastomeric spines are available. Their unique property is outstanding resistance to passage of gases and water vapor, making them useful in packaging applications.

Nitrile Rubber See ACRYLONITRILE-BUTADIENE COPOLYMER.

***o*-Nitrobiphenyl** (2-nitrobiphenyl, ONB) $O_2NC_6H_4–C_6H_5$. An involatile plasticizer for cellulose-ester polymers, and compatible with many others.

Nitrocellulose (guncotton, cellulose nitrate) In the first half of the twentieth century, this term was widely used in the plastics industry for CELLULOSE NITRATE, w s, and plastics based thereon.

Nitroethane $CH_3CH_2NO_2$. A colorless liquid used as a solvent for cellulosic, vinyl, alkyd, and other resins.

Nitrogen Nitrogen gas is the most widely used blowing agent for injection-molded, structural foams. It is less expensive than most chemical blowing agents, leaves no residue, is environmentally harmless, and is easy to handle. Nitrogen is added to the polymer melt by pumping it directly into the barrel of the injection machine. Nitrogen is also used in the same way in foam extrusion.

Nitromethane CH_3NO_2. A colorless liquid made by reacting methane with oxides of nitrogen or nitric acid under pressure, and used as a solvent for cellulosic, vinyl, alkyd, and other resins.

Nitropropane (1-nitropropane) $CH_3CH_2CH_2NO_2$. A colorless liquid boiling at 130°C, a good solvent for vinyl resins.

Nitroso Rubber A copolymer of tetrafluoroethylene and trifluoronitrosomethane.

N_{Ma} Symbol for MACH NUMBER, w s.

NMA (1) Abbreviation for METHYL NADIC ANHYDRIDE, w s. (2) Abbreviation for *N*-METHYL ACETAMIDE, w s.

NMR Abbreviation for NUCLEAR MAGNETIC RESONANCE, w s.

NODA Abbreviation for *n*-OCTYL-*n*-DECYL ADIPATE, w s.

Node (1) A characteristic of flax and hemp fibers, giving them a bamboo-like appearance under the microscope, seen in no other fiber. In flax, the nodes are fairly regularly spaced at intervals of about 0.5 mm. (2) In a vibrating body, a point, line, or surface which is wholly or mostly free of vibration and appears to be at rest.

NODP Abbreviation for *n*-OCTYL-*n*-DECYL PHTHALATE, w s.

NODTM Abbreviation for TRI(*n*-OCTYL-*n*-DECYL) TRIMELLITATE, w s.

NOL Ring Test A method of testing the tensile strength of circular rings (samples cut from tubes or vessels), originated at the former Naval Ordnance Laboratory, Silver Spring, MD. See SPLIT-DISK METHOD.

Nomogram (nomograph) A graph containing three (or more) lines, typically parallel, graduated for related variables. When known values of any two are located on their scales and connected by a straight edge, the value of the third may be read off at the point where the line intersects the third scale. Nomograms were very popular problem-solving devices in the engineering and plastics literature for many years but have largely been replaced by software and data banks available for personal computers.

Noncombustible Incapable of being ignited and burned in air; fire-resistant. See FLAMMABILITY.

Nondestructive Test (1) A test that yields information about failure under mechanical stress without actually stressing to failure. (2) More broadly, any test to evaluate a property of a material, part, or structure that does not significantly damage the part. Techniques used include ultrasound, magnetic inspection of metals and welds, X-ray inspection, infrared, nuclear magnetic resonance, and sonic analysis. Although an indentation-hardness test leaves a permanent mark, in many tests of parts it is nondestructive.

Nonflammable If combustible, burning without flame. Practically, whether or not a plastic material or part is "flammable" is a matter of its performance in a test—of which there are many—of FLAMMABILITY, w s. Note that the word "inflammable" has been deprecated by fire-safety authorities because of the ambiguity of the prefix "in-", and has long been superseded by "flammable" and "nonflammable".

Nonionic Pertaining to a material, atom, radical, or molecule that is incapable of being electrically charged. See also IONIC.

Nonisothermal See ISOTHERMAL.

NonNewtonian Designating liquids whose viscosities are dependent on the rate of shear (as well as temperature and pressure), or the flow of such a liquid. See NEWTONIAN LIQUID, PSEUDOPLASTIC FLUID, and VISCOSITY.

Nonpolar Having no concentrations of electrical charge on a molecular scale, thus incapable of significant dielectric loss. Examples of nonpolar resins are polystyrene and polyethylene.

Nonrigid Plastic (as given by ASTM D 883) For the purposes of general classification, a plastic that has a modulus of elasticity either in flexure or tension of not over 70 MPa (10,000 psi) at 23°C and 50% relative humidity when tested in accordance with ASTM D 638, D 747, D 790, or D 882. See also RIGID PLASTIC.

Nontoxic Materials The toxicity status of resins and additives used for food contact and packaging changes frequently, and in many cases percentages and conditions of end use are stipulated. Therefore, we do not attempt to list such materials here. The current statuses of resins, plasticizers, stabilizers, and other additives with respect to their permissible (in the US) use in contact with food ("indirect additives") are spelled out in great detail in the Code of Federal Regulations (CFR), Title 21 (Food and Drugs), Parts 173 through 184. The Code is reprinted annually in book form and updated weekly by the Federal Register, where one should check for the latest word on any particular substance. The parts referred to make up most of the 1991 volume: "Parts 170-199", available from the Government Printing Office, Washington, DC, 20402, for $17. Part titles are: Part 173, Secondary Direct Food Additives Permitted in Food for Human Consumption; Part 174, Indirect Food Additives: General; Part 175, Indirect Food Additives: Adhesives and Components of Coatings; Part 177, Indirect Food Additives: Polymers, of which Subpart B is Substances for Use as Basic Components of Single- and Repeated-Use Food-Contact Surfaces (58 sections) and Subpart C is Substances for Use Only as Components of Articles Intended for Repeated Use (21 sections); Part 178, Indirect Food Additives: Adjuvants, Production Aids, and Sanitizers, of which Subpart C is Antioxidants and Stabilizers (3 sections) and Subpart D is Certain Adjuvants and Production Aids (43 sections); Part 182, Substances Generally Recognized as Safe (9 subparts); and Part 184, Direct Food Substances Affirmed as Generally Recognized as Safe (no synthetic polymers listed).

Nonwoven Fabric A cloth formed from staple lengths of natural or synthetic fibers mechanically or hydraulically positioned into a random pattern, then lightly bonded with suitable resins to form sheets.

Nonwoven Mat A mat of glass fibers in random arrangement, lightly bonded so as to be able to be handled and cut, used in making reinforced-plastics structures.

Nonwoven Scrim An open-mesh glass fabric in which two or more layers of parallel yarns are bonded to each other by chemical or mechanical means, the yarns in adjacent layers lying at an angle to each other.

Nor- A prefix for organic compounds indicating the parent compound from which the subject compound may be derived, usually by removal of one or more carbon atoms and their attached hydrogens. Example: *norcamphor* is camphor from which three $-CH_3$ groups have been removed.

Norbornene-Spiro-Orthocarbonate A compound that, when added in small concentrations to matrix resins of carbon composites, apparently strengthens the fiber-matrix interfacial bond, as evidenced by improved strength properties and lower water absorption of the composites.

Normal Salt An ionic compound containing neither replaceable hydrogen nor hydroxyl groups.

Normal Solution (1-N solution) A solution containing a mass of the dissolved substance per liter equal to one mole divided by the hydrogen equivalence of the substance, i e, one gram-equivalent per liter.

Normal Stress (1) A stress directed at right angles to the area upon which it acts. (2) In a flowing viscoelastic liquid, tensile/compressive stresses at any point in the fluid in the principal coordinate directions, one of which will be the main direction of flow. Rheologists are usually concerned with *differences* between normal stresses acting in the flow direction and directions perpendicular to the flow.

Noryl® Tradename of the General Electric Co for a family of blends of polyphenylene oxide (PPO) with much less costly styrenic polymers. These blends have the processability, low water absorption, and good dielectric properties associated with polystyrene, while the PPO contributes heat resistance. Glass-reinforced grades are available.

Notch Sensitivity The extent to which a material's tendency to fracture under load, particularly an impact load, is increased by the presence of a surface inhomogeneity such as a notch or sharp inside corner, a sudden change in section thickness, a crack, or a scratch. Low notch sensitivity is usually associated with ductility, while brittle materials exhibit higher notch sensitivity. Most engineers and physical testers consider the notched IZOD and CHARPY IMPACT TESTs (w s) to be as much measures of notch sensitivity as they are of pure impact strength.

Novaculite A very fine-grained type of quartz found in Arkansas, Georgia, Massachusetts, North Carolina, Oklahoma, and Tennessee. A variety known as *altered novaculite*, typically about 99.5% quartz, is a solid crystalline substance with the basic hardness of quartz but more easily reduced to very fine particles. At their surfaces, these particles have high concentrations of ruptured Si–O bonds that readily combine with water to create surface hydroxides called *silanols*. Such novaculites are useful as semi-reinforcing fillers in silicone rubber, epoxy resins, urethane foams, and PVC.

Novolac (novolak) According to ASTM D 883, a novolac is a phenolic-aldehyde resin which, unless a source of methylene groups is added, remains permanently thermoplastic. For a preferred definition, see PHENOLIC NOVOLAC. However, the term is also used in connection with epoxies. See EPOXY-NOVOLAC RESIN.

Novoloid Fiber A phenolic fiber made by crosslinking a melt-spun novolac resin with formaldehyde. Novoloid fibers have good flame resistance, can serve at temperatures to about 220°C, and are used as reinforcement in a range of thermosetting matrices.

Nozzle In injection or transfer molding, the orifice-containing fitment at the delivery end of the injection cylinder or transfer chamber that contacts the mold's sprue bushing and conducts the softened resin into the mold. The nozzle is shaped to form a seal under pressure against the sprue bushing. Its orifice is tapered and sometimes contains a check valve to prevent flow reversal, or an on/off valve to interrupt the flow at any desired point in the molding cycle.

Nozzle Manifold A series of injection nozzles mounted on a common feed tube, each nozzle positioned so as to feed a single cavity in the mold. Such manifolds have been used to eliminate runners in molds when molding articles such as cups and when it is desired to gate the cavities at the centers of the cup bottoms.

Nozzle, Mold-Gating In injection molding, a nozzle whose tip is part of the mold cavity, thus feeding material directly into the cavity, eliminating the sprue and runner.

NR Abbreviation for *natural rubber.* See RUBBER, NATURAL and POLYISOPRENE.

N_{Re} Symbol for REYNOLDS NUMBER, w s.

Nuclear Magnetic Resonance (NMR) The spinning motion of atomic nuclei imparted by an alternating, high-frequency magnetic field. The phenomenon is the basis for an analytical method enabling identification and quantification of isotopes. NMR spectroscopy has been used to study the distribution of hydrogen in substituent groups, and the molecular structure of polymers, such as tacticity and occurrence of infrequent branches.

Nucleating Agent A chemical substance which, when incorporated in crystal-forming plastics, provide active centers (nuclei) for the growth of crystals as the melt is cooled through the melting range. In polypropylene, for example, a higher degree of crystallinity and more uniform crystalline structure is obtained by adding a nucleating agent such as adipic or benzoic acid or certain of their metal salts. Colloidal silicas are used as nucleating agents in nylon, seeding the polymer to produce more uniform growth of spherulites.

Number-Average Molecular Weight (M_n) The sum of the molecular weights of all the individual molecules in a given polymer sample divided by the total number of molecules. The defining equation is

$$M_n = \sum_i (N_i \cdot M_i) / \sum_i N_i$$

where N_i is the number of individual molecules in the sample having the molecular weight M_i. M_n relates to the colligative properties of polymer solutions, such as osmotic pressure. See also MOLECULAR WEIGHT, MOLECULAR-WEIGHT DISTRIBUTION, and WEIGHT-AVERAGE MOLECULAR WEIGHT.

Nutshell Flour Ground peanut or walnut shells, dried by heating or solvent extraction, have been used as low-cost fillers in polyethylene. Physical properties are comparable to those of PE filled with wood flour.

Nylon (polyamide) Generic name for all long-chain polyamides that have recurring amide groups (–CONH–) as an integral part of the main polymer chain. Nylons are synthesized from intermediates such as dicarboxylic acids, diamines, amino acids and lactams, and are identified by dual numbers denoting the number of carbon atoms in the polymer chain derived from specific constituents, that of the diamine being given first. Use of a single numeral signifies that the monomer was a lactam, as in *nylon 6*. The second number, if used, denotes the number of carbons derived from a diacid. For example, in nylon 6/6 the two numbers are the numbers of carbon atoms in hexamethylene diamine and adipic acid, respectively. However, in the literature these numbers may otherwise appear as 66, 6.6, 6,6, or 6-6, and sometimes precede, rather than follow, the word "nylon". The convention used here—numbers following, divided by slash mark—is almost universally used today. Nylon molding powders can be converted to useful products by injection molding, extrusion, and blow molding. Nylons are crystalline polymers. In nylon 6/6, a wide range of crystallinity is possible, depending on how quickly the melt is chilled and the presence or absence of nucleating agents. Injection-molded items normally have low-crystallinity skins with higher-crystallinity interiors. Very thin sections may have as little as 10% crystalline material. Finely powdered forms of nylon are available for fluidized-bed coating, rotational molding, and other powder processes. A casting process employs molten caprolactam monomer to which catalysts are added, polymerization occurring in the mold after pouring without additional heat or pressure.

Large solid castings and rotationally cast parts have been made by this method. In 1992, US production of nylon plastics was about 270 Gg (300,000 tons). Polyamides are now commercially available in so many filled and reinforced varieties that *Modern Plastics Encyclopedia*'s "Resins and Compounds" table for 1993 contained eight pages of polyamide listings, far more than for any other plastics family. Various types of nylons are described in the immediately following listings. See also INTERFACIAL POLYMERIZATION and POLYCYCLAMIDE.

Nylon 3 (polypropiolactam) A type of nylon that has been prepared and explored experimentally, but has not become commercial.

Nylon 4 (polypyrrolidinone) A polymer of 2-pyrrolidinone. Early attempts to commercialize nylon 4 failed because much of the material was of low molecular weight and decomposed at a relatively low temperature, making it unusable for melt spinning. Improved catalyst systems resulted in a polymer with a molecular weight (M_n) of about 400,000 and a melting point of 265°C. Today's nylon 4 has better heat stability than other nylons. Its moisture absorption is higher than that of nylon 6 and 6/6. It can be molded and extruded. Artificial leathers have been made from slurries of nylon-4 fibers.

Nylon 4/6 (polytetramethylenediamineadipamide) A condensation polymer of diaminobutane and adipic acid that melts higher than NYLON 6/6 (w s), so can be used at somewhat higher temperatures than the latter.

Nylon 6 (polycaprolactam) A type of nylon made by the polycondensation of CAPROLACTAM, w s, the second-most widely used polyamide in the US. Melting at about 228°C, it is used for fibers, including tire cord, and as a thermoplastic molding powder. Nylon 6 is as structurally sound as type 6/6 at room temperature, but it picks up moisture more rapidly and loses strength more rapidly as humidity and temperature increase. It is available in many grades, including glass-fiber-filled, and in a broad range of molecular weights, suitable for injection molding, extrusion, blow molding, and rotational molding. Parts can be machined, welded, and adhesive-bonded.

Nylon 6/6 (polyhexamethyleneadipamide) A type of nylon made by condensing hexamethylenediamine with adipic acid, first prepared by W. H. Carothers of Du Pont in 1936. It is the leading commercial polyamide, being used extensively for staple fibers and monofilaments, and is the most widely used type in other applications. The bulk polymer is a tough, white, translucent, crystalline material that melts rather sharply near 269°C. Nylon 6/6 is the strongest of the nylons over the widest range of temperature and humidity, but absorbs up to 2% water from air in the normal range of climatic humidity. Water acts as a plasticizer, reducing moduli but improving impact resistance and flex life in humid environments, with opposite effects in arid ones. During molding and extrusion, moisture content must be less than 0.1%.

Nylon 6/6 Salt (hexamethylenediammonium adipate) An intermediate in the manufacture of nylon 6/6, formed from one molecule each of adipic acid and hexamethylene diamine.

Nylon 6/10 (polyhexamethylenesebaçamide) The product of condensation of hexamethylenediamine with sebacic acid, used for brush bristles and monofilaments. It has lower water absorption and lower melting point than nylon 6 or 6/6. When a small amount of an alkyl-substituted hexamethylenediamine is added to the condensation mixture, a more elastic polymer known as *elastic nylon* is obtained.

Nylon 6/12 (polyhexamethylenedodecanamide) A nylon introduced by Du Pont in 1970, made from hexamethylenediamine and a 12-carbon dibasic acid. Nylon 6/12 is characterized by retention of physical and electrical properties over a wide humidity range, good dimensional stability, and low moisture absorption.

Nylon 6/T (polyhexamethyleneterephthalamide) A major member of an aliphatic-aromatic family of polyamides, none of which have gained commercial importance because they are difficult to prepare and to process.

Nylon 7 (polyenantholactam, polyheptanamide) A type of nylon known commercially in Russia as "Enant", but not yet commercial in the US. Its properties are similar to those of NYLON 6, w s.

Nylon 8 (polycaprylllactam, polyoctanamide) A nylon made by condensation polymerization from capryllactam. Its low melting temperature (200°C) and high cost of starting materials have limited the utilization of this polymer. It should not be confused with a type of nylon long marketed as "Type 8", which is actually a chemically modified nylon 6/6.

Nylon 9 (polypelargonamide, polynonanamide) A type of nylon made by melt condensation of aminopelargonic acid (9-aminononanoic acid). Nylon 9 has tensile yield and flexural strengths approaching those of nylon 6, while having low water absorption like those of nylons 11 and 12. It is, however, in limited use.

Nylon 10 An experimental polymer prepared, with difficulty, from aminodecanoic acid.

Nylon 11 (polyundecanamide) A type of nylon produced by polycondensation of the monomer 11-aminoundecanoic acid, a derivative of castor oil. It is available in the form of fine powders for rotational molding and other powder processes; and in pellet form for extrusion or molding. Like nylon 12, nylon 11 has properties intermediate between those of nylon 6 and polyethylene: good impact strength, hardness, and abrasion resistance, but other mechanical properties are lower than those of most other nylons. However, due to its exceptionally low water absorption, the dimensional stability of nylon 11 is high. A modified nylon 11 trade-named Rilsan N is flexible, transparent, and self-extinguishing.

Nylon 12 (polylauryllactam, polydodecanamide) A nylon made by the polymerization of lauric lactam (dodecanoic lactam) or cyclododecalactam, with 11 methylene groups between the linking –CONH– groups in the polymer chain. Its mechanical properties are intermediate between those of conventional nylons and polyethylene, and it is the lowest in water absorption (1.5%) and density (1.01 g/cm^3) of all the nylons.

Nylon 55 A blend of aliphatic-, cycloaliphatic-, and aromatic-based polyamides. The material is clear in thick cross sections, has low water absorption, good dimensional stability and solvent resistance, and can be processed economically by injection molding or extrusion.

Nylon Fiber Generic name for a manufactured fiber in which the fiber-forming substance is any long-chain synthetic polyamide having recurring amide groups (–CONH–) as an integral part of the polymer chain, with less than 85% of the amide groups bonded directly to aromatic rings. (If more than 85%, it's considered to be a POLYIMIDE, w s.) Nylon was the first fiber of major commercial importance to be made of wholly synthetic material. Carothers' pioneering research in 1929 culminated in Du Pont's introduction of nylon hosiery in 1940.

Nylon Monofilament Single strands much larger in diameter than those of staple fiber, used for fishing leaders and lines, brush bristles, racket strings, surgical sutures, and rope.

Nylon MXD/6 (poly-*m*-xylyleneadipamide) A type of nylon with lower elongation at break than nylon 6 or 6/6, but capable of attaining their properties by reinforcement with glass fibers. The resin has low melt viscosity, good flexural strength and modulus, and resists alkalies and hydrolytic degradation.

Nylon, Nucleated A nylon polymerized in the presence of a nucleating agent, e g, about 0.1% of finely dispersed silica, which promotes the growth of spherulites and controls their number, type, and size. Nucleated nylons have higher tensile strength, flexural modulus, abrasion resistance, and hardness, but lower impact strength and elongation than their unnucleated counterparts.

Nylon Salt Any of the intermediates in nylon synthesis formed by the combination of one molecule of diamine and one of diacid, such as NYLON 6/6 SALT, w s.

Nylon, Transparent Any of several nylon polymers based on aromatic ring units. The first such nylon, introduced in the early 1970s by Dynamit Nobel under the trade name Trogamid T, was made by polycondensation of terephthalic acid with 2,2,4-trimethylhexamethylenediamine. This crystal-clear, amorphous polyamide has excellent resistance to stress cracking and a glass-transition temperature comparable to those of polycarbonates.

Nylon "Zytel ST" A Du Pont rubber-toughened nylon, identified by the initials standing for *super-tough*, claimed to be the most rugged engineering resin then (1977) available. It is superior to polycarbonate in impact strength, with notched Izod = 9 J/cm (17 ft-lb$_f$/in).

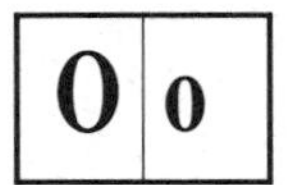

o- Abbreviation for prefix *ORTHO-*, w s, and ignored in alphabetizing compound names.

O Chemical symbol for the element oxygen.

OBP Abbreviation for OCTYL BENZYL PHTHALATE, w s.

OBSH Abbreviation for 4,4'-OXYBIS(BENZENESULFONYLHYDRAZIDE), w s.

Octobromobiphenyl (octabromodiphenyl) $(C_6H_2Br_4)_2$. A very dense, involatile liquid useful as a fire-retardant additive.

Octahedrite A term sometimes used for the anatase form of TITANIUM DIOXIDE, w s.

Octabis(2-Hydroxypropyl) Sucrose A viscous, straw-colored liquid used as a crosslinking agent for urethane foams and as a plasticizer for cellulosics.

1-Octene $C_6H_{13}CH{=}CH_2$. A comonomer, made from ethylene, and polymerized with ethylene to make linear, low-density polyethylene.

Octyl- The general term for all saturated, 8-carbon aliphatic radicals having the formula, C_8H_{17}–, often used imprecisely for the actual radical 2-ETHYLHEXYL–, w s.

Octyl Benzyl Phthalate (OBP) A plasticizer for PVC, cellulosics, polystyrene, and polyvinyl butyral. It is similar to butyl benzyl phthalate but has lower volatility. It resists oil extraction.

Octyl Biphenyl Phosphate A plasticizer for vinyl and other resins, with good permanence and low-temperature properties. It imparts flame resistance and is approved by FDA for use in food packaging.

***n*-Octyl-*n*-Decyl Adipate** (NODA, octyl decyl adipate, isooctyl isodecyl adipate) $C_8H_{17}OOC(CH_2)_4COOC_{10}H_{21}$. A plasticizer for cellulosics, synthetic rubbers, and vinyl resins. It imparts good low-temperature flexibility and resistance to extraction by water. NODA is also useful at low concentrations in polypropylene as a processing aid.

***n*-Octyl-*n*-Decyl Phthalate** (NODP, ethylhexyl decyl phathalate, octyl decyl phthalate) $C_8H_{17}OOCC_6H_4COOC_{10}H_{21}$. One of an important family of phthalate-ester plasticizers derived from C_6 to C_{10} alcohols. These plasticizers may be used interchangeably in PVC compositions, to which they impart somewhat better drape, flexibility, and low-temperature resistance than does dioctyl phthalate. They are also compatible with vinyl chloride-acetate copolymers, cellulose nitrate, ethyl cellulose, cellulose acetate butyrate, polystyrene, acrylic and butadiene-acrylonitrile resins, neoprene and chlorinated rubber.

***n*-Octyl-*n*-Didecyl Trimellitate** $C_8H_{17}OOCC_6H_3(COOC_{10}H_{21})_2$. An ester of trimellitic acid (1,2,4-benzene tricarboxylic acid), used as a plasticizer for vinyl chloride polymers and copolymers. The trimellitate plasticizers are used especially for nonfogging applications, and in adhesives or laminates where low migration is important.

Octyl Epoxy Tallate A monomeric plasticizer and heat and light stabilizer for vinyls and cellulosics. It imparts good low-temperature flexibility, has low volatility, and is used primarily in coated fabrics, garden hose, film and sheeting, and slush-molded parts.

Octyl Isodecyl Phthalate See ETHYLHEXYL ISODECYL PHTHALATE.

***n*-Octyl Methacrylate** $H_2C{=}C(CH_3)COOC_8H_{17}$. A comonomer for acrylic resins.

***p*-Octylphenyl Salicylate** $HOC_6H_4COOC_6H_4C_8H_{17}$. A white, crystalline powder, used as an ultraviolet absorber in polyolefins and cellulosics. It is reported to have increased the outdoor weathering life of polyethylene by 400%.

Octyl Phosphate See TRIOCTYL PHOSPHATE.

OD Abbreviation for *outside diameter* (of an annulus or spherical shell).

Offset Molding See JET MOLDING.

Offset Adapter In extrusion, a short, angled connector between extruder and die that orients the die axis on an axis different from, but sometimes parallel to that of the extruder axis.

Offset Printing A printing process in which the image to be printed is first applied to an intermediate carrier such as a rubber roll or plate, then is transferred (in reverse) to the surface to be printed.

Offset Yield Strength The stress at which the strain exceeds by a specified amount (the offset) an extension of the initial proportional portion of the stress-strain curve. It is expressed in force per unit area, usually psi or MPa (ASTM D 638).

-OH The hydroxyl radical, the characterizing group of inorganic bases, aliphatic alcohols, and phenols.

Ohm (Ω) The SI unit of electrical resistance, equal to one volt divided by one ampere. The SI definition for one ohm is the resistance between two points of a conductor when a constant difference of potential of one volt, applied between these two points, produces in this conductor a current of one ampere, this conductor not being the source of any electromotive force.

OIDP Abbreviation for *octyl isodecyl phthalate*. See ETHYLHEXYL ISODECYL PHTHALATE.

Oil Absorption The percentage increase in weight of a specimen after immersion in oil for a specified time. This property is important with respect to fillers used in plasticized thermoplastics, which can absorb plasticizer from the resin in a compound.

Oilcanning (1) A problem of distortion or buckling encountered in extrusion of rigid-vinyl sheet and shapes, usually attributed to uneven cooling of the extrudate. (2) A similar problem with dark-colored, rigid-vinyl house siding caused by solar heating and cured by adding to the extrusion compound a heat-resistant resin that elevates the glass-transition and heat-deflection temperatures of the siding.

Oil-Soluble Resin A resin that is capable of dissolving in or reacting with drying oils at moderate temperatures. Such resins are used for producing homogeneous coatings with modified characteristics.

Oleamide *cis*-$CH_3(CH_2)_7CH{=}CH(CH_2)_7CONH_2$. An ivory-colored powder used in low percentages as a slip agent for film-grade polyethylenes, as is its *trans*- isomer, ELAIDAMIDE, w s.

Olefin (1) Any of the class of monounsaturated, aliphatic hydrocarbons of the general formula C_nH_{2n}, and named after the corresponding paraffins by changing their *-ane* endings to *-ene* or *-ylene*. Examples are ethylene (ethene), propylene, and butenes. The class of polymers of olefins are called *polyolefins* or *olefin plastics*. (2) The term is sometimes taken to include aliphatics containing more than one double bond in the molecule such as a diolefin or diene. *Butadiene* is a typical member and an important comonomer for plastics.

Olefin Fiber Generic term for a manufactured fiber in which the fiber-forming substance is any long-chain synthetic polymer composed of at least 85% by weight of ethylene, propylene, or other olefin units (Federal Trade Commission).

Olefin Plastic See POLYOLEFIN.

Oleo Chemical A fatty acid, ester, amide, or a mixture of those compounded into thermoplastic resins, usually in fractional percentages, to enhance their release from injection molds and, in films, to reduce surface friction and stickiness.

Oleoresin A mixture of a natural resin obtained from a plant with an essential oil obtained from the same plant. They are usually semisolids and are sometimes called *balsams.*

Oligomer A substance composed of only a few monomeric units repetitively linked to each other, such as a dimer, trimer, tetramer, etc, or their mixtures (ASTM D 883). Other definitions in the literature place the upper limit of repeating units in an oligomer at about ten. The term *telomer* is sometimes used synonymously with oligomer.

ONB Abbreviation for *o*-NITROBIPHENYL, w s.

One-Shot Molding A process for molding polyurethane foam in which the reactants, usually an isocyanate, a polyol, and a catalyst, are fed in separate streams to a mixing head from which the mixed reactants are discharged into a mold. The polyol and catalyst are sometimes combined along with other additives prior to mixing, but the isocyanate is always fed separately to the mixing head.

On-Off Control A simple type of process control in which an instrument, the *controller*, senses a process state variable, such as temperature, and, when that variable passes its set point, shuts off or turns on a process factor that will tend to reverse the rise or fall of the state variable. Many older industrial ovens are controlled by a *thermostat*, i e, a temperature sensor that causes a heating circuit to be activated when the temperature falls below the setpoint by a small amount and turns off the circuit when the setpoint is exceeded by a small amount. Since even well insulated ovens lose heat to the surroundings, the rate of that loss, opposing the heater input, can provide an acceptable temperature band within the oven. In a more sophisticated variation used in process equipment, a cooling system may be turned on when the heater is turned off, and vice versa.

Opacity Of a material, the inability to transmit light; having a LIGHT TRANSMITTANCE (w s) of zero. With materials in thin layers, opacity depends on thickness. (Even normally opaque gold, rolled sufficiently thin, transmits some light.) Section 08 of ASTM Standards contains some nine tests dealing with optical properties of plastics. Test D 589, sec 15.09, is a method for determining the opacity of papers, some of which are used as backings for plastic sheets.

Opalescence (1) The limited clarity of vision through a sheet of transparent plastic at any angle, because of diffusion within or on the surface of the plastic. (2) Of a plastic material, the quality of having inner, tiny colored lights resembling those of opals.

Open-Cell Foamed Plastic A CELLULAR PLASTIC, w s, in which most of the cells are interconnected in a manner such that gases can flow freely from one cell to another.

Open-Mold Process Any technique for fabricating reinforced plastics in which a one-sided male or female mold is used, with no or low pressure being required. See HAND LAYUP, SPRAYUP, and BAG MOLDING.

Optical Brightener See BRIGHTENING AGENTS.

Optical Composite A composite that is transparent or nearly so. Making one requires that the refractive indices of the resin and reinforcing fiber, themselves transparent, be closely equal. This might be feasible with polymethyl methacrylate and a carefully chosen glass, since the refractive index of PMMA (n_D^{20}) is 1.49, that of fused quartz is 1.458, and most common glasses are above 1.51. Some fairly clear composites have been made with glass-reinforced polyurethanes.

Optical Density The degree of OPACITY (w s) of any translucent medium, sometimes expressed as the logarithm of the opacity.

Optical Dispersion Of a transparent material, the difference between the refractive indices of the material for two different wavelengths of light: in particular, the wavelengths of red and violet lights (ca 650 and 410 nm, respectively). Dispersion is an important property in the design of compound lenses.

Optical Distortion Any apparent alteration of the geometric shape of an object as seen either through a transparent material or as a reflection from mirror surface.

Orange Peel An uneven surface texture of a plastic article or its finish coating, somewhat resembling the surface of a navel orange. The condition is often caused by uneven wear of the mold surface due to overpolishing, overheating, or overcarburizing of the mold cavity. It may also be caused by overspraying with mold releases.

Organic Chemistry The chemistry of carbon compounds, usually containing hydrogen, and many containing oxygen, nitrogen, halogens, sulfur, phosphorus, and, occasionally, other elements. The term arose because such compounds were first obtained from living organisms.

Organic Peroxide See PEROXIDE.

Organic Pigment Any pigment derived from naturally occurring or synthetic organic substances, characterized by good brightness and brilliance and (usually) transparency. They are generally more resistant to chemicals than inorganic pigments, but are less resistant to heat, light, and solvents.

Organometallic Compound Any compound containing carbon, hydrogen and a metal, excluding ordinary metallic carbonates such as sodium bicarbonate and also excluding metallic salts of common organic acids. Many organometallic compounds are used as polymerization catalysts and stabilizers, most notably the ORGANOTIN STABILIZERs, w s.

Organopolysiloxane See SILICONE.

Organosol A suspension of a finely divided resin in a plasticizer together with a volatile organic liquid (ASTM D 883). A somewhat tighter definition requires that the volatile liquid comprise at least 5% of the total weight of the suspension. The resin used is most frequently PVC, but the term applies to such suspensions of any resin. An organosol can be prepared from a PLASTISOL, w s, merely by adding a volatile diluent or solvent that serves to lower viscosity and evaporates when the compound is heated. The diluents most often used in organosols are primary alcohols (methanol, propanol, butanol, pentanol, and octanol), also commercial mixed diluents such as Apco thinner, VM&P naphtha, Stoddard solvent, Varsol #2, Solvesso #100 and #150; and isoparaffins (Isopars).

Organotin Stabilizer Any of an important class of stabilizers for PVC, notable for their high efficiency, compatibility, and imparting of clarity. The family includes sulfides and oxides of tin-alkyls or -aryls, organotin salts of carboxylic acids, organotin mercaptides, and trialkyl or triaryl tin alcoholates. Certain dioctyltin mercaptides and maleate compounds have been approved for food-contact use. See also DI-*n*-OCTYLTIN MALEATE POLYMER and DI-*n*-OCTYLTIN-*S*,*S*'-BIS(ISOOCTYL MERCAPTOACETATE).

Orientation The process of stretching a hot plastic article to align the molecular chains in the directions(s) of stretching, thus improving modulus and strength in those direction(s). When the stretching is applied in one direction or two perpendicular directions, the process is called *uniaxial* or *biaxial orientation*, respectively. Upon reheating, an oriented film will shrink in the direction(s) of orientation. This property is useful in applications such as shrink packaging, and for improving the strength of formed or extruded articles such as lids of dairy-product tubs, pipe, filaments, film, and sheet.

Orientation-Release Stress The internal stress remaining in a plastic sheet after orientation, which can be relieved by reheating the sheet to a temperature above that at which it was oriented. This stress is measured by heating the sheet and determining the force per unit cross-sectional area exerted by the sheet as it attempts to revert to its preorientation dimensions.

Orifice (1) A small, usually cylindrical passage in a die, as in a strand die or an orifice-type rheometer. (2) A beveled, sharp-edged, usually circular hole in a thin metal plate that is inserted between flanges in a pipeline. By measuring the fluid pressure up- and downstream near the plate, one can calculate the rate of flow of the fluid. The method has long been used with simple gases and liquids and is useful for dilute polymer solutions, but not for melts.

Orr's White Synonym for LITHOPONE, w s.

Ortho (1) Short for ORTHOGONAL WEAVE, w s. (2) (*ortho-*) The relation of two adjacent carbon atoms in the benzene ring, abbreviated *o-* and ignored in alphabetizing lists of compounds.

Orthogonal Weave (1) A type of reinforcing cloth in which fibers are equally distributed in three principal directions (0, 60, and 120°) to make a sheet whose properties are nearly equal in all planar directions. (2) A type of three-dimensional reinforcement weave in which fibers are distributed equally in all three principal directions to generate a preform that is then impregnated with resin and cured.

Orthophthalate Plasticizer (phthalate-ester plasticizer) Any of a family of plasticizers derived by reacting phthalic anhydride with an alcohol. They include the widely used DOP, DIDP, and DTDP.

Orthorhombic One of the six crystal systems, in which the three principal axes are mutually perpendicular but the atomic spacings are different on all three.

Orthotropic Having three mutually perpendicular planes of elastic symmetry, as in composites having fibers running in two perpendicular directions, or biaxially oriented sheet. If the fibers or orientation are unidirectional, the material is still orthotropic but also isotropic in the two directions perpendicular to the fibers or oriented polymer chains.

Oscillating Die (1) A blown-film die that slowly rotates about its axis in one direction about 90°, then reverses to rotate as far in the opposite direction. The effect of the rotation is to distribute evenly over the width of the rolled-up film any slight differences in film thickness at different points around the die. (2) Flat or cylindrical strand dies in which the one die lip moves to and fro so as to cross the flow channels and produce nonwoven, knotless netting.

OSHA Abbreviation for *Occupational Safety and Health Administration*, the Federal Agency established by the Department of Labor, Bureau of Labor Standards, to enforce occupational safety and health standards. The standards are known as Part 1910 of amended Chapter XVII of Title 29 of the Code of Federal Regulations established on April 13, 1971 (36 F R 7006) and as amended thereafter.

Osmometer An instrument for measuring osmotic pressure (OP). The essential elements are a membrane which is permeable to solvents but impermeable to polymer molecules of a specific size range, reservoirs on each side of the membrane containing respectively the polymer solution and pure solvent, means for holding the temperature of the reservoirs constant, and means for measuring the differential osmotic pressure between the solution and the solvent. Osmotic pressure is proportional to the *number* of molecules per unit volume of the solution. From the mass per unit volume and the molar concentration estimated from the OP, one finds the value of the NUMBER-AVERAGE MOLECULAR WEIGHT, w s. The useful range of the method is for M_n from 20,000 to 1,000,000. Membrane materials include cellophane, cellulose acetate, polyvinyl alcohol, and polychlorotrifluoroethylene.

Osmosis The passage of solvent from a mass of pure solvent into a solution, or from a less concentrated to a more concentrated solution, through a membrane that is permeable to the solvent but not to the solute.

Osmotic Pressure (Π) The hydrostatic pressure at which the net flow of solvent through the membrane of an osmometer reaches zero. This pressure is related to the number of polymer molecules in dilute solution, and can be used to determine the molecular weight. See OSMOMETER.

Ostwald-deWaele Model See POWER LAW.

Ostwald-Fenske Viscometer See DILUTE-SOLUTION VISCOSITY.

Outgassing A term used in the vacuum-metallizing industry for the evaporation under vacuum of a volatile substance such as moisture, solvent, or plasticizer, from plastic articles to be coated with metal. Outgassing can cause pressure increase (loss of vacuum), also darkening and poor adhesion of the metal coating.

Outsert Molding A term coined to distinguish the process of molding small plastic parts in a large metal plate from the "insert molding", in which small metallic parts are incorporated into a larger plastic molding. The plate is indexed in the injection mold by a peg-and-hole system, with the plastic parts injected through blanks prepunched in the metal plate. The process makes it feasible to use plastics-metal combinations where economics formerly dictated all-metal components.

Ovaloid A surface of revolution symmetrical about the polar axis that forms the end closure for a filament-wound cylinder.

Overall Conductance (overall heat-transfer coefficient) In heat-transfer engineering, the reciprocal of the total THERMAL RESISTANCE, w s, for heat flow through plane walls or tube walls. It is defined by the equation: $U = q/A \cdot \Delta T$, where q = the rate of heat flow through (and normal to) the surface of area A, and ΔT is the fall in temperature through the layer in the direction of q. This is a modification of Fourier's law, invented to deal conveniently with heat flow through stagnant fluid films adjacent to walls, films whose thicknesses are difficult or impossible to measure. For curved surfaces such as pipes, U must be referred to either the inside or outside area, usually the latter (U_o). The SI unit for overall conductance is $W/(m^2 \cdot K)$.

Overcoating In extrusion coating, the practice of extruding a web beyond the edges of the substrate web.

Overcure A thermal decomposition in a thermosetting resin or vulcanizable elastomer due to overheating or excessive molding time.

Overflow Groove A small groove used in molds to allow material to escape freely to prevent flash and low part density, and to dispose of excess material.

Overlay Sheet (top sheet, surfacing mat) A nonwoven fibrous mat of glass or synthetic fiber used as the surfacing sheet in decorative laminates. Its function is to provide a smoother finish, hide the fibrous pattern of the laminate, and/or to provide a decorative motif when printed on the underside.

Overspray The roughness of a film of paint or lacquer due to dry particles deposited on, but not melded into a previously sprayed, semi-dried film. Overspray is encountered particularly with surfaces in more than one plane, such as auto bodies, television cabinets, etc. The remedy for overspray is a well-balanced solvent system containing enough high boiler to prevent the drying out of spray droplets before they land on an overshot surface.

Oxamide $H_2NCOCONH_2$. A white powder, used as a stabilizer for cellulose nitrate.

Oxetane Resin See CHLORINATED POLYETHER.

Oxidation Chemical reaction with oxygen. Any process that increases the proportion of oxygen or acid-forming element or radical in a compound. Oxidation is always exothermic. Rapid oxidation accompanied by flame is called *burning*.

Oxidative Coupling A process defined as a reaction of oxygen with active hydrogen atoms from different molecules, producing water and a dimerized molecule. If the hydrogen-yielding substance has two active hydrogen atoms, polymerization results. This process is used in the polymerization of phenols, particularly POLYPHENYLNE OXIDE, w s.

Oxidative Degradation Breaking down of a polymer or plastic product through the action of oxygen on the polymer itself or on other ingredients of the compound. The process may be signaled by change of color, visible deterioration of the part surface, or lowered performance in service.

Oxidative Dehydrogenation A chemical process used in making monomers such as styrene, butadiene, and vinyl chloride. Such "oxydehydro" processes involve either removal of hydrogen from a hydrocarbon by oxygen, forming water; or removal of hydrogen from a hydrocarbon by a halogen to form the hydrogen halide, then regeneration of the halogen with oxygen.

Oxirane A synonym for ETHYLENE OXIDE, w s.

Oxirane Group A synonym for EPOXY group, w s.

Oxirane Value (oxirane oxygen) The percent of oxygen absorbed by an unsaturated raw material during epoxidation; a measure of the amount of epoxidized double bond in a material. The oxirane value and the IODINE VALUE, w s, are used in evaluating epoxy plasticizers. A high oxirane value and low iodine value are considered to be essential for good performance, but these are not the only criteria.

Oxo Process A chemical process utilizing a reaction known as *oxonation* or *hydroformylation*, in which hydrogen and carbon monoxide are added across an olefinic bond to produce an aldehyde containing one more carbon atom than the olefin. The aldehydes produced by this process can be reduced to alcohols which are used for making many ester-type plasticizers.

Oxy- A prefix denoting the –O– radical or (primarily in Europe) the –OH radical.

***p*-Oxybenzoÿl Copolyester** Any of a family of readily moldable polyester copolymers consisting of mixtures of *p*-oxybenzoÿl with units from aromatic carboxylic acids and aromatic diphenols.

***p*-Oxybenzoÿl Polymer** A polymer based on *p*-hydroxybenzoic acid (derived from phenol and carbon dioxide). Technically a thermoplastic, the polymer retains good stiffness at temperatures up to 315°C and, at temperatures around 425°C, undergoes a second-order transition and becomes malleable so that it can be forged like ductile metals. Some other properties are high dielectric strength, elastic modulus, thermal conductivity, resistance to wear and solvents, self-lubricity, and good machinability. It has also been blended with metals to form alloys. Copolymers (see preceding entry) sacrifice some of the high-temperature performance to gain moldability.

4,4'-Oxybis(Benzenesulfonylhydrazide) (OBSH) The most important of the sulfonyl hydrazide family of blowing agents, a white crystalline solid melting at 164°C and yielding nitrogen upon decomposition. It is used widely in rubber-resin blends due to its ability to simultaneously foam the blends and act as a crosslinking agent, but it is also used in polyethylene, PVC, phenolics, and epoxy resins. Favorable properties are low odor, nontoxicity, and freedom from discoloration.

Oxygen-Index Flammability Test See FLAMMABILITY, ASTM D 2863.

Oxygen-Transmission Rate See PERMEABILITY.

Oxymethylene (1) Synonym for formaldehyde, w s. (2) (oxane) The group (–OCH_2–) the chain unit of ACETAL RESINs, w s.

Ozonation Chemical reaction with OZONE, w s.

Ozone An allotropic form of oxygen, O_3, a faintly blue, irritating gas and a powerful oxidant.

Ozone Resistance The ability of a plastic or elastomer to withstand, without diminution of useful properties, the chemical action (strong oxidation) of ozone.

P p

p (1) Abbreviation for SI prefix PICO-, w s. (2) *p*- Abbreviation for PARA-, w s.

P (1) Chemical symbol for the element phosphorus. (2) Abbreviation for SI prefix PETA-, w s. (3) Symbol for pressure or permeability.

Pa SI abbreviation for PASCAL, w s.

PA (1) Abbreviation for *polyamide*. See NYLON. (2) Abbreviation for PHTHALIC ANHYDRIDE, w s.

PAA Abbreviation for POLYACRYLIC ACID, w s.

PABM See POLYAMINOBISMALEIMIDE RESIN.

Packing In injection molding, the continuing slow flow of melt into a mold cavity after it has filled, but before the gate has frozen to cut off flow. The flow occurs because almost the full melt pressure is transmitted to the cavity while the initially injected melt is freezing and shrinking. Packing may cause problems traceable to high residual stresses near the gate and can also cause pieces to stick in the mold. However, it is sometimes done deliberately, with thick sections, to prevent formation of voids and sinks and to provide more intercavity and intershot consistency in dimensions.

Packing Braid A braid of reinforcing fiber having a fully filled, square cross section.

Paddle Agitator One of the simpler types of mixing equipment for plastics in the form of dispersions, pastes, and doughs. The most common form comprises a set of rotating blades driven by a vertical shaft and intermeshing with a set of fixed blades.

PAI Abbreviation for POLYAMIDE-IMIDE RESIN, w s.

Paint A dispersion of pigment in a liquid vehicle that may be applied to surfaces to form a thin adherent protective or decorative coatings. The liquid vehicle usually consists of a film-forming resin dissolved in a solvent, or a resin latex. Resins most frequently used in paints are phenolics, polyesters, ureas, melamines, cellulosics, acrylics, vinyls, alkyds, and epoxies. See also LACQUER and ENAMEL.

Painting of Plastics Plastics articles are painted not only to enhance their appearance, but also to provide desired surface properties lacking in the unpainted articles. For example, electrical properties and resistance to water, solvents, chemicals, and abrasion resistance may be improved by painting. Adhesion of paints to plastics is achieved by intermolecular attraction, solvent etching, or a combination of both. In the case of plastics of low surface energy, e g, polyethylene, an oxidative pretreatment is mandatory for good coating adhesion.. The methods used to apply paints to plastics are spraying (with or without masks), DIP COATING, FLOW COATING, ROLLER COATING, SCREEN PRINTING; and SPRAY-AND-WIPE PAINTING, all of which see.

PAK Abbreviation for *polyester alkyd resin.* See ALKYD RESIN.

PAN Abbreviation for POLYACRYLONITRILE, w s.

Paneling Distortion of a filled or partly full plastic container occurring during aging or storage, due to outward diffusion of solvent that causes reduced pressure inside the container.

PAPA Abbreviation for POLYAZELAIC POLYANHYDRIDE, w s.

Paper Chromatography The original CHROMATOGRAPHY, w s.

PAPI (PMPPI) Abbreviation for *polymethylenepolyphenylene isocyanate.* See DIISOCYANATE.

Para- (1) A chemical prefix from the Greek word meaning beside or beyond, denoting a relation "alongside" another compound such as a higher hydrated form of an acid or a polymeric form, as in paraldehyde. (2) (italicized, *p*-) Denoting the relation of opposite carbon atoms in the benzene ring, the 4-position in a singly substituted benzene. In this use, the prefix is ignored in alphabetizing compounds.

Paraffin (1) A synonym for ALKANE, w s. (2) A colorless, translucent wax obtained from petroleum-refining residues, a mixture of mostly saturated, straight-chain hydrocarbons melting between 49 and 63°C. (3) In Britain and its former possessions, kerosene.

Paraformaldehyde A low-molecular-weight, linear polymer of formaldehyde $HOCH_2(OCH_2)_nOCH_2OH$, a white solid that is easily depolymerized by mild heating to yield anhydrous formaldehyde gas. It is therefore a convenient form in which to handle and ship formaldehyde for industrial processes such as the manufacture of ACETAL RESINs (w s), its high-molecular-weight, stable homologs. See also 1,3,5-TRIOXANE.

Parallel-Laminated Pertaining to a laminate in which all layers of reinforcement are oriented approximately parallel with respect to the length or the direction of applied tensile stress.

Parallel-Plate Viscometer (1) An instrument consisting of two circular parallel plates, the lower one stationary, the upper one rotatable, the disk-shaped specimen being confined between the plates. Torque and rotational speed are measured; from these and the dimensions the viscosity may be estimated. A more elaborate development of this geometry is embodied in the MOONEY VISCOMETER, w s. (2) An instrument similar in geometry to that in (1) but neither disk is rotated. Instead, the upper plate is pressed downward at a controlled and measured rate, squeezing a thick specimen into a thin disk, and the force required is measured. This method is useful only for very viscous materials and low deformation rates.

Parallels (1) Spacers placed between the steam plate and press platen to prevent the middle section of a compression mold from bending under pressure. (2) Pressure pads or spacers between the steam plates of a mold when the land area is too small. The pads control the height when the mold is closed and thus prevent crushing parts of the mold.

Parameter (1) Loosely, a system factor or variable that may take on a range of values as decided by the observer or operator of the system. Example: hydraulic-line pressure and cylinder temperature are parameters in injection molding. (2) A defining constant of a statistical distribution, such as the mean or standard deviation of a normal distribution, and distinct from estimates of same calculated from sample measurements. (3) An independent variable through whose functions relations between other factors may conveniently be expressed.

Paraphthalate Plasticizer Any of a family of plasticizers derived by reacting terephthalic acid with an alcohol. They are similar in plasticizing capacity to the ORTHOPHTHALATE PLASTICIZERs, w s, while offering improved performance in areas such as volatility, low-temperature flexibility, electricals, and lacquer-marring. With the exception of dioctyl terephthalate (DOTP), a liquid plasticizer suitable for plastisols, most paraphthalates are solids when prepared from alcohols having an average chain length over six carbon atoms.

Paraxylylene See PARYLENE.

Parison The hollow tube or other preformed shape of molten thermoplastic that is inflated inside the mold in the process of blow molding. Most commonly, the parison is extruded immediately before blowing, but parisons are also injection molded and may also be chilled and stored, to be reheated before blowing. In the earliest application of blow molding, a pair of calendered sheets joined along the edges was used as the parison.

Parison Drawdown As a parison is extruded downward, the weight of that part which has already emerged pulls the regions above and stretches them, causing thinning of the wall, i e, drawdown. In modern blow-molding operations, drawdown effects are minimized both by programming the rate of extrusion of the parison and by extruding more rapidly, often with the aid of an ACCUMULATOR (1), w s.

Parison Programmer A device that varies the wall thickness of a parison in predetermined local areas while it is being extruded, in order to compensate for greater and lesser extents of blowing as the parison is blown to its final shape in the mold. For example, in a blow-molded bottle with a narrow waist portion, if the parison-wall thickness were the same all along its length, the waist would end up with a much thicker wall than the wider portions of the bottle. Programming provides less parison thickness at the waist, more in the shoulder and bottom of the bottle, with the final thickness being more uniform throughout. (Bottoms are often made thicker to provide resistance to dropping impact.) Programming is accomplished by varying the die gap, by moving either the die or the mandrel, the motions being controlled by a timer or an electromechanical synchronizer to assure that the programmed thicknesses always arrive at the correct heights in the mold.

Parison Swelling In blow molding, the momentary lateral enlargement of a parison immediately below the die as it emerges, a manifestation of released normal stress. It is expressed as the ratio of the cross-sectional area of the swollen parison to the cross-sectional area of the die opening. Parison swelling helps to offset PARISON DRAWDOWN, w s. See also EXTRUDATE SWELLING.

Parkesine The name given to the historic first (commercially unsuccessful) thermoplastic, made by plasticizing CELLULOSE NITRATE, w s. The polymer was dissolved in a solvent, castor oil was mixed in, and the solvent was evaporated. The product was developed by Alexander Parkes and was the forerunner of Celluloid, which was perfected in 1870 by John Wesley Hyatt, who used camphor as the plasticizer.

Particulate (1, adj) Existing as minute, separate particles. (2, n, plural) A particulate substance, a powder.

Particulate Composite A plastic filled with solid particles of one or more substances that do not melt during processing. See FILLER.

Parting Agent (release agent, mold lubricant, mold release) A lubricant, often wax, silicone oil, or fluorocarbon fluid or solid, used to coat a mold cavity to prevent the molded piece from sticking to it, and thus to facilitate its removal from the mold. Parting agents are often packaged in aerosol cans for convenience in application.

Parting Line (flash line) (1) A line marked on a three-dimensional model from which a mold is to be prepared, to indicate where the mold is to be split into two halves or several components. (2) The mark on a molded or cast article caused by slight flow of material into the crevices between mold parts. If the amount of material is sufficient that it must be removed, it is called FLASH, w s.

Partitioned Mold Cooling A large-diameter hole drilled into a mold or mold core and divided into two channels by a metal plate extending close to the bottom end of the channel. Water is introduced near the top of one channel, flows to the bottom of the hole, and is removed near the top of the other channel.

Parylene (poly-*p*-xylylene) Generic name for a group of film-forming thermoplastics introduced in 1965 by Union Carbide. The basic member, trade named Parylene N, is a polymer of *p*-xylylene, which has the structure shown below.

$$H_2C= \quad =CH_2$$

Parylenes C and D contain one and two chlorine atoms on the benzene ring, and other types have recently become available. The polymer is formed on a receiving surface by pyrolyzing the dimer of *p*-xylylene, a white powder, in vacuum (0.13 Pa abs) to its monomer vapor, which then flows to the room-temperature deposition chamber and forms a polymeric film with the structure $(-C_6H_4-CH_2CH_2-)_n$. Objects to be coated are usually mounted on rotatable racks to improve the uniformity of coating thickness. The films are tough, have excellent chemical resistance, low permeability, high thermal stability and dielectric strength, and have been used to protect electronic assemblies and other critical parts from atmospheric oxygen and moisture. Very thin films—down to 25 nm (1 μin)—can be formed without pinholes, and thickness uniformity is good even on irregular surfaces. The rather high cost of the material and processing still limits use.

Pascal (Pa) The SI unit of pressure and stress, equal to 1 newton per square meter (N/m^2). The pascal and its multiples are intended to supersede all other units of force per unit area such as pounds per square inch, atmospheres, torrs, etc. 1 psi = 6.894,757 kPa, 1 atm = 101.3250 kPa, and 1 torr = 133.322 Pa.

Pascal Second (pascal-second, Pa·s) The SI unit of dynamic (absolute) viscosity, equal to 1 $N \cdot s/m^2$. Some conversions of older viscosity units (of which there is a bewildering plethora) to Pa·s are given in the appendix. When shear stress τ assumes its alternate identity, MOMENTUM FLUX (w s), the pascal-second is interpreted as 1 kg/(m·s).

Pascal's Law Externally applied pressure on a confined fluid is transmitted equally in all directions.

Paste, PVC A term sometimes used for PLASTISOL, w s.

Paste Resin A term sometimes used for PVC resins used in making vinyl dispersions such as plastisols. See DISPERSION RESIN.

PAT Abbreviation for POLYAMINOTRIAZOLE, w s.

Pb Chemical symbol for the element lead (Latin: *plumbum*).

PB Abbreviation for *poly-1-butene*. See POLYBUTYLENE RESIN.

PBAN Abbreviation for *polybutadiene-acrylonitrile* copolymer. See NBR and ACRYLONITRILE-BUTADIENE COPOLYMER.

PBI Abbreviation for POLYBENZIMIDAZOLE, w s.

PBMA Abbreviation for POLY-*n*-BUTYL METHACRYLATE, w s.

PBS Abbreviation for *polybutadiene-styrene* copolymer. See STYRENE-BUTADIENE THERMOPLASTIC.

PBT See POLYBENZOTHIAZOLE and POLYBUTYLENE TEREPHTHALATE.

PBTP Abbreviation for POLYBUTYLENE TEREPHTHALATE, w s.

PC Abbreviation for POLYCARBONATE RESIN, w s.

PCL Abbreviation for POLYCAPROLACTONE, w s.

PCT see POLYCYCLOHEXYLENEDIMETHYLENE TEREPHTHALATE.

PCTFE See POLYCHLOROTRIFLUOROETHYLENE.

PDAP Abbreviation for *poly(diallyl phthalate)*. The abbreviation DAP is widely used for both the monomeric and polymeric forms of DIALLYL PHTHALATE, w s.

PDMS Abbreviation for POLYDIMETHYL SILOXANE, w s.

PE Abbreviation for POLYETHYLENE, w s, and sometimes used for PENTAERYTHRITOL, w s.

Peanut-Hull Flour See NUTSHELL FLOUR.

Pearlescent Pigment (pearl-essence pigment, nacreous pigment) A pigment with crystalline, transparent particles in the form of parallel platelets that impart an appearance of mother-of-pearl to plastics. The thin platelets have a high refractive index. Each crystal reflects only a portion of incident light reaching it, transmitting the remaining light to the crystal below. The simultaneous reflection of light from many parallel layers produces the characteristic pearly luster, the brilliance of which depends on the uniformity and parallelism of the crystals. Natural pearlescent pigments are composed primarily of guanine crystals derived from fish scales. They are expensive but nontoxic. The synthetic pearlescents are based on crystallized lead or bismuth compounds or platelets of mica coated with a dye or pigment.

Pearl Polymerization See SUSPENSION POLYMERIZATION.

Pear Oil See AMYL ACETATE.

Pebble Mill See BALL MILL.

Pectin A water-soluble plant polysaccharide, mainly *d*-galacturonic acid, but also containing other sugar units.

PEEK Abbreviation for POLYETHERETHERKETONE, w s.

Peeler A machine for slitting large rolls or blocks of foamed plastics into thin sheets, by rotating the blocks into a horizontally mounted bandsaw blade. Sheets as thin as 1.5 mm may be produced by this method.

Peel Ply The outside layer of a laminate that is removed or sacrificed to achieve improved bonding of additional layers.

Peel Test See "SCOTCH-TAPE" TEST.

PEG Abbreviation for POLYETHYLENE GLYCOL, w s.

PEI Abbreviation for POLYETHERIMIDE, w s.

PEK Abbreviation for POLYETHERKETONE, w s.

Pellets (molding powder) Granules or tablets of uniform size, consisting of resins or mixtures of resins with compounding additives, that have been prepared for molding and extrusion by shaping in a pelletizing machine or by extrusion into strands that are cut while hot or after solidifying in a water bath.

Peltier Effect The physical phenomenon occurring at the junctions, in electric circuits, of dissimilar metals that is the inverse of the Hall effect (see THERMOCOUPLE). When a current flows through the junction of two unlike metals it causes an absorption or liberation of heat, depending on the direction of flow and the metals.

Pendant-Perfluoryl Alkyl-Modified Acrylic Polymer A product introduced early in 1992 by Dow Chemical Company, suitable for spray application, and said to render surfaces to which it is applied extremely nonsticky and resistant to graffiti.

Pendulum Impact Strength See IMPACT RESISTANCE and IZOD IMPACT STRENGTH.

Pentacite An alkyd resin formed by using pentaerythritol as the polyhydric alcohol.

Pentaerythritol (tetramethylol methane) $C(CH_2OH)_4$. A white crystalline powder derived by reacting acetaldehyde with an excess of formaldehyde in an alkaline medium. It is used in the production of alkyd resins and chlorinated polyethers.

1,5-Pentanediol (pentamethylene glycol) $HOCH_2(CH_2)_3CH_2OH$. A colorless liquid used in the production of polyester and urethane resins.

Penton® Trade name for CHLORINATED POLYETHER, w s.

PEO Abbreviation for POLYETHYLENE OXIDE, w s.

Percentage Elongation 100 × ELONGATION, w s.

Perchloropentacyclodecane $C_{10}Cl_{12}$. A bridged bicyclic, saturated compound, a solid filler used as a flame retardant in epoxy resins, often in conjunction with antimony trioxide.

Perfluoroalkoxy Resin (PFA) A class of melt-processable fluoroplastics in which perfluoroalkyl side chains are connected to the fluorocarbon backbone of the polymer through flexible oxygen linkages. PFA resins have the desirable properties associated with fluoroplastics plus superior creep resistance, and are more easily processed by extrusion and injection molding.

Perfluoroelastomer (tetrafluoro-perfluoromethyl vinyl ether copolymer) Introduced by Du Pont in 1977 as Kalrez®, this elastomer combines the properties of a conventional fluoroelastomer, such as vinylidene fluoride-hexafluoropropylene copolymer, with those of a fluorocarbon resin such as polytetrafluoroethylene. It has found application for O-rings and seals that must withstand strong chemicals and solvents at high temperatures.

Perfluoroethylene Synonym for TETRAFLUOROETHYLENE, w s.

Perforating Any process by which a plastic film, sheet or tubing is provided with holes ranging from relatively large diameters for decorative effects (by means of punching or clicking) to very small, even invisible sizes. The latter are achieved by passing the material between rollers or plates, of which one of the pair is equipped with closely spaced, fine needles; or by spark erosion.

Perlite A siliceous lava which, when heated to 720–1090°C, expands to 10–20 times its original volume, forming tiny, hollow, spherical bubbles. Perlite is much used as an ingredient of lightweight concrete and as a density-lowering filler for plastics. See also MICROSPHERES.

Perm (1) A unit of permeability for seepage of fluids through particle beds; see PERMEABILITY (2). (2) A process involving ammoniacal chemicals and, usually, heat, by which straight strands of keratinous fiber are rendered into circular, spiral, and wavelike forms.

Permachor A concept of M. Salame (1961), who discovered that the permeation rates of many (though not all) organic liquids through polyethylene (PE) films could be estimated with the equation

$$\log Pf = 16.55 - 3700/T - 0.22\ \pi$$

in which Pf = the "permeability factor", T is absolute temperature, (K), and π is the *permachor* calculated for the permeating compound from a table of empirically determined, additive, atomic and structural contributions. Pf was measured by sealing the test liquid in a PE bag and measuring the loss of weight over a period of days, and was computed as (rate of weight loss, g/day)·(film thickness, mil)/(film area, 100 in^2), so has the units g·mil/(day·100 in^2), not the same as PERMEABILITY defined below. The concept worked well for some other film materials, too. Later (1967), Salame extended it to polymer permeation by oxygen, nitrogen, and carbon dioxide, this time computing the permachor for the polymer rather than the permeant.

Permanence The resistance of a plastic to changes in its properties and performance over extended time in the working environment.

Permanent Set The increase in length, expressed as a percentage of the original length, by which an elastic material fails to return to its original length after being stretched for a specified period of time.

Permanent White See BARIUM SULFATE.

Permeability (1) The ease with which a gas or vapor passes through a membrane, e g, a plastic sheet or film. It has been proved that permeability is equal to the product of *diffusivity* times the *solubility* of the gas or vapor in the plastic. *Coefficient of permeability* (*permeability coefficient*) ≡ the rate of permeation of a gas or vapor per unit cross-sectional area of the film, divided by the concentration gradient through the film. Because of the wide choice that has existed for units of permeation rate, film area, and concentration, many different "convenient" sets of units have been used. ASTM Test D 1434 (sec 15.09) defines permeability as the product of PERMEANCE (w s) times the thickness of a film. Since permeability can depend on material thickness, D 1434's authors suggest that it not be used unless it has been shown to be independent of thickness. The SI unit of permeability is mol/(m·s·Pa). However, this unit is not in use in US trade literature, where one is most likely (in 1992 and earlier) to see (for gases) the unit "cc-mil/100 sq in-24 hr-atm" at 23° or 37.8°C. And even in D 1434, reapproved in 1988, the results of a round-robin test were given in so-called "inch-pound units", i e, mL(STP)·mil/m^2·day·atm. The mil in both these last sets is 0.001 inch (of film thickness). D 1434 defines a permeability unit called the *barrer*, equal to 10^{-10} mL(STP)/(cm·s·cm Hg), = 0.3349 fmol/(m·s·Pa), and suggests using 10^{-18} mol (= 1 amol) or 10^{-15} mol (= 1 fmol) for reporting in the SI unit. A helpful table of conversions is provided there and in the Appendix of this Dictionary. Permeabilities of polymeric films to atmospheric gases and carbon dioxide vary from about 0.009 to 2.5 pmol/(m·s·Pa). See also WATER-VAPOR-TRANSMISSION RATE. (2) The proportionality constant in d'Arcy's equation (Darcy equation) for laminar flow of incompressible simple fluids through packed beds and porous solids (originally water through sand). The equation is

$$Q = \kappa \cdot A \cdot \Delta P/(\mu \cdot L)$$

in which Q = the volume flow rate, κ = the permeability, A = the cross-sectional area of the bed perpendicular to the net direction of flow, ΔP = the pressure drop (including any gravitational component) through the bed, μ = (Newtonian) dynamic viscosity, and L = the bed thickness in the flow direction. A permeability unit called the *perm*, equal to 1 (ft^3/d)·cp/[ft^2·(psi/ft)], corresponds to a flow of 1 ft^3 per day of water at 20°C (viscosity = 1.00 centipoise) through a packed bed in the form of a one-foot cube with sealed side faces. The analogous SI unit would have the form 1 (m^3/s)·(Pa·s)/[m^2·(Pa/m)], which reduces to m^2, providing one doesn't mind canceling shear-stress pascals with pressure-drop pascals and meters of three different meanings and directions. 1 perm = 1.5596×10^{-13} m^2. (3) In magnetic materials, the quotient of magnetic inductance and field strength, unity for most materials, slightly more or less for a few materials, and large for ferromagnetic materials (elements Fe, Co, Dy, Gd, Nd, and Ni and some alloys), and similarly large for plastics filled with these materials. The SI unit is henry per meter (H/m).

Permeance The ratio of the GAS-TRANSMISSION RATE, w s, to the difference in partial pressure of the gas on two sides of a sheet or film. The SI unit is mol/(m^2·s·Pa), but much smaller submultiples are convenient. The test conditions must be stated. See ASTM D 1434 (sec 15.09), also PERMEABILITY (1).

Permeation The passage of gas, vapor, or liquid molecules through a film or membrane, usually without physically or chemically changing it, except that permeation involves solubility of the vapor in the film. See PERMEABILITY.

Permittivity (specific inductive capacity, dielectric constant) For a plastic material, the ratio of the capacitance of a given configuration of two electrodes (most simply, parallel plates) separated by the material as the dielectric between the plates, to the capacitance of the same electrode configuration with air as the dielectric. See DIELECTRIC for some representative values for plastics.

Permselective Membrane A thin film that will preferentially permit gases of different kinds to pass through the film at different rates. For common gases such as hydrogen, oxygen, nitrogen, and carbon dioxide, silicone rubber is the most permeable polymer. Silicone rubber membranes as thin as 50 nm are 2.2 times as permeable to oxygen as to nitrogen. This property has been used in oxygen supplies for medical treatment.

Peroxide (peroxy compound) A compound containing at least one pair of oxygen atoms bonded by a single covalent bond. Organic peroxides, analogous to H_2O_2 in which either or both of the H atoms have been replaced by organic radicals, are thermally unstable and are widely used as initiators in polymerizations. As they decompose, they form free radicals that can initiate polymerization reactions and effect crosslinking. The rate of decomposition can be controlled by means of promoters or accelerators added to the system to increase the rate; or by inhibitors when it is desired to slow the rate. Peroxides are used in curing systems for thermosetting resins, and in polymerization reactions for many thermoplastics. Organic peroxides are often incorrectly labeled as catalysts.

Peroxyester (perester, *t*-alkyl peroxyester) Any of a family of liquid initiators used for crosslinking of polyethylene, polymerization of vinyls, high-temperature crosslinking of diallyl phthalate-modified polyesters, curing of styrene-modified polyesters, and styrenation of alkyd paints. They are aliphatic, not prone to yellowing and bleaching, and have good solubility and compatibility characteristics. One example of the numerous compounds in the family is *t*-butyl peroxypentanoate.

Persorption The adsorption of a substance in pores only slightly wider than the diameter of adsorbed molecules of the substance.

Perylene Pigment Any diimide of perylene-3,4,9,10-tetracarboxylic acid. Scarlet and vermilion varieties, resistant to bleeding, light, heat, and chemicals, are used in plastics.

PES Abbreviation for POLYETHERSULFONE, w s.

PET (PETP) Abbreviation for POLYETHYLENE TEREPHTHALATE, w s.

Peta- (P) The SI prefix meaning $\times 10^{15}$.

PETP (PET) Abbreviation for POLYETHYLENE TEREPHTHALATE, w s.

Petrochemical Any chemical derived directly or indirectly from petroleum or natural gas.

Petroleum Resin See HYDROCARBON PLASTICS.

PF Abbreviation for PHENOL-FORMALDEHYDE RESIN, w s.

PFA See PERFLUOROALKOXY RESIN.

Pfund Hardness A method of measuring INDENTATION HARDNESS (w s) of paints and related coatings in which a small hemispherical indenter made of quartz or sapphire is pressed into the coating (ASTM D 1474, sec 06.01).

pH A shorthand measure of the acidity or alkalinity of an aqueous solution or mixture, defined as the logarithm to base 10 of the reciprocal of the effective hydrogen-ion concentration (H^+) in gram-equivalents per liter. Pure water has a pH of 7 and is neutral. Acidic materials have pHs lower than 7, alkaline ones higher than 7. For example, vinegar and 0.1-N hydrochloric acid have pHs of 3.1 and 1.04, respectively, while 1% sodium carbonate and 0.1-N sodium hydroxide have pHs of 10.7 and 13, respectively. A 1-unit drop in pH corresponds to a 10-fold rise in H^+.

Phase Angle In many cyclic processes, particularly ones at high frequency, there is usually a time lag between the impulses driving the processes and the responses to those impulses. The lag time divided by the cycle period and multiplied by 2π is the phase angle (radians). This occurs in oscillatory rheometry, where the shear applied to the rotating element results in a lagging torque, which has two components, an elastic part, in phase with the displacement, and the viscous, out-of-phase component. The same situation exists in fatigue testing of plastics. An analogous phenomenon at the molecular and atomic levels in dielectric materials gives rise to the DIELECTRIC PHASE ANGLE, w s.

Phenol (1) Generic name for the class of organic compounds in which one or more hydroxyl (–OH) groups are attached directly to an aromatic ring. (2) The specific name for C_6H_5OH (carbolic acid, phenylic acid, hydroxybenzene). Phenol was first derived from coal tar but today is most commonly synthesized from benzene or toluene. About 60% of all phenol manufactured in the US is used for production of phenolic resins. Other applications in the plastics industry are for the production of bisphenol A, caprolactam, and the starting intermediates for nylon 6/6, adipic acid and hexamethylene diamine.

Phenol-Aralkyl Resin Any of several thermosetting resins produced by the condensation of aralkyl ethers and phenols. They are cured to hard, intractible resins by one of two methods: heating with hexamethylene tetramine or heating with selected epoxy compounds and an accelerator. They are available under the name Xylok® in ketonic solutions for manufacturing reinforced composites and as powders for the formulation of molding compounds. Having useful lives at 250°C about eight times that of phenolic resins, the phenol-aralkyl resins are used in high-performance, high-temperature electrical applications.

Phenol-Formaldehyde Resin (PF resin, phenolic resin) The most important of the PHENOLIC RESINs, w s. Made by condensing phenol with formaldehyde, these were the first synthetic thermosetting resins to be developed (L. H. Baekeland, 1907) and were marketed under the trade name Bakelite.

Phenol-Furfural Resin A type of FURAN RESIN, w s.

Phenolic Foam There are two basic types of phenolic foam—the *syntactic* type and the *reaction* type. The syntactic foams comprise hollow microspheres of phenolic resin mixed with a polyester or an epoxy resin to the consistency of putty, which can be applied to surfaces by troweling, or molded, or pressed into sandwich-core structures. The reaction-type foams are generated by heating a water-containing liquid phenolic resin, a blowing agent, an acid catalyst, and a surfactant. Heat liberated by the reaction of the acid with the phenolic resin vaporizes the blowing agent and water so that the foam structure is produced as the resin cures. The rising gas bubbles cause elongated cells, so the final foam has a grain structure somewhat like that of wood, with strength dependent on the direction of application of stress.

Phenolic Novolac (novolac, novolak) Thermoplastic, water-soluble resins obtained by reacting a phenol with an aldehyde, usually formaldehyde, in the proportion of less than one mole of the phenol with one mole of aldehyde, in the presence of an acid catalyst. When a source of methylene groups is added, linkage between the methylenes and the phenolic rings occurs, and the resins can react with diamines or diacids (e g, hexamethylenetetramine) to form thermosetting, insoluble resins. Absent a source of methylene groups, the resin remains permanently thermoplastic. See also RESINOID and EPOXY-NOVOLAC RESIN.

Phenolic Resin Any of a wide range of thermosetting resins made by reacting a phenol with an aldehyde, followed by curing and cross-linking. The phenols used commercially are phenol, cresols, xylenols, *p-t*-butylphenol, *p*-phenylphenol, bisphenols, and resorcinol. The aldehydes most commonly used are formaldehyde and furfural. In the uncured and semicured condition, phenolic resins are used as adhesives, casting resins, potting compounds, and laminating resins. Some are also used as vehicle resins in varnishes. Phenolic molding powders, compounded with reinforcements, fillers, and curing agents, e g, hexamethylene tetramine, are formed into a large variety of low-cost products by compression molding, transfer molding, and, to a limited extent, continuous extrusion. A modified type of injection molding, using a reciprocating screw rather than a ram, is also used for molding phenolic products. Phenolic moldings have good strength and modulus, good arc resistance and other electrical properties, resistance to solvents and high temperatures, and are bargain-priced.

Phenoplast Another name for PHENOL-FORMALDEHYDE RESIN, w s.

Phenoxy Resin (polyhydroxyether) Any linear thermoplastic resin made by reacting an exact stoichiometric equivalent of epichlorohydrin with bisphenol A and sodium hydroxide in dimethyl sulfoxide. The phenoxy resins are chemically similar to epoxies, but contain no epoxy groups, have higher molecular weights, and are true thermoplastics. However, the presence of many free hydroxyl groups permits cross-linking with isocyanates, anhydrides, triazines, and melamine. Their principal advantages are good processability, low mold shrinkage, excellent dimensional stability and creep resistance. Crystal-clear, water-white grades are available for extrusion, blow molding, and injection molding, which grades have been approved by FDA for food-contact use.

***m*-Phenylene Diamine** (1,3-diaminobenzene) 1,3-$(H_2N)_2C_6H_4$. A useful curing agent for epoxy resins, slower-reacting than aliphatic amines, thus providing longer pot life and milder exotherms.

Phenylene Oxide Resin See POLYPHENYLENE OXIDE.

Phenylethylene Synonym for STYRENE, w s.

Phenylformic Acid Synonym for BENZOIC ACID, w s.

Phenyl Group (phenyl radical) The group C_6H_5–, existing only in combination.

Phenylsilane Resin Any thermosetting copolymer of silicone and phenolic resins, available in solution form.

Phillips Process Polymerization of ethylene in hexane at 3 MPa and 130°C with a chromium oxide catalyst. The PE produced is highly linear and crystalline, with density between 0.962 and 0.968 g/cm^3.

Phosphate Plasticizer Any of a group of plasticizers derived from phosphoric acid and aliphatic alcohols and phenols, and used in conjunction with others to impart flame resistance. The *aromatic phosphates* (see TRICRESYL- AND CRESYL DIPHENYL PHOSPHATE) also impart good permanence and resistance to greases and oils, but have poor low-temperature properties. The *aliphatic phosphates* (see TRI-OCTYL-, TRIBUTOXYETHYL-, and OCTYL BIPHENYL PHOSPHATE) impart good low-temperature flexibility.

Phosphazene Polymer Any of a family of experimental resins built on long chains of alternating phosphorus and nitrogen atoms. The general structure is $(–PX_2{=}N–)_n$, where X may be a halogen or organic radical. They have been used in fuel hoses that remain flexible in subzero, arctic climates, prosthetics for reconstruction surgery, and fabric waterproofing. In solution these polymers can be reacted with various nucleophilic agents to form a range of thermoplastics and elastomers. An important potential use may be as flame retardants for textiles. They can also be foamed.

Phosphorescent Pigment One of a family of pigments, generally an inorganic sulfide crystal of fairly large and controlled size, that absorbs the energy of incident light then slowly re-emits it as radiation of a color specific to each pigment. The phosphorescence gradually dims in darkness, to be renewed by the next light restimulation.

Photoconductive Becoming electrically conductive when irradiated by light or ultraviolet light. See POLY(*N*-VINYLCARBAZOLE).

Photodegradation Breakdown of plastics due to the action of visible or ultraviolet light. Most plastics tend to absorb high-energy radiation in the ultraviolet portion of the spectrum, which elevates their elec trons to higher reactivity and causes oxidation, chain cleavage, and other destructive reactions. Prior to the rise of concern over permanence of discarded plastics in the environment, photodegradation was undesirable, to be overcome by protective additives. It later became desirable to induce delayed photodegradation of packaging materials and other articles that are discarded after use. The two most common methods of promoting degradation by UV light are incorporation of additives that act as photoinitiators or photosensitizers, and copolymerization, which forms weak links at designated sites along the polymer backbone. See also ULTRAVIOLET STABILIZER and STABILIZER.

Photoelasticity Changes in optical properties of isotropic, transparent materials when subjected to stress. Mainly, they become birefringent, a property that has been widely used to detect frozen-in stresses in sheets and other products, and in experimental stress analysis of models of products whose geometry is too complex for theoretical analysis.

Photopolymer A plastic compound containing an agent that undergoes a change, proportional to light intensity, upon exposure to light, so that images can be formed on its surface by a photographic process. Photopolymers play an important role in the manufacture of semiconductors.

Photopolymerization An addition reaction brought about by exposure of the monomer or mixture of monomers to natural or artificial light, with or without a catalyst. Methyl methacrylate, styrene, and vinyl chloride are examples of monomers that can be photopolymerized.

Phr Abbreviation for *parts* per *hundred* parts of *resin*, a fractional measure of composition, simpler than weight percentage, that facilitates making up multicomponent compounds. For example, as used in plastics formulations, 5 phr means that 5 kg of an ingredient would be combined with 100 kg of resin.

Phthalate Ester (*o*-phthalic ester) Any of a large class of plasticizers produced by the direct action of alcohols on phthalic anhydride. The phthalates are the most widely used of all plasticizers, and are generally characterized by moderate cost, good stability, low toxicity, and good all-round properties.

Phthalic Anhydride (phthalic-acid anhydride) $C_6H_4(CO)_2O$. A white, crystalline compound derived by oxidation of naphthalene or *o*-xylene, shipped in flake or molten form. Its major use is in the production of phthalate esters for plasticizing vinyl and cellulosic resins. It is also an important intermediate in the manufacture of alkyd and unsaturated polyester resins, and is a curing agent for epoxy resins.

Phthalocyanine Pigment Any of a group of pigments based on phthalocyanine (tetrabenzoporphyrazine), $(C_6H_4C_2N_2)_4H_2$, and modifications thereof. The metal-free parent compound is blue-green in color. Copper phthalocyanine, $(C_6H_4C_2N_2)_4Cu$, and relatives containing small amounts of chlorine, produce pure blues. Copper phthalocyanines in which 14 to 16 of the phenyl hydrogens are replaced by chlorine yield green shades. Certain insoluble forms of phthalocyanine blue and green pigments have been approved for use in food-contacting plastics. They are extremely lightfast, do not bleed in most vehicles, are high in tinctorial strength, and resistant to heat, acids, and alkalis.

Physical Catalyst Radiant energy capable of promoting or modifying a chemical reaction, as in PHOTOPOLYMERIZATION, w s.

PI Abbreviation for the *trans*-1,4- type of POLYISOPRENE, w s.

PIA Abbreviation for Plastics Institute of America, Inc, headquartered at 277 Fairfield Rd, Suite 100, Fairfield, NJ 07004-1932. PIA sponsors seminars in plastics technology and supports graduate research in plastics engineering and undergraduate scholarships.

PIB Abbreviation for *polyisobutylene*. See POLYBUTENE.

Pickup Groove See HOLD-DOWN GROOVE.

Pico- (p) The SI prefix meaning $\times 10^{-12}$.

Piggyback (adj) A word used to designate a system of two extruders in which one discharges to the other. Such an arrangement has occasionally been used to separate the melt-generating (plasticating) function from the pressure-developing and shaping function of the standard extruder, thus gaining more precise control over both functions and eliminating a main cause of surging at the die. For general extrusion this method of accomplishing that separation has been superseded by the cheaper and more compact combination of extruder with GEAR PUMP, w s. Also see EXTRUDER, TANDEM for a special application of the piggyback concept.

Pigment A general term for all colorants, organic and inorganic, natural and synthetic, that are insoluble in the medium in which they are used. Organic pigments are those that contain carbon as the basic part of the molecule. The inorganic pigments, many of which are derived from natural minerals, contain a metal oxide or salt as the basic part of the molecule. See also COLORANT.

Pill A term sometimes used for PREFORM, w s.

Pimelic Ketone A synonym for CYCLOHEXANONE, w s.

Pin Another name for MANDREL (2), w s.

Pinch-Off (1) In blow molding, a raised edge around the cavity in the mold that seals off the part and separates the excess material as the mold closes around the parison. (2) In making tubular film, the paired rollers (pinch rolls) at the top of the tower that flatten the tube and confine the air in the bubble.

Pinch-Off Blades In blow molding, the mating parts of the mold at the bottom that come together first, to pinch off the parison at the bottom and sever the tail, then, to help form the bottom of the part as the parison is inflated.

Pinch-Off Land The width of the PINCH-OFF BLADE (w s) that effects the sealing of the parison.

Pinch-Off Tail In blow molding, the bottom tip of the parison that is pinched off and severed as the mold closes. See the three preceding entries.

Pinene Resin See POLYTERPENE RESIN.

Pinhole Any very small hole in a film or coating that interrupts its continuity and reduces its integrity and efficiency as a barrier to vapor, gas, or electric current (as, for example, in wire insulation).

Pink Staining A pink-colored stain that sometimes appears on vinyl-coated fabrics of white and pastel colors when they have lain on earth for a long time. It has been attributed to growth of fungi of the genus *Penicillium*, and to the bacterium *Streptomyces rubrireticuli*. It can be prevented by treating the fabric with a fungicide, e g, *N*-(trichloromethylthio) phthalimide or an arsenic compound.

Pinpoint Gate (pin gate) In injection molding, a very small orifice, generally 0.75 mm (30 mils) or less in diameter (or maximum lateral dimension), connecting the runner and mold cavity, and through which molten plastic flows into and fills the cavity. Such a gate leaves a small, easily removed mark on the part, but due to the tendency of the melt to freeze early in the pinpoint gate as flow slows, its use is limited to small parts and to resins with good fluidity. In multicavity molds, the dimensions of pinpoint gates must be held within very tight tolerances in order to fill all cavities at the same time and to avoid differences in dimensions among the parts extracted from the several cavities. See also BALANCED GATING, GATE, and RESTRICTED GATE.

Pipe Die An extrusion die whose lands form a circular annulus used in extrusion of plastic pipe or tubing. The outer shell of the die is usually called the *die*, the core is called the *mandrel*. Pipe dies may be side-fed or end-fed, and the mandrel may be supported by a trio of legs called a *spider*, or it may be supported from the rear of a side-fed die. It is easier to achieve circumferential uniformity of wall thickness with a spider die, but the splitting of the melt stream at the legs has sometimes caused weak welds because of insufficient knitting time before the pipe emerges and is chilled.

Piperidine A heterocyclic, secondary amine with a six-membered ring, $C_5H_{10}NH$, a slow-acting curing agent for thick-section epoxy castings or laminates, where faster curing would cause exotherm prob– lems such as bubbling, distortion, or cracking.

Pipe Train A term used in pipe extrusion that denotes the entire equipment assembly, i e, extruder, die, external sizing means, cooling bath, haul-off, and coiler or cutter.

PIR Abbreviation for POLYISOCYANURATE, w s.

Piston See FORCE PLUG.

Pit An imperfection, a small crater in the surface of the plastic, with its width of about the same size as the depth.

Pitch (1) Of an extruder screw, the axial distance from a point on a screw flight to the corresponding point on the next flight. In a single-flight (single-start) screw, the pitch and LEAD, w s, are equal. In a screw having n parallel multiple flights, pitch = lead/n. In certain SOLIDS-DRAINING SCREWs (w s) with two flights, the lead of one flight is slightly larger than that of the other. In such a screw, pitch varies continuously along the two-flighted section. (2) Any of various black or dark semisolid to solid materials obtained as residues from the distillation of tars, and sometimes including natural bitumen.

Planar Helix Winding A winding in which the filament path on each dome lies on a plane that intersects the dome, while a helical path over the cylindrical section is connected to the dome paths.

Planar Winding A winding in which the filament lies on a plane that intersects the winding surface.

Planetary-Screw Extruder See EXTRUDER, PLANETARY-SCREW.

Planishing See PRESS POLISHING.

Plasma Etching A process in which a plastic surface to be metal-plated is exposed to a gas plasma in a vacuum, producing chemical and physical changes that yield bondability and wettability equivalent to those produced in the past by stringent and hazardous chemical pre-treatments. Although a variety of gases may be used, bottled oxygen has been found to be best. A radio-frequency source inside the high-vacuum chamber generates the plasma (an ionized gas consisting of an equal number of positive ions and electrons). The process has been effective on nylons, acrylonitrile-butadiene-styrene resins, and plastics based on phenylene oxide.

Plasma-Spray Coating A spray-coating process developed to apply sinterable plastics such as polytetrafluoroethylene to metals and ceramics. A special spraygun produces a rotating jet of hot, ionized gas particles (plasma) with laminar-flow characteristics. Plastic powder supplied to the gun is channeled within the gun so that it emerges as a layer on the periphery of the plasma jet where temperatures are lower than those in the center of the jet. The process is capable of producing coatings as thin as 2.5 μm on unprimed but clean substrates, without after-baking. Substrates must be capable of withstanding the sintering temperature of the polymer.

Plastic (n) A material that contains as an essential ingredient one or more high polymers, is solid in its finished state, and, at some stage in its manufacture or processing into finished articles, can be shaped by flow. However, this definition is supplemented by notes explaining that materials such as rubbers, textiles, adhesives, and paints, which may in some cases meet this definition, are not considered to be plastics. The terms *plastic*, *resin*, and *polymer* are somewhat synonymous, but resin and polymer most often denote the basic materials as polymerized, while the term *plastic* or *plastics* encompasses compounds containing plasticizers, stabilizers, fillers, and other additives. See also ELASTOMER.

Plastic (adj) (1) Indicating that the noun modified is made of or pertains to a plastic or plastics. The singular form is customarily used when the noun obviously refers to a particular, single plastic, as in "a plastic hose", and the plural form is often used when the noun could refer to several types of plastics, as in "the plastics industry". However, there has been a trend in Europe to use the plural form exclusively even when it results in ungrammatical phrases such as "a plastics hose". The intent of the ungrammatical pluralization is to distinguish between the synthetic polymers used in the plastics industry and other materials sometimes referred to as "plastic", such as hot glass, modeling wax, and clay in the wet, unfired state. The preference of most authors is to use the singular form when it is evident from context that the noun refers to a single material. (2) Capable of being deformed continuously and permanently without rupture at a stress above the yield value.

Plasticate (plastificate) (v) To render a thermoplastic more flexible, even molten, by means of both heat and mechanical working. Sometimes used imprecisely for PLASTIFY, w s, and incorrectly for PLASTICIZE, w s.

Plasticating Capacity Of an extruder or injection molder, the maximum rate at which the machine can melt room-temperature feedstock and raise it to the temperature suitable for extrusion or molding. This rate is determined mainly by the quotient of the available screw power, divided by the mean specific heat of the plastic of interest and the rise in temperature of the plastic from feed to die; and to lesser degrees by extruder length, screw design, and die design.

Plastic Deformation (1) A change in dimensions of an object under load that is not recovered when the load is removed. For example, squeezing a chunk of putty results in plastic deformation. The opposite of plastic deformation is *elastic deformation*, in which the dimensions return instantly to the original values when the load is removed, e g, as when a rubber band is stretched and released. (2) In tough plastics, deformation beyond the yield point, appearing on the stress-strain diagram as a large extension with little or no rise in stress. A part of the plastic deformation may be recovered when the stress is released; the remainder is PLASTIC FLOW, w s.

Plastic Flow (1) Irreversible flow above the yield point. (2) The flow of molten or liquid plastics during processing. (3) Deformation without change of stress.

Plastic Foam Synonym for CELLULAR PLASTIC, w s.

Plasticity The ability of a material to withstand, without rupturing, continuous and permanent deformation by stresses exceeding the material's yield value.

Plasticize To render a material softer, more flexible and/or more moldable by the addition and intimate blending in of a plasticizer. Should not be confused with PLASTICATE and PLASTIFY, w s.

Plasticizer A substance of low or even negligible volatility incorporated into a material (usually a plastic or an elastomer) to increase its flexibility, workability, or extensibility, while reducing elastic moduli. A plasticizer may also reduce melt viscosity and lower the glass-transition temperature. Most plasticizers are nonvolatile organic liquids or low-melting solids that function by reducing the normal intermolecular forces in a resin, thus permitting the macromolecules to slip past one another more freely. Some are polymeric. Plasticizers are classified in several ways according to: their compatibility (see PRIMARY and SECONDARY PLASTICIZERS); their general structure (monomeric or polymeric); their functions [flame-retardant, high-temperature, low-temperature, nontoxic (see NONTOXIC MATERIAL), stabilizing, crosslinking, etc]; and their chemical nature (see ADIPATE

PLASTICIZER, CHLORINATED PARAFFIN, EPOXY PLASTICIZER, PHOSPHATE PLASTICIZER, AND PHTHALATE ESTER). Many thousands of compounds have been developed as plasticizers, of which perhaps less than 200 are in widespread use today. The main facts about over 350 plasticizers are tabulated in the "Plasticizers" data table of the *Modern Plastics Encyclopedia* for 1993 (and most earlier years). About two-thirds of all plasticizers produced are used in vinyl compounds, in which field the three "workhorse" plasticizers are dioctyl phthalate, diisooctyl phthalate, and diisodecyl phthalate.

Plasticizer-Adhesive An additive, partly replacing plasticizers, that improves the adhesion of plastics coatings to substrates. For example, polymerizable monomers such as diallyl phthalate or triallyl cyanurate are added to PVC plastisols to improve their adhesion to metals, but these compounds also contribute to the plasticizing function.

Plasticizer Efficiency (1) The parts by weight of plasticizer per hundred parts of resin (phr) required to produce a plasticized PVC resin of a particular hardness on the Durometer A scale. (2) Taking dioctyl phthalate as the industry standard of comparison, one may express the efficiency (in percent) of another plasticizer as 100 (n_0/n_1), where n_0 is the phr of DOP required to achieve a particular Durometer value (or other desired property) and n_1 is the phr of the alternate plasticizer required to reach that same value.

Plasticizer Migration See MIGRATION OF PLASTICIZER.

Plasticizer, Polymerizable (reactive plasticizer) A special type of plasticizer, unique in that it functions as a plasticizer only before and during the processing step, consisting of a monomer added to a plastisol to increase its fluidity, which monomer cures in the presence of catalysts to become rigid in the fused plastisol article. Among such monomers are polyglycol dimethacrylates, dimethacrylates of 1,3-butylene glycol and trimethylolpropane, and some trade-named monomers whose compositions are proprietary. These polymerizable plasticizers enable one to liquid-cast very rigid articles that would otherwise have to be made, with very low plasticizer levels, by injection molding. Monomeric styrene, not ordinarily thought of as a plasticizer, performs in much the same way in polyester laminating formulations, lowering viscosity during wetting-out and the initial moments of pressure molding, then polymerizing to form crosslinks of the strong, stiff finished product. (2) Any of a new class of epoxy resins having the general structure

$$H_3C-C(CH_3)_2-\left[C(R_1)_2\right]_n-C(R_2)\overset{O}{-}C(R_3)-R_4$$

in which the R groups may be H, methyl, or ethyl, and n = 1 to 10.

These are very miscible with epoxy resins, they provide non-migrating internal plasticization after curing, and they are useful in coatings, adhesives, and sealants.

Plasticizer, Solid A plasticizer that is solid at room temperature but melts during processing to improve processability of the polymer in which it is incorporated. Upon cooling it resolidifies and thus does not soften the finished article. Solid plasticizers are used in rigid PVC, one of the most common being diphenyl phthalate (m p = 75°C).

Plastic Memory See MEMORY.

Plasticorder (plastograph) See BRABENDER PLASTOGRAPH.

Plastic Paper (synthetic paper) Paper-like products in which the skeletal structure is composed of synthetic resin. Three main types are SPUNBONDED SHEET (w s), *film paper*, and *synthetic pulp* (synpulp). Film papers are similar to thin films of oriented polystyrene or polyolefins but they are treated to obtain opacity and ink receptivity. Synpulps are papermaking pulps made usually from polyolefins by processes that produce fibrous pulps without the use of conventional spinning methods.

Plastic, Rigid See RIGID PLASTIC.

Plastic, Semi-Rigid See SEMIRIGID PLASTIC.

Plastics Recycling A term embracing systems by which plastics materials that would otherwise immediately become solid wastes are collected, separated, or otherwise processed and returned to the economic mainstream in the form of useful raw materials or products.

Plastic Strain Plastic flow above the yield stress expressed as a fraction or percent of the original dimension before applying stress.

Plastic Tooling A term designating structures composed of plastics, that are used as tools in the fabrication of metals or other materials including plastics. While they are usually made of reinforced and/or filled thermosets, flexible silicone or polyurethane tools are often used for casting plastics. Common applications of rigid plastics tooling are sheet-metal-forming dies, models for duplicators, drill fixtures, spotting racks, molds for thermoforming thermoplastic sheets, molds for layups or contact-pressure molding of reinforced plastics, and injection molds for short runs. The tools are formed by the usual processes used for thermosetting resins, such as laminating, casting, and sprayup.

Plastic Viscosity For a BINGHAM PLASTIC, w s, the difference between the shear stress and the yield stress, divided by the shear rate.

Plastic Welding See WELDING.

Plastify To soften a thermoplastic resin or compound by means of heat alone, as in sheet thermoforming. Should not be confused with PLASTICIZE or PLASTICATE, w s.

Plastigel A vinyl compound similar to a plastisol, but containing sufficient gelling agent and/or filler to provide a putty-like consistency. It may be molded to a shape-retaining form at room temperature, then heated and cooled to impart permanency.

Plastisol A suspension of a finely divided vinyl chloride polymer or copolymer in a liquid plasticizer which has little or no tendency to dissolve the resin at ambient temperatures but which becomes a solvent for the resin when heated. At room temperature the suspension is very fluid and suitable for casting. At the proper temperature, the resin is completely dissolved in the plasticizer, forming a homogeneous plastic mass which upon cooling is a more or less flexible solid. Additives such as fillers, stabilizers, and colorants are also usually present. A plastisol modified with volatile solvents or diluents that evaporate upon heating is known as an ORGANOSOL, w s. When gelling or thickening agents are added to produce a putty-like consistency at room temperature, the dispersion is called a PLASTIGEL, w s. The coined term *rigidsol* is used to denote a plastisol modified with polymerizable or cross-linking monomers so that the fused product is rigid rather than flexible. Products are made from plastisols by rotational casting, slush casting, dipping, spraying, film casting, and coating.

Plastograph See BRABENDER PLASTOGRAPH.

Plastomer (1) (Solprene®) Any of a family of thermoplastic-elastomeric, styrene-butadiene copolymers whose molecules have a radial or star structure in which several polybutadiene chains extend from a central hub, with a polystyrene block at the outward end of each segment. They are used in making footwear components, in adhesives and sealants, and are also blended with other resins to upgrade performance. (2) Late in 1992 this term was adopted as generic by Exxon Chemical and Dow Chemical for grades of VERY-LOW-DENSITY POLYETHYLENE w s, produced with so-called "exact" metallocene catalysts and offering the flexibility of rubber with the strength and processability of linear low-density polyethylene. Densities range from 0.880 to 0.905 g/cm^3.

Plastometer See RHEOMETER.

Plate Die An inexpensive and easily modified die for extruding a plastic profile, into which an orifice of the desired shape has been machined, typically by ELECTRICAL-DISCHARGE MACHINING (w s). The plate is bolted to the front of a universal die body.

Plate Dispersion Plug Two small, perforated, parallel disks joined by a central connecting rod. Such assemblies were sometimes inserted in the nozzles of ram-type injection-molding machines to improve the distribution of colorants in the resin as it flows through the nozzle.

Plate Mark Any imperfection in a pressed plastic sheet resulting from the surface of the pressing plate (ASTM D 883).

Platen Either of the sturdy mounting plates of a press, usually a pair, to which the entire mold assembly is bolted.

Plate-Out An objectionable coating gradually formed on metal surfaces of molds and calendering and embossing rolls during processing of plastics, and caused by extraction and deposition of some ingredient such as a pigment, lubricant, stabilizer, or plasticizer. In the case of vinyls, which are especially prone to this condition, plate-out can be reduced by using highly compatible stabilizers such as barium phenolates and cadmium ethylhexoate, or by incorporating silica in the formulation. Resins can play a role in the plate-out problem, although the degree and mechanisms of resin contributions to plate-out are controversial.

Platform Blowing A special technique for blow molding large parts. To prevent excessive sag of the massive parison, the machine employs a table that, after rising to meet the parison at the die, descends with the parison, but a little more slowly than the parison, so as to support its weight, yet not cause buckling.

Plating See ELECTROPLATING ON PLASTICS.

Plexiglas® Famous trade name of the Rohm and Haas Corporation for their acrylic resins and cast polymethyl methacrylate sheet.

Plug-and-Ring Forming A technique of SHEET THERMOFORMING, w s, in which a plug, functioning as a male mold, is forced into a heat-softened sheet held in place by a clamping ring.

Plug-Assist Forming (vacuum forming with plug assist) A sheet-thermoforming process in which a convex mold half presses the softened sheet into the concave half, accomplishing most of the draw, after which vacuum is applied, drawing the sheet onto the concave surface. The method provides more nearly equal bottom and side thicknesses than straight vacuum forming and permits deeper draws.

Plug Flow Pressure flow through a closed channel characterized by the local velocity being the same over the entire channel. This is an extreme seldom realized in practice, but can occur over the center of a Bingham-plastic stream or in a system where the fluid does not wet the bounding walls. As compared with Newtonian flow, the more pseudoplastic the plastic melt, the more nearly pluggish is its flow.

Plunger The part of a transfer-press or old-style injection machine that applies pressure on the unmelted plastic material to push it into the chamber, which in turn displaces the plastic melt in front of it, forcing it through the nozzle and into the mold. See also RAM, FORCE PLUG, and POT PLUNGER.

Plunger Molding A variation of TRANSFER MOLDING, w s, in which an auxiliary hydraulic ram is employed to assist the main ram. The auxiliary ram rapidly forces the material through a small orifice, thereby generating high frictional heat. The higher temperature speeds the cure of the material, which when transferred into the mold by the main ram, cures very soon after the mold is filled.

Ply A layer or lamina in a structure made up of two or more distinct layers bonded together, such as a reinforced-plastic laminate made from layers of preimpregnated cloth.

PMAC Abbreviation for POLYMETHOXY ACETAL, w s.

PMAN Abbreviation for POLYMETHACRYLONITRILE, w s.

PMC Abbreviation for *polymer-matrix composite.*

PMCA Abbreviation for *poly(methyl-α-chloroacrylate)*, a member of the acrylic-resin family.

PMDA Abbreviation for PYROMELLITIC DIANHYDRIDE, w s.

PMMA Abbreviation for POLYMETHYL METHACRYLATE, w s.

PMP Abbreviation for POLY(4-METHYLPENTENE-1), w s.

Pock Mark An imperfection on the surface of a blow-molded article, an irregular indentation caused by inadequate contact of the blown parison with the mold surface. Contributory factors are insufficient blowing pressure, air entrapment, and condensation of moisture on the mold surface. See also PIT for an ASTM-approved definition for synonymous term that is not specific to blow molding.

Poise (for J. L. Poiseuille, 1840) The cgs unit of viscosity, now deprecated but still widely used, equal to 0.1 Pa·s. See Appendix for other conversions.

Poiseuille Equation See HAGEN-POISEUILLE EQUATION.

Poiseuille Flow Laminar flow in a pipe or tube of circular cross section under a constant pressure gradient. If the flowing fluid is Newtonian, the flow rate will be given by the HAGEN-POISEUILLE EQUATION, w s.

Poisson's Ratio (for S. Poisson; symbol μ or ν) In a material under tensile stress, the ratio of the transverse contraction to the longitudinal elongation. For metals, Poisson's ratio is about 0.3, for concrete, 0.1. For many plastics, the ratio is in the range from 0.32 to 0.48 and for rubbers it is about 0.5. Mineral fillers in plastics reduce the ratio by 0.05 to 0.10. In nonisotropic structures such as uni- and bidirectional laminates, Poisson's ratio may be different in orthogonal directions; but the value perpendicular to the reinforcing fibers will be the one most likely to be relevant. ASTM Standards lists methods for determining Poisson's ratio for ceramics, concretes and structural materials, but none for plastics or rubbers. For isotropic, elastic materials, the ratio is related to the three principal moduli by the equations: $\mu = (E/2G) - 1 = 0.5 - (E/6B)$, where E, G, and B are the tensile modulus (Young's modulus), shear modulus, and bulk modulus, respectively. These relationships may be used to estimate the ratio when the strains are small. However, because plastics are not truly elastic, and because moduli may vary with temperature, they should be regarded as approximate when applied to plastics.

Polarization (1, dielectric) The slight shifting of molecular electric charges when a polymer is placed in a strong electric field, creating local electric dipoles. The shifting of charge takes finite time and generates friction. In a high-frequency field, the rapid shifting causes considerable dissipation of energy, which is the basis for DIELECTRIC HEATING (w s) of plastics. (2, light) Of the three types possible, the most useful and common is *plane polarization*. This occurs when ordinary, unpolarized light, having wave motions in all directions perpendicular to the ray, passes through nicol prisms or polarizing filters that deliver an exit ray whose vibrations lie in one plane.

Polarizing Microscope An optical microscope fitted above and below the specimen-holding stage with nicol prisms or polarizing filters. The lower filter (*polarizer*) imparts plane POLARIZATION (w s) to the incoming light. The upper one (*analyzer*) is rotatable, but is usually set so that its plane of vibration is at 90° to that of the lower one. With isotropic specimens, all light is blocked at the analyzer and the observer sees only darkness. With a birefringent specimen, if the original light source is white, the observer sees bands of colors related to the crystal structure and the specimen's refractive indices.

Polar Winding In filament winding, a winding in which the filament path passes tangent to the polar opening at one end of a chamber and tangent to the opposite side of the polar opening at the other end.

Polepiece In reinforced plastics, the supporting part of the mandrel used in filament winding, usually on one of the axes of rotation.

Polishing (1) Smoothing and imparting luster to a surface by rubbing with successively finer abrasive-containing compounds or by filling the minute low areas of the surface with a wax or polymeric finish. (2) Smoothing rough edges by applying a jet of hot gas (to plastics) or a flame (to glasses). Flame polishing of plastics is generally not recommended because of the likelihood of degrading the surface and/or leaving residual stresses, either of which can cause crazing.

Polishing Roll A roll, usually one of a set, that has a highly polished, chrome-plated surface that is mirrored on all sheet or film extruded onto the roll or calendered through it (them).

Poly- A prefix meaning many. Thus, the term *polymer* literally means *many mers*, a mer being the repeating unit of any high polymer.

Polyacetal See ACETAL RESIN.

Polyacetylene A polymer of acetylene, made with Ziegler-Natta catalysts and usually dark-colored, with the unusual property (for a polymer) of high electrical conductivity, achieved by doping the polymer with about 1% of ionic dopant such as iodine. It may become a useful solar-cell material because its absorption spectrum closely matches the solar spectrum, but mechanical properties and stability are poor. Also, practical processing methods have yet to be developed.

Polyacrylamide A nonionic, water-soluble polymer prepared by the addition polymerization of acrylamide (CH_2=$CHCONH_2$). The white polymer is readily soluble in cold water but insoluble in most organic solvents. It is used as a thickener, suspending agent, and as an ingredient in adhesives.

Polyacrylate A thermoplastic resin made by the polymerization of an acrylic ester such as methyl methacrylate. See ACRYLIC RESIN.

Polyacrylic Acid (PAA) A polymer of acrylic acid, used as a textile size.

Polyacrylonitrile (PAN) Made by free-radical polymerization of acrylonitrile (CH_2=$CHCN$) in solution or suspension, this highly polar polymer is the basis of large-volume acrylic and modacrylic fibers.

Polyaddition See ADDITION POLYMERIZATION.

Polyadipamide A polymer formed by the reaction of adipic acid with a diamine, NYLON 6/6 (w s) being the most important example.

Polyalcohol Synonym for POLYOL, w s.

Polyalkenamer A chlorine-containing elastomer developed by Goodyear, with properties similar to but somewhat better than those of neoprene rubber. It is a copolymer of the addition-reaction product of hexachlorocyclopentadiene or 1,5-cyclooctadiene and an olefin such as cyclopentene.

Polyalkylene Amide See AMINO RESIN.

Polyalkylene Terephthalate Any of a family of thermoplastic polyesters that are polycondensates derived from terephthalic acid, whose diol components may be any within a wide range. The principal members of the family are POLYETHYLENE TEREPHTHALATE and POLYBUTYLENE TEREPHTHALATE, w s.

Polyallomer A crystalline block copolymer produced from two or more different monomers, usually ethylene and propylene, by alternately polymerizing the monomers in the presence of anionic, coordination catalysts, resulting in chains containing polymerized segments of both monomers. The polymer chains exhibit degrees of crystallinity normally found only in stereoregular homopolymers of propylene and ethylene, and the copolymers possess properties different from those of either blends of the homopolymers or copolymers prepared by conventional polymerization processes. Among such properties are high impact strength, low density, and flexural-fatigue resistance. The name "polyallomer" is derived from *allomerism*, meaning a similarity of crystalline form with a difference in chemical composition.

Polyallyl Diglycol Carbonate A high-impact, transparent thermoplastic with excellent abrasion resistance, made from Pittsburgh Plate Glass Industries' CR-39 monomer (and hence sometimes called CR-39). It is widely used in eyeglasses.

Polyamic Acid [from *poly(amide-acid)*] A polymer containing both amide and acid groups. The aromatic varieties are precursors of POLYIMIDEs, w s.

Polyamide-Imide Resin (PAI) Any of a family of polymers based on the combination of trimellitic anhydride with aromatic diamines. In the uncured form (*ortho*-amic acid) the polymers are soluble in polar organic solvents. The imide linkage is formed by heating, producing an infusible resin with thermal stability up to 290°C. These resins are used for laminating, prepregs, and electrical components. Molding resins that behave as thermoplastics can be produced by thermally curing and modifying amide-imide polymers. These molding resins can be processed by compression molding, extrusion, and injection molding.

Polyamide Plastic (polyamide) See NYLON and other listings following same.

Polyamine Any of a family of compounds containing multiple amines—primary, secondary, or tertiary—or mixtures of these, useful as hardening agents for epoxy resins.

Polyamine-Methylene Resin A light-amber-colored resin derived from diphenylol and formaldehyde, used as an ion-exchange resin.

Polyamine Sulfone A water-soluble copolymer of diallylamine monomer and sulfur dioxide, used as a paint additive, antistatic agent, synthetic-fiber modifier, and polishing agent for metal platings.

Polyaminobismaleimide Resin (PABM) Any thermosetting resin of dark-brown color obtained by the addition reaction of an aromatic diamine and a bismaleimide. Typical prepolymers accept high percentages of fillers and can be cast and compression or transfer molded. PABMs have flow properties comparable to common thermosetting resins and thermomechanical properties exceeding those of some light alloys. They possess excellent dimensional stability and are flame- and radiation-resistant. They can be adapted to aircraft, electrical and electronic fabrication, ablation applications, and to chemical-process equipment where resistance to aromatic solvents, refrigerants, and acids is required.

Polyaminotriazole (PAT) A family of fiber-forming polymers made from sebacic acid and hydrazine with small amounts of acetamide.

Polyarylate A polyester made by the condensation of an aromatic diacid with a dihydroxy aromatic compound. Commercial resins are copolymers of iso- and terephthalic acid with bisphenol A, with properties comparable to those of polycarbonate and polyethersulfone.

Polyarylene Sulfide Any of a polymer family prepared by polymerization and reaction of polyhalogenated aromatics with sodium sulfide in a polar solvent at high temperature. The best known (and only commercial) resin is POLYPHENYLENE SULFIDE, w s.

Polyaryl Ether A polymer having both aromatic rings and ether links (–O–) in the chain. Sometimes used as generic for poly-2,6-dimethyl-1,4-phenylene oxide, of which several makers' brands are available.

Polyarylethersulfone Resin See POLYETHERSULFONE.

Polyaryloxysilane A family of polymers, resistant to high temperatures, made up of silicon atoms, oxygen atoms, and thermally stable aromatic rings, part organic and part inorganic in nature, like the SILICONEs, w s.

Polyazelaic Polyanhydride (PAPA) A carboxyl-terminated, low polymer of approximately 2300 molecular weight, used as a flexibilizing curing agent for epoxy resins.

Polybenzamide (poly-*p*-benzamide) A fiber-forming polymer made from *p*-aminobenzoÿl chloride by self-condensation polymerization. Fibers retain high modulus to high temperatures and have been used for composite reinforcement.

Polybenzimidazole (PBI) A family of high-molecular-weight, strong, and stable polymers containing recurring aromatic units with alternating double bonds. PBIs are produced mainly by the condensation of 3,3',4,4'-tetraaminobiphenyl (3,3'-diaminobenzidine) and diphenyl isophthalate. The polymers are brown, amorphous powders, exhibiting a high degree of thermal and chemical stability. They are used to make fibers and films with excellent resistance to high temperatures and flame. Principal applications have been in the aerospace field, as protective coatings on missiles, radar antennas, and supersonic aircraft; and in reinforced laminates for critical applications. In 1992, Hoechst Celanese and KMI Inc, announced a joint venture to injection-mold bearings and other parts from PBI, heretofore considered not to be melt-processable, using proprietary technology.

Polybenzothiazole (PBT) A family of resins, one type of which has been made by cooking a toluidine with sulfur, or by reaction of, e g, 3,3'-mercaptobenzidine with diphenyl phthalate. These polymers have outstanding thermal stability: glass-reinforced PBT composites have withstood 350°C for over ten days. Though they are technically thermoplastics, they aren't melt-processable. Applications are in coatings for high-temperature service and laminating.

Polybiphenylsulfone An engineering thermoplastic (Union Carbide's Radel®) with a deflection temperature of 205°C and notched-Izod impact strength approaching 8 J/cm (15 ft-lb_f/in), this resin is conventionally melt-processable.

Polyblend A colloquial term—shortened from *poly*mer *blend*—used for physical mixtures of two or more polymers, for example, polystyrene and rubber or PVC and nitrile rubber. Such blends usually yield products with favorable properties of both components, sometimes opening markets not available to either of the neat resins. The term *alloy* is sometimes used for blends.

Polybutadiene A synthetic rubber made from 1,3-butadiene ($H_2C=CH–CH=CH_2$). The *cis* type has superior abrasion resistance and resilience, while the *trans* type is similar to natural rubber.

Polybutadiene-Acrylic Acid Copolymer A binder used in solid propellants.

Polybutadiene-Type Resins Unsaturated, thermosetting hydrocarbons cured by a peroxide-catalyzed, vinyl-type polymerization reaction, or by sodium-catalyzed polymerization of butadiene or blends of butadiene and styrene. Liquid systems, curable in the presence of monomers, are used for casting, encapsulation, and potting of electrical components, and in making laminates. Molding compounds, often containing fillers and modified with other resins or rubbers, may be compression or transfer molded. Syndiotactic 1,2-butadiene, introduced in 1974 in Japan, is thermoplastic, with semicrystalline nature, with good transparency and flexibility without plasticization. In the presence of a photosensitizer such as *p,p'*-tetramethyl diaminobenzophenone, this polymer can readily be cured by ultraviolet radiation. Transparent films of the nontoxic polymer are used for packaging, and cellular forms for shoe soles. It is biodegradable.

Polybutene Any of a family of low-molecular-weight polymers of mixed 1-butene, *cis*-2-butene, *trans*-2-butene, and isobutene. Depending on molecular weight, these polymers range from oils through tacky waxes, crystalline waxes, and rubbery solids. See also BUTYL RUBBER and POLYBUTYLENE RESIN.

Polybutylene Adipate Glycol (PBAG) A polymeric diol used in the production of urethane elastomers.

Polybutylene Resin (PB, poly-1-butene) Any of a family of polymers consisting of isotactic, stereoregular, highly crystalline polymers based on 1-butene. Properties are similar to those of polypropylene and linear polyethylene, with superior toughness, creep resistance, and flexibility. PB has been used in pipe, wire coating, gaskets, and industrial packaging. PB pipe carries the highest design-stress rating of all flexible, thermoplastic piping materials, serving to temperatures of 90°C.

Polybutylene Terephthalate (PBTP, PBT, polytetramethylene terephthalate) A member of the polyalkylene terephthalate family, similar to polyethylene terephthalate in that it is derived from a polycondensate of terephthalic acid, but with butanediol rather than glycol. PBTP can be modified easily to overcome its relatively low operating-temperature limit, making it equivalent to plastics used in construction and appliances. Grades are available for injection and blow molding, extrusion, and thermoforming. Properties include high tensile and impact strength, dimensional stability, low moisture absorption, and good electricals; also resistance to fire and chemicals when suitably modified.

Poly-*n*-Butyl Methacrylate (PBMA) A rubbery polymer that enjoys some use as an adhesive and textile finish.

Polycaprolactam See NYLON 6.

Polycaprolactone (PCL) A low-melting (62°C) polyester resin with the linear structure $(-CH_2CH_2CH_2CH_2CH_2COO-)_n$, made by polymerizing ε-caprolactone. PCL is compatible with most thermosetting and thermoplastic resins and elastomers. It increases impact resistance and aids mold release of thermosets, and acts as a polymeric plasticizer with PVC. It is biodegradable, useful in containers for growing and transplanting trees and other plants. Unmodified PCL is completely consumed by soil microbes but the rate of degradation can be slowed by incorporating a non-biodegradable polymer. PCL is also useful in the production of polyurethane elastomers and foams, to which it imparts good low-temperature properties and water resistance.

Polycarbonate Resin (PC) Any of a family of special polyesters in which groups of dihydric phenols are joined through carbonate linkages. They can be produced by a variety of methods, of which the most commercially important are (1) phosgenation of dihydric alcohols, usually bisphenol A; (2) ester interchange between diaryl carbonates and dihydric phenols, usually between diphenyl carbonate and bisphenol A; and (3) interfacial polycondensation of bisphenol A and phosgene. Bisphenol A polycarbonates with molecular weights close to 33,000 can be processed by injection molding, extrusion, thermoforming, and blow molding. Melt-casting and solution-casting processes are also employed. Such polycarbonates have high impact strength (to 8 J/cm of notched width), good heat resistance, low water absorption, good electrical properties, and no toxicity. They are vulnerable to some common organic solvents. Crystal-clear grades have been developed for safety glazing, including multilayer glass-and-PC, bullet-proof structures. Other applications include dentures, food packages, electrical components, precision parts for instruments and household appliances and—a current large-volume use—compact-disc (CD) records and data disks.

Polycarboranesiloxane (SiB) A polymer whose chain consists of alternating carborane and siloxane groups. Commercial resins contain active end groups that may be vulcanized with peroxides to yield rubbers resistant to high temperatures (260°C in air).

Polycarboxane Synonym for ACETAL RESIN, w s.

Polychloroether Synonym for CHLORINATED POLYETHER, w s.

Polychloroprene Synonym for NEOPRENE, w s.

Polychlorostyrene Usually made from a mixture of *o*- and *p*-chlorostyrene isomers, this polymer has higher glass-transition temperature and fire resistance than polystyrene.

Polychlorotrifluoroethylene (CTFE, PCTFE) A family of polymers made by polymerizing the gas $ClFC{=}CF_2$ by mass, emulsion, or suspension polymerization. The polymers range from oils, greases, and waxes of low molecular weight to the tough, rigid thermoplastics most commonly used in industry. Unlike polytetrafluoroethylene, PCTFE may be processed by conventional thermoplastic methods. It is also available as dispersions in xylene or ketones for application by dipping or spraying. The polymers are nontoxic, resistant to heat, chemically inert, and have outstanding electrical properties.

Polycondensation See CONDENSATION POLYMERIZATION.

Polycyclamide (1) A polyamide containing a cycloalkane ring. (2) Any linear, high-molecular-weight polyamide formed by condensing cyclohexane-1,4-bis(methylamine) with one or more dicarboxylic acids. These polymers have high melting points, but are sufficiently stable to permit melt processing at temperatures above 300°C without thermal decomposition. Their excellent physical and chemical properties indicate their usefulness as fibers, films, and moldings. See also NYLON.

Polycyclic Of an organic compound, containing two or more rings, often with pairs of rings sharing two carbon atoms, as in the hydrocarbon anthracene ($C_{14}H_{10}$), whose structure is shown below.

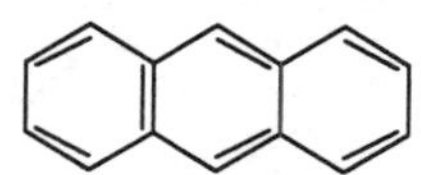

Polycyclohexylenedimethylene Terephthalate [PCT, poly(cyclohexane-1,4-dimethylene terephthalate)] The newest member of the commercial thermoplastic-polyester family, PCT is produced by reacting 1,4-cyclohexane dimethanol with dimethyl terephthalate. It is superior to its siblings (PET and PBT) in that it can serve at higher temperatures, with 1.8-MPa deflection temperatures to 260°C with 30% glass-fiber content. Moisture absorption is lower, too, and it has excellent chemical resistance. It is being used in automotive parts and dual-ovenable cookware. It can be compounded for high flame resistance.

Poly(Dichloro-*p*-Xylylene) (Parylene D) See PARYLENE.

Poly-2,6-Dimethyl-1,4-Phenylene Oxide See POLYPHENYLENE OXIDE.

Polydimethylsiloxane (dimethyl silicone) Any of a family of silicones of the composition $[-(CH_3)_2SiO-]_n$. Those of low molecular weight—several hundred to 10,000—are oils, some of which are widely used in aerosol mold releases for plastics that are not to be painted or printed. Polymers in the molecular-weight range near 10^5 are rubbers that are flexible at cryogenic temperatures but crystalline above –60°C. See also SILICONE.

Polydisperse Of a polymer, having a range of molecular weights as opposed to a single molecular weight (MONODISPERSE, w s), the usual state among commercial polymers. The broader the distribution relative to its center, the greater is its POLYDISPERSITY, w s.

Polydispersity The breadth of the molecular-weight distribution of a polymer. Two measures of polydispersity are in common use: (1) the ratio of the weight-average and number-average molecular weights (M_w/M_n), and (2) the g-INDEX, w s.

Polyelectrolyte (polyion, ionomer) Any of several classes of polymers having fixed ionizable groups, such as polyacids (e g, polyacrylic acid), polybases (e g, polyvinyl trimethylammonium chloride), and sodium- or potassium-salt complexes of such polymers as polyethylene oxide. They are much more electrically conductive than ordinary plastics, their conductivities generally rising with temperature. Some are finding use in battery separators, photoelectrochemical cells, and humidity sensors. An allied class is described under IONOMER, w s.

Polyester (alkyd) A general term encompassing all polymers in which the main polymer backbones are formed by the esterification condensation of polyfunctional alcohols and acids. The coined term *alkyd* (see ALKYD RESIN) is synonymous, chemically, with polyester. However, as more commonly used in the plastics industry, alkyd refers to polyesters modified with oils or fatty acids that are crosslinkable (see ALKYD MOLDING COMPOUND). The term polyester is explained further under POLYESTER, SATURATED and POLYESTER, UNSATURATED.

Polyester, Aromatic See POLY-*p*-HYDROXYBENZOIC ACID.

Polyester Fiber Generic name for a manufactured fiber in which the fiber-forming substance is any long-chain synthetic polymer composed of at least 85% by weight of an ester of a dihydric alcohol and terephthalic acid (Federal Trade Commission). The polyester fiber in widest use throughout the world is derived from polyethylene terephthalate. Polyester filaments are produced by forcing the molten polymer at a temperature of about 290°C through spinneret holes about 0.23 mm (9 mils) in diameter, followed by air cooling, combining the single filaments into yarns, and drawing. The major end use of polyester fibers is in blends with cotton or wool to enhance crease retention and reduce wrinkling of garment fabrics. It is also used in carpeting and tire cords.

Polyester Plasticizer Any of a broad class of plasticizers characterized by having many ester groups in each molecule. They are synthesized from three components: (1) a dibasic acid such as adipic, azelaic, lauric, or sebacic acid, (2) a glycol (dihydric alcohol), and (3) a monofunctional chain terminator such as a monobasic acid. Molecular weights are low—from 500 to 5,000. Polyester plasticizers are noted for their permanence and resistance to extraction.

Polyester, Saturated Any polyester in which the polyester backbone has no double bonds. The class includes low-molecular-weight liquids used as plasticizers and as reactants in forming urethane polymers; and linear, high-molecular-weight thermoplastics such as polyethylene terephthalate. Usual reactants for the saturated polyesters are (1) a glycol such as ethylene-, propylene-, diethylene-, dipropylene-, or butylene glycol, and (2) an acid or anhydride such as adipic, azelaic, or terephthalic acid or phthalic anhydride. Some saturated, branched polyesters are used in high-temperature varnishes and adhesives.

Polyester, Unsaturated A polyester family characterized by ethenic unsaturation in the polyester backbone that enables subsequent hardening or curing by copolymerization with a reactive monomer in which the polyester constituent has been dissolved. Unsaturated polyesters are made by agitating in a heated kettle a mixture of glycols, e g, propylene- or diethylene glycol; unsaturated dibasic acids or anhydrides, e g, fumaric acid or maleic anhydride; and, sometimes in order to control the reaction and modify properties, a saturated dibasic acid or anhydride, e g, isophthalic acid or phthalic anhydride. After removal of water and cooling, the fluid polyester may be dissolved in a reactive monomer in the same kettle, or it may be shipped to users who add the monomer and catalyst in their plants. Styrene is most widely used as the reactive monomer. Others sometimes used are diallyl phthalate, diallyl isophthalate, and triallyl cyanurate. A peroxide catalyst is generally used for the final copolymerization. These unsaturated polyesters are thermosetting and are most widely used in reinforced plastics for making boat hulls, trays, containers, and panels, and in potting of electrical assemblies. See also WATER-EXTENDED POLYESTER.

Polyether (1) Any polymer having the general structure $(-R-O-)_n$, where R may be simple or more elaborate. [Technically, polyoxymethylene, $(-CH_2-O-)_n$, is a polyether, though known as an ACETAL RESIN (w s) in the industry.] POLYPHENYLENE OXIDE (w s) is a well-known polyether. (2) A low-molecular-weight polymer containing hydroxyl end groups, used as a reactant in the production of polyurethane foams. One type of polyether, widely used for rigid foams, is obtained by reacting propylene oxide with a polyol initiator such as a glycol glycoside in the presence of potassium hydroxide as a catalyst.

Polyether, Chlorinated See CHLORINATED POLYETHER.

Polyetheretherketone (PEEK) An "advanced" polymer whose chain structure is shown below.

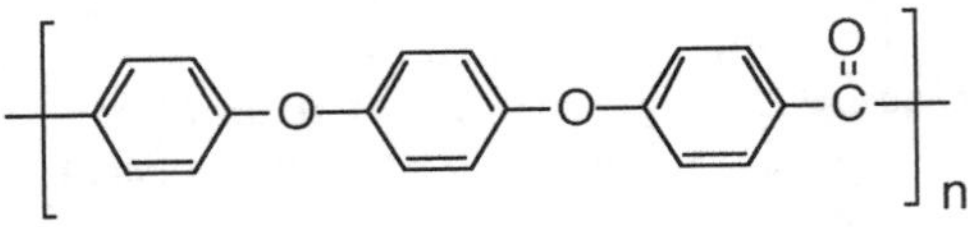

It has excellent temperature resistance among processable thermoplastics, with a melting temperature of 334°C, deflection temperature at 1.8 MPa of 160°C, and tensile yield strength of 91 MPa. Reinforcement with 30% glass fiber elevates the deflection temperature to about 300°C and almost doubles the yield strength.

Polyether Foam A type of POLYURETHANE FOAM, w s, that has been made by reacting an isocyanate with a polyether rather than a polyester or other resin component. For rigid foams, polyethers often used are the propylene oxide adducts of materials such as sorbitol, sucrose, aromatics, diamines, pentaerythritol, and methyl glucoside. These range in hydroxyl numbers from 350 to 600. For flexible foams, polyethers with hydroxyl numbers ranging from 40 to 160 are used. Examples are condensates of polyhydric alcohols such as glycerine, sometimes containing small amounts of ethylene oxide to increase reactivity.

Polyetherimide One of the "advanced" thermoplastics, having both ether links and imide groups in its chain, as shown below.

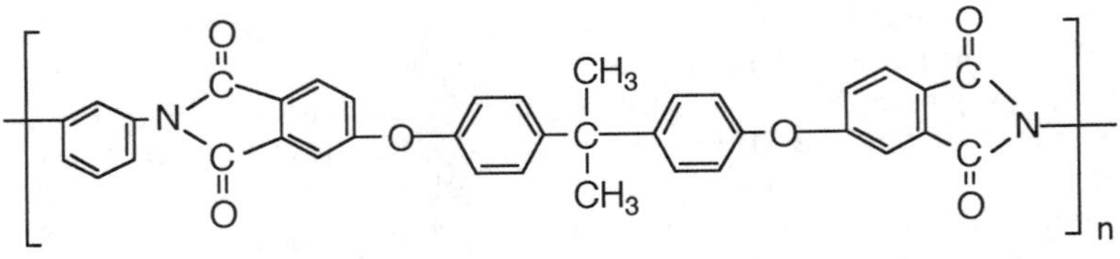

Deflection temperature at 1.8 MPa is 199°C, tensile modulus is 3.0 GPa, strength is 96 MPa, and the resin has good fire resistance.

Polyetherketone An "advanced" thermoplastic resin having both ether and ketone linkages in its chains, a close relative of POLYETHERETHERKETONE, above, and having the PEEK structure with the leftmost phenyl and ether oxygen deleted. This melt-processable polymer melts near 365°C, is fire-resistant, has good resistance to chemicals, and can be used at temperatures comparable to those for PEEK.

Polyethersulfone An "advanced" thermoplastic consisting of repeating phenyl groups (ϕ) linked by thermally stable ether and sulfone ($-SO_2-$) groups, its structure being like that of PEEK, above, but with the right-hand –O–ϕ–CO– section replaced by sulfone. The resin has good transparency and flame resistance, and has one of the lowest smoke-emission ratings among plastics. Both neat and reinforced grades are available in granule form for extrusion and molding. Unreinforced grades are used in high-temperature electrical applications, bakery-oven windows, and medical components. Reinforced grades are used for radomes, structural aircraft and aerospace components, and corrosion-resisting applications in packaging and chemical-plant hardware.

Polyethylene (PE, polyethene; in Britain, polythene) A huge family of resins obtained by polymerizing ethylene gas, $H_2C{=}CH_2$, and by far the largest-volume commercial polymer. Almost 10 Tg ($11 \cdot 10^6$ tons) was sold in the US in 1992, about one-third of all US resin sales. By varying the catalysts and methods of polymerization, properties such as density, melt-flow index, crystallinity, degree of branching and cross-linking, molecular weight and polydispersity can be regulated over wide ranges. Further modifications are created by copolymerization, chlorination, and compounding additives. Low-molecular-weight polymers of ethylene are fluids used as lubricants; medium-weight polymers are waxes miscible with paraffin; and the polymers with molecular weights over 10,000 (to which the above sales figure applies) are the familiar tough and strong resins, flexible or stiff, to make a myriad of products, both consumer and industrial. Polymers with densities ranging from about 0.910 to 0.925 g/cm^3 are called *low-density* polyethylene; those with densities from 0.926 to 0.940 are called *medium-density*; and those with densities from 0.941 to 0.965 and over are called *high-density* polyethylene. The low-density resins are polymerized at very high pressures and temperatures, and the high-density ones at lower pressures and temperatures, using special catalysts. Two newer types are *extra-high-molecular-weight* (EHMWPE) materials in the MW range from 150,000 to 1,500,000, and *ultra-high-molecular-weight* (UHMWPE) materials in the 1,500,000 to 3,000,000 range. Because UHMWPE does not melt and flow, it is processed by powder-molding and sintering techniques developed decades ago for polytetrafluoroethylene. Under carefully controlled conditions some EHMWPEs can be extruded, blow molded, and thermoformed on standard equipment. When fully crosslinked by irradiation or by the use of chemical additives, polyethylene is no longer a thermoplastic, and has superior strength, impact resistance, and electrical properties. A still newer member of the family, much used in grocery bags, is LINEAR LOW-DENSITY POLYETHYLENE, w s. Another new subfamily are the VERY-LOW-DENSITY POLYETHYLENEs, w s.

Poly(Ethylene-Chlorotrifluoroethylene) (PE-CTFE, ECTFE copolymer) A high-molecular-weight, 1:1 alternating copolymer of ethylene and chlorotrifluoroethylene. Available in pellet and powder form, PE-CTFE can be extruded, injection, transfer, and compression molded, rotocast and powder coated. It is a strong, highly impact-resistant material that retains useful properties over a wide temperature range. Good electrical properties and chemical resistance make it useful in electrical and chemical ware and in packaging applications requiring corrosion resistance.

Polyethylene Foam Low-density-PE foam, with foam densities as low as 0.03 g/cm^3, are made by thoroughly mixing a blowing agent with hot, molten polymer under pressure, then releasing the pressure and cooling. Foams are also made by extrusion, using pellets containing a heat-triggered foaming agent. Crosslinked PE foam is made by blending a peroxide crosslinking agent with the molten compound, then subsequently vulcanizing the molded shapes in a press. The denser foams have found application in packaging of electronic equipment.

Polyethylene Glycol Any of a family of polymers of ethylene glycol with molecular weights in the range from 200 to 8000, ranging from water-clear liquids to hard, waxy solids. They are used as plasticizers for polyvinyl alcohol, as intermediates, and in printing inks and mold releases.

Polyethylene Glycol (200) Dibenzoate $C_6H_5CO(OCH_2CH_2)_4OCO-C_6H_5$. A plasticizer compatible with cellulose acetate butyrate, ethyl cellulose, polymethyl methacrylate, polystyrene, and vinyl resins. Its major application is with phenol-formaldehyde resins in laminating applications, to improve flexibility without loss of electrical properties and high-temperature capability.

Polyethylene Glycol (600) Dibenzoate A plasticizer similar to the preceding one but with 13 $-OCH_2CH_2-$ groups, and with only partial compatibility with the resins listed for that one.

Polyethylene Glycol Di-2-Ethylhexoate A plasticizer for most cellulosic plastics, polymethyl methacrylate, polystyrene, and vinyls.

Polyethylene Glycol (400) Dilaurate A plasticizer for cellulose nitrate, PVC, and vinyl copolymers.

Polyethylene Glycol Terephthalate A longer name for POLYETHYLENE TEREPHTHALATE, w s.

Polyethylene Oxide (PEO) Low-molecular-weight polymers of ethylene oxide are viscous liquids or waxes. Those of high molecular weight are tough, highly crystalline, ductile thermoplastics that can be processed by molding, extrusion, etc. All PEO resins are soluble in water, and thus are used in the form of packaging film for powdered detergents, insecticides, and other household, industrial and agricultural products that are dissolved in water prior to use. The film is heat-sealable and permeable to gases.

Polyethylene-Propylene Adipate Glycol (PEPAG) A polymeric diol used in the production of urethane elastomers (Witco Corp, Formrez F 10-91).

Polyethylene Terephthalate (PET, polyethylene glycol terephthalate) A saturated, thermoplastic polyester resin made by condensing ethylene glycol and terephthalic acid, used for textile fibers, water-clear, biaxially oriented film (e g, Mylar®) and, more recently, for extruded, thermoformable sheet (TV-dinner trays), injection-molded parts, and large, blow-molded, soft-drink bottles. It is extremely hard, wear- and chemical-resistant, dimensionally stable, and has good dielectric properties. See also POLYESTER, SATURATED and CRYSTALLIZED POLYETHYLENE TEREPHTHALATE.

Poly(Ethylene-Tetrafluoroethylene) (PE-TFE) A crystalline resin in which the proportion of ethylene to tetrafluoroethylene (E/TFE) may range, for best combination of properties, between 2:3 and 3:2, modified with a vinyl copolymer for better toughness. It is stronger than either low-density polyethylene or polytetrafluoroethylene, has good electrical properties, high Izod-impact strength, and plastic memory that makes it useful for heat-shrinkable packaging.

Polyformaldehyde See ACETAL RESIN and PARAFORMALDEHYDE.

Polyglycidyl Polyepichlorohydrin Resin Any of a family of epoxy resins derived from epichlorohydrin and hydroxyl compounds, possessing flexibility and flame-retarding characteristics. They may be cured by themselves, or mixed with conventional epoxy resins to impart their favorable characteristics to laminates.

Polyglycol Distearate (polyethylene glycol distearate) The di(stearic acid) ester of polyglycol, used as a plasticizer.

Polyhexafluoropropylene A fully fluorinated polymer based on the gas $CF_3CF{=}CF_2$, not commercial. However, the copolymers of hexafluoropropylene and tetrafluoroethylene make up the family of FLUORINATED ETHYLENE-PROPYLENE RESINs, w s.

Polyhexamethyleneadipamide Explicit synonym for NYLON 6/6, w s.

Polyhexamethylenesebaçamide Explicit name for NYLON 6/10, w s.

Polyhexamethyleneterephthalamide Explicit name for NYLON 6/T, w s.

Polyhydric Alcohol Synonym for POLYOL, w s.

Poly-*p*-Hydroxybenzoic Acid A homopolyester of repeating *p*-oxyenzoÿl units with a high degree of crystallinity. It does not melt below its decomposition temperature, 550°C, but can be fabricated at 300 to 360°C by compression sintering and plasma-spray processes. Copolymers with aromatic dicarboxylic acids and aromatic bisphenols are processable by normal means. Applications include electrical connectors, valve seats, high-performance-aircraft parts, and automotive parts.

Polyhydroxyether Resin See PHENOXY RESIN.

Polyimidazopyrrolone (ladder pyrone, polypyrrolone) An aromatic, heterocyclic polymer that results from the reaction of an aromatic dianhydride with a tetramine. Due to the double-chain or ladder-like structure, these polymers have outstanding resistance to radiation, chemicals, and heat (no weight loss to 550°C!). However, this structure also makes them difficult to process. To overcome this difficulty pyrrone prepolymers in the form of solutions and salt-like powders have been made available. The powders can be molded under conditions that complete the cyclization or conversion to the ladder-like molecular structure during the molding cycle. The cyclization reaction generates water, which must be removed from the part.

Polyimide A polymer formed by the condensation of an organic anhydride or dianhydride with a diamine, in some cases followed by thermal dehydration (curing). The early polyimides from pyromellitic anhydride and aromatic diamines, when fully cured, had extremely high thermal stability but were unmeltable and required special processing. Later, addition-type polyimides based on reacting maleic anhydride and 4,4'-methylenedianiline were developed. These are processable by conventional thermoset molding, film casting, and solution-fiber techniques. Molding compounds filled with lubricating fillers or fibers produce parts with self-lubricating wear surfaces. Thermoplastic polyimide reinforced with glass, boron, or graphite fibers can be molded into high-strength structural components. Polyimide solutions are used as laminating varnishes to produce radomes, printed-circuit boards, and other components requiring fire resistance, good electrical properties, and strength at high temperatures. Printed-circuit boards of polyimide-glass laminate handily endure high-temperature soldering. Recently, the introduction of thermoplastic polyimides containing aromatic rings in the polymer backbone and trifluoromethyl side groups has opened these materials to a wider field of applications because of improved processability. Film has been used as insulation in electric motors, magnet wire and missile wiring, and in dielectric applications.

Polyimide Fiber Polyimides prepared by reacting diisocyanates with dianhydrides, developed by the Upjohn Co, have been used to spin fibers by the wet or dry process. They have good thermal-oxidative stability, flame resistance, and ultraviolet stability.

Polyimide Foam A family of polyimide-precursor powders enables the production of flexible and rigid polyimide-foam structures. These powders are poured into molds and heated until sufficient integrity for removal is attained, then they are subsequently cured at 300°C.

Polyisobutene See POLYBUTENE.

Polyisobutylene Any of a family of polymers of isobutylene, $(CH_3)_2C{=}CH_2$, for which the IUPOAC name is *2-methylpropene*. Depending on molecular weight, they range from oily liquids to elastomeric solids. The higher-molecular-weight polymers are used as impact-resistance improvers in polyethylene and other plastics. The liquid polymers are used as tackifying agents in adhesives.

Polyisobutylvinyl Ether (polyvinylisobutyl ether) Any polymer of isobutylvinyl ether. Some are liquids, others are solid and crystalline. They are used as adhesives, surface coatings, laminating agents, and filling compounds in electrical cables.

Polyisocyanate See ISOCYANATE.

Polyisocyanurate (PIR) A polymer containing isocyanurate rings,

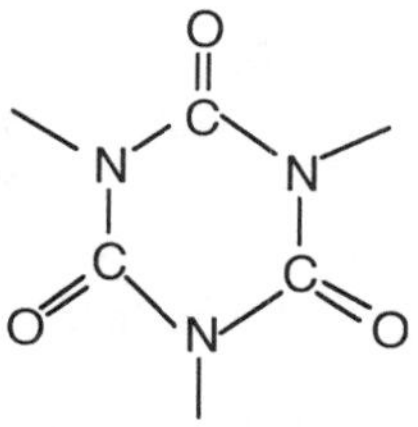

i e, isocyanate trimer, and forming foams that have better fire resistance than rigid polyurethanes, but are more brittle, so are often used in mixtures with the latter.

Polyisoprene A polymer of ISOPRENE, w s. The *cis*-1,4- type of polyisoprene occurs naturally as the major polymer in natural rubber, and is also produced synthetically. The *trans*-1,4- type resembles GUTTA-PERCHA, w s, and has in the past been used in golf-ball covers and shoe soles.

Polyisoprene, Deutero A polyisoprene in which heavy hydrogen (deuterium) atoms have replaced the ordinary hydrogen atoms. The *cis*-1,4-deuteropolyisoprene is more elastic than natural rubber.

Polylauryllactam See NYLON 12.

Polyliner A perforated, longitudinally ribbed sleeve that fitted snugly inside the cylinder of a ram-type injection-molding machine, replacing the conventional torpedo. It improved the heat transfer, plastifying rate, and uniformity of melt temperature at the nozzle.

Polymer Blend A physical mixture of two or more polymers and possible additives, achieved by kneading or by high-intensity mixing of fine powders. Because hot working can cause chain scission in some polymers, some grafting of the component polymers is likely to occur during such operations. Blends are made to take advantage of synergistic gains in properties, some of which may be better than the same properties of the blend components alone. Often, costly resins of outstanding properties are blended with cheaper, compatible ones to achieve blends of intermediate properties at a cost lower than that of the more expensive member. Melt viscosities of blends may lie, at a given shear rate and temperature, between those of the separate components, or above or below both of them.

Polymer Concrete A composite material consisting of graded aggregates with an organic binder or mixed organic and inorganic binders. Epoxy and other resins have been used, in contents between 8 and 20 percent. Compressive and flexural strengths are several times those of Portland-cement concretes, they are impervious to liquids, and can be made to look like granite or marble. However, cost is about five times that of Portland concrete, so polymer concretes, so far, have been limited to special uses.

Polymeric Modifier A term applied to any polymer that is blended with the principal polymer to alter the latter's characteristics. See also ELASTICIZER, IMPACT MODIFIER, and POLYMERIC PLASTICIZER.

Polymeric Plasticizer The term refers to plasticizers with molecules containing repeating mers and much larger than those of monomeric plasticizers that comprise virtually all other classifications of plasticizers. The two main types of polymeric plasticizers are the epoxidized oils of high molecular weight and POLYESTER PLASTICIZERs, w s. Polymeric plasticizers are noted for their permanence, which is due to the reduced tendency of the larger molecules to migrate and evaporate. However, the viscosity rises and the low-temperature properties of polymeric plasticizers decrease as their molecular weights increase. In cold weather, the high-molecular-weight polymerics may be difficult to handle and pump.

Polymeric Polyisocyanate A generic term for a family of isocyanates derived from aniline-formaldehyde condensation products, used as reactants in the production of polyurethane foams.

Polymeric Sulfur Nitride See SULFUR NITRIDE POLYMER.

Polymerizable Plasticizer See PLASTICIZER, POLYMERIZABLE.

Polymerization A chemical reaction in which the molecules of a simple substance (*monomer*) link together to form large molecules whose molecular weights are multiples (exact or nearly so) of that of the monomer. There are two general types of polymerizations, both with many variations: *addition polymerization*, which occurs when reactive, unsaturated monomers unite without forming any other products; and *condensation polymerization*, which occurs by combining of reactive end groups, accompanied by the elimination of a simple molecule such as water. Examples of condensation polymers are nylons and phenolic resins. The majority of thermoplastics, aside from polyamides and polyesters, and a few thermosets, are made by addition polymerization, in which a pair of shared electrons in each monomer molecule is utilized to link the separate molecules into long chains. Polymerization processes and related terms are defined under the following headings.

ADDITION POLYMERIZATION
AUTOACCELERATION
ALTERNATING COPOLYMER
BEAD POLYMERIZATION

BLOCK COPOLYMER
BRANCHING
BULK POLYMERIZATION
CHAIN-TRANSFER AGENT
CONDENSATION POLYMERIZATION
CROSSLINKING
EMULSION POLYMERIZATION
FREE-RADICAL POLYMERIZATION
GAS-PHASE POLYMERIZATION
GRAFT COPOLYMER
INTERFACIAL POLYMERIZATION
IONIC POLYMERIZATION
ISOTACTIC
NETWORK POLYMER
OXIDATIVE COUPLING
PHOTOPOLYMERIZATION
PRECIPITATION POLYMERIZATION
RADIATION POLYMERIZATION
RANDOM COPOLYMER
REDOX
SOLID-STATE POLYMERIZATION
SOLUTION POLYMERIZATION
STEREOBLOCK POLYMER
STEREOGRAFT POLYMER
STEREOREGULAR POLYMER
STEREOSPECIFIC
SUSPENSION POLYMERIZATION
SYNDIOTACTIC
THERMAL POLYMERIZATION

Polymer, Natural A substance of high molecular weight occurring naturally, consisting of molecules that are, at least approximately, multiples of simple units. Natural polymers are often regarded as organic, but many inorganic minerals such as quartz, feldspar, and asbestos are considered to be entirely or substantially polymeric. Examples of natural organic polymers are cellulose, natural rubber, lac, proteins such as collagen and keratin, and many natural fibers.

Polymerography (resinography) The use of microscopic and metallographic techniques in the study of polymers.

Polymer Structure (1) A general term referring to the relative positions, arrangements in space, freedom of motion of atoms in a polymer molecule, and orientation of chains. Such structural details have important effects on polymer properties such as the second-order-transition temperature, flexibility, and tensile strength. (2) The microstructure of a polymer, as observed by light- or electron-microscopic techniques, and including crystalline structure, birefringence, distribution of sizes of filler particles and spherulites, and distribution of reinforcement directions. These, too, have important influences on macroscopic properties and behavior.

Polymer, Synthetic The product of a polymerization reaction whose starting materials are one or more *monomers*. See POLYMERIZATION. When a single monomer is used, the product is called a *homopolymer*, *monopolymer*, or simply a polymer. When two monomers are polymerized simultaneously one obtains a *copolymer*. The term *terpolymer* is used for the polymerization product of three monomers. However, the term *heteropolymer* is also used for terpolymers as well as for products of more than three monomers. When no monomer is used, the product is known as a *nonomer*. The terms *polymer, resin, high polymer, macro-*

molecular material, and *plastic* are often used interchangeably, although *plastic* also refers to compounds containing major additives. Note: The definition approved by IUPAC and ISO for polymer is "a substance composed of molecules characterized by the multiple repetition of one or more species of atoms or groups of atoms (constitutional units) linked to each other in amounts sufficient to provide a set of properties that do not vary markedly with the addition or removal of one or a few constitutional units." A polymer may be amorphous or may contain crystalline structures up to and exceeding half its specific volume. In a given polymer, the crystalline regions are always more dense than the amorphous ones; thus, percent crystallinity can be estimated from density.

Polymethacrylate A polymer of a methacrylic ester, polymethyl methacrylate being the most important and useful member of the class.

Polymethacrylonitrile (PMAN) A thermoplastic obtained by the polymerization of methacrylonitrile, a vinyl monomer containing the nitrile group. The homopolymer has good mechanical strength and high resistance to solvents, acids, and alkalis, but discolors at molding temperatures.

Polymethoxy Acetal (PMAC) Any oligomer of methoxy dimethyl acetal with degree of polymerization in the range 3 to 10. These oligomers are high-boiling, yellowish liquids used as modifiers for phenolic resins, and as solvents and plasticizers.

Polymethyl Acrylate A polymer of methyl acrylate, having a glass-transition at 10°C, a leathery, tough material used in textile and leather finishing.

Polymethylene A polymer first made by polymerizing diazomethane (also called azomethylene) (CH_2N_2), with evolution of nitrogen gas. While this polymer has the same formula as polyethylene, it contains no side chains, so provides a standard with which branched ethylene polymers may be compared. Although long known in the laboratory, it is not a commercial resin.

Polymethyl Methacrylate [PMMA, poly(methyl 2-methyl propenoate)] The most important member of the family of acrylic resins, made by addition polymerization of the monomer, methyl methacrylate [$CH_2=C(CH_3)COOCH_3$]. Two outstanding characteristics of PMMA are its optical clarity (92% light transmission) and unsurpassed resistance to weathering. It also has good electrical properties, the ability to "pipe" light around bends, and is tasteless, odorless, and nontoxic. PMMA molding powders can be injection molded, extruded, and compression molded. The liquid monomer can be cast into rods, sheets, optical lenses, etc. Cast and extruded PMMA sheets are fabricated and thermoformed into many products such as aircraft canopies, skylights, lighting fixtures, and outdoor signs. See also ACRYLIC RESIN.

Poly(4-Methylpentene-1) (PMP) A polyolefin first introduced commercially in 1966 by Imperial Chemical Industries but now produced only by Mitsui Plastics ("TPX") and Phillips 66 Co ("Crystalor"). The monomer, 4-methylpentene-1, is produced by dimerization of propylene. Polymerization is conducted with a Ziegler-type catalyst. The polymers are supplied as free-flowing powders or as compounded granules, suitable for the usual thermoplastics processes. Properties of the resins are high light transmission (93%, better than many glasses), melting range near 240°C, rigidity and tensile properties similar to those of polypropylene, good electricals, high chemical resistance, and the lowest density of all commercial solid resins, 0.834 g/cm^3. It is approved for food contact. These properties account for its use in laboratory volumeware, sight glasses, high-frequency electrical components, coffee funnels, wire coatings, and microwave-safe cookware.

Poly(Monochloro-*p*-Xylylene) (Parylene C) See PARYLENE.

Polymorphism The ability of a crystalline substance to exist in two or more forms of crystalline structure.

Polyoctanoamide Synonym for NYLON 8, w s.

Polyol (*poly*hydric alcoh*ol*, *poly*alcoh*ol*) An organic compound having more than one hydroxyl (–OH) group per molecule. In the cellular plastics industry, the term includes monomeric and polymeric compounds containing alcoholic hydroxyl groups such as polyethers, glycols, glycerol, and polyesters, used as reactants in polyurethane foam.

Polyolefin Any of the largest genus of thermoplastics, polymers of simple olefins such as ethylene, propylene, butenes, isoprenes, and pentenes, and copolymers and modifications thereof. The two most important are polyethylene and polypropylene, which, together, accounted for just under half of all US resin sales in 1992. Polyolefin plastics are most usually processed into end products by extrusion, injection molding, blow molding, and rotational molding. Thermoforming, calendering, and compression molding are used to a lesser degree. An inherent characteristic common to all polyolefins is a nonpolar, nonporous, low-energy surface that is not receptive to inks, lacquers, etc, without special oxidative pretreatment. See ETHYLENE-PROPYLENE RUBBER, ETHYLENE-VINYL ACETATE COPOLYMER, IONOMER RESIN, POLYALLOMER, POLYBUTENE, POLYETHYLENE, POLYISOPRENE, POLY(4-METHYLPENTENE-1), and POLYPROPYLENE.

Polyorganophosphazene A polymer obtained by reaction of phosphorus pentachloride and ammonium chloride. A cyclic trimer, $(NPCl_2)_3$, or tetramer, $(NPCl_2)_4$, is formed which can be converted to polyorganophosphazenes, $(-N{=}PR_2-)_n$ where R represents an organic side group. The polymers have found some use in hose, gaskets, and seals in aviation-fuel-handling equipment. They have better solvent resistance and low-temperature elasticity than siloxane-carborane polymers, and are less costly.

Polyoxamide Generic name for nylon-type materials made from oxalic acid and diamines. Their extremely high melting temperatures have kept them out of commerce.

Polyoxetane See CHLORINATED POLYETHER.

Poly-*p*-Oxybenzoÿl See POLY-*p*-HYDROXYBENZOIC ACID.

Polyoxymethylene (POM) Linear polymers of formaldehyde or oxymethylene glycol with the formula $(-OCH_2-)_n$, in which n is above 100. Those in the range $100 < n < 300$ are brittle solids used as intermediates. Those in the range $500 < n < 5000$ are ACETAL RESINs, w s.

Polyoxypropylene Glycol Any low-molecular-weight polymer with the structure $H[-OCH(CH_3)CH_2-]_nOH$, derived from propylene oxide and used in the production of polyurethane foams.

***Trans*-Polypentenamer** An elastomer obtained by the polymerization of cyclopentene, using complex catalysts. Its structure is highly linear and the molecular weight has a wide range. Its properties are similar to those of natural rubber and *cis*-polybutadiene.

Polyphenone A phenolic-like material developed in the early 1970s by Union Carbide, but still not commercial. Unlike phenolic, it was to be available in a range of light colors, with good moldability and electrical and physical properties equal to those of mineral-filled phenolics.

Polyphenylene Benzobisthiazole (PBT, PBZ) A liquid-crystalline polymer from which very strong and heat-resistant fibers are made.

Poly-1,3-Phenylenediamine Isophthalate A high-temperature fiber, trade named Nomex® by Du Pont. This fiber resists common flame temperatures around 500°C for a short time and thus is suitable for fire-protective clothing and insulation of motors and transformers.

Polyphenylene Oxide (PPO, poly-2,6-dimethyl-1,4-phenylene oxide) A thermoplastic, linear, noncrystalline polyether obtained by the oxidative polycondensation of 2,6-dimethylphenol in the presence of a copper-amine-complex catalyst. The resin has a wide useful temperature range, from below −170° to +190°C, with intermittent use to 205°C possible. It has excellent electrical properties, unusual resistance to acids and bases, and is processable on conventional extrusion and injection-molding equipment. Because of its high cost, PPO is also marketed in the form of polystyrene blends (see NORYL®) that are lower-softening (T_g of PS is about 100°C vs 208°C for PPO), and have working properties intermediate between those of the two resins.

Polyphenylene Sulfide (PPS) A crystalline polymer having a symmetrical, rigid backbone chain consisting of recurring *p*-substituted benzene rings and sulfur atoms. A variety of grades suitable for slurry coating, fluidized-bed coating, electrostatic spraying, as well as injection and compression molding are offered (Phillips 66 Co's Ryton® and others). The polymers exhibit outstanding chemical resistance, thermal stability, and fire resistance. Their extreme inertness toward organic solvents, and inorganic salts and bases make for outstanding performance as a corrosion-resistant coating suitable for contact with foods. Doping with arsenic pentafluoride imbues the resin with usefully high electrical conductivity.

Poly-(*p*-Phenylene Sulfone) (PPSU) Chemically similar to the polysulfones, this high-performance polymer has better impact resistance. It also has excellent heat resistance, low creep and good electrical properties. It is difficult to process, however, which has limited its commercial acceptance.

Polyphenylquinoxaline (PPQ) Any of a family of high-performance thermoplastics that have potential for use as functional and structural resins in applications demanding high chemical and thermal stability. The most attractive synthesis is by copolycondensation of an aromatic bis(*o*-diamine) powder and a stirred solution or slurry of bis(1,2-dicarbonyl) monomer in an appropriate solvent such as a mixture of *m*-cresol and xylene. In solution form, the polymers can be used directly for prepreg and adhesive-tape formulations, film casting, etc. If desired, the polymer can be isolated from solution and compression molded. It is convertible to a thermoset form by rigidizing the linear polymer backbone with reactive latent groups and by crosslinking.

Polyphosphazene A family of inorganic-base polymers having phosphorus-nitrogen backbones joined with fluorine or chlorine. Depending on which organic side groups are linked to the backbones, a wide variety of polymers can be made with properties ranging from rigid and flexible thermoplastics and elastomers to glass-like thermosets. Some grades outperform silicones in biomedical uses. See also PHOSPHAZENE POLYMER.

Polyphosphazene Fluoroelastomer Any of a family of elastomers developed primarily for fuel tanks to be used in the Arctic, having the typical chain-unit configuration $(CF_3CH_2O)_2PN(CHF_2C_3F_6CH_2O)_2PN$. These elastomers are inert to aviation fuel and remain flexible to –60°C, lower than other elastomers previously used in this application.

Polyphthalamide (PPA) A polyamide in which the residues of terephthalic or isophthalic (or mixed) acid components are part of the mer unit of the chain. PPA is an advanced engineering polymer first commercially offered in 1991 by Amoco Chemical Co under the trade name Amodel®.

Polypivalolactone A crystalline thermoplastic polymerized by ring opening from the cyclic monomer, $(CH_3)_2CCOOCH_2$. It is useful for making high-strength fibers, also as a high-crystallinity (75%) matrix resin with carbon fibers.

Polypropylene (PP, polypropene) Any of several types of a large family of thermoplastic resins made by polymerizing propylene with suitable catalysts, generally solutions of aluminum alkyl and titanium tetrachloride. Its density (approximately 0.905 g/cm^3) is among the lowest of all plastics. PP and copolymers enjoyed the third largest sales in the US in 1992, 3.8 Tg ($4.2 \cdot 10^6$ tons), about 13% of all US resins sales. As with the POLYETHYLENEs, w s, properties of the polymers vary widely according to molecular weight, method of production, and copolymers involved. The grades used for molding have molecular weights of 40,000 or more, are 90 to 95% isotactic with about 50% crystallinity. They have good resistance to heat, chemicals, and solvents, and good electrical properties. Properties can be improved by compounding with fillers, e g, mica or glass fibers; by blending with synthetic elastomers, e g, polyisobutylene; and by copolymerizing with small amounts of other monomers. Fibers are the single largest use of polypropylene. See SPLIT-FILM FIBER.

Polypropylene Adipate (polypropylene glycol adipate) A polymeric plasticizer for vinyl chloride polymers and copolymers formed by reacting propylene glycol and adipic acid.

Polypropylene Glycol (PPG, polypropylene oxide) A family of nonvolatile liquids with the general formula $HOCH(CH_3)[-CH_2CH-(CH_3)O-]_nCH_2OH$. They are similar to the polyethylene glycols, but are more oil-soluble and less water-soluble. They are polyols used in producing polyurethane foams, adhesives, coatings, and elastomers.

Polypyromellitimide (PPMI) The original polyimide family of polymers, having enhanced heat resistance and formed from polyamide carboxylic acids derived by reacting pyromellitic dianhydride with 4,4'-diaminophenyl ether. Grades of the polymer are used for forming films, paint components, and are processable as thermoplastics under special conditions. More easily processed copolymers have enjoyed greater commercial success.

Polysilane A polymer whose backbone is composed of covalently linked silicon atoms with organic side groups that may be aliphatic, aromatic, or mixed, and not to be confused with *polysiloxane* (see SILICONE). Polysilanes have been used as lithographic resists.

Polysiloxane Synonym for SILICONE, w s.

Polystyrene An important family of workhorse plastics, the polymer of styrene (vinyl benzene), which has been commercially available for more than half a century. In 1992 polystyrene, neat and modified, accounted for about 10% of US resins sales, i e, 3.0 Tg ($3.2 \cdot 10^6$ tons). The homopolymer is water-white, has excellent clarity and sparkle, outstanding electrical properties, good thermal and dimensional stability, is hard, stiff, and resistant to staining, and is inexpensive ($1/kg in 12/92). However, it is somewhat brittle, and is often modified by blending or copolymerization to a desired mix of properties. High-impact grades (HIPS) are produced by adding rubber or butadiene copolymers. Heat resistance is improved by including some α-methyl styrene as a comonomer. Copolymerization with methyl methacrylate improves light stability, and copolymerization with acrylonitrile raises resistance to chemicals. Styrene polymers and copolymers possess good flow properties at temperatures safely below degradation ranges, and can easily be extruded, injection molded, or compression molded.

Polystyrene Foam (expanded polystyrene, EPS) A low-density, cellular plastic made from polystyrene by either of two methods. Extruded foam is made in tandem extruders, the first for plasticating the resin, the second to homogenize the blowing agent, which may be a gas or volatile liquid, such as nitrogen or pentane, and reduce the temperature of the melt before it reaches the die. As it emerges from the die the large drop in pressure frees the blowing agent and the mass expands to form a low-density "log" conveyed through a long cooling tunnel. The cooled slab is usually sliced into a large range of shapes marketed through building-materials dealers. In the other basic method, a volatile blowing agent, e g, isopentane, is incorporated into the tiny PS beads as they are polymerized, or afterward. The beads are first pre-expanded, allowed to "rest" for about a day, then molded in a closed, steam-heated mold, and finally cooled with water in the mold members. This method, which generally produces closed-cell foams, is used to mold finished products such as coffee cups, packaging components, and life-preserver rings. Beads are also used to generate very large, thick slabs ($6 \times 1.2 \times 0.6$ m) by blowing live steam into an expandable, low-pressure mold charged with measured weight of beads. After cooling, these slabs are sliced with multiple hot-wire cutters to produce foam "lumber", as with the extruded foam.

Polystyrol A rarely used term for POLYSTYRENE, w s.

Polystyrylpyridine (PSP) A thermosetting resin, resistant to high temperatures, formed by condensation of 2,4,6-trimethylpyridine and 1,4-benzyldialdehyde, and a useful matrix for carbon-fiber composites.

Polysulfide Rubber (T) A family of sulfur-containing polymers prepared by condensing organic polyhalides with sodium polysulfides in aqueous suspension. They range from liquids to solid elastomers. The first commercial polysulfide was Thiokol®A, polyethylene tetrasulfide, made from sodium tetrasulfide and ethylene dichloride. This elastomer had outstanding solvent resistance, but its poor mechanical properties and unpleasant odor limited its use to plasticizing acid-resistant cements. Modern T materials have the general structure $(-R-S_m-)_n$ where m is usually 2 to 4 and R is $(CH_2)_2$ or an ether group. These elastomers are used in hose, printing rolls, gaskets, and gas-meter diaphragms. Polysulfide products have excellent resistance to oils, solvents, oxygen, ozone, light, and weathering, and low permeability to gases and vapors.

Polysulfonate Copolymer (sulfonate-carboxylate copolymer) A family of transparent, thermoplastic polyesters, moldable at 250° to 300°C, and formed by reaction of a diphenol, generally bisphenol A, with an aromatic disulfonyl chloride and an aliphatic disulfonyl chloride or carboxylic acid chloride. These copolymers have good electrical and mechanical properties, and excellent resistance to hydrolysis and aminolysis.

Polysulfone (PSU, PPSU) A family of sulfur-containing thermoplastics, closely akin to POLYETHERSULFONE, w s, made by reacting bisphenol A and 4,4'-dichlorodiphenyl sulfone with potassium hydroxide in dimethyl sulfoxide at 130° to 140°C. The structure of the polymer is benzene rings or phenylene units linked by one or more of three different chemical groups—a sulfone group, an ether link, and an isopropylidene group. Each of these three linking components acts as an internal stabilizer. Polysulfones are characterized by high strength, very high service-temperature limits, low creep, good electrical characteristics, transparency, self-extinguishing ability, and resistance to greases, many solvents, and chemicals. They may be processed by extrusion, injection molding, and blow molding.

Polyterephthalate See POLYESTER, SATURATED, and TEREPHTHALATE POLYESTER.

Polyterpene Resin Any of several thermoplastic resins of low molecular weight obtained by polymerizing turpentine or its derivatives, e g, α- or β-pinene, in the presence of catalysts such as aluminum chloride or mineral acids. The amber-colored resins, ranging from viscous liquids to solids, are used as tackifiers, wetting agents, and modifiers in the manufacture of adhesives, paints and varnishes, and caulking and sealing compounds. They are compatible with natural and synthetic rubbers, polyolefins, alkyd resins, other hydrocarbon resins, and waxes. See also HYDROCARBON PLASTICS.

Polytetrafluoroethylene (PTFE) The oldest of the fluorocarbon-resin family, discovered in 1938 by R. J. Plunkett, developed by Du Pont and marketed under the trade name Teflon®. It is made by polymerizing tetrafluoroethylene, $F_2C{=}CF_2$, and is available in powder and aqueous-dispersion forms. PTFE, inert to virtually all chemicals, has a crystalline melting point of 327°C, though it does not truly liquefy. Molecular weights of commercial PTFE powders are very high, on the order of 10^6. It has a very low coefficient of friction on most surfaces, resists adhesion to almost any material unless strenuously pretreated, and has excellent electrical properties. Its nonstick character has long been evidenced by its everyday use as an interior coating in cooking utensils. Its inability to form a true melt long ago forced the development of special extrusion, molding, and calendering processes in which the PTFE powder is pressed, then sintered with heat. PTFE tape and film are made by skiving pressed (or extruded) and sintered rods. Continuous extrusion is accomplished by alternating strokes of two ram extruders feeding a single die block. PTFE's low modulus and tendency to creep under load can be substantially improved by addition of inorganic fillers or chopped glass fiber. See also FLUOROCARBON RESIN.

Polytetrahydrofuran (polytetramethylene ether, PTHF) A type of polyol made from tetrahydrofuran by ring opening, having the mer [$-CH_2(CH_2)_3-O-$] and $-OH$ end groups, with low to moderate molecular weights. These polymers have long been used as prepolymers for polyurethane elastomers.

Polytetramethyleneadipamide Synonym for NYLON 4/6, w s.

Poly(Tetramethylene Terephthalate) Synonym for POLYBUTYLENE TEREPHTHALATE, w s.

Polythene The British name for POLYETHYLENE, w s.

Polythiazyl See SULFUR NITRIDE POLYMER.

Polytrifluorostyrene A clear, thermoplastic material introduced in 1965 and said to combine the oxidation resistance of polytetrafluoroethylene with the mechanical and electrical properties and ease of processing of polystyrene, but still not commercially available in 1992.

Polyurethane (PUR) A large family of polymers with widely ranging properties and uses, all based on the reaction product of an organic isocyanate with compounds containing a hydroxyl group, and having the $-RNHCOOR'-$ group in their chains. The reaction product of an isocyanate with an alcohol is called a *urethan* according to the rules of chemical nomenclature, but the terms *urethane* and *polyurethane* are almost universally used in the plastics industry. The types and properties of polyurethanes are so varied that they have been dubbed the "erector set" of the plastics industry. They may be thermosetting or thermoplastic, rigid and hard or flexible and soft, solid or cellular; and the properties of any of these types may be varied within wide limits to suit the desired application. Members of this family are described under POLYURETHANE ELASTOMER, POLYURETHANE RESIN, SPANDEX, URETHANE COATING, POLYURETHANE FOAM, IONIC POLYURETHANE, and INSTANT-SET POLYMER.

Polyurethane Elastomer (PU) Any condensation polymer made by reacting an aromatic diisocyanate with a polyol that has an average molecular weight greater than about 750. The aromatic diisocyanates usually employed are toluene diisocyanate (TDI) and diphenylmethane diisocyanate (MDI). The polyol component is either a polyester or polyether. In the "one-shot" system the elastomer is prepared directly in the mold that shapes the final product. The diisocyanate, polyol, and catalyst are rapidly and intimately mixed, then immediately poured or pumped into the mold. Prepolymers, in the form of liquids or low-melting solids, are used when a longer working time is desired. The prepolymers are mixed with catalysts, heated when necessary, degassed, and poured into the molds. Also available are millable gums and pellets containing all components, which have been reacted to a degree that permits further processing by methods used for rubber, including injection molding, compression molding, and transfer molding. The polyurethane elastomers most widely used are harder than natural rubber, and possess excellent resistance to flexural fatigue, abrasion, impact, oils and greases, oxygen, ozone, and radiation, but are susceptible to hydrolysis.

Polyurethane Foam (urethane foam, isocyanate foam) This family of foams differs from other cellular plastics in that the chemical reactions causing foaming occur simultaneously with the polymer-forming reactions. As in the case of POLYURETHANE RESINs (w s), the polymeric constituent of urethane foams is made by reacting a polyol with an isocyanate. The polyol may be of the polyester or polyether type. When the isocyanate is in excess of the amount that will react with the polyol, and when water is present, the excess isocyanate will react with water to produce carbon dioxide which expands the mixture. The hardness of the cured foam is governed by the molecular weight of the polyol used. Low-molecular-weight polyols (approximately 700) produce rigid foams, and high-molecular-weight polyols (3000 to 4000) produce flexible foams. Polyols with molecular weights around 6000 are used for the so-called "cold-cure", highly resilient foams. They are usually capped with ethylene oxide to provide terminal primary hydroxyl groups that increase the polyols' reactivity about threefold. Crosslinked foams are rigid or semirigid. Auxiliary blowing agents are often used, especially in rigid foams where they improve the insulation values. Other ingredients often incorporated in urethane foams are catalysts to control the speed of reaction, and a surfactant to stabilize the rising foam and control cell size. Three basic processes are used for making urethane foams: the prepolymer technique, the semi-prepolymer technique, and the one-shot process. In the prepolymer technique, a polyol and an isocyanate are reacted to produce a compound that may be stored and subsequently mixed with water, catalyst, and, in some cases, a foam stabilizer. In the semi-prepolymer process about 20% of the polyol is prereacted with all of the isocyanate, then this product is later reacted with a masterbatch containing the remainder of the ingredients. See also ONE-SHOT MOLDING, ISOCYANATE, POLYOL, POLYETHER FOAM, RETICULATED POLYURETHANE FOAM, and INTEGRAL-SKIN MOLDING.

Polyurethane/Imide Modified Foam A polyaryl polyisocyanate (PAPI) is reacted with a 3,3',4,4'-benzophenone tetracarboxylic dianhydride (BTDA) to form an isocyanate prepolymer. This prepolymer can be compounded with a polyol, a blowing agent, a catalyst, and a cell stabilizer to form the modified foam. Such a foam containing 5% BTDA in the prepolymer has better thermal properties than conventional polyurethane foams.

Polyurethane Resin (isocyanate resin) A family of resins produced by reacting diisocyanates with organic compounds containing two or more active hydrogen atoms to form polymers having free isocyanate groups. These groups, under the influence of heat or certain catalysts, will react with each other, or with water, glycols, etc, to form a thermosetting material.

Polyvinyl Acetal The general name for resins formed by partially or completely replacing the hydroxyl groups of polyvinyl alcohol with aldehydes, by means of a condensation reaction. Two commercially important members of the class are POLYVINYL BUTYRAL and POLYVINYL FORMAL, w s. Polyvinyl acetal resins are thermoplastics that can be cast, extruded, molded, and coated onto substrates, but their main uses are in adhesives, lacquers, coatings, and films.

Polyvinyl Acetate (PVA, PVAc) A colorless, transparent thermoplastic, prepared by the polymerization of vinyl acetate. The homopolymers are available as beads, powder, solutions, emulsions, and latex. The major use is in water-based latex paints, adhesives, fabric finishes, and lacquers. In the plastics industry, the *copolymers* of vinyl acetate, particularly with vinyl chloride, are of most interest.

Polyvinyl Alcohol (PVA, PVAL, PVOH) A water-soluble thermoplastic prepared by partial or complete hydrolysis of polyvinyl acetate with methanol or water. Although it can be extruded and molded, its principal uses are in packaging films, fabric sizes, adhesives, emulsifying agents, etc. The packaging films are impervious to oils, fats, and waxes, and have very low transmission rates of oxygen, nitrogen, and helium. Thus, they are often used as barrier coatings on other thermoplastics or coextruded with them. The water solubility of polyvinyl alcohol films can be regulated to some degree. The "standard" type, made from higher-molecular-weight polymers and plasticized with glycerine, is only weakly soluble in cold water. The other type, known as CWS (cold-water-soluble), is made from internally plasticized or lower-molecular-weight resins. See also ETHYLENE-VINYL ALCOHOL COPOLYMER.

Polyvinyl Butyral (PVB, polyvinyl butyral acetal) A member of the POLYVINYL ACETAL (w s) family, made by reacting polyvinyl alcohol with butyraldehyde, with some unreacted PVAL groups retained in the polymer. It is a tough, sticky, colorless, flexible solid, used primarily as the interlayer in automotive safety glass. Other applications include adhesive formulations; base resins for coatings, toners, and inks; solutions for rendering fabrics resistant to water, staining and abrasion; and crosslinking with resins such as ureas, phenolics, epoxies, isocyanates, and melamines to improve coating uniformity and adhesion, increase toughness, and minimize cratering.

Poly(*N*-Vinylcarbazole) (PVK) A thermoplastic resin, brown, obtained by reacting acetylene with carbazole. It has excellent electrical properties and good heat and chemical resistance, and is used as an impregnant for paper capacitors. It is photoconductive, a property that has found use in xerography.

Polyvinyl Chloride (PVC) A polymer made by the catalytic polymerization of vinyl chloride, CH_2=CHCl, and a leader in the mass and value of products containing PVC, and in the variety of its uses. US sales of PVC and copolymers in 1992 came to 4.6 Tg ($5.0 \cdot 10^6$ tons), accounting for 15% of all US resin sales in 1992, and third in volume after the polyethylenes. "PVC" also includes copolymers that contain at least 50% vinyl chloride. The neat homopolymer is hard and horny, brittle and difficult to process, but it becomes flexible when plasticized. Processing of the rigid grades is eased by copolymerization with vinyl acetate and by compounding PVC resin with acrylic-polymeric processing aids, lubricants, and heat stabilizers. Applications for rigid PVC include pipe, house sidings, and bottles. PVC is copolymerized with many monomers and blended with other polymers to obtain a wide variety of properties. PVC resins, especially those copolymerized with minor amounts of vinyl acetate, may be dissolved in volatile solvents to make lacquers, coatings and cast films. A special class of PVC resin of fine particle size, often called *dispersion-grade resin*, can be dispersed in liquid plasticizers to form PLASTISOLs, w s. PVC molding compounds can be extruded, injection molded, compression molded, calendered, and blow molded to form a huge variety of products, flexible and rigid, according to the amount and type of plasticizers incorporated. Far more compounding recipes exist for PVC than for any other polymer, with thousands of US patents issued, making it uniquely versatile among plastics. Rigid PVC is strong, difficult to burn, has excellent resistance to strong acids and bases, to most other chemicals, and to many (but not all) organic solvents. Plasticized compounds range from fairly stiff to very flexible and find applications as diverse as garden hose, flooring, women's shoes and purses, shower curtains, automotive upholstery, and thousands more. Finally, PVC is one of the least expensive plastics: the average bulk price for pipe-grade resin during 1992 was 59¢/kg.

Polyvinyl Chloride-Acetate An important copolymer family of vinyl chloride and vinyl acetate, usually containing 85% to 97% vinyl chloride. These copolymers are more flexible and more soluble in solvents than PVC, and are used in solution coatings as well as in most of the processes and applications employing PVC.

Polyvinyl Dichloride See CHLORINATED POLYVINYL CHLORIDE.

Polyvinyl Fluoride (PVF) The polymer of vinyl fluoride (fluoroethylene, $H_2C=CHF$). The fluorine atom forms a strong bond along the hydrocarbon chain, accounting for properties such as high melting point, chemical inertness, and resistance to ultraviolet light. In the form of film, PVF is used for packaging, glazing, and electrical applications. Laminates of PVF film with wood, metal, and polyester panels are being used in building construction. Although it cannot be dissolved in ordinary solvents at room temperature, coating solutions can be made by dissolving PVF in hot "latent solvents" such as dimethyl acetamide and the lower-boiling phthalate, glycolate, and isobutyrate esters. Such solutions are used to protectively coat the insides of rigid metal containers for chemicals and industrial compounds.

Polyvinyl Formal (PVFO, PVFM) A member of the POLYVINYL ACETAL family, w s, made by condensing formaldehyde in the presence of polyvinyl alcohol or by the simultaneous hydrolysis and acetylation of polyvinyl acetate. It is used mainly in combination with cresylic phenolics for wire coatings and impregnating, but can also be molded, extruded, or cast. It is resistant to greases and oils and to moderately high temperatures.

Polyvinyl Halide A term sometimes used (almost exclusively in patents) for polymers and copolymers of vinyl chloride. Aside from polyvinyl fluoride, which is more similar structurally to polyethylene, and brominated butyl rubber, which has enjoyed some use in the automobile-tire industry, no polymers containing the other halogens (bromine, iodine, and astatine) exist in commerce.

Polyvinylidene Chloride (PVDC) A thermoplastic polymer of 1,1-dichloroethylene, $H_2C=CCl_2$. The homopolymer is of limited commercial value due to difficulty of processing. Copolymers with vinyl chloride (15% or more) are widely used as packaging and food-wrapping films under the name SARAN, w s.

Polyvinylidene Fluoride (PVDF, PVF_2) A member of the fluorocarbon-resin family, made by polymerizing 1,1-difluoroethylene, $H_2C=CF_2$, a colorless gas. The resin is thermally stable to high temperatures, is stronger and more abrasion-resistant than other fluoroplastics, and is easier to process on conventional thermoplastics equipment. Its oriented film is piezoelectric, a property useful in loudspeakers. Major applications of PVDF are in the fields of electrical insulation for high-temperature service, pipe and other chemical-process equipment, and coatings for industrial and commercial buildings.

Polyvinylisobutyl Ether See POLYISOBUTYLVINYL ETHER.

Poly(Vinylmethyl Ether) [PVME, PVM, poly(methylvinyl ether)] A family of polymers polymerized from vinylmethyl ether, $H_2C{=}CHO{-}CH_3$. They range from viscous liquids to stiff rubbers. The liquids, soluble in cold water but not in hot water, are used in pressure-sensitive and hot-melt adhesives for paper and polyethylene, and as a tackifier in rubbers. PVM also designates copolymers of vinyl chloride and vinylmethyl ether.

Poly(1-Vinylpyrrolidone) [PVP, poly(*N*-vinyl-2-pyrrolidone)] A highly water-soluble polymer prepared by the addition polymerization of 1-VINYL-2-PYRROLIDONE, (w s for structure). Molecular weights range from 10,000 to 360,000. Solutions of the polymer are used as protective colloids and emulsion stabilizers, and it has been used as a substitute for human blood plasma. PVP films are clear and hard, but can be plasticized.

Polyvinyl Resin See VINYL RESIN.

Polyvinyl Stearate A wax-like polymer of vinyl stearate, of limited use in the plastics industry. However, the monomer is copolymerized with vinyl chloride, acting as an internal lubricant.

Polywater In the early 1960s, it was reported that the Soviet physicist, Boris Derjaguin, had discovered a polymeric form of water, formed by condensing ordinary water on the inside quartz capillary tubing of very fine bore. Properties of the polymer, dubbed *polywater*, were said to be thermal stability up to 500°C, density equal to 1.4 g/cm^3, i e, 40% greater than that of ordinary water, and solidification to a glass-like state at –40°C. There were subsequent reports that US scientists had confirmed the existence of polywater. Later, however, the discoverer of "polywater" admitted that the substance he had created was actually impurities dissolved from the quartz tubes used in the experiment.

Poly-*p*-Xylylene (Parylene N) See PARYLENE.

Poly-*m*-Xylyleneadipamide Synonym for NYLON MXD/6, w s.

POM Abbreviation for POLYOXYMETHYLENE, w s. Also see ACETAL RESIN.

Pony Mixer See CHANGE-CAN MIXER.

Poromeric (from micro*poro*us and poly*meric*) A material that has the ability to transmit moisture vapor to some degree while remaining essentially waterproof. The first plastic material of this type was Du Pont's "Corfam", introduced in 1963 and vigorously marketed as a leather substitute in shoe uppers, at which task, alas! it enjoyed only mediocre success. It was a composite of urethane polymers and polyester fibers. The most successful poromerics are the fabrics known as Gore·Tex®, developed by W L Gore Associates, and applied widely to raincoats, sport garments, and camping gear.

Porosity The ratio of the volume of voids contained within a sample of material to the total volume (solid matter plus voids), expressed as a fraction (*void fraction*) or percentage (*percent voids*).

Porous Mold A mold that is made up of bonded or sintered aggregates (powdered metal, pellets, etc) in such a manner that the resulting mass contains numerous connected interstices of regular or irregular shape and size through which air may escape as the mold is filled.

Positive Mold A compression mold in which the pressure is applied wholly on the material, and which is designed to prevent the escape of any molding material.

Post-Curing Completing the cure of a thermosetting casting or molding after removal from the mold in which a partial cure has been accomplished. Post-curing usually involves heating, for example, in a circulating-air oven.

Postforming (1) The heating and reshaping of a fully or partially cured laminate. On cooling, the formed laminate retains the contours and shape to which it has been postformed. (2) Operations applied to still warm extrudates, particularly some types of profile extrusions, in which limbs of the extrudate pass through fixtures that bend or curl them into their final shapes.

Pot (1, n) A chamber to hold and heat molding material for a transfer mold. (2, v) See POTTING.

Potassium Titanate Fiber $K_2O(TiO_2)_n$ wherein n = 4 to 7. Highly refined, single crystals, approximately 6 μm long by 0.1 μm in thickness, used as reinforcing fibers in thermoplastic composites. The fibers melt at 1370°C; density is 3.2 g/cm^3. They also act as white pigments.

Pot Life (working life) The period after mixing the ingredients of an active compound during which the viscosity remains low enough to permit normal processing. Dissatisfied with the vagueness of this definition, Breitigam and Ulrich (August, 1990) cautiously defined pot life for their epoxy compounds as the time, at room temperature, for the viscosity to reach double its initial value.

Pot Plunger A plunger used to force softened molding material from the pot into the closed cavity of a transfer mold.

Pot Retainer A plate channeled for passage of a heat-transfer medium (e g, hot oil) and used to hold the pot of a transfer mold.

Potting The process of encasing an article or assembly in a resinous mass, performed by placing the article in a container that serves as a disposable mold, pouring a liquid resin into the mold to completely submerge the article, then curing the resin. The container remains attached to the potted article. The main difference between potting and ENCAPSULATION, w s, is that in the latter the mold is removed from the encapsulated article and reused. These processes are widely used in the electronics industry.

Potting Syrup See CASTING SYRUP.

Poundal An obsolete, but still occasionally seen, unit of force (the force required to accelerate a 1-pound mass 1 ft/s^2), analogous to the dyne in the cgs system, both created many years ago to perpetuate the illusion that Newton's law of momentum change needs no proportionality constant. See FORCE.

Powder Blend See DRY BLEND.

Powder Compact A molding material in the form of dry, friable pellets prepared by compacting dry-blended mixtures of resin (typically PVC) with plasticizers and other compounding ingredients. The powder compacts are about as easy to handle and process by extrusion as PELLETS, w s, and offer the advantages of lower heat history and somewhat lower cost than equivalent materials in the form of fused pellets. See also DRY BLEND.

Powder Density See BULK DENSITY.

Powdered Plastic A resin or plastic compound in the form of extremely fine particles, for use in fluidized-bed coating, rotational molding, and various sintering techniques.

Powder Molding A general term encompassing rotational molding, slush molding, compression molding, and centrifugal molding of dry, sinterable powders such as polyethylene, nylon, PVC, polytetrafluoroethylene and ultra-high-molecular-weight polyethylene. The powders are charged into molds that are heated, and manipulated or pressed, according to the process being used. These actions cause the powders to sinter or fuse into a uniform layer or molding against the mold surfaces, which are then cooled. See also FLUIDIZED-BED COATING, ROTATIONAL MOLDING, CENTRIFUGAL MOLDING, SLUSH MOLDING, and SINTER MOLDING.

Powell-Eyring Model (Eyring-Powell model) A complex rheological equation containing three parameters that must be evaluated by fitting experimental flow data. It has the form:

$$\tau_{xz} = C\left(\frac{dv_z}{dx}\right) + \frac{1}{B}\sinh^{-1}\left[\frac{1}{A}\left(\frac{dv_z}{dx}\right)\right]$$

where τ_{xz} is the z-directed shear stress perpendicular to x, v_z is the velocity in the z-direction, dv_z/dx is the shear rate at x, and A, B, and C are temperature-dependent constants characteristic of the flowing medium. The model calls for Newtonian-flow regions at both very low and very high shear rates, a type of behavior seen in a few polymer solutions but rarely in melts. The limiting viscosity at low shear rate is given by C + 1/AB, while the high-shear limiting viscosity is C. Applying this model to even simple flow geometries is cumbersome, especially when it is to be used to find shear rates, velocities, and flow rates from known stresses, so it hasn't seen much use. Compare EYRING MODEL.

Power The rate at which work is being done or energy expended. The SI unit is the watt (W), equal to 1 joule/second (J/s). Some conversions of other units are given in the Appendix. In purely resistive, direct-current electric circuits, power is given by the product of voltage drop times current, ($\Delta e \times i$) and the watt is equal to 1 volt-ampere (V·A). In sinusoidally alternating circuits, power is given by $\Delta e \cdot i \cdot \cos \phi$, in which ϕ is the phase angle between the current and the voltage.

Power Factor (1) The ratio of actual power (wattage) being used in an alternating circuit to the product of voltage drop (Δe) times current, (i), usually expressed as a percentage. When the load in the AC circuit is purely resistive, as with ovens and incandescent lamps, the wattage equals $\Delta e \cdot i$ and the power factor is 100%. When the load includes inductive elements such as motors and transformers, the current lags the voltage by a phase angle ϕ, which can range from 0 for a purely resistive load to 90° for a load that is wholly inductive, and the power factor is referred to as a *lagging* power factor. When the load in the AC circuit contains capacitive elements, the current *leads* the voltage by the phase angle and the term *leading* power factor is used. In a circuit containing all three types of elements, the net effect will generally be current lagging or leading, but with a relatively smaller phase angle. The difference between actual power and $\Delta e \cdot i$ is called *reactive power*, which increases as the power factor decreases. Reactive power does no useful work but costs the same as actual power used. Thus, low power factors increase power costs, cause overloading of motors and transformers, and reduce load-handling capacity of plant electrical systems. (2) In testing the behavior of plastics as dielectrics, power factor is the cosine of the phase angle when the voltage across the capacitor varies sinusoidally. In a perfect dielectric (pure capacitance), the current would lead the voltage and the phase angle (ϕ) would be 90°, its cosine 0. When loss occurs, the phase angle is $90 - \delta$, where δ = the *loss angle*, hence $\cos \phi = \sin \delta$. In the literature, $\tan \delta$ is often called the power factor. In capacitor applications, δ is usually very small, so the difference between sine and tangent is negligible. This might not be so in dielectric heating, say, of phenolics or vinyls, where power factors are higher. Dielectric loss depends on frequency. Because it is generated by oscillatory movement of molecular and atomic dipoles within the material, the loss spectrum over the frequency range of many decades will usually show one or more maxima and minima.

Power Law (Ostwald-deWaele model) The simplest representation of pseudoplastic flow, and characteristic of most polymer melts over several decades of shear rate. One versatile form of the model is

$$\tau_{xz} = -m \left| \frac{dv_z}{dx} \right|^{n-1} \left(\frac{dv_z}{dx} \right)$$

where τ_{xz} = the z-directed shear stress on a surface perpendicular to x, v_z = the velocity in the z-direction, dv_z/dx is the shear rate at x, and m and n are constants peculiar to the liquid. n is called the *flow-behavior index* and has a value between 0.25 and 0.9 for most polymer melts. The quantity m is analogous to viscosity and is temperature-dependent. For n = 1, the power law reduces to Newton's law of flow and $m = \mu$, the Newtonian viscosity. Over the limited range of shear rates occurring in a given process, the power law can often provide a sufficiently accurate approximation to the actual flow behavior. Chemical engineers often cast the power law into the simpler form:

$$\frac{\Delta P \cdot D}{4L} = K \left(\frac{8V}{D} \right)^n$$

in which the left side is the shear stress at the wall of a pipe of diameter D and length L, ΔP is the pressure drop over that length of pipe, K is a viscosity-like property (temperature-dependent), V = the average liquid velocity, and n is the flow-behavior index of the liquid. The quantity $(8V/D)$ is the apparent Newtonian shear rate at the tube wall.

PP Abbreviation for POLYPROPYLENE, w s.

ppb Abbreviation for *parts per billion.*

PPG Abbreviation for POLYOXYPROPYLENE GLYCOL, w s.

PPI Abbreviation for POLYMERIC POLYISOCYANATE, w s.

ppm Abbreviation for *parts per million.*

PPMI Abbreviation for POLYPYROMELLITIMIDE, w s.

PPO Abbreviation for POLYPHENYLENE OXIDE, w s.

PPOX Abbreviation for *polypropylene oxide.* See POLYPROPYLENE GLYCOL.

PPS Abbreviation for POLYPHENYLENE SULFIDE, w s.

PPSU Abbreviation for POLY(p-PHENYLENE SULFONE), w s.

Prandtl Number (Pr, N_{Pr}) A dimensionless group important in the analysis of convective heat transfer, defined as (in consistent units) $C_P\mu/k$, where C_P = the specific heat of a fluid at constant pressure, μ =

its viscosity, and k = its thermal conductivity. The Prandtl number is also the ratio of the KINEMATIC VISCOSITY to the THERMAL DIFFUSIVITY (see both entries).

Precipitation Polymerization A polymerization reaction in which the polymer being formed is insoluble in its own monomer or in a particular monomer-solvent combination and thus precipitates as it is formed.

Precure A partial or full state of cure existing in an elastomer or thermosetting resin prior to its use as an adhesive or in a forming operation.

Precursor One who or that which precedes and suggests the course of future events. A compound or polymer that is later or transformed into another material or polymer by chemical reaction.

Predrying The drying of a resin or molding compound prior to its introduction into an extruder, a mold, or molding machine. Many resins and plastics compounds are hygroscopic and require this treatment, to prevent formation of bubbles in the product, particularly after exposure to a humid atmosphere. Predrying for extrusion or injection molding is usually accomplished by passing heated, bone-dry air up through the bed of pellets in an enclosed feed hopper. This has a bonus of reducing the heat input required from the extruder drive and can boost extruder output. The exit, moistened air is recycled through a dryer packed with silica gel or other drying agent.

Preform (n) (1) A compressed, shaped mass of plastic material or fibrous reinforcing material or a combination of both, prepared in advance of a molding operation for convenience in handling or for accuracy of loading by weighing the mass. The term also applies to tablets and biscuits of thermoplastic and thermosetting compounds. (2) In the reinforced-plastics industry, a preform is a mat of chopped strands bonded together by a resin in approximately the shape of the end product, for use in processes such as matched-die molding. Or it may be a complex shape made by two- or three-dimensional weaving or braiding that, when wet out with resin and cured, will become the finished product. See NEAR-NET-SHAPE CONFIGURATION.

Preform (v) (1) To make plastic molding powders into pellets, tablets, or biscuits of known mass that facilitate accuracy in compression molding. (2) To prepare by hand cutting of reinforcing cloth or mat, or by blowing chopped fibers onto a contoured screen, the reinforcement for a fiber-reinforced molded object. The reinforcement, which has a shape close to that of the final molded object, is placed into or onto the mold along with the required amount of resin, then wet out and cured.

Preform Binder A light application of resin applied to a mat or screened preform that provides enough shape stability to permit handling the preform into the mold without tearing the mat or shifting the fiber distribution.

Preform Molding See MATCHED-METAL-DIE MOLDING.

Pregel An unintentional, prematurely cured, or partially cured layer of resin on part of the surface of reinforced plastic prior to molding. Should not be confused with GEL COAT, w s.

Pregl Method See COMBUSTION ANALYSIS.

Preheating Heating of feedstock or material to be processed prior to the main processing step. In extrusion of coated wire, the wire is resistively heated just before entering the die to ease melt flow through the die and maintain coating quality at high wire speeds. (Compare PREHEAT ROLL.) In compression molding with preforms, the preforms are commonly preheated electronically before being loaded into the mold, thus improving the flow, reducing curing time in the mold, and shortening the cycle. In some extrusion and injection-molding operations, pellet feedstocks are dried in the hopper with hot air. This predrying of the feed not only precludes splay marks, bubbles, etc, caused by moisture in the melt, but significantly reduces the amount of energy per unit mass of plastic that must be furnished by screw action, permitting higher throughputs with lower screw-energy input per unit of product delivered.

Preheating Hopper See HOPPER DRYER.

Preheat Roll In extrusion coating, a heated roll installed between the pressure roll and unwind roll, the purpose of which is to heat the substrate before it is coated, thereby providing better adhesion of the coating and permitting lower melt temperature and, possibly, a higher production rate.

Preimpregnation A method of preparing fiber-reinforced molding material by forcing thermoplastic resin, or thermosetting resin advanced only to the B-stage, into mats or cloths of fiber reinforcement. The product, called a PREPREG (w s), is ready for molding, but storable for periods up to several months and is shippable.

Premix ("gunk") A term originally applied to mixtures of polyester resin with sisal or glass fiber reinforcement and fillers, usually prepared by molders shortly before use. The ASTM definition (D 883) specifies that the premix should not be in web or filamentous form. The term premix is now often used by molding compounds of any thermosetting resin mixed with fillers, reinforcements, and catalysts. The terms *sheet molding compound* (*SMC*) and *bulk molding compound* (*BMC*) have some currency for premixes containing thickening agents such as Group II oxides (e g, magnesium hydroxide and calcium hydroxide), and for thermoplastic polymers, used respectively for sheet molding and injection molding.

Premix Molding ("gunk" molding) A variation of matched-die molding in which the ingredients, usually chopped roving, resin, pigment, filler and catalyst, are premixed and divided into accurately weighed charges for molding.

Preplasticization In plunger-type injection molding, the technique of premelting molding powders in a separate chamber, then transferring the melt to the injection cylinder. The technique shortened molding cycles and provided a more homogeneous melt entering the mold. With the widespread adoption of screw-injection machines, the need for separate preplasticization has fallen sharply to a few special circumstances.

Prepolymer A polymer of relatively low molecular weight, usually intermediate between those of the monomer or monomers and the final polymer or resin, that may be mixed with compounding additives, and that is capable of being hardened by further polymerization during or after a forming process.

Prepolymer Molding In the polyurethane-foam industry, a system whereby a portion of the polyol is prereacted with the isocyanate to form a liquid prepolymer with a viscosity suitable for pumping or metering. This component is supplied to end-users with a second premixed blend of additional polyol catalyst, blowing agent, etc. When the two components are vigorously mixed, foaming and crosslinking occurs. For a contrasting method, see ONE-SHOT MOLDING.

Prepreg In the reinforced-plastics industry, a mat or shaped mass of reinforcing fibers, typically glass strands, impregnated with a thermosetting resin advanced in cure only through the B-stage. Such prepregs may be stored until needed for a molding or laminating operation. A prepreg containing a chemical thickening agent is called a MOLD-MAT, w s. The term "prepreg" also includes fabrics such as jute, coconut fiber, or rayon yarn impregnated with a thermoplastic resin, e g, vinyl, acrylonitrile-butadiene-styrene, or acrylic. For sheet forms, the term "prepreg" is being displaced by the more specifically descriptive *sheet molding compound* (*SMC*).

Prepreg Molding A type of MATCHED-METAL-DIE MOLDING, w s, in which the fibrous mat is preimpregnated with a partially cured, thermosetting resin.

Preprinting In sheet thermoforming, the inversely distorted printing of sheets before they are formed. During forming, the stretching of the sheet brings the print into its proper size and spacing.

Preservative A chemical incorporated in a material to prevent deterioration, mainly by living organisms, but more generally, also by heat, oxidation or weather. See also ANTIOXIDANT, FUNGICIDE, and STABILIZER.

Press Polishing (planishing) A finishing process used to impart high gloss and improved clarity and mechanical properties to sheets of vinyl, cellulosic, and other thermoplastics. The sheets are hot-pressed against thin, highly polished metal plates.

Pressure Force exerted over an area, expressed as force per unit area. The SI unit is the pascal (Pa), equal to 1 newton per square meter, the same as the unit of stress. Since our pervading atmosphere keeps us all under a pressure of about 101 kPa, many pressure-sensing devices detect and indicate the *difference* between a process pressure and atmospheric, called *gauge* (or *gage*) *pressure*. Pressure referred to total vacuum is *absolute pressure*. When exerted by solid contact, as by a ram on an elastic surface, pressure may vary over the contact area. Pressures of confined gases at rest are equal everywhere within the container while liquid pressures can depend significantly on depth because of density and gravity. Conversions of traditional pressure units are listed in the Appendix.

Pressure-Bag Molding See BAG MOLDING.

Pressure Break As applied to a defect in a laminated plastic, a break apparent in one or more outer sheets of the paper, fabric, or other base visible through the surface layer of resin that covers it.

Pressure Flow (1) In general, any flow that is driven by a pressure gradient along a flow path, including any vertical component due to gravity's action on the fluid density. Flows through orifice-type rheometers and extrusion dies are pressure flows. (2) Specifically, in the metering section of an extruder screw, the rearward flow ("back flow") that would occur *if* the screw were not rotating and the pressure gradient were unaltered. In the rotating screw, pressure flow opposes the productive drag flow, reducing net output, but can never exceed it. In Newtonian flow equations for extruders the pressure flow is subtracted from the drag flow to obtain the net flow (throughput). In the actual plastics extruder, the nonNewtonian character of the melt invalidates the algebraic summing, yet because pressure flow is usually a third or less of the drag flow in a well designed system, the errors of this algebraic summing are seldom serious. If the simplified flow equation overstates the actual output, the difference is more likely to be due to insufficient feeding or poor plasticating action than to equation errors. On the other hand, there have been instances, with screws of high compression ratio, where the pressure of the melt entering the metering section was as high or higher than that at the die. In that case, there may be little or no pressure flow or even *positive* (forward) pressure flow. See also DRAG FLOW, LEAKAGE FLOW, and NET FLOW.

Pressure Forming A variant of sheet thermoforming in which pressure above atmospheric is used to push the heat-softened sheet against the mold surface, as opposed to using only a vacuum to suck the sheet against the mold. The cycle may be shortened. Pressure forming has been effectively used to form container lids from biaxially oriented polystyrene sheet without losing the orientation and good strength properties accruing therefrom. See TRAPPED-SHEET FORMING.

Pressure Pad A reinforcement of hardened steel, several of which may be distributed around the dead areas in the faces of a mold to help the land absorb the final pressure of closing without collapsing.

Pressure Roll In extrusion coating, a roll that presses the coating and substrate together to form a strong bond, continuous over the entire interface.

Pressure-Sensitive Adhesive (PSA) An adhesive that develops a strong bond to most surfaces by applying only a moderate pressure.

Pressure Transducer An instrument that converts a sensed fluid pressure into an electrical signal that in turn can be converted to a pressure reading and recorded. Transducers for extruders presented a difficult problem of temperature compensation and need for extreme ruggedness in service, and were pioneered by Dynisco in the 1950s. Several reliable makes are now available.

PRI Abbreviation for Plastics and Rubber Institute (United Kingdom).

Primary (adj) In chemistry, a functional group at the end of a molecule's chain (or branch) in which only one of the hydrogen atoms has been replaced by some other link, as a primary alcohol, $-CH_2OH$, or primary amine, $-NH_2$. See SECONDARY and TERTIARY.

Primary Plasticizer A plasticizer that, within reasonable compatibility limits, may be used as the sole plasticizer, is completely compatible with the resin, and is sufficiently permanent to produce a composition that will retain its desired properties under normal service conditions throughout the expected life of the article. See also PLASTICIZER and SECONDARY PLASTICIZER.

Primer A coating applied to a substrate to improve the adhesion, gloss, or durability of a subsequently applied coating. For example, vinyl copolymer, acrylic, phenolic, epoxy, and polyester resins are used as primers for adhering vinyl coatings to metals.

Primrose Chrome See CHROME-YELLOW PIGMENT.

Primrose Yellow See CHROME-YELLOW PIGMENT.

Printing on Plastics Many methods commonly used on paper and other materials are also used for printing on plastics, with slight modifications such as the use of special inks. Such processes are letterpress, offset, silk screen, electrostatic, and photographic methods. See also ELECTROSTATIC PRINTING, FLEXOGRAPHIC PRINTING, GRAVURE PRINTING, HOT STAMPING, SPANISHING, and VALLEY PRINTING. Polyolefins are normally oxidatively treated before printing so as to make them receptive to inks. See CASING, CORONA-DISCHARGE TREATMENT, FLAME TREATING, and ULTRAVIOLET PRINTING.

Probit A translation of origin of the scale of standard normal deviates (not a contradiction!) to avoid the inconveniences of negative signs. For a given percentage point of the standard normal distribution, the probit = the corresponding standard normal deviate +5, i e, $z + 5$. There is available normal probability paper that has a probit scale alongside the probability scale. The device is useful in plotting and discussing the results of testing by the UP-AND-DOWN METHOD, w s.

Processability The ease with which a polymer, elastomer, or plastic compound can be converted to high-quality, useful products with standard melt-processing techniques and equipment. Some quantitative tests of processability have been devised; for example, see MOLDING INDEX and THERMOFORMABILITY.

Processing Aid A substance added to a compound to improve its behavior during processing. Many processing aids have been tried with rigid PVC because the neat resin is heat-sensitive, decomposing autocatalytically with evolution of toxic hydrogen chloride gas (HCl) at temperatures near 215°C. Processing aids may include heat stabilizers, lubricants, and other resins, even plasticizers.

Process Variation The degree to which measurements of the same process parameter, or characteristic or dimension of successive parts or products are different. See STANDARD DEVIATION and RANGE.

Producer's Risk In quality control and acceptance sampling, the probability, under a given sampling plan, of making a Type I error, that is, of rejecting a lot whose true quality is at the desired acceptable level.

Profile (1) Any extruded product but those of the simplest cross sections, such as film, sheet, rod stock, pipe, and coated substrates. Examples of profiles are angle-stock and channels; square, triangular, and trapezoidal solids and annuli; house siding and refrigerator-door gaskets. (2) The lineal variation of the smoothness/roughness of a finished surface. See PROFILOGRAPH. (3) The pattern of variation of some process parameter over time, or more usually, distance. Examples are the channel-depth profile of an extruder screw and the temperature profile along an extruder cylinder.

Profile Die A die used to form an extruded PROFILE (w s). Two basic types are used: PLATE DIEs (w s) and streamlined dies. The former are cheaper to make and alter; the latter are essential when extruding rigid PVC and other heat-sensitive plastics, and are apt, with *any* compound, to permit higher extrusion rates of good product.

Profilograph (profilometer) An instrument that measures the roughness of a surface, usually expressed as the local root-mean-square average in nm (or μin). The profile taken in any direction can be magnified and displayed graphically.

Progressive Aging In a heat-aging test, stepwise raising of the temperature at preset time intervals.

Progressive Bonding A method of curing thermosetting-resin adhesives in laminates or plywood slabs that are larger in area than the press platens in which they are being bonded. A partial area, say, a quarter of the laminate, is cured by application of heat and pressure. The press is then opened, and a different quarter of the laminate is moved between the platens and cured, and so on, until the entire laminate has been cured.

Projected Area In molding, the area of a cavity, or all the cavities, or cavities and runners, perpendicular to the direction of mold closing force and parallel to the parting plane. In injection molding and blow molding, this area must be safely less than the quotient of the force applied to hold the mold closed divided by the maximum melt pressure or blowing pressure within the mold. In transfer molding, it must also be about 15% less than the cross-sectional area of the pot.

Promotor (promoter) A chemical substance that, in very small concentration, increases the activity of a CATALYST, w s. The promotor may itself be a weak catalyst. Examples in the curing of polyester resins are cobalt octoate used as the promotor with methyl ethyl ketone peroxide, and *N*-alkyl anilines used with benzoÿl peroxide.

Proof Resilience (energy to break) The work required to stretch an elastomeric test specimen from no elongation to its breaking point, expressed in J/cm^3 of specimen volume.

Propagation (chain propagation) The middle phase of any polymerization process during which monomers are extending polymer chain lengths by addition or condensation reactions.

2-Propanone Synonym for ACETONE, w s.

Propeller Mixer A device comprising a rotating shaft with a propeller at its end, used for mixing relatively low-viscosity dispersions and holding contents of tanks in suspension. The propellers, of which there may be two or three on a single shaft, resemble boat propellers, having two to four broad, curved lobes. See also PADDLE AGITATOR.

Propenal Synonym for ACROLEIN, w s.

Propenenitrile Synonym for ACRYLONITRILE, w s.

Propenoic Acid Synonym for ACRYLIC ACID, w s.

Proportional Control A method of controlling processes in which the control action taken is proportional to the difference (process error) between the sensed state variable of a process and the desired target level of that variable. See ON-OFF CONTROL.

Proportional Limit The greatest stress a material is capable of sustaining without deviating from direct proportionality (linearity) between stress and strain (Hooke's law). See also ELASTIC LIMIT and YIELD POINT.

***n*-Propyl Acetate** (propyl acetate) $C_3H_7OOCCH_3$. A clear, colorless liquid with a pleasant odor, used as a solvent for cellulosics, vinyls, acrylics, polystyrene, alkyds , and coumarone-indene resins.

Propylene (propene) $CH_3CH{=}CH_2$. A colorless gas produced mainly by cracking propane, butane, or other refinery off-gases, or by cracking hydrocarbons during the production of ethylene. It is the monomer from which polypropylene is made, and also has many uses as an intermediate.

Propylene Glycol (1,2-propanediol) $CH_3CH(OH)CH_2OH$. An important starting material for unsaturated polyester resins that are compatible with styrene.

1,2-Propylene Glycol Monolaurate $C_{11}H_{23}COOCH_2CH(OH)CH_3$. A plasticizer for cellulosics, polystyrene, and vinyl resins.

1,2-Propylene Glycol Monooleate $C_{17}H_{33}COOCH_2CH(OH)CH_3$. A plasticizer for cellulose nitrate and ethyl cellulose.

Propylene Oxide (1,2-propylene oxide, 1,2-epoxypropane) $CH_3CH(O)CH_2$. A low-boiling, liquid epoxide compound derived from the intermediate propylene chlorohydrin, which is itself produced by reacting propylene with chlorine and water. Propylene oxide is an important intermediate for the manufacture of polyglycols used for polyurethane foams and resins, and polyester resins.

Propylene Plastic See POLYPROPYLENE.

Propylene-Vinyl Chloride Copolymer Any of a family of copolymers ranging from 2% to 10% by weight of propylene, that provide the application-properties advantages of PVC homopolymers plus the processing advantages attributable to the introduction of stable hydrocarbon structures as end groups. The copolymers are easy to mold and extrude, and have high thermal stability and low melt viscosity.

***n*-Propyl Oleate** $C_{17}H_{33}COOC_3H_7$. A monounsaturated fatty ester, and a plasticizer for ethyl cellulose, polystyrene, and, with limited compatibility, some vinyl and acrylic resins.

Protein Resin A generic term for resins derived from proteins, constituting CASEIN PLASTICS and ZEIN, w s.

Proton A nuclear particle that used to be considered elementary, having a positive charge equal to the negative charge of an electron but possessing a mass approximately 1837 times that of an electron at rest, and slightly less than that of a neutron. The proton is in effect a hydrogen-atom nucleus.

Prototype Mold A temporary or experimental mold used to make a few samples to test product designs or obtain market reactions. Such a mold is often made from a low-melting metal-casting alloy or from a filled and reinforced epoxy resin.

Protrusion Any raised area on a molded or painted surface, such as a blister, bump, or ridge.

PS Abbreviation for POLYSTYRENE, w s.

Pseudoplastic Fluid A solution or melt whose apparent viscosity decreases instantaneously and reversibly with increasing shear rate and stress. There is no yield stress. Most polymer solutions and melts are pseudoplastic. See also POWER LAW and ELLIS MODEL. Pseudoplastic behavior is often confused with, and mistakenly labeled as THIXOTROPY, w s.

Pseudoplasticity Time-independent shear thinning with no yield stress.

PSP Abbreviation for POLYSTYRYLPYRIDINE, w s.

PSU Abbreviation for POLYSULFONE, w s.

PTFE Abbreviation for POLYTETRAFLUOROETHYLENE, w s.

PTHF Abbreviation for POLYTETRAHYDROFURAN, w s.

PTMT See POLY(TETRAMETHYLENE TEREPHTHALATE).

PU Abbreviation sometimes used in Europe for POLYURETHANE, w s.

Pulforming A postforming technique applied to pultruded plastics to form mildly curved shapes such as leaf springs.

Pulled Surface Imperfections in the surface of a laminated plastic, ranging from a slight breaking or lifting of its surface in spots to pronounced separation of its surface from its body (ASTM D 883).

Puller Any device used to pull an extrudate away from the extruder and through the cooling tank, playing a role in determining the dimensions of the product's cross section. See CATERPILLAR for a description of the kind most used.

Pull-out Strength Of threaded inserts in plastics moldings, the force required to pull the insert out of the molding. It may be expressed as the force per unit area of the engaged outside surface. (As of 1991, ASTM listed no test of this strength in plastics.)

Pull Strength The bond strength of an adhesive joint, obtained by pulling in a direction perpendicular to the plane of the bond. This is an uncommon mode of test for adhesive bonds: the usual mode is to pull apart the ends of lap-joined specimens, thus testing the joint in shear. See TENSILE-SHEAR STRENGTH.

Pulp Molding A process by which a resin-impregnated pulp material is preformed by application of a vacuum and subsequently oven cured or molded. The pulp is first mixed with water and pumped into a tank wherein a mold, usually of wire mesh shaped like the finished article, is positioned. Air is evacuated from the mold to attract the pulp fibers, forming a preformed layer in contact with the screen. The mold is then removed from the vacuum tank, the pulp deposit is stripped off and dried, then the preform is molded to final form by fluid pressure or conventional compression methods.

Pulsed Positive/Negative-Ion Chemical Mass Spectrometry See MASS SPECTROMETRY.

Pultrusion A reinforced-plastics technique for continuously producing profiles of constant cross section, both solid and annular. Strands of reinforcing material are conveyed through a tank of resin—usually polyester but silicone and epoxy are also used—from which they are pulled through a long, heated steel die shaped to impart the desired profile. Both gelling and curing of the resin are sometimes accomplished entirely within the die length, which can be as much as 75 cm. In other variations of the process, preheating of the resin-wet reinforcement is effected by dielectric energy prior to its entering the die, or heating may be continued in an oven after emergence from the die. In the past, the pultrusion process has mainly yielded continuous lengths of material with high unidirectional strengths, used for building siding, fishing rods, golf-club shafts, etc, but recent advancements in the technique permit multidirectional reinforcement and strengths.

Pumice A highly porous, igneous rock, used in pulverized form as an abrasive and a filler for plastics.

Pumicing A finishing method for molded plastics parts, consisting of the rubbing off of traces of tool marks and surface irregularities by means of wet pumice stones.

Pump Ratio In single-screw extrusion with two-stage screws (as in vented operation), the ratio of the drag-flow capacity of the forward (final) pumping section to that of the rear (first) pumping section. This ratio is approximately equal to (but slightly less than) the ratio of the two pump depths, providing the lead angle is constant throughout, and is usually in the vicinity of 1.5.

Punching A method of producing components, particularly electrical parts, from flat sheets of rigid or laminated plastics by cutting out shapes with a matched punch and die in a punch press.

Puncture Resistance The ability of a plastic film or sheet to resist being penetrated by pointed objects. The most nearly relevant ASTM tests are two in which a specimen of films or sheets are punctured by not very pointy objects. One is D 1709, the free-falling dart method, in which a variably weighted dart having a hemispherical nose is dropped on a clamped specimen, a new specimen being used with each weight change and drop. By one of two testing techniques (see UP-AND-DOWN METHOD) the mean weight required for penetration is determined. In the other, more sophisticated test, D 3763, an instrumented plunger, also round-nosed, is forced at high speed through the clamped film or sheet specimen and a load-vs-displacement trace is developed.

PUR The preferred (in US) abbreviation for POLYURETHANE, w s.

Purging In extrusion or injection molding, the cleaning of one color or type of material from the machine by forcing it out with the new color or material to be used in subsequent production, or with another compatible purging material. The operation goes faster when the purg*er* is more viscous than the purg*ee*. See also DRY PURGE and PURGING COMPOUND.

Purging Compound A plastic compound especially designed to quickly purge most other plastics from an extruder or molder. It may contain organic fibers that help to scour the cylinder, and some purging compounds contain percentages of ultra-high-molecular-weight polyethylene which, because it doesn't actually melt in the extruder, also tends to be an efficient purger.

Pushback Pin See RETURN PIN.

Pushing Flight Synonym for LEADING FLIGHT FACE, w s.

Push-Pull Molding An injection-molding technique that uses twin injection units to fill a mold through well separated gates. By oscillating the advance and retraction of the injection screws or rams, the material in the mold is sheared and oriented, breaking up weld lines. It is particularly suited to molding of liquid-crystalline polymers. See also MULTI-LIVE-FEED MOLDING.

Pushup In the packaging industry, a container bottom with sufficient concavity to prevent rocking of the container when it is filled and placed on a flat surface.

PVA Abbreviation for either POLYVINYL ALCOHOL or POLYVINYL ACETATE, w s.

PVAC Abbreviation for POLYVINYL ACETATE, w s.

PVAL Abbreviation for POLYVINYL ALCOHOL, w s.

PVB Abbreviation for POLYVINYL BUTYRAL, w s.

PVC (1) Abbreviation for POLYVINYL CHLORIDE, w s. (2) In the paint industry, abbreviation for *pigment volume concentration*.

PVCA An abbreviation for copolymers of vinyl chloride and vinyl acetate.

PVD A rarely used abbreviation for polyvinyl dichloride. See CHLORINATED POLYVINYL CHLORIDE.

PVDF Abbreviation for POLYVINYLIDENE FLUORIDE, w s.

PVF See POLYVINYL FLUORIDE. The possibility of confusion exists because this abbreviation has been used in some literature for POLYVINYL FORMAL, for which the alternative abbreviations PVFM and PVFO have been employed.

PVFM, PVFO Abbreviations for POLYVINYL FORMAL, w s.

PVI Abbreviation for POLYISOBUTYLVINYL ETHER, w s.

PVK Abbreviation for POLY(*N*-VINYLCARBAZOLE), w s.

PVM, PVME Abbreviation for POLY(VINYLMETHYL ETHER), w s.

PVOH Abbreviation for POLYVINYL ALCOHOL, w s. The abbreviation PVA is more commonly used.

PVP Abbreviation for POLY(1-VINYLPYRROLIDONE), w s.

PX Abbreviation for *p*-XYLYLENE, w s.

Pycnometer A small flask for measuring the density of a liquid. It is usually of glass and one type is provided with a standard-taper, ground-glass stopper carrying a thermometer and a capped, capillary sidearm so that the pycnometer can be filled to its known volume with high precision. The same instrument may be used to measure the density of particulate matter, such as plastic pellets, by immersing it in a liquid that is inert to, and significantly less dense than the solid matter. A *dilatometer* is a special pycnometer equipped with instruments to study SPECIFIC VOLUME (w s) as a function of temperature.

Pyranyl Foam A type of rigid, pour-in-place, thermosetting foam similar to a polyurethane foam, but with superior resistance to high temperatures. It is formed in the same manner as polyurethane foams, using as the monomer a pyranyl (radical) derived from polypropylene by heating and oxidation to form an acrolein dimer, which ultimately forms the pyranyl.

Pyrazolone Red A metal-free diazo pigment based on a pyrazolone.

Pyrogenic Silica See FUMED SILICA.

Pyrogram A chromatogram (see CHROMATOGRAPHY) obtained from the pyrolysis products of a sample.

Pyrolysis The decomposition of a complex organic substance to ones of usually simpler structure by heating in vacuum or an inert atmosphere. Some polymers will depolymerize at high temperatures, either to polymers of lower molecular weight, or, in some cases, back to the monomers from which they were derived. Thermosets tend to give up hydrogen and other elements and small compounds and retain their carbon, taking on a char-like character. The fragments generated by pyrolysis are often amenable to chromatographic analysis, which can yield insights to structure of the polymer or a compound's ingredients.

Pyromellitic Dianhydride (PMDA, 1,2,4,5-benzenetetracarboxylic anhydride) A triple-ring heterocycle with the structure shown below,

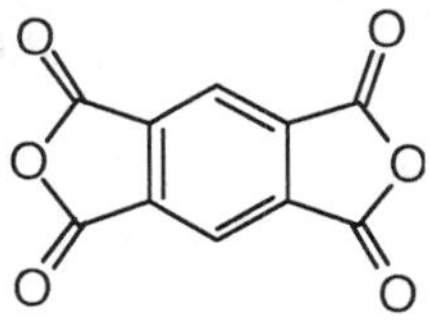

O O
O O
O O

PMDA is a curing agent for epoxies giving cured products of high deflection temperatures. It is the least costly starting material, reacted with diamines, for producing polyimides and related high-temperature-resistant polymers.

Pyrometer (1) A hand-held instrument for measuring temperatures of melts and surfaces, usually furnished with two or three probes for different applications, and consisting essentially of an iron-constantan thermocouple (in each probe) connected to a milliammeter whose dial indicates temperature rather than current. Reference-junction compensation is built into the circuitry. (2) An INFRARED PYROMETER, w s.

Pyrophoric Igniting spontaneously in air.

Pyroxylin Lower-nitrated cellulose containing less than 12% nitrogen, flammable but less explosive than guncotton. See CELLULOSE NITRATE. Its solutions are known as COLLODION, w s.

Pyrrone See POLYIMIDAZOPYRROLONE.

Qq

Q Symbol, in electronics, for the ratio of the reactance to the resistance of an oscillatory circuit, and often called the *quality factor* of the circuit. $Q/2\pi$ = the ratio of energy stored to energy dissipated per cycle. A closely analogous measure applies to mechanical oscillating systems and, when the system is oscillating at or near its resonant frequency, Q is proportional to that frequency.

QA Abbreviation for QUALITY ASSURANCE, w s.

QC Abbreviation for QUALITY CONTROL, w s.

Quadripolymer (tetrapolymer) A rarely used term for the product of simultaneous polymerization of four monomers.

Quadrupole Spectrometer A type of mass spectrometer with two dipoles that provide better separation of ionic masses than can a single dipole. Carrying the idea even further are modern triple-quadrupole spectrometers. See MASS SPECTROMETRY.

Quality Assurance (QA) A system of activities whose purpose is to provide assurance, with documentation, that the overall quality-control function for any product, operation, service, or entire organization is in fact being accomplished.

Quality Characteristic Any dimension, property, aspect of appearance, surface finish, or performance specification that helps to determine the acceptability of a product or its ability to perform particular design functions. Most quality characteristics are measurable and therefore objective, but some, such as odor or texture, may be subjective and may be determined by the judgment of an expert or a panel of potential consumers of the product.

Quality Control (QC) The techniques, measurements, and other activities that monitor and maintain product-quality characteristics within stated limits. These means require sampling of the product, measurement of important quality characteristics, statistical analysis, continuous presentation of the data (typically by means of *control charts*), and taking of decisions on whether lots are to be accepted, reworked, or scrapped.

Quantitative Analysis A branch of chemistry, encompasssing very many methods and techniques, whose scope is to determine the amounts of the different elements in substances or the percentages of molecular entities in a mixture of gases, liquids, or solids.

Quantitative Differential Thermal Analysis DIFFERENTIAL THERMAL ANALYSIS (w s) in which the equipment used is designed to produce quantitative results in terms of energy and/or other physical parameters (ISO).

Quartile Any of the three values that divide a data set that has been ordered from smallest to largest into four equal parts by number. The quartiles are also the 25th, 50th, and 75th percentiles. The second quartile, i e, the 50th percentile, is more familiarly known as the MEDIAN, w s.

Quartz The most common of minerals, of the rhombohedric crystal habit, occurring in a myriad of minerals and colors, but in its purest form, silicon dioxide (SiO_2), colorless, clear, very hard and transparent to both visible and ultraviolet light. The term is also used for synthetically produced, amorphous fused quartz or vitrified SILICA, w s.

Quartz Fiber Fiber produced from natural quartz crystals of high purity (99.95% SiO_2). Quartz melts at 1610°C and is immune to thermal shock. However, high-purity crystals are rare, and in many applications one can settle for high-silica fiber, made from ordinary sand. Quartz- and silica-fiber-reinforced composites are used in jet aircraft, rocket nozzles, and reentry nose cones. Quartz whiskers are also in use where their high cost is justified. They are among the strongest and stiffest of all fibers, comparable with graphite whiskers, with strength of 21 GPa and modulus of 700 GPa. Density is 2.65 g/cm^3.

Quaterpolymer The IUPAC term for a COPOLYMER (w s) derived from four species of monomers.

Quench Bath The cooling medium used in QUENCHING, w s.

Quenching (1) A process of shock cooling thermoplastic materials from the molten state, usually done experimentally with thin films of crystal-forming polymers in order to minimize the crystalline content and to study the nearly amorphous material. (2) Following the heat treating of steel tools, such as mold-cavity inserts and extrusion-die parts after machining, immersing the parts in an oil bath or sometimes just rapidly cooling in air to increase temper and hardness.

Quench-Tank Extrusion An extrusion process wherein the extrudate is conducted through a water bath for rapid cooling.

Quick-Burst Pressure (of a pipe, tube, or pressure vessel) See HYDROSTATIC STRENGTH.

Quicklime Synonym for CALCIUM OXIDE, w s.

Quinacridone Pigment A family of organic pigments based on substituted and unsubstituted forms of linear *trans*-quinacridones. Colors available include several shades of red, violet, gold, orange, magenta, and maroon. These pigments have good lightfastness, intensity of hue, resistance to bleeding and chemical attack, good transparency, and heat resistance.

Quinol See HYDROQUINONE.

Quinone See *p*-BENZOQUINONE.

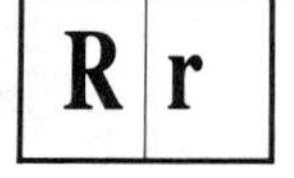

r Symbol for radius of a circle or sphere, or radial coordinate in cylindrical and spherical coordinate systems, or product-moment correlation coefficient in statistics.

R (1, often –R) In organic chemistry, symbol for a general attached group or radical, frequently an aliphatic or aromatic hydrocarbon group, that may take on any of various specific identities. (2) Abbreviation for ROENTGEN, w s. (3) Abbreviation for generalized or multiple correlation coefficient, the square root of the COEFFICIENT OF DETERMINATION, w s. (4) (°R) Abbreviation for degrees Rankine. See RANKINE TEMPERATURE. (5) Symbol for electrical resistance.

R^2 Symbol for COEFFICIENT OF DETERMINATION, w s.

Rabinowitsch Correction The ingenious correction derived by B. Rabinowitsch (1929) that, when applied to the Newtonian shear rate at the wall of a circular tube through which a nonNewtonian liquid is flowing, gives the true shear rate at the wall. For pseudoplastic liquids, such as most polymer melts, the correction is always an increase. If the fluid obeys the POWER LAW, w s, it reduces to a simple correction factor, $(3n + 1)/4n$, where n is the flow-behavior index of the liquid.

Rad (1) A deprecated, but still widely used, unit of energy absorbed by a material, including living matter, from exposure to ionizing radiation. One rad = 0.01 gray (Gy) = 0.01 J/kg. Also see ROENTGEN. (2, rad) The abbreviation for the SI unit of plane angular measure, the *radian*, the angle intercepting a circular arc of length equal to its radius ($= 360°/2\pi = 57.3°$).

Radiant Heating The net transfer of heat from a hotter body to a cooler one by (usually infrared) radiation. Radiant transfer is one of the three basic mechanisms of HEAT TRANSFER, w s, requiring no contact or fluid between the bodies. The net rate is proportional to the difference between the fourth powers of the *absolute* temperatures of the hotter and cooler bodies ($T_1^4 - T_2^4$), and depends also on the thermal "color" of the bodies (their emissivities), their geometries, and their positioning relative to each other. The principal use of radiant heating in the plastics industry is in SHEET THERMOFORMING, w s.

Radiation Compatibility The ability of a plastic to maintain its properties when exposed to X-ray, gamma, electron, or other ionizing radiation.

Radiation Crosslinking The formation of chemical links between polymer chains through the action of high-energy radiation, commonly gamma radiation from a cobalt-60 source or electrons from an electron gun. The treatment has improved the modulus and raised the use temperature of polyethylene wire coatings and some polymer films. Exposure must be accurately controlled if the crosslinking is to be achieved without degrading the resin.

Radiation Degradation Breakdown of a plastic caused by too long exposure to radiation, or to radiation of too high energy levels, or both. The radiation may be X-ray, electron, gamma, or neutron beams. The mechanism is ionization and chain scission.

Radiation Polymerization A polymerization reaction initiated by exposure to radiation such as ultraviolet or gamma rays rather than by means of a chemical initiator.

Radiation Processing See IRRADIATION.

Radiation Sterilization The process of sterilizing the contents of sealed packages or cans containing foods or medical materials by exposing them to controlled levels of high-energy radiation, usually either gamma from a cobalt-60 source or electrons from guns. The process has long been known to provide 100% bacteriological action, to be safe, to have flavor advantages over some other methods of sterilization, to give long shelf life for processed foods, and to generate zero induced radioactivity in the treated materials, but its use has been held back by unfavorable—and irrational—publicity.

Radical A group of atoms, normally part of a molecule, that may replace a single atom (frequently H in organic compounds) and remain unchanged during reactions of the compound. Some examples are the *ethyl* radical, $-C_2H_5$, the *acetate* radical, CH_3COO-, and the *phenyl* radical, $-C_6H_5$. Many chemical-reaction mechanisms postulate the transitory existence of unattached ("free") radicals as intermediates, which, because of their charge, are extremely reactive. FREE RADICALs (w s) play important roles in addition polymerizations. A few free radicals are known that are sufficiently stable to permit their identification and quantitative determination as chemical entities.

Radioactive Tracer A chemical compound or other material in which one or more of the ordinary atoms have been replaced by their radioactive isotopes. Carbon-14, tritium (hydrogen-3), and iodine-131 are among the isotopes that have been used in this way. The tracer isotopes have been useful in elucidating chemical-reaction mechanisms and in tracking human and animal metabolisms.

Radio Frequency (RF) A frequency of electromagnetic radiation within the broad range of radio and radar transmission, e g, from about 300 kHz to 20 GHz.

Radio-Frequency Heating See DIELECTRIC HEATING.

Radio-Frequency Preheating (RF preheating) A method of preheating used for thermosetting molding materials to facilitate the molding operation or shorten the molding cycle. The frequencies most commonly used are near 20 or 40 MHz.

Radio-Frequency Welding (dielectric welding, high-frequency welding) See DIELECTRIC HEAT SEALING.

Ram (1) In compression and matched-die molding, the press member that enters the cavity block and exerts pressure on the molding compound, designated by its position in the assembly as the *top force* or *bottom force.* (2) In older injection machines predating the development of screw injection, the plunger that forced the feed pellets through the annulus between cylinder and torpedo, and that also accomplished, in most such machines, the injection of melt into the mold. Very few of these machines are still (1993) in commercial use. (3) The piston of a melt accumulator such as may be used in blow-molding large objects, or in special injection-molding techniques. (4) The piston (plunger) of a *ram extruder.* See EXTRUDER, RAM.

RAM Acronymic abbreviation for *random-access memory*, computer memory for storing and working with programs and data, and erasable by the operator. Compare ROM.

Ram Extruder See EXTRUDER, RAM.

Ramie A natural vegetable fiber obtained from the stems of the hemp *Boehmeria nivea*, used as a reinforcement.

Ram Injection Molding See RAM (2) and INJECTION MOLDING.

Ramped Temperature The heating of a sample (as in thermogravimetric analysis) at a closely controlled, linear rate.

Ram Travel The distance the injection ram (or screw) moves in filling the mold in injection or transfer molding.

Random Copolymer A copolymer consisting of alternating segments of two different monomeric units of random lengths, including single mer units. A random copolymer usually results from the copolymerization of two monomers in the presence of a free-radical initiator, for example the so-produced rubbery copolymer of ethylene and propylene.

Range (*R*) In sampling of product dimensions and properties, the difference betrween the largest and smallest values in the sample. Range chartts for small samples have long been used in quality-control work. The range is simple to calculate and is almost as efficient as the STANDARD DEVIATION, w s, in samples of 2 to 5 items.

Rankine Temperature The absolute temperature scale, now deprecated, derived from the Fahrenheit scale, having its zero at –459.67°F. To convert Rankine to SI's kelvin (K), multiply by 5/9.

RAPRA Former acronym for Rubber and Plastics Research Association, since 1985 renamed RAPRA Technology, an industry-supported organization headquartered at Shawbury, Shrewsbury, SLAOP, SY4 4NR, England. Among many contributions by RAPRA people to plastics and rubber processing is the CAVITY-TRANSFER MIXER, w s.

Rate of Shear See SHEAR RATE.

Rate-Process Theory A general theory, derived from statistical mechanics, applicable to both chemical reactions and creep phenomena in plastics. For the latter, the theory relates time-to-rupture to stress and the reciprocal of the absolute temperature. An equation recommended for pipe in ASTM D 2837 is

$$\log t = A_0 + A_1/T + (A_2/T)\cdot\log S$$

where the A_i are empirical coefficients, different for each plastic, t = the time to failure under sustained hoop stress S, and T = the absolute temperature.

Rayon The definition established by the Federal Trade Commission in 1951 is: "Generic name for a manufactured fiber composed of regenerated cellulose, as well as manufactured fibers composed of regenerated cellulose in which substituents have replaced not more than 15% of the hydrogens of the hydroxyl groups." Prior to that date, going back to 1924 when the name rayon was first used (inspired by its sheen invoking the brilliance of a ray of sunlight), the term was used for all man-made fibers derived from cellulose, including cellulose acetate and cellulose triacetate. Rayon is the oldest of the synthetic fibers, having been produced commercially since 1855. All methods of producing rayon are based on treating fibrous forms of cellulose to make them soluble, extruding the solution through the tiny orifices of a spinneret, then converting the filaments into solid cellulose. Most rayon fibers are produced from the intermediate VISCOSE, w s. See also CUPRAMMONIUM RAYON.

Re (N_{Re}) An alternate, older symbol for REYNOLDS NUMBER, w s.

Reaction Injection Molding (RIM) This term is usually applied to the process of injection molding of urethane reactants in which the two primary constituents, isocyanate and polyol, are pumped by a metering device into a mixing head from which the intimately mixed reactants are quickly injected into a closed mold. The injection pressure is much lower than in conventional injection molding of molten plastics, enabling the use of inexpensive, light-weight molds. However, the mixing head is a high-pressure impingement mixer in which pressures may reach 14 to 21 MPa. One mixing head may be used to feed up to ten separate molding presses. One of the largest-volume applications of RIM is the production of exterior automotive parts such as body panels and bumpers. Furniture is another big use. The term LIQUID INJECTION

MOLDING (LIM), w s, is usually applied to the similar process of molding other thermosetting resins such as polyesters, epoxies, silicones, alkyds, and diallyl phthalate resins. The terms *liquid reaction molding* (*LRM*) and *high-pressure injection molding* (*HPIM*) have sometimes been used for either or both processes. If reinforcing fibers are included in the reaction mix, the process is called *reinforced reaction injection molding* (*RRIM*). *Structural reaction injection molding* (*SRIM*) is a variation in which there is some foaming of the polyurethane in the core of the molding, with a solid skin on the outside. This technique reduces part weight with little loss of stiffness or strength and is widely used in the commodity-furniture industry.

Reactive Plasticizer See PLASTICIZER, POLYMERIZABLE.

Reactive Processing A molding or extrusion operation in which chemical reactions are carried out. Extruders, mainly specialized twin-screw machines, have successfully carried out partial and complete polymerizations on a large scale. Transfer and compression molding of thermosets have always been reactive processes, but see REACTION INJECTION MOLDING.

Reagent Resistance (chemical resistance) The ability of a plastic to withstand exposure to acids, alkalis, oxidants, and solvents.

Ream (1) Layers of inhomogeneous material parallel to the surface in a transparent or translucent plastic article. (2) A quantity of paper, 472 to 500 sheets, depending on the type of paper.

Reciprocating-Screw Injection Molding In this process the screw serves to both plasticate the feedstock and inject the melt into the mold. During part of the cycle, the screw rotates rapidly, moving backward as it accumulates a charge (*shot*) of melt in the forward end of the cylinder. A limit switch stops the rotation and two hydraulic rams, one on either side, push the screw forward, forcing the melt into the mold and holding melt pressure until the gates freeze. The rams are withdrawn, screw rotation recommences, the mold opens to eject the parts, the mold closes, and a new cycle begins. Most new injection machines sold today (1993) are of this type.

Reclaim (v) (recycle) To salvage plastics from discarded products such as milk and soda bottles, automobiles, and packaging films.

Recycle (1) (regrind) In a processing plant, to recover trim scrap and faulty parts by granulating them and blending the ground material with virgin feed. (2) To RECLAIM, w s.

Recycled Plastic (reclaimed plastic) A plastic prepared from discarded articles that have been cleaned and ground (ISO). This material may or may not be reformulated by the addition of stabilizers, plasticizers, fillers, pigments, etc.

Red 2B Pigment (permanent red 2B, pigment red 48) Any of the calcium, strontium, or barium precipitates of the coupling from diazotized 2-chloro-4-aminotoluene-5-sulfonic acid with 2-hydroxy-2-naphthoic acid. These pigments have high tinctorial strength and good resistance to bleeding and heat, but are poor in chemical resistance and fair in lightfastness. Colors range from orange-red to ruby.

Redox A contraction (and reversal) of the term "oxidation-reduction". A *redox catalyst* is one entering into an oxidation-reduction reaction. A polymerization initiator comprising a mixture of a peroxide and a reducing agent is called a *redox initiator*. Polymers formed by such reactions are sometimes called *redox polymers*.

Reduced Viscosity (IUPAC: viscosity number) Of a dilute polymer solution, the ratio of the specific viscosity to the concentration. Reduced viscosity is a measure of the specific capacity of the polymer to elevate viscosity. See DILUTE-SOLUTION VISCOSITY.

Reduction (1) Any chemical process that increases the proportion of hydrogen or base-forming elements or radicals in a compound. (2) The gaining of electrons by an atom, ion, or element, thereby reducing the positive valence of that which gained the electron(s). The reverse of OXIDATION, w s.

Reentrant Mold A mold containing an undercut that tends to impede withdrawal of the molded product. If the undercut is more than slight, the mold will probably be designed with a side draw that retracts from the undercut region as the mold opens and relieves the undercut.

Reference Material (reference standard) In analytical chemistry, a gas mixture, pure liquid, solution, pure solid, or alloy whose composition is certified with a stated, high degree of accuracy. Reference materials are used to calibrate analytical procedures and instruments. They are available from private companies, also from the National Institute of Standards and Technology (the former National Bureau of Standards).

Refiner A machine similar to a ROLL MILL (w s), but operated with the rolls very close together so as to crush lumps and hold them in the roll nip until reduced. Refiner rolls are shorter and of larger diameter than mill rolls, and are operated at higher differential speeds to provide more intense shear.

Reflectance See LIGHT REFLECTANCE.

Reflectometer An instrument that measures the total luminous flux from a surface and reports it as a percentage of the incident flux on the surface.

Refractive Index See INDEX OF REFRACTION.

Refractivity The INDEX OF REFRACTION (w s) minus 1. *Specific refractivity* is given by the expression $(n - 1)/d$ where n = the index of refraction and d = the density of the material.

Refractometer An instrument used to measure the index of refraction of transparent materials, both solid and liquid. The Abbé design is convenient and is used worldwide. See INDEX OF REFRACTION and DIFFERENTIAL REFRACTOMETER.

Regenerated Cellulose A transparent cellulosic plastic made by mixing cellulose xanthate with a dilute sodium hydroxide solution to form a VISCOSE, w s, extruding the viscose into film, sheeting, or fiber form, then treating the extrudate with acid to effect regeneration. In fiber form, the material is called RAYON (w s). The term CELLOPHANE (w s) is used for films and sheets.

Regression Analysis (method of least squares) A family of statistical techniques for fitting equations to data based on the principle that the best-fitting parameters are those that minimize the sum of squares of the differences between the original observations and the corresponding equation values. A simple case is fitting a straight line to a set of data (x_i, y_i) to obtain the equation $\hat{y} = a + b \cdot x$. *Multiple linear regression* includes the application of the least-squares principle to any form of relationship between a dependent variable y and several "independent" factors $x_{1i}, x_{2i}, \ldots x_{ki}$, or explicit functions of the x_{ji}, that is linear with respect to the constants to be fitted. *Nonlinear regression*, which requires one or another iterative-search method, deals with relationships that are not linear in the constants sought. Powerful programs for all these techniques are now available for personal computers. Regression equations provide the vehicle for estimating the outcomes of future experiments in the systems investigated, with known errors of estimate. They have been a powerful tool in polymer science, plastics processing, compound development, and in estimating future performances of plastics products in service.

Regrind (n) Thermoplastic waste material such as sprues, runners, excess parison material, sheet trimmings, and rejected parts from molding, extrusion, and thermoforming operations, that has been reclaimed by shredding or granulating. Regrind is usually mixed with virgin compound at a predetermined percentage for remolding, etc.

Regular Block In the chemical structure of polymers, a block that can be described by only one species of mer in a single sequential arrangement (IUPAC, slightly modified).

Regular Polymer A polymer whose molecules can be described by only one species of mer in a single sequential arrangement (IUPAC).

Regulator In polymerization reactions, a chain-transfer agent used at low concentration to limit the molecular weight of the polymer.

Reinforced Molding Compound A compound containing reinforcing fibers and supplied by the raw-material producer in the form of ready-to-use material, as distinguished from PREMIX, w s.

Reinforced Plastic (RP) A plastic composition in which are embedded fibers that are much stronger and typically much stiffer than the matrix resin. The reinforcements are usually fibers, rovings, fabrics, or mats, or mixed forms of glass, carbon, asbestos, metals, ceramics, paper, sisal, cotton, or nylon. Resins most commonly used are polyesters, phenolics, aminos, silicones, epoxies and various thermoplastics. The term *reinforced plastic* includes some forms of LAMINATE (w s) and molded parts in which the reinforcements are not in layered form. When the resin is thermoplastic, the term *reinforced thermoplastic* is often used. Methods of forming reinforced-plastics articles from thermosetting resins are defined under the entries listed below.

AXIAL WINDING
BAG MOLDING
CENTRIFUGAL CASTING
CERAPLAST
CONTACT-PRESSURE MOLDING
FIBERFILL MOLDING
FILAMENT WINDING
FURAN PREPREG
IMPREGNATION
LAMINATE
LAP WINDING
LAYUP MOLDING
LOW-PRESSURE LAMINATE
MATCHED-METAL-DIE MOLDING
PREFORM
PREMIX
PREPREG
PREPREG MOLDING
PULFORMING
PULP MOLDING
PULTRUSION
REACTION INJECTION MOLDING
RESIN-TRANSFER MOLDING
REVERSE HELICAL WINDING
SHEET-MOLDING COMPOUND
SLURRY PREFORMING
SPRAYUP

Reinforced Thermoplastic (RTP) A reinforced structure in which the bonding resin is a thermoplastic rather than a thermoset. Over the past two decades, applications for RTP have grown rapidly, mainly based on nylons, polycarbonates, acetal resins, polystyrene, polypropylene, and ADVANCED RESINs (w s). The tensile strength and modulus of a thermoplastic can be at least doubled by the addition of glass reinforcement, and creep under load is greatly decreased. Because thermoplastics are remeltable, RTPs are most commonly produced as pelletized molding compounds for injection molding.

Reinforcement A strong, inert, fibrous material incorporated in a plastic mass to improve mechanical properties. Typical reinforcements are ASBESTOS, BORON FIBER, CARBON FIBER, CERAMIC FIBER, FLOCK, GLASS-FIBER REINFORCEMENT, GRAPHITE, JUTE, MICA, SISAL, and WHISKERS, all of which see. Others sometimes used are chopped paper, macerated fabrics, synthetic fibers, and metal wires. To be effective, a reinforcement must form a strong adhesive bond with the matrix resin, to which end adhesion-promoting substances known as coupling agents are often preapplied to the fibers. Reinforcements differ from FILLERs, w s, in that they markedly improve modulus and strength, whereas fillers do not.

Reinforcing Pigment A pigment that also serves to improve the strength of the finished product. An example is carbon black.

Relative Density Synonym for SPECIFIC GRAVITY, w s.

Relative Humidity (RH) The ratio, always expressed in percent, of the quantity of water vapor present in the ambient atmosphere to the quantity that would saturate it at the prevailing temperature. It is also the ratio of the partial pressure of water vapor present to the vapor pressure of water at the prevailing temperature. High relative humidity, which occurs in summer in many manufacturing-plant atmospheres, can cause condensation of water on chilled surfaces such as injection molds, with attendant mold defects. It can also cause blushing problems in the painting of plastics and metals. See also HUMIDITY.

Relative Standard Deviation See COEFFICIENT OF VARIATION.

Relative Viscosity (IUPAC: viscosity ratio) The ratio of the kinematic viscosity of a dilute polymer solution to that of the pure solvent. See also DILUTE-SOLUTION VISCOSITY.

Relaxation A gradual decrease in stress in a structure under sustained constant strain. See STRESS RELAXATION.

Relaxation-Map Analysis (RMA) A technique used on the results of a series of THERMALLY STIMULATED CURRENT (w s) experiments in which the TSC data are transformed into relaxation times and plotted versus reciprocal absolute temperature to estimate enthalpy and entropy of activation for the molecular relaxations.

Relaxation Time Of a viscoelastic material under constant strain (specifically, one behaving as a Maxwell element), the time required for the stress to diminish to 1/e (= 0.368) of its initial value. Compare RETARDATION TIME.

Release Agent See PARTING AGENT.

Release Paper A layer of paper that can be readily separated from the surface of a plastic article to which it has been applied or against which the plastic article has been formed. The term applies to papers used to protect the surfaces of plastic sheets, to temporary backings for pressure-sensitive adhesives, and to papers used as temporary carriers in film- and foam-casting processes. They may also be preprinted with an ink that is transferred to the cast film.

Relief Angle In an injection or blow mold, the relief angle is the angle between the narrow pinch-off land and the cutaway portion adjacent to the pinch-off land.

Rennet Casein A type of casein precipitated from milk by means of rennet, the dried extract of stomach secretions from calves or other ruminants containing the enzyme rennin. See also CASEIN and CASEIN PLASTIC.

Reprocessed Plastic Thermoplastic material that has been left over from molding, extrusion, or thermoforming, such as sprues and runners, sheet trim, trim between thermoformed parts, and rejected parts, then molded, extruded, or thermoformed again into useful articles by other than the original processors. See also RECYCLED PLASTIC.

Residual Monomer The unpolymerized monomer that remains incorporated in a polymer after the polymerization reaction has been completed. Polystyrene and vinyls are examples of resins in which residual monomer has been found, although manufacturers of vinyls now take steps to eliminate residual monomer (because it is a known liver carcinogen) from their resins.

Residual Solvent Solvent, usually polymerization solvent, remaining in an unfinished resin, or a pelletized resin ready for market, or solvent remaining in a solvent-cast film after drying. Either is usually expressed as a weight percent.

Residual Strain Strain remaining in a part that has been chilled while undergoing plastic deformation or immediately thereafter. Often, if the part is reheated, some or nearly all of the strain may be recovered. Usually associated with RESIDUAL STRESS, w s.

Residual Stress (frozen-in stress) Stress remaining in a part that has been chilled quickly during or after molding, extrusion, or forming. It remains because there was too little time for the stress to relax while the material was soft. Over time, high residual stress can cause parts to warp and shrink. It can be relieved and rendered harmless by annealing residually stressed parts while restraining them in fixtures.

Resilience (1) The degree to which a body can quickly resume its original shape after removal of a deforming stress. When the body is a standard test specimen, the resilience, expressed as the percentage recovery from a stated maximum strain, may be attributed to the material from which the specimen was made. ASTM Tests D 926 and D 945 (sec 09.01) describes compression and shear tests for resilience of rubber and foam rubber. (2) The fractional return, to an impacting body, of the energy with which it strikes a resilient test specimen. ASTM D 1054 details a pendulum-rebound test, while D 2632 and D 3574 describe drop-weight-rebound tests, all employing this principle and all in section 09.01.

Resin The term *resin* is defined by ASTM D 883 as a solid or pseudosolid material, often of high molecular weight, that exhibits a tendency to flow when subjected to stress, usually has a softening or melting range, and usually fractures conchoidally. A note appended to this definiton explains that in a broad sense, the term is used to designate any polymer that is a basic material for plastics. However, common uses of the term in the plastics industry do not always conform to this definition. The term is also used for uncured fluid thermosetting materials, some chemically modified natural resins, and often synonymously with the terms *plastic* and *polymer*. See also RESIN, NATURAL; RESIN, SYNTHETIC; and CELLULOSIC PLASTIC.

Resin Applicator In filament winding, a device that deposits the liquid resin onto the reinforcement band.

Resin-Bonded Laminate See LAMINATE and REINFORCED PLASTIC.

Resin Content The percentage, by weight or volume, of resin in a laminate or filled or reinforced thermoplastic molding.

Resinoid (n) A term that has sometimes been used for a thermosetting resin, either in its initial, temporarily fusible state or in the final cured state.

Resin, Natural When certain trees and other plants are wounded, either by natural accident or by tapping, they exude liquids that harden or partially harden upon exposure to air to form resinous or balsam-like products. Deposits of such secretions undergo chemical changes—oxidation and polymerization—when buried for long periods, forming solid or semisolid products that are soluble in oils and organic liquids but insoluble in water. Such water-insoluble resins are generally known as the natural resins, sometimes called *varnish resins*. Those that are water-soluble are known as gums or essential oils. Examples of natural resins are accroides (acaroid resin), amber, Canada balsam, Congo copal, dammar, elemi, Kauri copal, and sandarac. They are used in varnishes and lacquers, also as modifiers for plastics. See also RESIN, SYNTHETIC; and AMBER.

Resinography See POLYMEROGRAPHY.

Resin Pocket An apparent accumulation of excess resin in a small localized area between laminations in a laminated-plastic article, visible on a cut edge or a molded surface.

Resin-Rich Area A region in a reinforced-plastic article in which there is an objectionable excess of resin.

Resin-Starved Area (dry spot) An area of a reinforced-plastic article that has an insufficient amount of resin to wet out the reinforcement completely, evidenced by low gloss, dry spots, or exposure of fiber.

The condition may be caused by improper wetting or impregnation, or by excessive molding pressure.

Resin Streak A long, narrow surface imperfection on the surface of a laminated plastic caused by local excess of resin.

Resin, Synthetic (synthetic polymer) A RESIN, w s, that has been produced from simple materials, or intermediates made from such chemicals, by either addition or condensation polymerization. Of the commercial plastics, all but the cellulosics are based on synthetic resins. Among commercial elastomers, only natural rubber, refined from the sap of the *Hevea* tree, is natural.

Resin-Transfer Molding (RTM) A variation of MATCHED-METAL-DIE MOLDING in which, after placing the preformed reinforcement in the heated mold, premixed, quick-curing resin is injected while the mold is closing or after it has closed. The technique has been used to make body parts for specialty cars and aircraft components.

Resist (n) (1) In preparing zinc printing plates, a material, such as wax, that covers the areas of the plate that are not to be etched by acid (*acid resist*); or, in electroplating, a material that covers areas that are not to be plated. (2) In photolithography and microlithography, widely used in making solid-state electronic devices and printed circuits, a thin film, applied over a substrate, whose solubility in a developer solvent is reduced (*negative resist*) or enhanced (*positive resist*) by exposure to UV or other radiation. In polymeric resists, the mechanism of solubility reduction is crosslinking and of solubility enhancement, chemical reaction or chain scission.

Resistance (electrical resistance) The property of a body that limits or opposes the passage of a steady electric current. In a direct-current circuit element, the resistance equals the quotient of the voltage drop across the element and the current passing through it (Ohm's law). If the dimensions of the conducting element and the material's VOLUME RESISTIVITY (w s) are known, the resistance may be calculated from them. The SI unit of resistance is the OHM, w s.

Resistance Heating Heating by means of dissipation of electrical energy in resistive circuit elements. Resistive heating is the principle of HEATER BANDs and CAST-IN HEATERs (w s) for extruders and injection machines, and of CARTRIDGE HEATERs for dies. It is the way that wire is preheated in wire-coating operations, and has been used in welding plastic components by placing between them flattened wires that are cut off and left in the part after welding. Also, it is the principle used in heating the radiant heaters used in sheet thermoforming. The rate of heating (watts, W) is given by the quotient of the square of the voltage drop across the heater (V^2) divided by its at-temperature resistance (Ω), also by the product of the voltage drop and the current ($V \cdot A$).

Resistance Thermometer A temperature-measuring device consisting of an encapsulated, fine coil of platinum wire whose resistance increases substantially and nearly linearly with rising temperature. See THERMISTOR. The change in resistance is sensed and converted to a temperature reading. Resistance thermometers have found considerable use in plastics processing.

Resistance Welding See THERMOBAND WELDING.

Resistivity (specific resistance) When this word stands alone, it usually means VOLUME RESISTIVITY, w s. But also see SURFACE RESISTIVITY.

Resite A term that has sometimes been used for a thermosetting resin in the fully cured or C-STAGE, w s.

Resitol See RESOLITE.

Resol A thermosetting resin composition in its unformed and uncured state, but containing all the necessary ingredients for hardening upon heating. See A-STAGE.

Resolite Synonym for a resin at its B-STAGE, w s.

Resonance (1) In chemistry, the periodic cycling of electrons from one atom of a molecule or ion to another atom of the same molecule or ion. Thus, given atoms remain in a fixed spatial arrangement with their electrons oscillating between atoms so as to satisfy two (or more) possible structural formulas. Resonance was first conceived to account for the outstanding stability of the benzene ring and it took almost a century for researchers to prove its reality. (2) A large-amplitude vibration of a mechanical or electrical system caused by a relatively small periodic stimulus applied at the natural frequency of the system. It is a phenomenon to be avoided in most structures. That is done by designing the natural frequency and the first few harmonics to be very different from the loading frequencies expected during use.

Resorcinol (resorcin, *m*-dihydroxybenzene, 3-hydroxyphenol) $C_6H_4{=}(OH)_2$. A highly reactive phenol, which, when reacted with formaldehyde, produces resins suitable for cold-setting adhesives.

Resorcinol Monobenzoate $HOC_6H_4OOCC_6H_5$. A white, crystalline solid used as an ultraviolet screener in plastics. It is particularly useful in applications requiring a high degree of transparency, and can be used with cellulosics, vinyls and certain polyesters.

Restricted Gate A small orifice between runner and cavity in an injection or transfer mold. Such a gate freezes (or cures) quickly when the cavity has filled and flow stops, preventing packing of the region near the gate in the cavity. When the piece is ejected, this gate breaks cleanly, simplifying separation of the runner from the molded item. It may be so small (see PINPOINT GATE) that its tiny stub need not be removed from the piece.

Restrictor Bar Synonym for CHOKER BAR, w s.

Restrictor Ring A ring-shaped part protruding from the torpedo surface within the cylinder of a plunger-type injection machine. It was claimed that the ring provided higher injection pressure on the melt, higher rate of filling the cavities, and improved weld strength.

Retainer Plate In injection molding, a plate that reinforces the cavity block against the injection pressure, and also serves as an anchor for the cavities, ejector pins, guide pins, and bushings. The retainer plate is usually cored for circulating water or steam.

Retaining Pin A pin in an injection or transfer mold on which an insert is placed and located prior to molding. The term is sometimes used to mean *guide pin* or *dowel pin.*

Retardation Time Of a stressed viscoelastic material (specifically, one behaving like a Voigt element), the time required after release of stress for the strain to decrease to 1/e (= 0.368) of its original value.

Retarder Synonym for INHIBITOR, w s.

Reticulated Polyurethane Foam A urethane foam of extremely low density, characterized by a three-dimensional skeletal structure of strands with few or no membranes between the strands, and containing up to 97+% of void space. Such foams are made by treating an open-cell foam structure with a dilute aqueous sodium hydroxide solution under controlled conditions so that the thin membranes are dissolved, leaving the strands substantially unaffected. Ultrasonic vibrating is sometimes used to assist the solution process. These foams are used in filters for air conditioners, automobile carburetors, and air-cleaning systems; and in acoustical panels, humidifiers, and various household products.

Return Pin In injection molding, any of the set of pins that return the ejector mechanism to its molding position.

Reverse-Flighted Screw A short section of a screw in a twin-screw extruder in which the flights have a helical direction opposite to that of the main screw sections, thus opposing the forward flow of the heat-softened plastic. The purpose is to improve mixing and dispersion of compound ingredients such as pigments, and sometimes to heat the melt just before it enters a vacuum-extraction zone. Reverse-flight sections have rarely been used in single-screw machines.

Reverse Helical Winding In filament winding, a pattern in which a continuous helix is laid down, reversing direction at each of the polar ends. It differs from biaxial, compact, or sequential winding in that the fibers cross each other at definite equators, the number depending on the helical lead, with the minimum number of crossovers being three.

Reverse Impact Test A test for sheet material in which one side of the specimen is struck by a pendulum or falling object and the reverse side is inspected for damage.

Reverse-Roll Coating A method of coating wherein the coating material is premetered between two rolls, one of which deposits the coating on a substrate. The thickness of the coating is controlled by the gap between the rollers and also by the speed of rotation of the coating roll.

Reworked Material A thermoplastic from a processor's own production that has been reground, pelletized, or solvated, after having been previously processed by molding, extrusion, etc. Note: In many specifications the use of reworked material is limited to clean plastic that meets the requirements specified for the virgin material, and yields a product essentially equal in quality to one made from only virgin material (ASTM D 883). See also REGRIND.

Reynolds Number (for Osborne Reynolds, 1883; Re, N_{Re}) A dimensionless ratio that relates to frictional pressure drop in fluid flow in pipes, defined by $D \cdot V \cdot \rho / \mu$ for gases and Newtonian liquids, where D is the inside pipe diameter, V is the average velocity of the fluid (= flow rate/cross section), ρ is the fluid density, and μ is the mass-based viscosity [Pa = kg/(m·s) in SI units]. When N_{Re} = 2100 to 4000, the character of the flow changes from streamline (laminar) to turbulent. For nonNewtonian liquids the criterion for flow transition by the Reynolds number is redefined. For a power-law liquid it becomes $D^n \cdot V^{2-n} \cdot \rho / (g_c \cdot K \cdot 8^{n-1})$, where n = the flow-behavior index of the liquid, K = its temperature-dependent consistency index, and g_c = the proportionality constant in Newton's law of momentum change. See POWER LAW. The Reynolds number is also applicable to other flow geometries, such as packed beds, with appropriate modifications.

R-Factor See THERMAL RESISTANCE.

RF Curing Hastening the final crosslinking of thermosetting compounds and laminates by application of radio-frequency energy.

RF Heating Synonym for DIELECTRIC HEATING, w s.

RF Shielding Enclosing equipment that generates radio-frequency radiation in a conductive housing that prevents the radiation from being broadcast and causing interference in nearby electronic devices that operate in the same frequency range. Required by law for all RF-heating equipment except that operating at 27.12 MHz, the frequency band reserved for industrial use.

RH Abbreviation for RELATIVE HUMIDITY, w s.

Rhe In the now deprecated cgs system, the rhe was the unit of fluidity, equal to 1 poise^{-1} [10 (Pa·s)$^{-1}$].

Rheogoniometer See WEISSENBERG RHEOGONIOMETER.

Rheology The study of deformation and flow of matter. The term rheology, derived from the Greek *rheos* meaning "something flowing", was proposed by Bingham in 1929 for the rapidly growing science of flow and deformation properties of materials, including liquids, solids, and even powders, in terms of stress, strain rates, and time. See also CONSISTENCY, DILATANCY, NEWTONIAN FLOW, NONNEWTONIAN, POWER LAW, THIXOTROPY, VISCOELASTICITY, VISCOSITY, and YIELD VALUE.

Rheomalaxis Synonym for SHEAR DEGRADATION, w s.

Rheometer (plastometer) An instrument for measuring the flow behavior of high-viscosity materials such as molten thermoplastics, rubbers, pastes, and cements. The most widely used principle is that of the CAPILLARY RHEOMETER, w s, of which a variety of makes and models are in daily use. Instruments for measuring the flow properties of less viscous fluids, e g, dilute polymer solutions, are called VISCOMETERs, w s, but the terms *rheometer* and *viscometer* are often used interchangeably. Currently (1993), computerized, on-line capillary rheometers linked to a single control station can simultaneously monitor melt viscosity in ten or more extruders in a resin-finishing plant.

Rheometer, Foam A special type of capillary rheometer developed by W. Kostyrzewski to measure the density and viscosity of foaming polyurethane immediately after mixing. Creaming time is also measured. It consists of a three-section, vertical acrylic tube, adapted at the bottom to a reaction-injection-molding mixer ("RIM machine"). The bottom chamber is large enough to contain the mix charge delivered by the RIM machine, and the mixture rises through an instrumented and temperature-controlled "capillary" as the foam forms and expands. The instruments are connected to an analog-to-digital converter and computer. The foam passes up through the capillary into a larger-diameter collection chamber and raises a floating disk that can be weighted to adjust the pressure on the foam as it rises. Flow rate is inferred from foam density; rate of change of density, and shear rate at the capillary wall are computed from flow rate. Shear stress is inferred from the pressure-transducer readings.

Rheopexy The inverse of THIXOTROPY, w s. The viscosity of a rheopectic material increases with time under an applied constant stress, approaching a constant value. When the stress is removed or reduced, the viscosity diminishes toward its original value.

Rhodamine Any of a class of organic dyes that exhibit bright orange to red fluorescent colors when viewed under ultraviolet light. An interesting characteristic is that, in vinyl compounds, the color shade is dependent on the degree of fusion. Thus, they can be used as indicators of fusion completeness.

Ribbon Blender A type of low-intensity mixer comprising helical, ribbon-shaped blades supported by spokes from the horizontal shaft, with the blade edges fitting fairly closely to the lower half of the U-shaped shell of the mixer. The shell is usually jacketed for temperature control and the shaft, spokes and blades may also be cored for circulation of heat-transfer liquid (usually water). They are used in the plastics industry mostly for cooling dry blends such as PVC extrusion and calendering compounds discharged from high-intensity mixers before storage and/or shipment.

Ricinus Oil Synonym for CASTOR OIL, w s.

Ridge Forming A variation of SHEET THERMOFORMING, w s.

Rigidity The ability of a structure to resist deformation under load. It is a function of both the material's modulus of elasticity and, often more critically, of the geometry of the structure. In a loaded beam, whatever the load distribution or type of beam supports, the maximum deflection is inversely proportional to the product, $E \cdot I$, of the material's elastic modulus and the moment of inertia of the beam's cross section about its neutral axis. See also SECTION MODULUS. The term *rigidity* is often applied loosely to materials themselves without reference to a particular structure when what the speaker actually has in mind is the elastic modulus.

Rigidity, Modulus of The slope of the linear portion of the stress-strain graph of a specimen tested in shear. See SHEAR MODULUS.

Rigid Plastic For the purposes of general classification, a plastic that has a modulus of elasticity either in flexure or in tension greater than 700 MPa (100 kpsi) at 23°C and 50% relative humidity when tested in accordance with ASTM methods D 747, D 790, D 638, or D 882 (ASTM D 883). This simple ASTM criterion has not always been adequate, especially with respect to vinyls whose impact strengths and other properties can vary widely while elastic modulus remains fairly constant. Vinyls are classified as rigid if their moduli are 1.4 GPa or higher, semirigid from 0.4 to 1.4 GPa, and flexible below 0.4 GPa.

Rigid PVC See RIGID PLASTIC and POLYVINYL CHLORIDE.

Rigidsol A coined term for a plastisol that forms an article of very high Durometer hardness. Such hardness is obtained by means of compounding techniques that permit the use of relatively small amounts of plasticizer, and/or by the incorporation of monomers that serve as diluents at room temperature but crosslink or polymerize upon heating.

RIM Abbreviation for REACTION INJECTION MOLDING, w s.

Ring A polymeric structure resulting from the reaction of one end of a molecule with the other end, forming a ring structure that may be likened to a snake biting its own tail. The stability of ring molecules formed from carbon-carbon chains is generally greatest in 5- to 6-membered rings, and least in 9- to 11-membered rings. The probability of ring formation decreases rapidly as the length of the molecule increases. Thus the presence of a few small rings in the polymer is usually insignificant.

Ring Gate A gate for molding tubular objects in which the top of the cavity is encircled by a thick runner connected to the cavity through a thin web (the ring gate) along its entire circumference. This promotes filling the mold without objectionable weld lines.

Ring Tensile Test See SPLIT-DISK METHOD.

RMA Abbreviation for RELAXATION-MAP ANALYSIS, w s.

Rocker A colloquial term for a blown container that is defective by reason of a bulged or deformed bottom causing the container to rock when placed upright on a flat surface.

Rockwell Hardness The net depth of penetration of one of several spherical indenters as the load on the indenter is increased from the *minor load* of 98 N to the *major load* of 588 or 981 N, then returned to the minor load. The test for plastics is described in ASTM D 785. One chooses the most appropriate of the five available scales, which have the following combinations of indenter and load. In order of increasing hardness they are R, L, M, E, and K scales.

Scale	R	L	M	E	K
Indenter diameter, mm	12.7	6.35	6.35	3.175	3.175
Major load, N	60	60	100	100	100

In an interlaboratory test, the overall variability for the six resins and 12 laboratories tested was about 10%. See also INDENTATION HARDNESS and SCRATCH HARDNESS.

Rodent Resistance The ability of a plastic to withstand or repel attacks by rodents. Some plastics require additives to prevent rodents from chewing objects such as cable insulation. One such additive is TETRAMETHYLTHIURAM DISULFIDE, w s.

Rod Mill A grinding machine consisting of a cylindrical, horizontal shell containing a number of free steel rods and rotated about its axis at a speed such that each rod is lifted almost to the top of the cylinder before it falls to impact the charge of material being ground. Compare BALL MILL.

Roentgen (R) The former international unit, now being phased out, of quantity or dose of X rays or gamma rays, a measure of the ionization induced by these rays, equal to 2.58 × 10^{-4} coulomb per kilogram (C/kg).

Roll Bending In calendering of sheet, the practice of applying a bending moment to the ends of the calender rolls that opposes the bending caused by the pressure forces as the plastic is squeezed between the rolls. The object is to produce sheet whose thickness varies minimally across its width. See also CROWN and ROLL CROSSING.

Roll Crossing In calendering, a method of compensating for the slight, but significant, bending apart of the rolls that occurs under working pressure, in which the roll axes are set slightly askew of each other. The effect is to make the thickness of the calendered sheet more nearly uniform across the sheet's width. See also CROWN and ROLL BENDING.

Roller Coating (roll coating) The process of coating substrates with liquid resins, solutions, or dispersions by contacting the substrate with a roller on which the fluid material has been spread. The process is often used to apply a contrasting color or raised lettering or markings. See also GRAVURE COATING, KISS-ROLL COATING, and REVERSE-ROLL COATING.

Roll Forming See COLD FORMING.

Roll-Leaf Stamping See HOT STAMPING.

Roll Mill An apparatus for mixing a plastic material with compounding ingredients, comprising two horizontal rolls placed close together. The rolls turn at speeds that differ by about 10% to produce a shearing action in the material being compounded. Mixing plows and slitting knives are sometimes used to work the stock across the rolls, thus improving the uniformity of additive distribution in the compound. See also REFINER.

ROM Acronymic abbreviation for *read-only memory*, the permanent part of computer memory not alterable by the user.

Root Diameter The smaller (*minor*) diameter of an extruder screw at any point along its length. In most industrial extruder screws, the increase in root diameter from the feed section to the metering section is the most important factor creating the COMPRESSION RATIO, w s.

Rosin An important natural resin, obtained from pine trees of several species by collection of exudate from cuts through the bark or by extraction from stumps (*wood rosin*). Rosin is brittle and friable, and is used mostly for sizing paper. Modified rosins are used in printing inks, pressure-sensitive adhesives, and chewing gum.

Rossi-Peakes Tester An instrument that tests a peculiar combination of properties of a thermoplastic molding compound that may relate to its moldability. Its use is the subject of ASTM Test D 569. A small, premolded test slug is inserted in a preheated charge chamber beneath an orifice (also preheated to the chosen temperature) 3.18 mm in diameter by 38 mm in length. In one procedure, a pressure of 10.3 MPa is immediately applied by a ram, forcing the charge, as it warms, to flow up into the orifice. A follower rod resting on top of the melt indicates the extent of flow of the material into the orifice. For each material tested, it is recommended that tests be run at three temperatures for which the flow is in the range 12.7 to 38 mm, the *flow temperature* being the temperature at which the flow is 25.4 mm.

Rotameter A simple flow-measuring device for gases and low-viscosity liquids consisting of a vertical glass tube whose internal diameter increases gradually from bottom to top, graduated on the outside, and containing a float that reaches an equilibrium position that rises in proportion to the flow rate. By calculation or calibration, one can infer the flow rate from the float position and the pressure and temperature of the fluid at the discharge end. Floats of small rotameters may be simple spheres interchangeable with others of different density to expand the instrument's range. The useful range of a good-quality rotameter and float is about tenfold. They are made in sizes ranging over many orders of magnitude of flow rate, from tubes not much larger than a pencil with a 2-mm ball float to others several inches in diameter with floats of elaborate conical or spindle shapes.

Rotary Joint A plumbing fixture used with cored extruder screws and chilled rolls that permits circulation of liquid within the screw or roll, without leakage at the point of connection of the inlet and return lines.

Rotary Molding A term sometimes used to denote a type of injection, transfer, compression, or blow molding utilizing several mold cavities mounted on a rotating table or dial. Not to be confused with ROTATIONAL MOLDING, w s.

Rotary-Vane Feeder A device for conveying and metering dry materials, comprising a cylindrical housing containing a concentric shaft with blades or flutes attached, rotating at a rate selected to feed the material at a desired rate.

Rotating Die See OSCILLATING DIE.

Rotating Spreader A type of torpedo used in plunger-type injection machines that consisted of a finned torpedo rotated by a shaft extending through the cored injection ram behind it. This was soon superseded by the screw-injection machine.

Rotational Casting The process of forming hollow articles from liquid materials by rotating a mold containing a charge of the material about one or more axes at relatively low speeds until the charge is distributed on the inner walls of the mold by gravity and hardened by heating, cooling, or curing. Rotation about one axis is suitable for cylindrical objects. Rotation about two axes and/or rocking motions are employed for completely closed articles. The process of rotational casting of plastisols comprises placing a measured charge of plastisol in the bottom half of an opened mold, closing the mold, rotating the mold about one or more axes while applying heat until the charge has been distributed and fused against the mold walls, cooling the mold until the deposit has gained strength, opening the mold, and removing the article. See also CENTRIFUGAL CASTING and ROTATIONAL MOLDING.

Rotational Injection Molding A modified injection-molding process applicable to hollow, axisymmetrical articles such as cups and beakers, in which the male half of the mold is rotated during the molding cycle until the material has hardened to a prechosen degree. The rotation produces orientation and increased crystallinity of some polymers, resulting in improved toughness and stress-craze resistance.

Rotational Molding (rotomolding) The preferred term for a variation of ROTATIONAL CASTING, w s, utilizing dry, finely divided, sinterable powders, such as polyethylene, rather than liquid slurries. The powders are first distributed, then sintered against the heated mold walls, the mold is cooled, and the product stripped from the mold.

Rotational Rheometer An instrument for measuring the viscosity of molten polymers (and many other fluid types) in which the sample is held at a controlled temperature between a stator and a rotor. From the torque on either element and the relative rotational speed, the viscosity can be inferred. The most satisfactory type for melts is the *cone-and-plate* geometry, in which the vertex of the cone almost touches the plate and the specimen is situated between the two elements. This provides a uniform shear rate throughout the specimen. It may be operated in steady rotation or in an oscillatory mode.

Rotational Viscometer An instrument for measuring the viscosity of pourable liquids, slurries, plastisols, and solutions. Most are of the *bob-and-cup* type. In these, the bob is a polished, accurate cylinder that is immersed in the liquid contained in the cup. Either the bob or cup is rotated and the torque on one or the other is measured, as is the rotational velocity. From these and the dimensions, the viscosity can be inferred, either directly by calculation from principles or indirectly by calibration with standards of known viscosity. An instrument widely used in the plastics industry is the BROOKFIELD VISCOMETER, w s.

Rotomolding A contraction of the term *rotational molding*, sometimes used indiscriminately for both the processes of ROTATIONAL CASTING and ROTATIONAL MOLDING, w s.

Roughing Pump In high-vacuum work, e g, vacuum metallizing, a mechanical pump that removes most of the air from the chamber, leaving the remainder to a secondary pump, usually a diffusion pump, capable of reducing the pressure to about 0.13 Pa absolute.

Roughness See FINISH (3).

Roving A form of fibrous glass comprising from 8 to 120 single filaments or strands gathered together in a bundle. When the strands are twisted together, the term *spun roving* is used. Roving is used in continuous lengths for filament winding, chopped into short lengths for use in reinforced-plastic molding compounds and in sprayup, and woven into skeins or mats for use in laminates. See REINFORCEMENT.

RP Abbreviation for REINFORCED PLASTIC, w s.

rpm Abbreviation for *revolutions per minute*, a convenient industrial unit of rotational speed. One rpm equals 0.10472 radian per second.

RRIM Abbreviation for *reinforced reaction injection molding*. See REACTION INJECTION MOLDING.

RTM Abbreviation for RESIN-TRANSFER MOLDING, w s.

RTP Abbreviation for REINFORCED THERMOPLASTIC, w s.

RTV Abbreviation for *room-temperature vulcanizing*, a characteristic of some elastomers that do not require heating to cure.

Rubber Hydrochloride A nonflammable thermoplastic material containing about one-third chlorine, obtained by treating a solution of rubber with anhydrous hydrogen chloride (HCl) under pressure at low temperature. The packaging film Pliofilm® is a representative product.

Rubber, Natural (India rubber, caoutchouc) An amorphous polymer consisting essentially of *cis*-1,4-polyisoprene, obtained from the sap (latex) of certain trees and plants, mainly the *Hevea brasiliensis* tree. The material is shipped from tropical plantations in one of two primary forms: latex, usually stabilized and preserved with ammonia and centrifuged to remove part of the water; or sheets made by milling the coagulum from the latex. Natural rubber has very high molecular weight and is usually masticated to reduce the molecular weight and improve processability. A major use is sidewalls of automotive tires.

Rubber-Plate Printing A marking method sometimes employed for intricate parts such as molded terminal blocks. Numbers, instructions, etc are stamped with conventional rubber stamps or printing plates.

Rubber-Plunger Molding A variation of matched-die molding that employs a deformable rubber plunger and a heated metal concave mold. The process enables the use of high fiber loadings.

Rubber, Synthetic See ELASTOMER.

Rubber, Synthetic Natural An awkward term sometimes used for synthetic elastomers having the composition of natural rubber, *cis*-1,4-polyisoprene.

Rubber Toughening The practice of compounding into a brittle plastic 5–20% of a rubber in the form of spherical particles, in order to improve the plastic's resistance to impact. The process has been used with both thermoplastics and thermosets, e g, polystyrene and epoxies. Some users of this term include toughening achieved by copolymerization with elastomer-forming monomers.

Rubber Transition (rubbery transition, gamma transition) See GLASS TRANSITION.

Rule of Mixtures See LAW OF MIXTURES.

Run (1) In experimental or test work, performance of the experiment or test at one set of experimental factors. (2) Resin flow down the vertical or sloping surface of a sprayed-up or laid-up reinforced-plastic fabrication before it can be cured (usually unwanted).

Runner In an injection or transfer mold, the feed channel, often branched to serve multiple cavities, and usually of semicircular or trapezoidal cross section, that connects the sprue with the cavity gates. The term is also used for the plastic piece formed in this channel.

Runnerless Injection Molding (1) In molding thermoplastics, a process in which the runners are insulated from the cavities and kept hot, so that the molded parts are ejected with only small gates attached. See also HOT-RUNNER MOLD. (2) In thermoset molding, the same as COLD-RUNNER INJECTION MOLDING, w s.

Runner System This term is sometimes used for the entire mold-feeding system, including sprue, runners, and gates, in injection or transfer molding.

Rupture (fracture) Failure by cracking or tearing of, or in, a test specimen or working part.

Rupture Disk A thin metal disk, contained in a *rupture-disk fitting*, that, when subjected to a known high fluid pressure, will tear, permitting outflow of fluid and rapid relief of pressure. Rupture disks are now used routinely on extruders and gear pumps to protect both personnel and equipment against the dangers of excessive melt pressure.

Rutile One of the crystalline forms of TITANIUM DIOXIDE, w s.

R-Value See THERMAL RESISTANCE.

S s

s (1) Abbreviation for SI's basic time unit, the second. (2) Symbol for sample STANDARD DEVIATION, w s.

S (1) Abbreviation for the SI unit of conductance, the SIEMENS, w s. (2) Chemical symbol for the element sulfur. (3) Symbol for ENTROPY, w s.

σ Symbol for STANDARD DEVIATION (2), w s.

SABRA Abbreviation for *surface activation beneath reactive adhesives*, a method of bonding plastics, such as polyolefins and polytetrafluoroethylene, that are normally unreceptive to adhesives without pretreatment. The method consists of mechanical abrasion of the surfaces to be joined to roughen their outer layers, scission of bonds with creation of free radicals, and further reaction with primers in the liquid, vapor, or gaseous phase. An adhesive such as an epoxy is then applied.

SACMA Acronymic abbreviation for SUPPLIERS OF ADVANCED COMPOSITE MATERIALS ASSOCIATION, w s.

SAE Abbreviation for SOCIETY OF AUTOMOTIVE ENGINEERS, w s.

Safety Factor See FACTOR OF SAFETY.

Sag (1) In blow molding, the local reduction in parison diameter, or *necking down*, caused by gravity. It is usually greatest on the portion nearest the die, and increases as the parison grows longer. (2) In thermoforming, sag is the downward bulge in the heat-softened sheet.

SAIB Abbreviation for SUCROSE ACETATE ISOBUTYRATE, w s.

Salt (1) In inorganic chemistry, an ionic substance formed by the reaction of an acid with a base. Polybasic acids can form acid salts and, when dissolved in water, generally yield acidic solutions, i e, their pH is less than 7. Similarly, polyacidic bases can form basic salt whose solutions are generally basic. (2) By analogy with the above, some organic reaction products of diacids and diamines that have ionic character are called salts. See NYLON SALT.

Sample Mean See ARITHMETIC MEAN.

SAN Abbreviation for STYRENE-ACRYLONITRILE COPOLYMER, w s.

Sanding A finishing process employing abrasive belts or disks, sometimes used on thermosetting-resin parts to remove heavy flash or projections, or to produce radii or bevels that cannot be formed during molding.

Sand Mill An apparatus used for preparing pigment dispersions, consisting of a vertical cylinder with a centrally mounted agitator shaft on which are mounted several flat, annular disk impellers. The mill is charged with coarse natural sand or high-silica ceramic beads as the grinding medium. The pigment slurry is pumped into the bottom of the mill and becomes mixed with the grinding medium. As the mixture is forced upward through the mill, it passes through several zones of agitation and finally flows through a screen at the top that retains the sand or beads but allows the much smaller pigment particles to pass.

Sandwich Heating (two-sided heating) A method of heating a plastic sheet for thermoforming in which the sheet is positioned between two radiant heaters. This practice cuts heating time to about one-fourth the time required with a single heater and also provides more uniform temperature through the sheet, so it is almost universally used.

Sandwich Structure A term often employed for a laminate comprising at least three layers; for example, a cellular-plastic or honeycomb core sandwiched between two layers of glass-reinforced laminate. See also LAMINATE.

Saran (1) Generic name for thermoplastics consisting of polymers of VINYLIDENE CHLORIDE (PVDC), w s, or copolymers of same with lesser amounts of other unsaturated compounds. (2) Saran® is also Dow Chemical's trade name for its copolymer films. PVDC's very low permeability to gases and vapors makes it an excellent barrier material for packaging and food wrapping.

Saran Fiber Generic name for a manufactured fiber in which the fiber-forming substance is any long-chain synthetic polymer composed of at least 80% by weight of vinylidene chloride units ($–CH_2CCl_2–$) (Federal Trade Commision).

Saturable Reactor (saturable-core reactor, SC, SCR) A component of some temperature-control systems, it is an AC coil of variable impedance which is limited by saturation of its iron core by means of a separate DC excitation coil.

Saturated Compound An organic compound in which there are no double or triple bonds between pairs of carbon atoms and which, therefore, cannot engage in addition reactions.

Saturated Polyester See POLYESTER, SATURATED.

Saturator A machine designed to impregnate paper, fabrics, and the like with resins. The web to be saturated is conveyed by rollers through a pan containing a solution of the resin, then through metering devices such as squeeze rolls, scraper blades, or suction elements that control the amount of resin retained.

SAXS Abbreviation for SMALL-ANGLE X-RAY SCATTERING, w s.

Saybolt Viscosity The time in seconds required to fill a 60-cm^3 flask with an oil specimen preheated to 38, 54, or 99°C and draining through a standard orifice. The measurement is convertible to kinematic viscosity by the equation: $\nu = 0.00222\ t - 1.83/t$, where ν = kinematic viscosity, cm^2/s and t = Saybolt seconds. Several ASTM tests make use of the Saybolt viscometer but none are related to plastics.

Sb Chemical symbol for the element antimony (Latin: stibnium).

SB Abbreviation for copolymers of styrene and butadiene.

SBR Abbreviation for STYRENE-BUTADIENE RUBBER, w s.

SBS Abbreviation for *styrene-butadiene-styrene* block terpolymer.

Scale of Segregation In mixing, the average distance between regions of the same component. See also INTENSITY OF SEGREGATION.

Scanning Electron Microscope (SEM) An electron microscope that uses electrons reflected from the sample surface to form images. The surface must first be made opaque and reflective, usually by sputtering onto it a very thin layer of gold. The microscope has more depth of field than the transmission-type instrument and can produce very sharp and insightful pictures. See ELECTRON MICROGRAPH.

Scarf Joint A joint made by cutting away congruent acute-angular segments on two pieces to be joined, then bonding the cut surfaces to make the adherends coplanar. See also BUTT JOINT and LAP JOINT.

Scatter Diagram A graph showing the closeness and character of the relationship between two measured variables, e g, tensile yield strength vs penetration hardness of the grades of a plastic, or elastic modulus vs glass content of a laminate. If the scatter diagram appears to be linear, one may then compute the CORRELATION COEFFICIENT, w s.

Scintillometer (scintillation counter) An instrument used to detect the presence of and measure the concentration of beta-emitting radioactive isotopes. When the emitted beta particle is intercepted by a molecule of the scintillant material in solution, a tiny flash of light is given off. By counting the flashes over a period of hours, one obtains an estimate of concentration of the known beta emitter.

Scleroscope An instrument for measuring impact resilience by dropping a ram with a flattened-cone tip from a specified height onto the specimen, then noting the height of rebound.

Score (v) To mark a line on a flat surface using a sharp-pointed instrument (*scriber*) and a straightedge. With a brittle material, the scoring may be deepened by repeated strokes, sometimes on both sides of the sheet, making it possible to cleanly break the sheet along the line of scoring. The same can be accomplished with brittle tubing with a triangular file, scoring only the outside surface.

"Scotch-Tape" Test A method of evaluating the adhesion of a lacquer, paint, or printed label to a plastic substrate. Pressure-sensitive adhesive tape is applied to an area of the painted plastic article, which may first be cross-hatched with scored lines. Adhesion is considered to be adequate if no paint adheres to the tape when it is peeled off.

SCR Abbreviation for either *saturable-core reactor* (see SATURABLE REACTOR) or *silicon-controlled rectifier* (see SCR DRIVE).

Scrap All products of a processing operation that are not present in the primary finished articles. This includes flash, runners, sprues, excess parison, and rejected articles. Scrap from thermosetting molding is generally not reusable. That from most thermoplastic operations can usually be reclaimed for reuse in the molder's own plant or can be sold to a commercial reclaimer.

Scrap Grinder See GRANULATOR.

Scrapless Thermoforming See SHEET THERMOFORMING.

Scratch Hardness The resistance of a material to scratching by another material. The test most often employed with plastics is the Bierbaum test, in which the specimen is moved laterally on the stage of a microscope under a loaded diamond point. The standard load is 0.0294 N. The width of the scratch is measured with a micrometer eyepiece and the hardness value is expressed as the quotient of the load divided by the scratch width. See also MOHS HARDNESS.

SCR Drive A variable-speed motor drive widely used today on new extruders in which a silicon-controlled rectifier (SCR) converts alternating current to run a direct-current motor. Unlike AC motors, DC motors have good torque characteristics over a wide speed range.

Screen Changer A device bolted to the head end of an extruder, between extruder and die adapter, that allows the operator to quickly replace a dirty screen pack with a clean one, usually without having to rethread the downstream line. The most common type consists of a two-position slide that can be shifted horizontally, in some cases by hand, in others by action of a hydraulic piston. As the clean pack moves smartly into position, the loaded one is exposed for removal and renewal. See also EXTRUDER SCREEN PACK.

Screen Pack See EXTRUDER SCREEN PACK.

Screen Printing (silk-screen printing) A printing process widely used on plastic bottles and other articles, employing as a stencil a taut woven fabric secured in a frame, the fabric being coated in selected areas with a masking material that is not penetrated by the ink being used. The stencil fabric is commonly called a "silk screen" even though silk is rarely used today. Nylon is most often used, and screens of copper, stainless steel and many other materials are suitable. The screen is placed above the part to be decorated, and a flexible squeegee forces ink through the openings in the screen onto the surface of the plastic article. Multicolor work requires multiple screens and impressions.

Screw In extrusion, the shaft provided with a helical channel that conveys the material from the feed hopper through the barrel, working it vigorously and causing it to melt, then developing pressure through pumping action to force the molten plastic through the die. See EXTRUDER SCREW for more details.

Screw Characteristic The relationship between volumetric flow rate in a melt-metering extrusion screw and the head pressure. For a melt screw operating isothermally, it is given by the equation:

$$Q = \alpha N - \beta \Delta P/(\eta L)$$

where N = the screw rotational speed, ΔP = the rise in pressure along the screw, equal to the head pressure, L = the axial length of the screw, η = the melt viscosity based on temperature and shear rate in the screw channel, and α and β are constants related to the screw dimensions. See also DIE CHARACTERISTIC.

Screw Conveyor A device for moving and metering the flow of solid particles at a fairly well controlled rate, consisting of a trough or sometimes a closed tube within which rotates a deep-flighted screw. Unlike an extruder screw, a conveyor screw normally operates only partly full and is a low-friction, low-torque device. A conveyor can be jacketed for heating or cooling. Screw conveyors are sometimes used as feeders for extruders and injection machines when it is desired to operate them in the starved mode.

Screw Extruder See EXTRUDER.

Screw Injection Molding See INJECTION MOLDING.

Screw Lead See second LEAD.

Screw-Piston Injection Molding See INJECTION MOLDING.

Screw Pitch See PITCH (1).

Screw-Plasticating Injection Molding See INJECTION MOLDING.

Scrim A low-cost, nonwoven, open-mesh reinforcing fabric made from continuous-filament yarn.

Sealant A liquid, paste, or coating, or tape that fills small gaps between mating parts, e g, pipe-thread sealant, or plugs small holes, stopping fluid leaks.

Sealing See HEAT SEALING.

Sebacic Acid (sebacylic acid, decanedioic acid) $HOOC(CH_2)_8COOH$. White leaflets derived from butadiene or castor oil, used as an intermediate in the production of plasticizers, alkyd resins, and certain nylons.

SEBS Abbreviation for block terpolymer *styrene-ethylene/butene-styrene*.

Secant Modulus The ratio of stress to corresponding strain at any specific point on the stress-strain curve. It is expressed in force per unit area, MPa or kpsi. The modulus must be labeled with its associated strain, e g, the 1% modulus. This definition of modulus is useful for many plastics whose stress-strain graphs are curved even at very low strains, exhibiting little or no Hooke's-law range.

Second (1) (s) The SI basic unit of elapsed time, the duration of 9,192,631,770 periods of the radiation corresponding to the transition between the two hyperfine levels of the ground state of the cesium-133 atom. All other measures of time intervals are defined in terms of the second. (2) A finished product containing slight imperfections but without serious functional faults and therefore usable, but usually sold at a discount.

Secondary (1, adj) In organic chemistry, a functional group in which two of its hydrogen atoms have been replaced by other groups, as a secondary alcohol, >CHOH, or a secondary amine, >NH. Examples are isopropyl alcohol, $(CH_3)_2CHOH$, and dimethyl amine, $(CH_3)_2NH$.

Secondary Plasticizer (extender plasticizer) A plasticizer that is less compatible with a given resin than is a PRIMARY PLASTICIZER, w s, and thus would exude or cause surface tackiness if used in excess of a certain concentration. Secondary plasticizers are used in conjunction with primaries to reduce cost or to obtain improvement in electrical or low-temperature properties.

Second-Order Transition See GLASS TRANSITION.

Second-Surface Decorating A decorating process used with transparent plastics in which the decoration is applied to the back of the part so that it is visible from the front but is not exposed to possible damage.

Section Modulus (Z) In a beam under load, the quotient of the moment of inertia of the beam's cross section about its neutral axis divided by the distance from the neutral axis to the outermost surface of the beam (I/c). The bending moment divided by the section modulus gives the maximum stress in the beam at any point along it.

Seebeck Effect The electrical phenomenon responsible for the action of a THERMOCOUPLE, w s. If a circuit consists of two metals, one junction being hotter than the other, a current flows in the circuit. The magnitude and direction of the current depend on the metals chosen and on the difference in temperature between the junctions. If the cold junction is held at a known constant temperature, e g, 0°C, the current becomes a measure of the temperature at the other junction.

Segregation (1) A close succession of parallel, rather narrow and sharply defined, wavy lines of color on the surface of a plastic, said color differing in shade from surrounding areas, and creating the impression that components of the plastic have separated. (2) In 1952, P. V. Danckwerts introduced the terms (w s) SCALE OF SEGREGATION and INTENSITY OF SEGREGATION to define the state of a heterogeneous mixture and to quantify the effectiveness of mixing processes.

Self-Extinguishing A somewhat loosely used term denoting the ability of a material to cease burning after the source of flame has been removed. PVC, vinyl chloride-acetate copolymers, polyvinylidene chloride, some nylons, and casein plastics are examples of self-extinguishing materials. See also FLAMMABILITY and FLAME RETARDANT.

Self-Ignition Temperature See AUTOIGNITION TEMPERATURE.

Semi-Automatic Molding Machine A machine in which only part of the operation is controlled by the direct actions of a person. The automatic segments are controlled by the machine's instruments or by a computer according to a predetermined program.

Semicrystalline Describing a polymer having both amorphous and crystalline regions in the range 20 to 80 percent. Most so-called "crystalline polymers" are, in most articles made from them, actually semicrystalline.

Semipermeable Membrane See OSMOMETER and OSMOSIS.

Semipositive Mold A mold with a plunger that fits loosely within the cavity as the mold begins to close, allowing excess material to escape as flash. As the mold nears complete closing, the plunger fits more closely to exert full molding pressure on the material. The semipositive mold combines the free flow of material inherent in a flash-type mold with the high density of moldings obtained with a positive mold.

Semirigid Plastic For the purposes of general classification, a plastic that has a modulus of elasticity in either flexure or tension of between 70 and 700 MPa at 23°C and 50% relative humidity when tested in accordance with ASTM D 747, D 790, D 638, or D 882 (ASTM D 883). See also RIGID PLASTIC.

Sequential Arrangement The arrangement of mer units in a polymer. A HEAD-TO-TAIL POLYMER, w s, exemplifies one possible sequence.

Sequential Winding See BIAXIAL WINDING.

Sequestering Agent A chemical that prevents metallic ions from precipitating from solutions of anions that would normally, without the sequestering agent being present, precipitate those ions. See also CHELATE.

Serpentine A type of asbestos containing CHRYSOTILE, w s.

Service Factor (1) See FACTOR OF SAFETY. (2) The fraction (or percentage) of planned operating time during which a device or system satisfactorily accomplishes its anticipated mission.

Service Life The time over which a part or system will continuously and satisfactorily perform its designed functions under stated service conditions. Service life may be determined by actual life testing or may be estimated by extrapolation from shorter-term testing, or from short-term testing under more severe conditions, as at higher temperature via TIME-TEMPERATURE EQUIVALENCE, w s.

Set (1, v) To become at least partly fixed or hardened by chemical or physical action, such as condensation polymerization, oxidation, vulcanization, gelation, hydration, or evaporation of volatile constituents. See also CURE. (2, n) See PERMANENT SET.

Setpoint The value set in a control instrument at which a control action will be taken when the instrument's sensor signals that the quantity controlled has just passed above (or below) the set value.

Setting Agent Synonym for CURING AGENT, w s.

Setting Time Of adhesives, castings, or hand layups and sprayups, the time from application until the material has become firm and handleable and will no longer flow by gravity, usually a time less than that required to dry completely or to reach a full cure or full strength.

S Glass A specialty glass composition containing 64% silica, 25% alumina, and 10% magnesia that provides high strength—4.5 GPa in fibers—sometimes used as reinforcement for laminates where high specific strength is wanted. See also GLASS-FIBER REINFORCEMENT.

Shank The section of an extruder screw to the rear of the flighted sections. The forward part of the shank contains a milled keyway holding the key that engages the tubular drive shaft and may act as a shear safety protecting the screw from overtorquing, while the rear part of the shank engages the radial and thrust bearings.

Sharkskin An irregularity of the surface of an extrudate in the form of finely spaced sharp ridges perpendicular to the extrusion direction, believed to be caused by a relaxation effect on the melt at the die exit.

Shaw Pot A name used in the early years of the industry for the original transfer-molding machine. It consisted of a conventional hydraulic press with a pot suspended above the mold. Material was charged into pot, then forced into the mold by the closing of the press.

Shear The movement, in a fluid or solid body, of a layer parallel to adjacent layers. See also SHEAR STRAIN.

Shear Degradation (rheomalaxis) Chain scission of a polymer caused by subjecting it to an intense shear field, such as exists in the close clearances of extruders and internal mixers.

Shear Flow Flow caused by the relative parallel or concentric motion of the surfaces confining a liquid, as in an extruder screw; or caused by a pressure drop, in the direction of flow, from the entrance of a flow passage to its exit, as in a die. Sometimes the two basic driving modes coexist, as in the metering sections of most extrusion screws and in wire-covering dies. In the direction crosswise to any laminar flow, successive layers slide past each other in shear.

Shear Heating Synonym for VISCOUS DISSIPATION, w s.

Shear Modulus (G, modulus of rigidity) The ratio of shear stress to shear strain within the proportional limit of a material. If the material exhibits no proportional region, a SECANT MODULUS, w s, is used.

Shear Rate (shear-strain rate, velocity gradient) The rate of change of shear strain with time. In concentric-cylinder flow where the gap between the cylinders is much smaller than the cylinder radii, shear rate is almost uniform throughout the fluid and is given by $\pi(R_1 + R_2)N/(R_2 - R_1)$, where R_1 and R_2 are the radii of the cylinders, one rotating, the other stationary, and N is the rotational speed in revolutions per second. The universally used unit of shear rate is s^{-1}. In tube flow, the shear rate varies from zero at the center to its maximum at the tube wall where, for a Newtonian liquid, it takes on the value $4Q/\pi R^3$. Q = the volumetric flow rate and R = the tube radius.

Shear Strain The amount of movement of one layer relative to an adjacent layer divided by the layer thickness. This may be expressed as an *angle of shear*, in radians.

Shear Strength The maximum SHEAR STRESS (w s) required to shear the specimen in such a manner that the moving portion completely clears the stationary portion.

Shear Stress Force per unit area acting in the plane of the area to which the force is applied. In an elastic body, shear stress is equal to shear modulus times shear strain. In an inelastic fluid, shear stress is equal to viscosity times the shear rate. In viscoelastic materials, shear stress will be a function of both shear strain and shear rate.

Sheet (sheeting) Sheet is distinguished from film in the plastics and packaging industries according to thickness: a web under 0.25 mm thick is usually called film, whereas thicker webs are called sheet. Sheeting is most commonly made by extrusion, casting, and calendering, but some, such as decorative laminates, are compression molded.

Sheet Die A heavy-walled, extremely rigid steel structure, bolted to an extruder head, whose inner passages form the molten plastic leaving an extruder screw into the shape of a flat sheet. Most modern sheet dies are of the coathanger type with multi-zone temperature control, and contain adjustable choker bars and die lips for close control of lateral variation in the sheet thickness. See COATHANGER DIE, CHOKER BAR, and FLEXIBLE-LIP DIE.

Sheeter Lines Parallel scratches or projecting ridges distributed over considerable area of a plastic sheet such as might be produced during a slicing operation.

Sheet Line The entire assembly necessary to produce plastic sheet, including the extruder, die, polishing-roll stand, cooling conveyor, pull rolls, and winder, or cutter and stacker, and all associated controls.

Sheet-Molding Compound (SMC) A fiberglass-reinforced thermosetting compound in sheet form, usually rolled into coils interleaved with plastic film to prevent autoadhesion. This term was chosen to replace the term PREPREG, w s, which was deemed to be confusing and insufficiently definitive. SMC can be molded into complex shapes with little scrap, and is low in cost.

Sheet Thermoforming (thermoforming) The process of forming a thermoplastic sheet into a three-dimensional shape by clamping the sheet in a frame, heating it to render it soft and flowable, then applying differential pressure to make the sheet conform to the shape of a mold or die positioned below the frame. When the pressure is applied entirely by vacuum drawn through tiny holes in the mold surface, the process is called *vacuum forming*. When above-atmospheric pressure is used to partially preform the sheet, the process becomes *air-assist vacuum forming*. In another variation, mechanical pressure is applied by a plug to preform the sheet prior to applying vacuum (*plug-assist forming*). In the *drape-forming* modification, the softened sheet is lowered to drape over the high points of a male mold prior to applying vacuum. Still other modifications are *plug-and-ring forming* (using a plug as the male mold and a ring matching the outside contour of the finished article); *ridge forming* (the plug is replaced with a skeleton frame); *slip forming* or *air-slip forming* (the sheet is held in pressure pads that permit it to slip inward as forming progresses); and *bubble forming* (the sheet is blown by air into a blister, then pushed into a mold by means of a plug). In *free forming* the sheet is formed entirely by gentle inflation with air and touches no mold, a method used to make cockpit canopies. The term sheet thermoforming also includes methods employing only mechanical pressure, such as *matched-mold forming*, in which the hot sheet is formed between registered convex and concave molds. Corrugated sheets are produced by this method. In *scrapless forming*, designed to reduce the amount of scrap inherent in other thermoforming methods, the process starts with a blank, usually square, of the material to be thermoformed. This blank is forged in a press to a circular preform, which can be thermoformed to a bowl or other shape without trimming. Millions of dairy-tub lids are formed every year by a variation called *trapped-sheet forming*, in which biaxially oriented polystyrene sheet is clamped tightly around the perimeter of each lid in a sheet while it is heated, so as to prevent relaxation of the orientation, then pressure formed with air at about 300 kPa. See also BLISTER PACKAGING and SKIN PACKAGING.

Shelf Life (storage life) The length of time over which a product will remain suitable for its intended use during storage under specific conditions. The term is applied to some finished products as well as to raw materials. Premixes and prepregs, being reactive materials, enjoy considerably longer shelf lives if refrigerated

Shell Flour See NUTSHELL FLOUR.

Shell Molding In metal foundries, a process of casting metal objects in thin molds made from sand or a ceramic powder mixed with a thermosetting-resin binder. Some authors have misused the term by equating it to plastics processes such as dipping and slush casting.

Shift Factor (a_T) The amount by which the logarithm of the modulus (or compliance) of a plastic, measured at temperature T (K) must be shifted along the time axis to bring it onto a single curve with the modulus measured at T_g, the glass-transition temperature. See TIME-TEMPERATURE EQUIVALENCE and WILLIAMS-LANDELL-FERRY EQUATION.

Shish-Kebab Structure A term borrowed from the broiling kitchen to designate a polymeric microstructure in which the random-coil chain of an amorphous polymer (shish) has been interlaced with crystalline cross segments (kebab) produced by shearing the polymer in the molten condition or in solution. The string-of-lumps microstructure so resembles the edible delicacy that its borrowed name seems quite apt.

Shoe See CHASE.

Shore Hardness See INDENTATION HARDNESS.

Short In reinforced plastics, an imperfection caused by an absence of surface film in some areas, or by lighter, unfused particles of material showing through a covering surface film, accompanied possibly by thin-skinned blisters.

Short-Beam Shear Strength Shear strength of reinforced-plastics materials as measured in three-point bending of a specimen whose length between end supports is about five times the specimen depth. ASTM D 2344 describes one such test for flat specimens and short arcs cut from relatively large rings, while ASTM D 4475 describes a similar test for pultruded round rod.

Short-Chain Branching The presence of 2- to 4-carbon side chains along the backbone of a polymer molecule. In an average low-density polyethylene molecule, both ethyl and butyl side chains, mostly the latter, are believed to be present, with a total of 50 in a typical molecule, in addition to one much longer branch.

Short Fibers A somewhat relative term: inorganic fibers from 12 to 38 mm long are referred to as "short staple", those longer than 76 mm, "long staple". In fiber-reinforced thermoplastics sold in pellet form, the fibers are less than 12 mm long, yet provide substantial reinforcement. Studies have shown that most of the possible strength gain is achieved with an aspect ratio of about 200 or more. Since glass-fiber diameters are about 10 μm, a length of 2 mm should be sufficient. For *modulus* improvement, greater lengths are beneficial.

Short Shot In injection molding, failure to fill all cavities of the mold completely, caused by too low melt temperature, too low injection pressure, insufficient plastication time, too constricted gates, too viscous resin, inadequate venting of cavities, etc. Short shots are often made deliberately when testing a new multicavity mold to reveal the pattern of runner flow and the sequence of cavity filling.

Shortstopper A term used for an agent added to a polymerization-reaction mixture to inhibit or terminate polymerization.

Shot (1) One complete cycle of a molding machine. (2) Imprecise synonym for SHOT WEIGHT, w_s.

Shot Capacity The maximum weight of plastic that can be delivered to an injection mold by one stroke of the ram or screw. In the case of screw-injection molding machines that are not equipped with backflow-preventing valves at the end of the screw, slippage of material may occur in the screw flights and may not be reckoned with in calculations of shot capacity that are based on cubic displacement.

Shot Weight (shot) In injection and transfer molding, the entire mass of plastic delivered in one complete filling of the mold, including the molded parts, sprue, runners, cull, and flash.

Shrinkage See MOLDING SHRINKAGE.

Shrinkage Allowance The dimensional allowance that must be made in molds to compensate for shrinkage of the plastic compound on cooling. The ASTM method for determining shrinkage from molded bars or disks and mold dimensions is D 955. This method does not provide for additional shrinkage that may occur as molded materials age beyond the first 48 h after removal from the mold.

Shrinkage Pool An irregular, slightly depressed area on the surface of a molding caused by uneven shrinkage before hardening is complete.

Shrink Film The prestretched or oriented film used in SHRINK PACKAGING, w s. See, too, HEAT-SHRINKABLE FILM.

Shrink Fit A method of joining circular and annular parts in which the outer member, having a slightly smaller inside diameter than the inner member's outside diameter, is heated, causing it to expand, then slipped into place over the inner member and allowed to cool. Alternatively, one can chill the inner member in liquid nitrogen, slip it into the outer member, and let it warm. Care must be taken in designing the joint to have the final stresses in both members well below yield values so as not to lose the joint to creep over time. Compare SNAP FIT.

Shrink Fixture Synonym for COOLING FIXTURE, w s.

Shrink Mark See SINK MARK.

Shrink Packaging A method of wrapping articles utilizing prestretched (oriented) films that are warmed to cause them to shrink tightly around the enclosed articles. First, the article is placed in a loose envelope of two layers of film, usually in the form of a V-folded strip. This envelope is heat sealed around the edges and detached from the strip, both of which operations can be done with an L-shaped, thermal-impulse sealer and cutter. The package is then conveyed through a hot-air oven or other heating device to shrink the film.

Shrink Tunnel An oven in the form of a tunnel mounted over, or containing a continuous conveyor belt, used to shrink oriented films in the shrink-packaging process.

Shrink Wrap (stretch wrap) The use of plastic films for unitizing several boxes or items loaded on a pallet. Film may simply be stretched over the materials to be protected, pulled tight and secured; or shrunk by application of heat. ASTM D 4649 (sec 15.09) is a guide to the selection of such materials. See also SHRINK PACKAGING.

Si Chemical symbol for the element silicon.

SI (1) Abbreviation for SILICONE, w s, or POLYDIMETHYLSILOXANE, w s. (2) Abbreviation for "International System of Units", derived from the official French name, *Le Systèm International d'Unités*. The system is a modern version of the MKSA (meter-kilogram-second-ampere) system published by an international treaty organization that is attempting to have the system adopted throughout the world. To date (1993) it has become official everywhere but in the US and Liberia. On December 23, 1975, President Ford signed the "Metric Conversion Act", which established a board to coordinate the voluntary conversion to the SI system in the US. "Practice for Use of the International System of Units (SI)", available from both ASTM and ANSI, provides an excellent summary of the system, as does "NBS Guidelines for the Use of the Modernized Metric System", available from the National Institute of Standards and Technology. In this Dictionary, SI units are emphasized, with equivalents in common working units given in occasional entries. A conversion table for selected quantities forms the Appendix of the Dictionary. Many professional organizations, including ASTM, American Chemical Society, American Institute of Chemical Engineers, and the Society of Plastics Engineers now require authors of papers submitted to their publications to use SI units.

Siamese Blow A colloquial term denoting the process of blow molding two or more objects or parts of objects in a single blowing mold, then cutting them apart.

SiB Abbreviation for POLYCARBORANESILOXANE, w s.

Side Bar A loose piece used to carry one or more molding pins, and operated from outside the mold (seldom seen today).

Side-Draw Pin A projecting mold element used to core a hole in a direction other than the direction of mold closing, and which must be retracted before the mold is opened and the part ejected.

Siemens (S) The SI unit of electrical conductance, equal to and replacing the *mho*, and the reciprocal of the OHM, w s. 1 Siemens = 1 ampere/volt (A/V).

Sieve Analysis (screen analysis) The separation of particulate solids into sequentially finer size fractions by placing a weighed sample into the topmost of a stack of graded standard sieves, mechanically shaking and tapping the stack for ten minutes, then weighing the material collected on each sieve and the pan beneath the lowest, finest sieve. The procedure is described in ASTM D 1921.

Sieve Fraction The mass fraction of a sieve-analyzed powder found between two successive screens in a SIEVE ANALYSIS, w s. For example, one might say, "The –0.420-, +0.250-mm fraction was 15.27 percent." See also MESH NUMBER.

Sigma-Blade Mixer A type of INTERNAL MIXER, w s, having blades that are (roughly) S-shaped.

Silane Coupling Agent Any silane or oxysilane that has the ability to bond inorganic materials such as glass, mineral fillers, metals, and metallic oxides to organic resins. The adhesion mechanism is due to two groups in the oxysilane structure. The $Si(OR_3)$ portion reacts with the inorganic reinforcement, while the organofunctional (vinyl–, amino–, epoxy–, etc) group reacts with the resin. The coupling agent may be applied to the inorganic materials (e g, glass fibers) as a pretreatment and/or added to the resin. Examples of silane coupling agents are:

N-β(AMINOETHYL)-γ-AMINOPROPYLTRIMETHOXY SILANE
γ-AMINOPROPYLTRIETHOXY SILANE,
BIS(β-HYDROXYETHYL)-γ-AMINOPROPYLTRIETHOXY SILANE
β-(3,4-EPOXYCYCLOHEXYL) ETHYLTRIMETHOXY SILANE
γ-GLYCIDOXYPROPYLTRIMETHOXYSILANE
γ-METHACRYLOXYPROPYLTRIMETHOXY SILANE
SULFONYLAZIDOSILANE
VINYLTRICHLOROSILANE
VINYLTRIETHOXYSILANE
VINYL-*tris*-(β-METHOXYETHYL)SILANE.

A newer class of silane coupling agents is known as *silyl peroxides*, represented by the general formula: $R'_mR''_{4-n-m}Si(OOR)_n$. A typical member of this family is vinyl-tris-(*t*-butylperoxy) silane. The coupling mechanism of the silyl peroxides, effected by heat only, is free-radical in nature. The conventional silanes require an external free-radical source and couple via an ionic mechanism initiated by hydrolysis.

Silica (silicon dioxide) SiO_2. A substance occurring widely in minerals such as quartz, sand, flint, chalcedony, opal, agate, and many more. In powdered form it is used as a filler, especially in phenolic compounds for ablative nose cones of rockets. Synthetic silicas, made from sodium silicate or by heating silicon compounds, are useful in preventing PLATE-OUT, w s. See also FUMED SILICA.

Silica Gel A form of colloidal silica consisting of grains having many fine pores and capable of adsorbing, and firmly retaining at room temperature, substantial quantities of water and some other compounds. It is used to dry gas streams and organic liquids to very low moisture levels. It can be reactivated by heating to temperatures above 100°C. Compare MOLECULAR SIEVE.

Silicon Carbide (SiC) In the form of crystals, produced in an electric furnace by reaction of carbon with sand, silicon carbide is a dense, extremely hard filler, used in some plastics to increase abrasion resistance, elastic modulus, and thermal conductivity.

Silicon Carbide Whiskers These high-modulus SiC fibers are made by pyrolysis of organosilanes in hydrogen at 1500 to 2000°C and by other methods. The whiskers are very fine, with diameters of about 2.5 μm, density of 3.2 g/cm^3, ultimate strength of 20 GPa (3 Mpsi), and modulus of 480 GPa (more than twice that of steel).

Silicone One of a large family of semi-organic polymers (*polyorgano siloxanes*) comprising chains of alternating silicon and oxygen atoms, modified with various organic groups attached to the silicon atoms. Depending on the nature of the attached organic groups, molecular weight, and the extent of crosslinking between chains, the polymers may be oily fluids ranging in viscosity from 0.001 to over 1000 Pa·s, or elastomers, or solid resins. The earliest silicones were dimethyl polysiloxanes, made by treating silicon derived from sand with methyl chloride in the presence of a catalyst to form a chlorosilane, hydrolyzing the chlorosilane to form a cyclic trimer of siloxane, then polymerizing the siloxane to form a dimethyl polysiloxane. Although this type of silicone is still in widespread use, many modifications have been made such as by the incorporation of phenyl groups, halogen atoms, alkyds, epoxides, polyesters, and other organic compounds containing –OH groups. The silicone fluids are used as lubricants, mold-release agents, heat-transfer fluids, and water-repellent coatings. The elastomers, often called silicone rubbers and reinforced with inorganic fillers or fibers, are vulcanizable and offer superior resistance to high temperatures and weathering. The silicone resins, possessing good electrical properties and strength at high temperature, are widely used for encapsulating and potting electrical components and in reinforced laminates. They have been alloyed with other resins, e g, polystyrene, which alloy has many of the good silicone attributes but costs less than neat silicone resin. Silicone solutions are also available for coating and varnishing.

Silicone Foam Foam based on fluid silicone resins is made by mixing the resins with a catalyst and blowing agent, pouring the mixture into molds, and curing at room temperature for about 10 h or at elevated temperatures for shorter periods. Silicone-foam sponge is made by mixing unvulcanized silicone rubber with a blowing agent and heating to the vulcanizing temperature.

Silicone-Polycarbonate Copolymer Introduced in 1969 by the General Electric Co, these thermoplastic copolymers vary from strong elastomers to rigid engineering plastics, depending on composition. They can be extruded, cast, or molded into optically clear films.

Silicone Rubber A synthetic rubber made by vulcanizing an elastomeric silicone gum such as dimethyl silicone. A free-radical-generating catalyst such as benzoÿl peroxide is usually used as the vulcanizing agent. The tensile strength of unreinforced silicone rubber is low, about 350 kPa. Higher strengths are obtained by adding reinforcing fillers such as finely divided or fumed silica, or by putting crystallizing segments such as silphenylene into the polymer. See also SILICONE.

Silicon Nitride Whiskers Very fine fibers of Si_3N_4 prepared by vapor-phase reaction of silicon and a silicate in nitrogen and hydrogen at 1400°C. Density is 3.2 g/cm^3, ultimate strength is 4.8 GPa (700 kpsi), and modulus is 276 GPa. They are used as reinforcements in specialty laminates.

Silk-Screen Printing See SCREEN PRINTING.

Siloxane (1) A chemical group with the structure shown below:

$$-\overset{\displaystyle R_1}{\underset{\displaystyle R_2}{\overset{|}{\underset{|}{Si}}}}-O-$$

in which the Rs are usually alike and can be alkyl radicals or even just hydrogen. (2) A compound containing siloxane links. A simple representative is hexamethyldisiloxane, $(CH_3)_3SiOSi(CH_3)_3$. Siloxane links are common in silicone resins. See SILICONE.

Silver Filler Particles or flakes of silver in the 1- to 15-μm size range, compounded with plastics to provide electrical conductivity. Flakes are mechanically flattened to provide layers of contiguous flakes and high conductivity. Silver oxide, sulfate, and carbonate are also good conductors and are used for the same purpose.

Silver-Spray Process A metalizing process based on glass-mirror art. The plastic article is first cleaned and lacquered as in vacuum metalizing, then the lacquer is sensitized in an oxidizing aqueous solution of sulfuric acid and potassium dichromate. A silver-forming solution, e g, silver nitrate and an aldehyde, is sprayed on the article with a two-nozzle spray gun so that the components mix at the surface. After rinsing and drying, a final topcoat of lacquer is applied over the silver.

Silyl Peroxide See SILANE COUPLING AGENT.

Simplified Flow Equation An equation giving the delivery rate of a single-screw extruder as a function of the screw diameter, screw speed, channel depth in the metering section, and room-temperature density of the plastic. Underlying assumptions include: the screw is single-flighted in the metering section, with typical dimensional proportions and shape factors, and a lead angle of 17.7°; the feeding and plasticating capabilities of the screw (and drive!) are sufficient to keep the metering section filled with melt; the feedstock is not preheated; and the total resistance to melt flow of screens and die does not entail excessive head pressure at the desired rate. See NET FLOW.

Single-Circuit Winding A winding in which the filament path makes a complete traverse of the chamber, after which the following traverse lies immediately adjacent to the previous one.

Single-Screw Extruder See EXTRUDER, SINGLE-SCREW.

Single-Stage Resin (single-step resin) See RESOL.

Sinking (hobbing) In mold making, pressing a hardened hob into annealed, soft mold steel or beryllium copper. See HOB.

Sink Mark (shrink mark) A shallow depression or dimple on the surface of an injection-molded article, usually in a thick section, caused by local internal shrinkage after the gate seals, or by a slightly short shot. Sinks can be diminished or prevented by reducing melt and mold temperatures, opening gates, filling faster, increasing pressure-holding time, and/or raising injection pressure.

Sinter Coating A coating process in which the article to be coated is preheated to sintering temperature and immersed in a plastic powder, then is withdrawn and heated to a higher temperature to fuse the adhering sintered powder into a continuous skin on the article.

Sintering The welding together of powdered plastic particles at temperatures just below the melting or fusion range. The particles are fused together to form a relatively strong mass, but the mass as a whole does not melt and may retain some porosity.

Sinter Molding The process of compacting fine thermoplastic particles by applying pressure at temperatures a little below the melting range and holding pressure until the particles fuse together, often followed by further heating and postforming. Porous nylon bearings capable of absorbing lubricants are made by this method, and sinter molding is a main molding method for powders of polytetrafluoroethylene or ultra-high-molecular-weight polyethylene, which do not form true melts.

Sisal A strong, durable, white fiber obtained from the agave plant grown in India, Indonesia, Mexico, and the West Indies. Chopped sisal enjoys some use as a reinforcement in thermosetting resins.

Size In the plastics industry, a synonym for COUPLING AGENT, w s.

Size-Exclusion Chromatography (SEC, gel-permeation chromatography) A column-chromatography technique employing as the stationary phase a swollen gel made by polymerizing and crosslinking styrene in the presence of a diluent that is a nonsolvent for the styrene polymer. The polymer to be analyzed is introduced in dilute solution at the top of the column and then is eluted with pure solvent. The polymer molecules diffuse into the gel, and out of it, at rates depending inversely on their molecular size. As they emerge from the bottom of the column they are detected by a differential refractometer connected to a computer or recorder, where a plot of concentration vs retention time is developed. This is converted, through calibration information, to a molecular-weight distribution whose parameters are calculated and printed out by the computer. See also CHROMATOGRAPHY.

Sizing (1) The process of applying a material to a surface to fill pores to smooth it and reduce absorption of a subsequently applied adhesive or coating, or to otherwise modify the surface. (2) Determining dimensions during design or production.

Sizing Plate In tubing and pipe extrusion, a plate with a central hole that may form the entrance to the water bath, and through which the still warm extrudate is passed to bring the outside diameter closer to its final dimension. At high line speeds, several plates having successively slightly smaller, smoothly finished openings may be used in sequence. Compare SIZING RING.

Sizing Ring (calibrator) In pipe extrusion, a hollow ring, slotted around its polished inside circumference, used to reduce the slightly oversize outside diameter of the warm pipe toward the desired final value. The core and slot of the ring are connected to a vacuum line. As the pipe passes through the ring, vacuum is applied to the core, sucking the pipe circumference against the slot and the ring's inside surface. Having made that seal, the operator raises the water level to submerge the pipe and cool it. Several rings of slightly decreasing inside diameters may be used in sequence for better control of pipe outside diameter and increased production rate. Compare SIZING PLATE.

Skin A relatively dense layer at the surface of a cellular polymeric material (ASTM D 883).

Skin Packaging A variation of the thermoforming process in which the article to be packaged serves as the mold. The article is usually placed on a printed card prepared with an adhesive coating or mechanical surface treatment to seal the plastic film to the card. See also BLISTER PACKAGING.

Skiving Slicing off a thin layer. Skiving is the method by which veneers are cut from logs and by which polytetrafluoroethylene film is produced from cylindrical bars of ram-extruded or sinter-molded resin. PTFE film is made in this way because, unlike most thermoplastics, it cannot be directly extruded into film.

Slab Stock Large, thick sheets of plastics, usually formed by casting from syrup or melt and used for fabrication of larger structures such as tanks.

Slate A fine-grained metamorphic rock of varied composition, used in powdered form as a filler, especially in flooring compounds.

Sleeve Ejector A bushing-type KNOCKOUT, w s.

Slip (1) With reference to adhesives, slip is the ability to move or reposition the adherends after an adhesive has been applied to their surfaces. (2) Of plastic film, having low surface friction and sliding easily over another layer of film or over machine surfaces in film-fabricating and packaging equipment.

Slip Agent (slip additive) A modifier (e g, oleamide) that acts as an internal lubricant by exuding to the surface of a plastic film during and immediately after processing. In other words, an invisible coating blooms to the surface, providing the necessary lubricity to reduce the coefficient of friction and thereby improve slip characteristics. Some experts feel that slip agents also act as ANTIBLOCKING AGENTs, w s.

Slip Forming (slip-ring forming) A variation of SHEET THERMOFORMING, w s, employing a sheet-clamping frame provided with tensioned pressure pads that permit the plastic sheet to slip inward as the part is being formed. This controlled slippage provides more uniform wall thickness in the formed article.

Slit Die (slot die, strip die) A PROFILE DIE, w s, with a nearly rectangular opening, used to produce an extrudate with a thin, moderately wide, rectangular cross section, but too narrow to be called sheeting. Sheet and film dies, which are more elaborate and have adjustable, removable lips, are not considered to be slit dies. A die producing a blocky rectangular cross section is called a *bar-stock die*.

Slitting The conversion of a given width of plastic film or sheeting to several smaller widths by means of knives. The operation can be performed as the material emerges from a production unit such as a calender, film-casting unit, or an (extrusion) roll stand (*in-line slitting*); by unwinding, slitting, then rewinding of rolls; or by slitting of rolls without unwinding (*roll slicing*). Slitting knives may be actual straight-edge knives, or razor blades, or circular knives.

Slug Molding A process for making thin-walled containers of 200- to 300-ml capacity. Melt from an extruder is fed into a metering head that delivers a slug of precise mass into a cylindrical bushing. From the bushing the slug is propelled into a single-cavity mold by a high-speed ram that passes through the bushing. The finished part is removed by a mechanical arm and transferred to an air conveyor that takes it to an automatic stacking unit. Trimming is unnecessary.

Slurry Preforming A method of preparing reinforcement preforms by wet-processing techniques similar to those used in PULP MOLDING, w s. For example, glass fibers suspended in water are directed onto a shaped screen that retains the fibers while allowing the water to pass through, thus forming a mat that has the shape of the object to be molded.

Slush Casting A method of forming hollow objects, widely used for doll parts and squeeze toys, in which a hollow mold provided with a closable opening is filled with a fluid plastic mixture, usually a vinyl plastisol. Heat, applied to the mold before and/or after filling, causes a layer of material to gel against the inner mold surface. When the layer has attained the desired thickness, the excess fluid is poured out, and additional heat is applied to fuse the gelled layer. After cooling, the article is stripped from the mold. Molds for slush casting are thin-walled for rapid heat transfer and are made of electroformed copper or cast aluminum.

Slush Molding The preferred term for the process resembling SLUSH CASTING, w s, but employing dry, sinterable powders.

SMA Abbreviation for copolymers of styrene and maleic anhydride.

Small-Angle X-ray Scattering (SAXS) A technique using high-intensity X-ray sources for determining a wide range of information on submicroscopic structures larger than 2 nm. Some applications with polymers are particle size and macromolecular structure of particles, SCALE OF SEGREGATION (w s) in blends, void structures, and SPECIFIC SURFACE, w s.

Smart Skin (smart composite) A composite containing molded-in sensors and microtransmitters that enable aerospace engineers to detect in-flight changes in temperature, strain, ice thickness, and cracks.

SMC Abbreviation for SHEET-MOLDING COMPOUND, w s.

Smectic (adj) See LIQUID-CRYSTAL POLYMER.

Smoke Suppressant An additive that, compounded into poorly burning polymers such as polyvinyl chloride, reduces smoke generation in fires. ASTM 2843 is a test for smoke generation, described under FLAMMABILITY.

SMS Abbreviation for copolymers of styrene and α-methylstyrene.

Sn Chemical symbol for tin (Latin: *stannum*).

Snap-Back Forming See VACUUM SNAP-BACK FORMING.

Snap Fit A method of reversibly joining plastic parts in which one part has a concave element—usually a diameter—slightly smaller in an inside dimension than the outside dimension of the mating convex element of the second part. Behind the circle of interference the convex part will be relieved to a smaller diameter, the concave part to a larger one. To assemble, one forces (snaps) the two parts together. Once the zone of interference is passed, the momentarily high stress is partly or wholly relieved. Disassembly is easily accomplished by pulling the two members apart, but is unlikely to occur spontaneously. A ubiquitous application of the principle is seen in caps for polyolefin half- and whole-gallon milk bottles; another is the Tupperware® family of reusable food containers.

S/N Curve Abbreviation for *stress at failure vs number of cycles* (curve). See FATIGUE CURVE.

$(SN)_x$ See SULFUR NITRIDE POLYMER.

Soap, Metallic Any product derived by reacting a fatty acid with a metal. Metallic soaps are widely used as stabilizers for plastics. The fatty acids commonly used are lauric, stearic, ricinoleic, naphthenic, octanoic (2-ethylhexanoic), rosin, and tall oil. Typical metals are aluminum, barium, calcium, cadmium, copper, iron, lead, magnesium, tin, and zinc.

Society of Automotive Engineers (SAE) Address: 400 Commonwealth Dr, Warrendale, PA 15906. Its 66,000 members are engineers, managers, and scientists in the field of self-propelled land, sea, air, and space vehicles. Among its publications are an annual handbook of standards, many of them shared with ANSI (w s). The AMS 3000 series includes many standards on elastomers, casting resins, plastics extrusions, moldings, potting compounds, and reinforcing fibers.

Society of Plastics Engineers (SPE) The leading international organization devoted to plastics engineering and technology, with 35,000 members, headquartered at 14 Fairfield Dr, Brookfield, CT 06804-0403. In 1993, SPE had 86 geographic sections, including 14 outside the US, and 20 special-interest divisions. SPE publishes several journals, holds many regional and national technical meetings each year, including its annual technical conference (ANTEC), has sponsored publication of scores of books on all aspects of plastics, supports plastics education at all levels, and founded the Plastics Institute of America, now a separate organization.

(The) Society of the Plastics Industry (SPI) An organization whose members include companies from all areas of plastics: materials producers, plastics sales people, processors and fabricators, designers and users of plastics products, and a few individual members. SPI's main address is 1275 K Street NW, Washington, DC 20005. SPI's Composites Institute is located at 355 Lexington Ave, New York, NY 10017. SPI sponsors a triennial international trade show. It has groups that interact with AMERICAN NATIONAL STANDARDS INSTITUTE (w s) to promulgate standards for safe operation of plastics processing equipment and machinery, and for certain plastics products such as pipe. The Composites Institute sponsors an annual technical conference on reinforced plastics. SPI lobbies for sensible legislation affecting plastics, and established and supports the Plastics Hall of Fame.

Sodium Aluminum Hydroxycarbonate $NaAl(OH)_2CO_3$. Named Dawsonite (after the mineral of the same nominal composition) by Alcoa, this material is produced in the form of microfiber crystals useful for upgrading physical properties of thermoplastics. In PVC compounds it also acts as a smoke suppressant and HCl scavenger.

Sodium Borohydride $NaBH_4$. A white crystalline powder, used as a blowing agent for foamed plastics such as rigid PVC and polystyrene, and for elastomers. The material decomposes at room temperature in the presence of water and an acidic medium, releasing hydrogen.

Sodium Hydroxide (lye, caustic soda) NaOH. A strong, cheap alkali, completely ionic, useful both as a reactant and a catalyst in polymer chemistry.

Sodium Polyacrylate $(-CH_2CH-)_nCOONa$. A thickening agent, the sodium salt of polyacrylic acid.

Sodium Stearate $NaOOCC_{17}H_{35}$. A water-soluble white powder, a soap, used as a nontoxic stabilizer.

Softening Range A temperature interval over which a plastic changes from a rigid to a soft state or undergoes a rather sudden and substantial change in hardness (ISO). ASTM D 1763 describes the Durrans' softening-point procedure for fully reactive epoxy resins. See also BALL-AND-RING TEST, MELTING POINT, and VICAT SOFTENING POINT.

Solid Casting The process of forming solid articles by pouring a fluid resin or dispersion into an open mold, causing the material to solidify by curing or by heating and cooling, then removing the formed article. Compare CASTING.

Solid-Phase Forming This term includes the shaping of plastic sheets or billets into three-dimensional articles either at room temperature (see COLD FORMING) or at higher temperatures up to the softening or melting range (see WARM FORGING) by processes resembling those used in the metals-working industry. These processes include forging with closed metal dies, rubber-pad forming (using one metal die and a rubber pad as the matching die), diaphragm forming, stamping, drawing, rolling, cold heading, threading, and coining. An advantage of solid-phase forming is that working of the material below its melt temperature orients it, making the thin sections stronger than the thicker sections. The opposite occurs in thermoforming processes employing temperatures above the melting range of the material: thin sections tend to be weaker than heavier sections. Among plastics suitable for at least some of the solid-phase processes are acrylonitrile-butadiene-styrene resins, acetals, cellulosics, polyolefins, polycarbonates, polyphenylene oxides, and polysulfones. Brittle materials such as acrylics and polystyrene cannot be formed by solid-phase processes.

Solids-Draining Screw (barrier screw) Any of a half-dozen designs of screws for single-screw extruders whose intent is to separate generated melt from the bed of solids as the melt is formed, the two main goals being better contact of the remaining bed with the hot barrel and preventing unmelted solids from passing into the metering section and die. The first of these, and still in use, was the MAILLEFER SCREW, w s. Some others bear the names Barr, Dray and Lawrence, Ingen Housz, Kim, SPR (for Scientific Process & Research), and Uniroyal.

Solid-State Controls Control instruments or motor-drive controls whose circuitry employs transistors and kindred elements rather than mechanical or vacuum-tube devices. Practically all modern process instruments are of this type.

Solid-State Polymerization A chain-growth polymerization initiated by exposing to ionizing radiation a crystalline monomeric substance. A large number of olefinic and cyclic solid monomers have been so polymerized, the crystalline monomer converting directly to the polymer with no obvious change in appearance of the solid.

Solprene See PLASTOMER.

Solubility (1) The mass of a substance that will dissolve in a given volume or mass of a solvent to form a solution that is homogeneous to the molecular level, sometimes expressed as weight percent. See also MISCIBILITY. (2) When no solvent and conditions are specified, the solubility in water at room temperature is usually meant.

Solubility Parameter (SP, δ) The square root of the COHESIVE ENERGY DENSITY (w s) of a polymer, solvent, or adhesive. For solvents, SP equals the square root of the heat of vaporization per unit volume $(J/cm^3)^{0.5}$. Values for polymers (which do not vaporize) are found by indirect methods. If the solubility parameters of a polymer and solvent differ by 3 $(J/cm^3)^{0.5}$ or less, the polymer will probably dissolve in the solvent. SPs for organic solvents range from 13 for neopentane to 30 for methanol; for polymers, from 13 for polytetrafluoroethylene to 32 for polyacrylonitrile.

Solute That constituent, usually a solid substance, of a solution which is considered to be dissolved in the solvent. The solvent is usually present in larger percentage than the solute.

Solution A single-phase mixture of two or more components, homogeneous at the molecular level, such as a resin dissolved in a liquid, that forms more or less spontaneously, will not separate, and has no fixed proportions of the components. Polymer solutions are used in the plastics industry to apply coatings, for film casting, and for spinning fibers. The term also includes gas/gas, gas-in-liquid, liquid/liquid, and solid/solid solutions.

Solution Casting See FILM CASTING.

Solution Coating Any coating process employing a solvent solution of a resin, as opposed to a dispersion, hot-melt, or uncured thermosetting resin. See also SPREAD COATING.

Solution Polymerization A polymerization process in which the monomer, or mixture of monomers, and the polymerization initiators are dissolved in a nonmonomeric solvent at the beginning of the polymerization reaction. The liquid is usually also a solvent for the resulting polymer or copolymer. Solution polymerization is most advantageous when the resulting polymeric solutions are to be used for coatings, lacquers, or adhesives. Vinyl acetate, olefins, styrene, and methyl methacrylate are the monomers most often employed.

Solution Viscosity See DILUTE-SOLUTION VISCOSITY.

Solvation The process of swelling, gelling, or solution of a resin by a solvent or plasticizer as a result of chemical compatibility. See SOLUBILITY PARAMETER.

Solvent (1) Broadly defined, a liquid with the ability to dissolve other substances. Solvents are used by the plastics industry in three main ways. As intermediates, solvents are used in the production of many monomers and resins. In plastics processing, solvents are used in etching, welding, polishing, film casting, fiber spinning, and making laminates. Finally, solvents are widely used in adhesives, printing inks, and surface coatings for plastics, as well as those based on plastics and used on other materials. The major types of solvents used in all these applications are alcohols, esters, glycol ethers, ketones, aliphatic and aromatic hydrocarbons, chlorinated hydrocarbons, and nitroparaffins. (2) The constituent of a solution that is (usually) present in larger percentage; or, in the case of solutions of solids or gases in liquids, the constituent that is liquid in the pure state at room temperature.

Solvent Bonding See SOLVENT WELDING.

Solvent Casting A process for forming thermoplastic articles by dipping a male mold in a solution or dispersion of the resin and drying off the solvent to leave a layer of plastic film adhering to the mold.

Solvent Cement See ADHESIVE.

Solvent Cementing (solvent bonding, solvent welding) The process of joining articles made of thermoplastic resins by applying a solvent capable of softening the surfaces to be joined, and pressing the softened surfaces together until the bond has gained strength. Adhesion is attained by evaporation of the solvent, absorption of it into adjacent material, diffusion of liquefied polymer molecules or chain segments across the interface, and/or polymerization of the solvent cement. Plastics joined by this method include acrylonitrile-butadiene-styrene resin, acrylics, cellulosics, polycarbonates, polystyrenes, and vinyls.

Solvent Diffusion The migration of solvent molecules into or out of a polymer as driven by the concentration gradient. Diffusion is the rate-limiting process in drying plastics and is also important in EXTRACTION EXTRUSION, w s.

Solvent Polishing A method for improving the gloss of thermoplastic articles by immersion in, or spraying with a solvent that will dissolve surface irregularities, followed by evaporation of the solvent. The method is used primarily for cellulosics, for which dipping is suitable. Plastics that are subject to crazing, such as polystyrene, are usually sprayed rather than dipped.

Solvent Resistance The ability of a plastic to withstand, unchanged, exposure to solvents. Plastics vary widely in their resistance to specific solvents.

Solvent Welding See SOLVENT CEMENTING.

Sonic Modulus The tensile/compressive modulus (E) estimated by measurement of sound-wave propagation in a material. ASTM Test C 769 (sec 15.01) describes such a method.

Sonic Velocity The speed of sound in a material. In air at 20°C and 50% relative humidity, the velocity of a plane longitudinal sound wave is 344 m/s. In polymethyl methacrylate, polystyrene, and polyethylene it is 2680, 2350, and 1950 m/s, respectively.

Soot-Chamber Test (ASTM D 2741) A test evaluating the relative effectiveness of antistatic agents in blown polyethylene bottles. After conditioning at 23°C and 50% RH, each bottle is rubbed ten times with a paper towel to generate static charge, then placed in a test chamber at 15% RH for 2 h. Filter paper wetted with toluene is ignited and the smoke is blown into the test chamber. Fifteen minutes later the bottles are removed from the chamber and examined. Soot accumulation is judged to be clean (i e, none), slight, moderate, or severe.

Sorption The process of one substance taking up and holding another by physical or chemical action. Usually the sorbed substance is mobile, a gas or vapor, and the sorbing phase is dense, a liquid or solid. However, components of liquids can also be sorbed by solids, as by a MOLECULAR SIEVE, w s. See also ABSORPTION, ADSORPTION, CHEMISORPTION, and PERSORPTION.

Soybean Meal The product of grinding soybean residue after extraction of oil, sometimes treated with formaldehyde to reduce moisture absorption. It is used as a filler, often in conjunction with wood flour, in thermosetting resins.

Soybean Oil A pale yellow oil extracted from soybeans, used in epoxidized form as plasticizers and stabilizers for vinyl resins.

Spandex Generic name for a manufactured fiber in which the fiber-forming substance is a long-chain synthetic polymer comprised of at least 85% of a segmented polyurethane (Federal Trade Commission). These fibers are used in garments to enhance stretchability and resilience. Compared to its competitors for this purpose, spandex is stronger and stiffer, with better resistance to heat and oxidation.

Spangles See GLITTER.

Spanishing A printing process similar to VALLEY PRINTING, w s. Ink is deposited on the bottoms and sides of depressed areas of an embossed plastic film.

Spark Machining See ELECTRICAL-DISCHARGE MACHINING.

SPC Abbreviation for STATISTICAL PROCESS CONTROL, w s.

SPE Abbreviation for SOCIETY OF PLASTICS ENGINEERS, w s.

Specific Adhesion Adhesion between two surfaces that are held together by valence forces of the same type as those that give rise to cohesion, as opposed to mechanical adhesion in which the adhesive holds the parts together by interlocking action.

Specific Gravity (sp gr) The ratio of the density of a liquid or solid material to the density of water at 4°C, or other specified temperature.

The notation $d_{t_2}^{t_1}$ is often used to specify the two temperatures, the subscript t_2 being that of water. For gases, specific gravity is ratioed on that of air at standard conditions, usually 101.325 kPa and 0°C. When buoyancy corrections are made to mass determinations of the densities, the term *absolute specific gravity* is used. Specific gravity is often misused as a synonym for density but the error isn't as serious as the parallel error with specific heat because, in the density unit still in most common use, g/cm^3, water's density is very close to 1.000 at room temperature. The ASTM test for specific gravity (and density) of plastics by displacement is D 792. See also DENSITY and BULK DENSITY.

Specific Heat Strictly, specific heat is the ratio of the HEAT CAPACITY, w s, of a substance to that of water at 15°C. In traditional cgs and English units, the heat units (calorie and British thermal unit) were defined by the heat capacity of water, making that of water at room temperature closely equal to 1.00 in either system. Thus, for other materials, specific heat and heat capacity were numerically equal. This fact led to the use, still ongoing, of "specific heat" when the property meant was heat capacity. In the SI system, the heat capacity of water is 4.18 J/g·K, whereas specific heats are dimensionless and not affected by changes in units , so are no longer equal to heat capacities. Soon, it is hoped, this confusing and now useless term will fade out of the language of science and engineering.

Specific Inductive Capacity Synonym for PERMITTIVITY, w s.

Specific Insulation Resistance See VOLUME RESISTIVITY, w s.

Specific Modulus Elastic modulus (usually the tensile modulus) divided by density, the SI unit being Pa/(kg/m^3) (technically reducible to J/kg, which blurs its derivation and true nature). For example, the specific moduli of neat acetal resin (homopolymer) and glass-fiber-reinforced epoxy are, respectively, 2.3 and 11 MPa/(kg/m^3).

Specific Strength Tensile strength (if some other is not specified) divided by density. The SI unit is Pa/(kg/m^3). For example, the specific strengths of neat acetal resin (homopolymer) and 30-33%-glass-reinforced nylon 6/6 are, respectively, 48 and 118 kPa/(kg/m^3). Specific strength, like SPECIFIC MODULUS, w s, is a more useful criterion than strength alone for material selection when both strength and minimum weight are important in the design of a structure.

Specific Surface Of porous solids, massed fibers, and particulate materials, the total surface area per unit of bulk volume or per unit mass. Specific surface is usually measured by gas adsorption or estimated from mercury-porosimetry measurements.

Specific Viscosity Of a dilute polymer solution, the specific viscosity is equal to the relative viscosity (*viscosity ratio*) minus 1. It represents the fractional increase in viscosity that can be attributed to the polymer solute. See DILUTE-SOLUTION VISCOSITY.

Specific Volume The volume of a unit mass of a material; the reciprocal of its density. The SI unit of specific volume is m^3/kg, but for polymers, cm^3/g (= 1000 m^3/kg) is more convenient.

Spectrophotometer An instrument for measuring the brightness of various portions of spectra. One useful application of this instrument is in the formulation of colorants to match a given sample under all types of illumination. The instrument produces a curve representing the amounts of light energy the specimen to be matched absorbs over a wide range of wavelengths. Matching this curve assures that the developed compound will appear to be the same color as the specimen under any lighting condition. Another important application of the principle in a different instrumental format is in the field of quantitative chemical analysis. See also COLORIMETER.

Spectroscopy The study of electromagnetic waves that are absorbed or emitted by substances when excited by an arc, a spark, X rays, or a magnetic field. Each element emits light of characteristic wavelengths, by which minute quantities can be detected and estimated. See also NUCLEAR MAGNETIC RESONANCE.

Specular Gloss The luminous fractional reflectance of a specimen in the specular direction. The ASTM method for measuring specular gloss is D 523.

Specular Transmittance (regular transmittance) Of a transparent plastic, the ratio of the light flux transmitted without diffusion to the flux incident (ASTM D 883). See also LIGHT TRANSMITTANCE.

Speed Reducer (gear reducer) The gearbox between motor and screw coupling of an extruder that reduces the high speed of the motor shaft to the much lower speed, with inversely higher torque, of the extruder screw. The reduction of about 12 to 1 normally requires two stages. The gears may be spur gears (cheapest and least desirable), helical gears, or double-helical (herringbone) gears. Where the extruder is expected to handle a wide variety of plastics and products using an assortment of screws, a change-gear reducer is recommended. It enables the operator to quickly accomplish large changes in the reduction ratio, while also permitting safe and efficient operation of the variable-speed motor-drive and gears. One may still find a worm-gear reducer in an antique extruder long past its rightful retirement age.

Spew Groove Synonym for FLASH GROOVE, w s.

Spew Line Synonym for PARTING LINE, w s.

Spherulite A rounded aggregate of radiating crystals with a fibrous appearance under the microscope. Spherulites have been observed in most crystalline plastics. They originate from a nucleus such as a particle of contaminant, residual catalyst particle, or chance fluctuation in density. They may grow through stages: first needles, then bundles and sheaf-like aggregates, and finally the spherulites. Spherulites may range in diameter from a few tenths of a micrometer to several millimeters. They are birefringent, displaying a characteristic maltese-cross insignia when viewed through crossed polarizers.

SPI Abbreviation for (THE) SOCIETY OF THE PLASTICS INDUSTRY, w s.

Spider (1) In a molding press, that part of an ejector mechanism that operates the ejector pins. (2) Within an end-fed extrusion die making a tubular section, the three or four legs extending from die to mandrel and supporting the mandrel at the center of the die. The spider legs may themselves be cored to pass a temperature-control fluid through the mandrel and into an extended calibrating mandrel attached to the die mandrel. Also called *spider fins*. (3) In rotational casting, the grid-work of metallic members supporting cavities in a multicavity mold.

Spider Lines In blow molding or pipe extrusion, visible marks parallel to the parison or pipe axis and corresponding to the positions of the spider legs. They are due to incomplete welding of the divided stream downstream of those legs. These lines are the exterior traces of weld surfaces that go through the annular wall, surfaces that are sometimes weaker than the material between them.

Spider Runners (spider gating) A design for melt distribution to multiple cavities in an injection mold in which the cavities are arranged in a circle and fed by runners radiating from a central sprue.

Spin Dyeing (mass dyeing, dope dyeing) The process of coloring fibers or yarns by incorporating pigments or dyes in the material prior to spinning, either during or after polymerization of the material.

Spinneret (1) An extrusion die consisting of a plate with many tiny holes, through which a plastic melt or solution is forced, to make fine fibers and filaments. Early spinneret holes were round and thus produced fibers of circular cross section. Today, spinneret holes have many different shapes, even annular ones, to produce fibers of corresponding cross sections. One purpose is to decrease the fiber-bundle density, giving added warmth, moisture permeability, and enhanced dye receptivity to the textile fabric. An important application of hollow fibers is in artificial kidneys for dialysis. Filaments emerging from the spinneret may be hardened by cooling in air or water, or by chemical action of solutions. (2) A spinneret hole.

Spinning The process of forming staple fibers by extruding polymers. There are three main variations of the process: *melt spinning*, *dry* *spinning*, and *wet spinning*. All employ extrusion dies with from one to thousands of tiny orifices, called *jets* or *spinnerets*. In melt spinning, molten polymer is pumped first through sand-bed filters, then through the die orifices by small gear pumps operating at extremely high pressures. In both wet and dry spinning the polymer is dissolved in a solvent prior to extrusion. In dry spinning, the extrudate is subjected to a hot atmosphere that removes the solvent by evaporation. In wet spinning, the spinneret is immersed in a liquid that either diffuses the solvent or reacts with the fiber composition, precipitating it from solution. However produced, the fibers are then oriented to realize their optimal strength and modulus, four times or more that of the unoriented fibers. With larger fibers (see MONOFILAMENT) this is done by reheating the fiber in a carefully controlled oven, and while it is warm, stretching it with godet stands operating at a large speed differential. Much staple fiber is also oriented in this way. With melt-spun staple, the trend is toward accomplishing most, if not all of the orientation by high-speed windup and drawing as the extruded filaments are leaving the spinnerets.

Spin Welding (friction welding) A process for joining thermoplastic articles at circular mating surfaces by rotating one part in contact with the other until sufficient heat is generated by friction to produce a thin interfacial layer of melt. Rotation is stopped and pressure is maintained while the melt solidifies, completing the weld. The process can be performed manually in a drill press, or can be fully automated with devices for feeding, timing, controlling stroke and pressure of the press, and ejection. In a version of the process that is applicable to some noncircular joints, rotation is oscillatory through small, reversing arcs.

Spiral-Flow Test A method of evaluating the molding flow of a resin in injection or transfer molding in which the melt is injected into a spiral runner of constant trapezoidal cross section with numbered and subdivided centimeters (or inches) marked along the runner. The mold is filled from a sprue at the center of the spiral and pressure is maintained until flow stops, the number just aft of the molded-spiral tip giving the flow distance. The spiral-flow test has been widely used since it was introduced in the early 1950s but has been standardized in the US only for thermosetting resins, in ASTM D 3123.

Spiral-Mandrel Die The most popular die design for extrusion of blown film, in which the melt, fed into a manifold at the bottom of the die, flows up to the die lip via multiple spiral grooves. The design is capable of handling a range of resins, provides even distribution to the lip, thus helping to minimize circumferential variations in film thickness, and has relatively low pressure drop.

Spiral Mold Cooling A method of cooling injection molds (or heating transfer molds) wherein the heat-transfer medium flows through a spiral passage in the body of the mold. In injection molds, the cooling medium is introduced at the center of the spiral, near the sprue section, because more heat is concentrated in this area.

Splash See SPLAY.

Splay (splay marks, silver streaking) Scars or surface defects on injection molding arising from two main mechanisms. One is the injection of a high-velocity stream of molten material into the mold ahead of the general flow front. The prematurely injected melt, especially in the case of crystalline polymers with a sharp freezing point, cools and solidifies before the mold cavity is completely filled. These defects occur most frequently in the gate area, but may be washed into other areas of the cavity. Remedies are: increasing the mold temperature, local heating of the gate area, reduction of injection rate, and increasing the gate cross section. The second mechanism is release of moisture or residual monomer or solvent in the melt in the form of fine bubbles at the part surface. The remedy for this type of splay is predrying to remove volatiles from the feedstock before it enters the molder and/or careful handling of regrind to prevent moisture pickup during recycling. These precautions are especially important with hygroscopic resins such as nylon 6/6.

Split-Disk Method (NOL ring test) A method of testing the tensile strength of tubular plastics and reinforced plastics embodied in ASTM D 2290. The standard test specimen is a ring whose internal diameter (ID) = 146.05 mm, width = 6.35 mm, and thickness = 1.52 mm. The ring is slipped over a wider steel disk split along its diameter and only slightly smaller in diameter than the ID of the test ring. The disk halves have holes bored through them to accommodate pins in ⊃-shaped fixtures that can be gripped by the testing machine. Thermoplastic specimens may be molded or cut from pipe, using suitably sized disk halves, but reinforced-plastic rings are specially filament-wound in accordance with ASTM D 2291. Because the disk halves fit so closely, the very small sections of the specimen at the ends of the horizontal slit between the disk halves are pulled in pure tension.

Split-Film Fiber A type of polypropylene fiber made by extruding a thin film, applying a high lengthwise draw to strongly orient it, at the same time weakening it in the transverse direction, pounding it to cause lateral fibrillation into narrow ribbons about 1 mm wide, and further orienting the ribbon fibers. These may be woven into a variety of products such as sacking, artificial turf, and indoor-outdoor carpet.

Spray-and-Wipe Painting (fill-in marking) A decorating process for articles with depressed letters, figures, or designs. A lacquer or enamel is applied by spraying either the article's entire surface or a restricted area including the depression, then removing the excess wet paint by buffing or wiping the raised areas.

Spray Coating The application of a plastic coating to a substrate by means of a spray gun. The process is used for coating any material with a plastic, and for coating plastics for decorative purposes. In the latter application, masks are usually employed to apply the coating only to selected areas. See also ELECTROSTATIC SPRAY COATING, FLAME-SPRAY COATING, and PLASMA-SPRAY COATING.

Spray Drier A cylindrical or conical-bottomed chamber through which hot air rises and into which a resin emulsion is sprayed in small droplets. The solvent evaporates from the droplets as they fall and the dried-resin powder is removed from the chamber bottom by scraping or air blasting. Spray drying has been employed for emulsion-polymerized vinyl resin, and for amino and phenolic resins.

Sprayed-Metal Mold A mold made by spraying molten metal onto a master form to obtain at shell of desired thickness, which may subsequently be backed with plaster, cement, casting resin, etc. Such molds are used most commonly in sheet-forming processes.

Spray Molding See SPRAYUP.

Sprayup A term coined in February, 1958, by the Engineering Editor of *Modern Plastics* to identify a new technique of reinforced-plastics molding that employed a gun that chops roving and impels the chopped strands through the focus of two resin sprays and onto the mold or mandrel. One resin stream contains catalyst, the other promoter, formulated to set quickly at room temperature. An advantage of the process is the ease with which section thickness may be varied in the molding to meet local strength and stiffness needs. The term sprayup was later extended to the foamed-plastics field, denoting the spraying of fast-reacting polyurethane or epoxy-resin systems onto surfaces where they react to foam and cure. In both areas the external mixing of streams by impingement avoids pot-life problems in spray equipment and tanks.

Spread (n) In the adhesive-bonding trade, the quantity of adhesive applied to a unit area of a material to be bonded to another. It has customarily been expressed in lb/1000 ft^2 of joint area.

Spread Coating A process for coating fabrics, sheet metals, etc, with fluid dispersions such as vinyl plastisols. The substrate is supported on a carrier, and the fluid material is applied to it just ahead of a blade or "doctor knife" that regulates the thickness of the coating. The deposit is then heated to fuse the coating to the substrate, often followed by embossing to impart texture.

Spreader (1) See TORPEDO (2). (2) Any device, such as a knife or roller, a part of spread-coating equipment that helps to produce an evenly thick coating.

Spring-Box Mold A type of compression mold equipped with a spacing fork that prevents the shifting of bottom-loaded inserts or loss of fine details, and which is removed after partial compression.

Sprue (1) In injection and transfer molding, the main feed channel connecting the machine nozzle with the runners leading to the various cavities. The sprue is usually conical, widening slightly toward the mold, so that, as the mold opens, the plastic within the sprue remains attached to the runners and the sprue is cleared for the next shot. (2) The conical plastic stub that is formed within the sprue with each shot and is removed with the runners.

Sprue Bushing (British: sprue bush) A hardened steel insert in an injection mold that contains the tapered sprue passage and has a suitable seat, usually hemispherical, making a seal with the nozzle of the injection cylinder.

Sprue-Ejector Pin Synonym for SPRUE PULLER, w s.

Sprue Gating (direct gating) In injection molding with single-cavity molds, particularly those with circular symmetry, filling from the center with the sprue connected directly to the gate and mold cavity.

Sprue Lock In injection molding, a portion of the plastic composition that is held in the cold-slug well by an undercut, used to pull the sprue out of the sprue passage as the mold is opened. The sprue lock itself is pushed out of the mold by an ejector pin. When the undercut occurs on the cavity-block retainer plate, this pin is called the *sprue-ejector pin.*

Sprue Puller A pin having a Z-shaped slot undercut in its end, by means of which it pulls the molded sprue out of the sprue passage.

Spunbonded Sheet A sheet structure resembling paper or felted fabric, made by heat sealing webs of randomly arranged, continuous thermoplastic fibers. Three such materials, of polyolefin and polyester fibers, were introduced by Du Pont in 1968. Good qualities are all-directional tensile strength, high tear resistance, good flex life, and puncture resistance. Applications include wall coverings, book covers, tags, labels, mailing envelopes, industrial clothing, and filter media.

Spun Roving A glass-fiber strand, repeatedly doubled back on itself to form a roving, sometimes reinforced by one or more straight strands (ISO).

Spur A term sometimes used to mean SPRUE (2), w s.

Sputtering See VACUUM METALIZING.

SQC Abbreviation for (*statistical*) QUALITY CONTROL, w s.

Squeegee A soft, flexible blade or roller used in wiping operations, particularly in SCREEN PRINTING and LAYUP MOLDING, w s.

Squeeze Molding A process for making prototypes from sheet molding compounds (SMC) with inexpensive tooling and very low molding pressure, to develop designs for parts that will be produced by injection molding or from metal by die casting. An epoxy two-piece mold is prepared, details such as ribs, gussets, and bosses are positioned, then the mold is filled with reinforced SMC and pressed at 140 to 210 kPa until cured.

SRIM Abbreviation for *structural reaction injection molding*. See REACTION INJECTION MOLDING.

SRP Abbreviation for *rubber-reinforced polystyrene*. See RUBBER TOUGHENING.

SS (1) Abbreviation for *single-stage* (resin). See RESOL. (2) Abbreviation for STAINLESS STEEL, w s.

Stabilizer A chemical added to some plastics to assist in maintaining the physical and chemical properties of the compound at suitable values throughout the processing and service life of the material and articles made therefrom. An *emulsion stabilizer* serves to keep emulsions and suspensions from separating. A *viscosity stabilizer* is used in vinyl dispersions to retard viscosity increase on aging. An agent used primarily to protect plastics and rubbers from deterioration by oxidation is called an ANTIOXIDANT, w s. The remaining, and most important, types of stabilizers are those that protect plastics from the effects of heat and light. Such effects are evidenced by a change of color, ranging from slight yellowing to blackening; a progressive decrease in mechanical properties; a decrease in electrical properties; or undesirable surface conditions such as blisters, spew, or exudation of ingredients rendered incompatible by heat or light. Many resins are vulnerable to ultraviolet light because they are good absorbers of UV and because its photonic energy is high. Stabilizers that function primarily by absorbing UV are described under ULTRAVIOLET STABILIZER. Thousands of compounds have been proposed as heat stabilizers and as combination heat and light stabilizers for various plastics. The principal classes of such compounds are: (1) Group II metal salts of organic acids (primarily the barium, cadmium, and zinc salts of fatty acids and phenols, the most important group); (2) ORGANOTIN STABILIZERs, w s; (3) EPOXY STABILIZERs, w s; (4) salts of mineral acids, e g, carbonates, sulfates, silicates, phosphites, and phosphates; (5) other organic compounds of metals and metalloids, e g, alcoholates and mercaptides. Heat stabilizers are used nearly exclusively with vinyl resins. See also ZINC STABILIZER.

Stainless Steel A generic term embracing over 70 standard ferrous alloys produced in either wrought or cast forms, containing from 12 to 30 percent chromium and from 0 to 22 percent nickel (18 Cr, 8 Ni is typical). Also present may be small percentages of other elements such as carbon, columbium, copper, molybdenum, tantalum, and titanium. Stainless and semi-stainless steels have been used in plastics processing equipment where corrosion is a problem. A stainless steel has sometimes been chosen instead of chrome plating in environments where the plating has to be frequently repeated. Stainless type 17-4 PH, containing 17% Cr, 4% Ni, 4% Cu and 0.35% Co, which is hardenable, has served well in extrusion dies for PVC.

Stain Resistance The ability of a plastic material to resist staining caused by traffic, coffee, tea, blood, waxing compounds, grease deposits, and other staining agents. In the case of plasticized PVC, the most severe staining is caused by shoe polish, tobacco smoke, lipstick, nail polish, ketchup, and mustard. The degree of staining can be reduced to some extent by use of certain plasticizers, e g, butyl benzyl phthalate, diethylene glycol dibenzoate, dipropylene glycol dibenzoate, and 2,2,4-trimethyl-1,3-pentanediol monoisobutyrate benzoate.

Staircase Method See UP-AND-DOWN METHOD.

Staking A term sometimes used for the process of forming a head on a protruding portion of a plastic article for the purpose of holding a surrounding part in place. Ultrasonic heating of the protrusion facilitates the staking operation.

Stalk (n) A European term for SPRUE (2), w s.

Stamping See DIE CUTTING.

Stamp Molding A compression-molding variation in which the mold is closed so suddenly that the plastic is impacted rather than pressed, as in forging.

Standard (secondary standard) (1) An amount of a substance whose content of specified elements or compounds is known within error limits that are narrower than those of the analytical method or instrument to be calibrated with the standard. (2) An object having one or more measurable properties whose values have been certified by an appropriate standards-issuing authority to be within error limits that are usually narrower than those of the measuring method or instrument that the object will be used to calibrate. Examples are gauge blocks ("Jo blocks") for micrometers, standard resistors and capacitors, standard cells (DC emf sources) for potentiometers, NIST-calibrated thermocouples and thermistors, balance weights, wavelength (or frequency) standards, tensile- and compressive-force standards, and time standards. Secondary standards that are traceable to government-maintained primary standards are the ultimate basis of all quantitative scientific and engineering work, of manufacturing-quality control, and of commerce. Standards and calibration services are available from the National Institute of Standards and Technology, and from many private companies. See also CALIBRATION.

Standard Deviation (1) (s) The most used and most efficient measure of the dispersion of values in a sample of measurements drawn from a population of possible measurements of a particular quantity or dimension. Its definition is

$$s = \left[\frac{1}{n-1} \sum_{i=1}^{n} (x_i - \bar{x})^2 \right]^{0.5}$$

where x_i = any individual measurement, $\bar{x}$ = the sample mean, and n = the number of measurements in the sample (*sample size*). s^2 is called the *sample variance* and $(n - 1)$ is known as the *degrees of freedom* for the sample. (2) (σ) The standard deviation of the conceptually very large population of possible measurements from which a sample is drawn.

Standardized Measurement In a sample of n measurements of a property or dimension of a part, material, or product, the difference between any individual measurement and the sample mean, divided by the sample standard deviation, i e, $(x_i - \bar{x})/s$. Standardizing helps in making tests of normality of a sample, as by plotting on normal probability graph paper. See also STUDENT'S *t*.

Standard Normal Deviate (z) The difference between any member of a normal distribution of measurements and the distribution (true) mean, divided by the standard deviation, i e, $(x_i - \mu)/\sigma$. Compare this with the identically structured STANDARDIZED MEASUREMENT for a sample.

Stannous 2-Ethylhexanoate (stannous octanoate) A polymerization catalyst for urethane foam.

Staple Fiber Short, spinnable fibers between 1 and 13 cm in length.

Starch, Permanent An aqueous emulsion of a synthetic resin for application to fabrics which when ironed become stiff as if starched.

Starved Area See RESIN-STARVED AREA.

Starve Feeding (starved feeding) In extrusion, regulating, with an auxiliary feeding device such as a weigh feeder or screw conveyor, the rate at which feedstock enters the feed port of the extruder so that the screw flights are less than full.

Static Eliminator A mechanical or electrical device for draining off static electrical charge from plastics articles by creating an ionized atmosphere in close proximity to the surface. Types include static bars, ionizing blowers and air guns, and radioactive elements. All except latter operate on the principle that a high-voltage discharge from the applicator to ground creates an ionized atmosphere. A newer type employs ceramic microspheres containing radioisotopes that emit alpha particles. A layer of the microspheres is bonded to a substrate with a resinous binder, and the laminate is installed in an aluminum housing. By ionizing air this device provides a conductive path through which charge is drained.

Static Mixer (motionless mixer) Any of several types of devices, used widely in the process industries and in plastics extrusion, that contain no moving parts but instead accomplish mixing by repeatedly splitting the melt (or other) stream and braiding or intertwining the streamlets. The scale of segregation of the different constituents in the flow or of hotter and cooler regions, is exponentially reduced in proportion to the number of stages. They are mainly used between extruder and die to insure the thermal homogeneity of the melt reaching the die lip, with consequent close control of extrudate thickness across the lip.

Stationary Platen In an injection molder, the large front plate to which the front plate of the mold is bolted. This platen does not normally move.

Statistical Process Control (SPC) The application of statistical methods, both simple and advanced, to identifying and reducing sources of product-quality variation and output-quality limiters in production processes, the goal being to build quality ("zero defects") into the process rather than to remove defective items by inspection, as in (statistical) QUALITY CONTROL, w s.

Steady Flow (steady-state flow) Any flow in which velocities throughout the stream do not vary with time.

Steady State A condition of processes or parts of processes in which the state variables describing the process, e g, temperature, pressure, compositions and velocities of streams, and amounts of materials residing in various process equipment, do not change with time. Most extrusion operations closely approximate the steady state except during startup and shutdown, whereas injection molding and sheet thermoforming are unsteady, *intermittent* processes.

Steam Molding A process for molding plastic-foam parts from pre-expanded beads of polystyrene that contain a volatile hydrocarbon, e g, isopentane, as a blowing agent. The steam is usually in direct contact with the beads but, with thin-wall moldings such as coffee cups, may be used indirectly to heat mold surfaces that contain the beads. The process is widely used for molding packaging elements for the electronic-equipment industry, but it can also make huge "logs" that are subsequently sliced into foam "lumber". See POLYSTYRENE FOAM.

Steam Plate See FORCE PLATE.

Stearic Acid (octadecanoic acid) $CH_3(CH_2)_{16}COOH$. A saturated organic fatty acid obtained by the hydrolysis with sodium hydroxide (*saponification*) of beef tallow. The salts, amides, and esters of this acid have long played important roles in the realm of plastics as lubricants, slip agents, and plasticizers, and stearic acid itself is present in numerous plastics compounds.

Stearyl Methacrylate A group name for compounds of the general formula $CH_2{=}C(CH_3)OOC(CH_2)_nCH_3$, in which n is from 13 to 17. It is a polymerizable monomer for acrylic plastics.

Steel-Rule Die A sharp-edged knife fashioned from thin steel strip, flexible enough to be shaped to complex outlines. It is used as the cutting element in DIE CUTTING, w s.

Stellite® Trade name of the Haynes Stellite Co for a family of hard and corrosion-resistant metal alloys. Stellites 1 and 6 are used for hard-facing extruder screws by first machining a groove in what will become the flights' outer surfaces, then welding the alloy into the groove and grinding to final diameter and finish after machining the main screw channel. Both alloys are mainly cobalt, Stellite 1 containing about 31% chromium and 13% tungsten (W), while Stellite 6, the choice for use in Xaloy-306-lined extrusion cylinders, contains about 28% Cr and 3–6% W. Hardness of Stellite 1 is 52–55 on the Rockwell C scale; that of Stellite 6 is a little less.

Stereoblock A regular block (in a polymer) that can be described by one species of stereorepeating unit in a single sequential arrangement (IUPAC).

Stereoblock Polymer A polymer whose molecules consist of stereo-blocks connected linearly (IUPAC). See also STEREOSPECIFIC.

Stereograft Polymer A polymer consisting of chains of an atactic polymer grafted to chains of an isotactic polymer. For example, atactic polystyrene can be grafted to isotactic polystyrene under suitable conditions.

Stereoisomerism A kind of isomerism in organic compounds arising from the fact that a carbon atom linked to four different groups can exist in two spatially different forms that, though chemically identical, are not superimposable. A simple example is lactic acid, $CH_3CH(OH)$–$COOH$.

Stereoregular Polymer A regular polymer whose molecules can be described by only one species of STEREOREPEATING UNIT, w s, in a single sequential arrangement. A stereoregular polymer is always a tactic polymer, but a tactic polymer need not have every site of stereoisomerism defined (ISO).

Stereorepeating Unit A configurational repeating unit having defined configuration at all sites of stereoisomerism in the main chain of a polymer molecule (IUPAC). In stereoregular polypropylene, the two simplest possible stereorepeating units are:

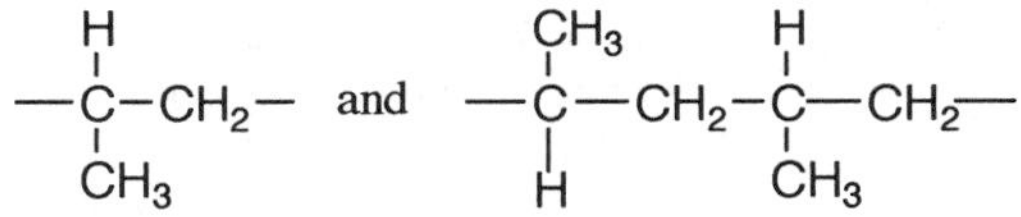

$$\begin{array}{c} H \\ | \\ -C-CH_2- \\ | \\ CH_3 \end{array} \quad \text{and} \quad \begin{array}{c} CH_3 \qquad\quad H \\ | \qquad\qquad | \\ -C-CH_2-C-CH_2- \\ | \qquad\qquad | \\ H \qquad\quad CH_3 \end{array}$$

which would be the mers of the corresponding stereoregular polymers, the first being isotactic, the second, syndiotactic.

Stereoselective Polymerization Polymerization in which a polymer molecule is formed from a mixture of stereoisomeric monomer molecules by incorporation of only one stereospecific species (IUPAC).

Stereospecific Of polymerization catalysts, implying a specific or definite order of spatial arrangement of molecules in the polymer resulting from the catalyzed polymerization. This ordered regularity of molecules (TACTICITY, w s), in contrast to the branched or random structure found in other plastics, permits close packing of the molecular segments and leads to high crystallinity, as in polypropylene. The adjective is sometimes applied imprecisely to polymers to mean tactic.

Stereospecific Polymerization A polymerization in which a tactic polymer is formed (IUPAC).

Stiffness (1) The capacity of a structure to resist elastic deformation under stress. (2) Loosely applied to materials as a synonym for any elastic modulus. Compare RIGID PLASTIC.

Stir-in Resin (dispersion resin, paste resin) A vinyl resin that does not require grinding or extremely high-shear mixing to effect dispersion in a plasticizer or to form a plastisol or organosol.

Stitching (stitch welding) The progressive joining of thermoplastic film and sheets by successive applications of two small, mechanically operated electrodes, connected to the terminals of a radio-frequency generator, using a mechanism somewhat like that of an ordinary sewing machine.

Stock Temperature In plastics processing, the temperature of the plastic (as opposed to temperatures of metal parts of the equipment) at any point. Often taken to mean, if not otherwise qualified, the temperature of the melt within an extruder head or leaving the nozzle of an injection molder. See MELT TEMPERATURE. Stock temperatures may be measured by sturdy thermocouples or thermistors inserted into the plastic stock, or by infrared instruments pointed at emerging extrudates, sheets being thermoformed, etc.

Stoddard Solvent A petroleum distillate comprising 44% naphthenes, 39% paraffins, and 17% aromatics, used as a diluent in PVC organosols.

Stoichiometric Pertaining to a mixture of chemical reactants, each ion or compound of which is present in the exact amount necessary to complete a reaction with no excess of any reactant. For example, each ingredient of a urethane-foam formula should be present in its stoichiometric quantity in order to assure a high-quality product.

Stoke The deprecated cgs unit of kinematic viscosity, equal to 10^{-4} m^2/s.

Storage Life A synonym for SHELF LIFE, w s. See also WORKING LIFE.

Storage Modulus In dynamic mechanical measurements, the part of the COMPLEX MODULUS, w s, that is in phase with the strain, with the symbol G' if the testing mode is shear, E' if it is tension or compression.

Stormer Viscometer A rotational-type instrument in which the test liquid fills a stationary, baffled cup and the rotating element may be a concentric cylinder or one of several paddle designs. The rotor is powered by adjustable falling weights through step-up gearing. Because of the complex geometry with any of the Stormer rotors, viscosity cannot be directly calculated from the weight and the time required for the rotor to make 100 revolutions. For each configuration, the instrument must be calibrated with liquids of known viscosity to establish the instrument constant for use with unknown liquids or slurries. Operation of the instrument, used mostly in the paint industry, is described in ASTM D 562 (sec 06.01).

STP Abbreviation for *standard conditions of temperature and pressure*. In scientific work these are 0°C and 101.325 kPa (1 atm). American gas industries and some others often choose 70°F (21.1°C) as their standard temperature.

Strain In tensile and compression testing, the ratio of the elongation to the gauge length of the test specimen, that is, increase (or decrease) in length per unit of original length. The term is also used in a broader sense to denote a dimensionless number that characterizes the change in dimensions of an object during a deformation or flow process. In shear deformation, strain is the shear angle in radians. See also SHEAR STRAIN and TRUE STRAIN.

Strain Birefringence Double refraction in a transparent material subjected to stress and accompanying strain. One of the techniques of experimental stress analysis is based on this phenomenon. See also FLOW BIREFRINGENCE.

Strain Energy The recoverable, elastic energy stored in a strained body and recovered quickly upon release of stress. Strain energy in a perfectly elastic material is equal to the area beneath the stress-strain curve up to the strain being considered. For a Hooke's-law material, it is equal to $0.5 \cdot \text{modulus} \cdot \text{strain}^2$. This area, which appears to have the dimensions of stress (Pa), is actually the strain energy per unit volume (J/m^3).

Strain Gauge A small electrical element consisting of a very fine wire of many short runs and reverse turns embedded in a tape-like matrix. The strain gage is adhered to an object whose deformation under stress it is desired to measure. With complex objects, several gages may be attached at different locations and oriented in different directions. The gage undergoes the same strain as the object when stress is applied and the strain stretches the wire, which, through the many doublings-back, magnifies the change in length. The increase in resistance, proportional to object strain, is measured with a bridge circuit, of which the gage is one branch.

Strain Relaxation A misnomer for CREEP, w s. What may be mistakenly meant by this term is STRESS RELAXATION, w s.

Strand A primary bundle of continuous filaments combined in a single compact unit, without twist. The number of filaments in a strand is usually 52, 102, or 204.

Strand Chopper (1, pelletizer) A type of cutter into which multiple extruded and chilled strands are fed for cutting into short lengths about equal to the strand diameter. The strands are fed perpendicular to the edge of a thick stationary knife and are sheared off by several rotating knives bolted to a massive cylindrical head. The pellets so formed are discharged into a shipping container or conveyed into mass storage bins. In this way, molding powders are produced for extrusion or molding into finished products. Widely used by many smaller compounders and by reclaimers, this method of producing pellets has largely given way, in the plants of the large resin producers, to UNDERWATER PELLETIZING, w s. (2) A device for chopping strands of fibrous reinforcement into lengths suitable for blowing onto a preform screen, compounding into premix, or combining with sprayed resin in SPRAYUP, w s.

Strapping (1) Thin, flat, continuous-strip material available in widths from 6 to 25 mm, usually in coils, and designed to be used with hand-operated machines that dispense, tighten, and clamp the strip around a package or bundle of packages. The material may be steel, unidirectionally fiber-reinforced plastic, or oriented nylon or polypropylene. (2) High-strength hoisting straps woven from aramid fiber, replacing chains in many industries for handling loads with hoists.

Stress (σ or τ) The force producing or tending to produce deformation in a body, divided by the area over which the force is acting. If the stress is tensile or compressive, the area is perpendicular to the stress; in shear it is parallel to the stress. The SI unit of stress is the pascal (Pa) equal to 1 newton per square meter (N/m^2). Practical stresses are usually in the range from 1 kPa to 1 MPa. In theoretical mechanics, the stress tensor in a body has nine possible components, three of which are tensile/compressive, the others shear. In many cases, all but one or two of the components are zero or not relevant to the behavior question of interest.

Stress Concentration The magnification of applied stress in the vicinity of a notch, hole, inclusion, or inside corner. Minimizing stress concentrations is an important aspect of plastics product design.

Stress Corrosion The preferential attack of areas under stress in a corrosive environment, when environment alone would not have caused corrosion.

Stress Cracking (stress crazing) External or internal cracking in a plastic caused by tensile stresses less than the short-time tensile strength. See ENVIRONMENTAL STRESS CRACKING.

Stress-Intensity Factor (K) A fracture-mechanics parameter that describes the magnifying of stress caused by a flaw in a material. It is defined by the equation:

$$K = f \cdot S(\pi \cdot a)^{0.5}$$

where f = a factor dependent on the flaw geometry and the structure in which it is contained, S = the nominal or average stress caused by the load, and a = half the length of the flaw measured normal to the direction of stress.

Stress Relaxation The decay of stress with time at constant strain. If a plastic specimen is strained and the recovery of the strain prevented, the chain segments of the molecules will tend to realign so as to lower the free energy of the system, sometimes by breaking covalent bonds. The elastic and retarded strains which would usually be recovered upon release of the stress are instead converted into unrecoverable strains when chain segments have been rearranged. The decrease in stress is often plotted against time in order to estimate when it will have fully decayed. The strains induced during processing of molten thermoplastics, particularly injection molding, often do not completely recover before cooling and become frozen into the material. At ambient temperatures these strains may slowly recover, as evidenced by warping and excessive shrinkage of parts. For this reason, some moldings with high molded-in strain are annealed while being securely held in shape-retaining fixtures, to permit stress relaxation. ASTM D 2991 describes a standard practice for testing stress relaxation of plastics.

Stress Rupture The sudden, complete failure of a plastic member held under load. In laboratory testing, the temperature and rate of loading or, in longer-term tests, the time for which the load was sustained, should be stated along with the load and corresponding type and mode of stress. The mode of loading may be tensile, flexural, torsional, biaxial, or hydrostatic.

Stress-Strain Diagram The plot of stress on a test specimen, usually in tension or compression, versus the corresponding strain, usually carried to the point of failure. The test is usually carried out at constant crosshead speed, i e, at a constant rate of nominal elongation. The stress plotted is usually the *nominal stress*, i e, the measured force at any time divided by the original cross-sectional area, normal to the force, over which the force is distributed. The strain plotted is usually also the nominal quantity, i e, the increase in specimen gage length divided by the original length. See also TRUE STRESS and TRUE STRAIN.

Stress Whitening Whiteness seen in some plastics and rubbers that are subjected to extreme stretching, thought to be due to the formation of light-scattering microvoids (*microcavitation*) within the material.

Stress Wrinkles Distortions in the face of a laminate caused by uneven web tensions, slowness of adhesive setting, selective absorption of the adhesive, or by reaction of the adherends with materials in the adhesive.

Stretch-Blow Molding A blow-molding variant for making bottles from polyethylene terephthalate in which shaped preforms (as distinct from simple tubular parisons) are injection molded, reheated through several zones, then inflated and stretched lengthwise to essentially the preform shape magnified many times, cooled, and released from the mold. The process may be done in a single stage, or in two essentially separate operations, or in an integrated two-stage method in a single machine (Husky, 1992). See also BLOW MOLDING.

Stretched Tape Strong tape made of a crystalline plastic such as polypropylene or nylon, that has been unidirectionally oriented by warm stretching followed by cooling while under tension. Used mainly for STRAPPING, w s.

Stretch-Film Wrapping See SHRINK WRAP.

Stretch Forming A sheet-forming technique in which a heated thermoplastic sheet is stretched over a mold and subsequently cooled. See also DRAPE FORMING and SHEET THERMOFORMING.

Stretching See ORIENTATION.

Stretch Ratio In making uniaxially oriented films and filaments, the length of a sample of stretched film or filament divided by its length before stretching. Also see BLOW-UP RATIO and DRAW RATIO.

Stretch Wrap See SHRINK WRAP.

Striae Surface or internal thread-like inhomogeneities in a transparent plastic.

Striation (1) In blow molding a rippling of thick parisons caused by a local weld-line effect in the melt, imparted by a spider leg. See SPIDER LINES. (2) In the mixing of contrasting colors of resin by viscous shear in the melt, as in extrusion, a layer of one color adjacent to one of the contrasting color. The thickness of the layer is called the *striation thickness* and the thinning of the layer is a measure of the effectiveness of mixing, corresponding to Danckwerts' SCALE OF SEGREGATION, w s.

Strippable Coating A temporary coating applied to finished articles to protect them from abrasion or corrosion during shipment and storage, which can be removed when desired without damaging the articles. Vinyl plastisols, applied by dipping, spraying, or roll coating, are often used for this purpose. See also RELEASE PAPER.

Stripper Plate In molding, a plate that strips a molded piece from core pins or a force plug. The plate is actuated by opening the mold.

Stripping Fork (comb) A tool, usually of brass or laminated sheet, used to remove articles from molds.

Stripping Torque (1) Of a self-tapping screw, the twisting moment in N·m required to strip the threads formed by the screw in a softer material. (2) Of a molded-in insert, the torque required to break the mechanical bond between the plastic and the insert's knurled surface.

Stroke (1) The movement of a hydraulic piston, press ram, or other reciprocating machine member. (2) The distance between the extremes of movement of such a member in normal operation.

Structural Foam A term originally used for cellular thermoplastic articles with integral solid skins having high strength-to-weight ratios and also used for high-density cellular plastics that are strong enough for structural applications. Today the term is more apt to mean the products of SRIM (w s at REACTION INJECTION MOLDING). In the original Union Carbide process called *structural-foam molding*, pellets of resin containing a blowing agent are fed into an extruder provided with an accumulator, where the melt is maintained above the foaming temperature but at a pressure high enough to preclude foaming. A piston in the accumulator forces a measured charge of molten resin into the mold, the volume of the charge being only about half of the mold volume, which is quickly filled, however, by the expansion of the gas. As the foam contacts the mold surfaces, the cells at those surfaces collapse to form a solid skin. Parts produced by this process are from three to four times as rigid as solid injection moldings of the same weight (because the latter are so much thinner). Many variations of the process have been developed, most employing injection molding or extrusion, with emphasis on elimination of swirls in the products.

Student's t (t) The difference between the mean of a sample of measurements and the presumed true mean of the parent population from which the measurements were drawn, divided by the standard error of the sample mean, i e,

$$t = \frac{\bar{x} - \mu}{s/\sqrt{n}}$$

in which s is the sample standard deviation and n is the sample size. If the individual x_i are normally distributed, then t will have the t-distribution first formulated by W. S. Gosset in 1908 under the pseudonym, "Student". The t-statistic and its distribution are very useful in testing hypotheses about population means and in setting up confidence intervals for same. Tables of the percentage points of the t-distribution are widely available in handbooks and computer software.

Styrenated Alkyd See ALKYD MOLDING COMPOUND.

Styrene (vinyl benzene, phenylethylene, cinnamene) $C_6H_5CH{=}CH_2$. A colorless liquid with a strong, sharp odor, produced by the catalytic dehydrogenation of ethylbenzene. Styrene monomer is easily polymerized by exposure to light, heat, or a peroxide catalyst, and even spontaneously, so a little inhibitor is added if it is to be stored. Styrene is a versatile comonomer, polymerizing readily with many other monomers, and is the active crosslinking monomer in most polyester laminating resins.

Styrene-Acrylonitrile Copolymer (SAN) Any of a group of copolymers containing 70–80% styrene and 30–20% acrylonitrile and having higher strength, rigidity, and chemical resistance than straight polystyrene. These monomers may also be blended with butadiene to make a terpolymer or the butadiene may be grafted onto the SAN, either method producing ABS resin.

Styrene-Butadiene Rubber (SBR, Buna-S, GR-S) A group of widely used synthetic rubbers comprising about 3 parts butadiene copolymerized with 1 part of styrene, with many modifications yielding a large variety of properties. The copolymers are first prepared as latices, in which form they are sometimes used. The latices can be coagulated to produce crumb-like particles resembling natural crepe rubber. SBR has better abrasion resistance than natural rubber and has largely supplanted it in tire treads, though not in sidewalls.

Styrene-Butadiene Thermoplastic (S/B, SB) A group of thermoplastic elastomers introduced in 1965 (Shell Chemical Co, Thermolastic®) They are linear block copolymers of styrene and butadiene, produced by lithium-catalyzed solution polymerization, with a sandwich molecular structure containing a long polybutadiene center surrounded by shorter polystyrene ends. The materials are available in pellet form for extrusion, injection molding, and blow molding, and S/B sheets are thermoformable.

Styrene-Maleic Anhydride Copolymer An alternating copolymer useful as a textile size and emulsifier.

Styrene Plastic See POLYSTYRENE.

Styrenic Plastic (styrenic) A term that encompasses all the wide variety of thermoplastics in which styrene is the monomer or comonomer or in which polystyrene is a member of a polymer blend.

Styrene-Rubber Plastic See HIGH-IMPACT POLYSTYRENE.

Styrol The name given to styrene by the chemist who first observed the monomer in 1839. The name was changed to styrene by German researchers around 1925.

Submarine Gate (tunnel gate) In injection molds, a type of edge gate where the opening from the runner into the mold is located below the parting line or mold surface, as opposed to conventional edge gating where the opening is machined into the surface of the mold or mold cavity. With submarine gates, the item is broken from the runner system on ejection from the mold. See also GATE.

Submicron Reinforcement Very fine WHISKERS, w s, that, when used as reinforcements for thermoplastic resins, are short enough to flow through restricted gates without breaking, yet have aspect ratios well over 200, adequate for good isotropic strength development in the composite. However, they are costly, so are used only for special applications.

Substrate A surface or sheet on which another material such as an adhesive, coating, or paint is deposited.

Sucrose Acetate Isobutyrate (SAIB) A modifying extender for lacquers and finishes based on resins such as cellulose acetate butyrate, acrylics, alkyds, and polyesters.

Sucrose Octaacetate $C_{12}H_{14}O_3(OOCCH_3)_8$. A plasticizer for cellulosic resins and polyvinyl acetate.

Sucrose Octabenzoate $C_{12}H_{14}O_3(OOCC_6H_5)_8$. A plasticizer for polystyrene, cellulosics, and some vinyls.

Sulfide Staining Discoloration of a plastic caused by the reaction of one of its constituents with a sulfide in a liquid, solid, or gas to which the plastic article has been exposed. Stabilizers based on salts of lead, cadmium, antimony, copper, or other metals sometimes react with external sulfides to form a staining metallic sulfide.

Sulfonate-Carboxylate Copolymer See POLYSULFONATE COPOLYMER.

Sulfonyl An organic radical of the form RSO_2–, in which R may be aliphatic or aromatic. If the bond is filled by an –OH group, the compound is called a sulfonic acid. The sulfonyl chlorides are highly active reagents used in many organic syntheses.

Sulfonylazidosilane Any of a family of organofunctional coupling agents that, in contrast to most other SILANE COUPLING AGENTs (w s), enter into direct chemical reaction with organic polymers. They function by insertion into carbon-hydrogen bonds, which avoids generation of free radicals and degradation of radical-sensitive polymers such as polypropylene, polyisobutylene, and polystyrene.

Sulfonyldianiline (aminophenyl sulfone) $(NH_2C_6H_4)_2SO_2$. A curing agent for epoxy resins.

Sulfur Nitride Polymer (polysulfur nitride, polythiazyl) $[-(SN)-]_n$. First synthesized in 1910 but ignored for six decades, this covalent polymer has been restudied and found to have the physical and electrical properties of a metal. It is formed by passing the vapor of $(SN)_4$ over a catalyst that cracks it to $(SN)_2$, which is condensed on a cold surface where it spontaneously polymerizes into crystalline form. The crystals are malleable and can be cold-worked into thin sheets or fibers under pressure, having an electrical conductivity similar to that of mercury, and superconducting near 0 K.

Sunlight Resistance See LIGHT RESISTANCE.

Superposition Principle See BOLTZMANN SUPERPOSITION PRINCIPLE.

Suppliers of Advanced Composite Materials Association (SACMA) A recently formed industry group headquartered at 1600 Wilson Blvd, Ste 1008, Arlington, VA 22209.

Surface Characteristics All those properties attributable to surfaces, such as roughness, adsorptivity, coefficient of friction, surface energy, and chemical activity.

Surface Conductance The direct-current conductance (A/V) between two electrodes in contact with a specimen of solid insulating material when the current is passing only through a thin film of moisture on the surface of the specimen.

Surface Energy An alternate aspect of SURFACE TENSION, w s.

Surface Pin Synonym for EJECTOR-RETURN PIN, w s.

Surface Resistivity The electrical resistance between two parallel electrodes in contact with the specimen surface and separated by a distance equal to the contact length of the electrodes. The resistivity is therefore the quotient of the potential gradient, in V/m, and the current per unit of electrode length, A/m. Since the four ends of the electrodes define a square, the lengths in the quotient cancel and surface resistivities are reported in "ohms per square". For reproducibility of results, specimens must be carefully cleaned and dried, and protected from contamination.

Surface Tension (free-surface energy, surface energy) The work required to increase the surface of a solid or liquid (in contact with air) by one unit of area. Surface energy and surface tension are equivalent terms, but with liquids it is possible to measure directly the surface tension. The unit still in common use is dynes per centimeter, equal to 0.001 N/m, or, in surface-energy terms, 0.001 J/m^2. Surface tension is measured with a *tensiometer* or by capillary rise. Surface energies of plastics are determined indirectly by observing the angles of contact of a graded series of increasingly polar liquids of known surface tension on the plastic surface (see ASTM D 2578), then applying regression analysis to determine the polar and nonpolar components of surface energy. The surface energy is the sum of the two components.

Surface Treating Any method of treating a plastic surface to render it more receptive to adhesives, paints, inks, lacquers, or to other surfaces in laminating processes. The two methods in widest use are CORONA-DISCHARGE TREATING and FLAME TREATING, w s. See also CASING, ION PLATING, IRRADIATION, and PLASMA ETCHING.

Surfacing Mat A very thin mat, usually 0.2 to 0.5 mm thick, of highly filamentized glass fiber used primarily to produce a smooth, strong surface on a reinforced-plastic laminate.

Surfactant A chemical compound that reduces the SURFACE TENSION (w s) of a liquid (typically water) in which it is dissolved, making it easier for the solution to wet solid surfaces and penetrate pores.

Surging Any irregularity in the output rate of an extruder. Extrusion engineers have recognized two kinds: long- and short-period surging. Long-period surging has been traced to diurnal changes in electrical supply and environmental conditions, while short-period surging, which has sometimes been so serious as to interrupt flow, is attributed to many causes such as bridging of pellets in the feed throat or compression section of the screw, too abrupt transition between deep feed flights and shallow metering flights, inadequate melting capability for the rate attempted, improper profile of barrel temperatures, voltage surges because of large loads coming on or going off the line, and changing character of feedstock loaded from bags into the hopper. In tracking down and correcting the causes of surging, it is imperative for the extruder to be equipped with instruments that sense melt temperature and pressure at the head, screw speed, extrusion rate, and either screw torque or motor kilowattage, with recorders or computer data acquisition to record these quantities.

Surlyn® Trade name for an IONOMER, w s.

Suspension A fluid medium with fine particles of any solid more or less stably dispersed therein. The particles are called the *disperse phase*, and the suspending medium is called the *continuous phase*. When the particles do not settle out and are small enough to pass through ordinary filters, the suspension is called a *colloid* or *colloidal suspension*. Solid particles suspended in air or other gas, particularly solid particles arising from burning, are called a *smoke*. In the plastics field, a suspension is essentially synonymous with DISPERSION, w s. See also EMULSION.

Suspension Polymerization (pearl, bead, or granular polymerization) A polymerization process in which the monomer or mixture of monomers is dispersed by mechanical agitation in a second liquid phase, usually water, in which both monomer and polymer are essentially insoluble. The monomer droplets are polymerized while maintained in dispersion by vigorous agitation. Polymerization initiators and catalysts used in the process are generally soluble in the monomer. According to the type of monomer, emulsifier, protective colloid, and other modifiers used, the resulting polymer may be in the form of pearls, beads, soft spheres, or irregular fine granules that are easily separated from the suspending medium when agitation is stopped. Suspension polymerization is used primarily for PVC, polyvinyl acetate, polymethyl methacrylate, polytetrafluoroethylene, and polystyrene.

Sustained-Pressure Test In testing pipe or tubing, subjecting the specimen to constant internal pressure over a long time, up to 1000 h or more, during and after which changes in diameter, wall thickness, and appearance are noted. It is a special-purpose creep test.

Sweating Exudation of small drops of liquid, usually a plasticizer or softener, on the surface of a plastic part.

Swelling (1) Volume expansion of a material specimen or manufactured article due to a rise in temperature (see COEFFICIENT OF THERMAL EXPANSION) or absorption of water or other liquid. It is usually expressed as a percentage of the original volume or, sometimes, as percentage change in lineal dimensions. (2) EXTRUDATE SWELLING, w s.

Swirl A term applied to visual and tactile surface roughness sometimes observed in the structural-foam-molding process. It results from jetting at the gate of the mold, causing surface wrinkling as the polymer melt flows along the wall of the mold. The condition can be alleviated by measures such as reducing the filling speed and raising the temperature of melt and/or mold.

Sylvic Acid A synonym for ABIETIC ACID, w s.

Sym- (*s*-) Prefix abbreviation for *symmetrical*, referring to the relative location of atoms, chain groups, or substituent groups in a cyclic compound. Two examples are *sym*-trioxane, 1,3,5-$(CH_2O)_3$, and 1,3,5-trinitrobenzene (the corresponding 1,2,4-isomer in each case being unsymmetrical). The prefix is usually ignored in alphabetizing lists of compound names.

Syndiotactic Derived from the Greek words *syndio*, meaning "every other" and *tatto*, meaning "to put in order", the term refers to polymers having alternating, different substituent groups along the chain capable of exhibiting mirror-image symmetry (see ENANTIOMER). The polymer structure has such groups attached to the backbone chain in an order a-b-a-b- on one side of the chain and b-a-b-a- on the other. See also ISOTACTIC and TACTICITY.

Syndiotactic Polymer See preceding entry. A regular polymer whose molecules can be described in terms of alternation of configurational base units that are enantiomeric (ISO).

Syneresis The contraction of a gel upon standing, usually accompanied by the separation of a liquid from the gel.

Synergism A phenomenon wherein the effect of a combination of two ingredients (or other experimental factors) is greater than the sum of their individual effects, or, in the case of polymer blends or composites, than would be estimated by a simple LAW OF MIXTURES, w s. For example, some stabilizers and some fire retardants for plastics have a mutually reinforcing effect and are thus termed synergistic. Similarly, some plastic alloys have higher strength than either of the neat resins.

Syntactic Foam (cellular mortar) A term applied to composites of tiny, hollow spheres in a resin or plastic matrix. The spheres are usually of glass, although phenolic microspheres were used in the early years of the art. The resin most used is epoxy, followed by polyesters, phenolics, and PVC. Syntactic foams of the most common type—glass microspheres in a binder of high-strength thermosetting resin—are made by mixing the spheres with the fluid resin, its curing agent, and other additives, to form a fluid mass that can be cast into molds, troweled onto a surface or incorporated into laminates. After forming, the mass is cured by heating. These foams are characterized by densities lower than those of the matrix resins, ranging from 0.57 to 0.67 g/cm^3, and very high compressive strengths. Their first applications were for deep-submergence buoys capable of withstanding depths of 6000 m. When both gas bubbles and hollow glass spheres are used in the same mixture, the resulting composite has been called a *diafoam*.

Synthetic Fiber (man-made fiber) A class name for various fibers (including filaments), distinguished from natural fibers such as wool and cotton, produced from fiber-forming substances which may be: (1) modified or transformed natural polymers, e g, alginic and cellulose-based fibers such as acetates and rayons; (2) polymers synthesized from chemical compounds, e g, acrylic, nylon, polyester, polyurethane, polyethylene, polyvinyl, and carbon/graphite fibers; or (3) fibers of mineral origin, e g, glass, quartz, boron, and alumina

Synthetic Resin See RESIN, SYNTHETIC.

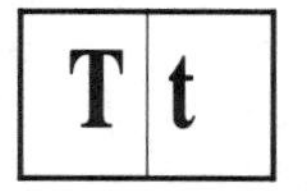

t Symbol for elapsed time, for thickness, or for STUDENT'S t, w s.

T (1) Symbol for torque or ABSOLUTE TEMPERATURE, w s. (2) SI abbreviation for TERA-, w s.

Tab Gate In injection molding, a small removable tab of approximately the same thickness as the molded item, usually located perpendicularly to the item. The tab is used as a site for edge-gate location, typically on items with large flat areas.

TAC Abbreviation for TRIALLYL CYANURATE, w s.

Tack (1) The immediate stickiness of an adhesive, measurable as the force required to separate an adherend from it by viscous or plastic tensile flow of the adhesive. (2) The tendency of a soft material to stick to itself but not to other materials; *tackiness*.

Tackifier A substance such as a rosin ester that is added to synthetic resins or elastomeric adhesives to improve the initial tack and extend the tack range of the deposited adhesive film.

Tack Range The period of time during which an adhesive will remain tacky after application to an adherend, under specified conditions of temperature and humidity.

Tactic Block In a polymer, a regular block that can be described by only one species of configurational repeating unit in a single sequential arrangement (IUPAC).

Tactic Block Polymer A polymer whose molecules consist of TACTIC BLOCKs (w s) connected linearly (IUPAC).

Tacticity (1) The orderliness of the succession of configurational repeating units in the main chain of a polymer molecule (ISO, IUPAC). (2) Any type of regular or symmetrical molecular arrangement in a polymer structure, as opposed to random positioning of substituent groups along a polymer backbone. See also STEREOSPECIFIC.

Tactic Polymer (1) A regular polymer whose molecules can be described by only one species of configurational repeating unit in a single sequential arrangement (ISO, IUPAC). (2) A polymer whose chains have either an isotactic, syndiotactic, or tritactic arrangement. These structures are most commonly produced in polyolefins by means of catalysts derived from a transition-metal halide and a metal alkyl, etc.

Take-off The mechanism for drawing extruded or calendered material away from the extruder or calender. The most common form of extrusion take-off is a pair of endless caterpillar belts with resilient grip pads conforming to the section being extruded, driven at a speed synchronized with that of the extrudate.

Talc (steatite, talcum) $Mg_3Si_4O_{10}(OH)_2$. A natural hydrous magnesium silicate, sometimes used as a filler.

Tandem Extruder See EXTRUDER, TANDEM.

Tangent Modulus The slope of the curve at any point on a static stress-strain graph ($d\sigma / d\varepsilon$) expressed in pascals per unit of strain. This slope is the tangent modulus in whatever mode of stress the curve has arisen from—tension compression, or shear. [Since strain is dimensionless, the unit given for modulus is normally just stress (Pa).]

Taper (1) In a CONICAL TRANSITION (w s) section of an extruder screw, the vertex angle of the axial cone defined by the increasing root diameter of the screw. Compare with HELICAL TRANSITION. (2) An often used synonym for DRAFT (w s) in molds.

Taper Pin A slightly conical, hard steel dowel driven into mating holes drilled into the contact faces of adjacent major machine components, after the components have been aligned, to preserve alignment. At least two pins are normally used at each such interface.

TBT Abbreviation for TETRABUTYL TITANATE, w s.

TCE Abbreviation for TRICHLOROETHYLENE, w s.

TCEF Abbreviation for *trichloroethyl phosphate*, a plasticizer.

TCP Abbreviation for TRICRESYL PHOSPHATE, w s.

TDI Abbreviation for *toluene diisocyanate*, an 80-20 mixture of the 2,4- and 2,6- isomers. See also TOLUENE-2,4-DIISOCYANATE and DIISOCYANATE.

T-Die A center-fed, slot die for extrusion of film whose horizontal cross section, together with the die adapter, resembles the letter T.

Teardrop Die See MANIFOLD.

Tear Strength The force or stress required to start or continue a tear in a fabric or plastic film. See ELMENDORF TEAR STRENGTH.

Teflon® Du Pont's trade name for its fluorocarbon resins, including polytetrafluoroethylene, perfluoropropylene resin and copolymers.

Telechelic Polymer A polymer having purposely introduced end groups of a particular chemical type, e g, acetal homopolymer that has been "end-capped" by treatment with acetic anhydride.

Telescopic Flow A picturesque name for laminar flow in a circular tube, derived from visualizing successively smaller cylindrical shells of liquid, from the tube wall toward the center, each moving faster than the next outer one, sliding like the tubes of a sectional telescope. See POISEUILLE FLOW.

Telomer An OLIGOMER, w s, formed by addition polymerization in the presence of excess chain-transfer agent (polymerization stopper) whose free radicals become the end groups of the telomer.

Temperature Profile (1) In extrusion or injection molding, the sequence of barrel temperatures from feed opening to head, sometimes presented as a plot of temperature vs longitudinal position, hence *profile*. (2) The sequence of metal temperatures across a sheet or film die, or around a large blown-film die. (3) The sequence of temperatures across the width of a slab of newly extruded or cast plastic foam, as indicated by temperature sensors placed laterally at the same depth in the foam. (4) In analysis of nonisothermal, laminar flow of very viscous liquids (e g, polymer melts) within tubes and dies, the sequence of temperatures from one sidewall through the center to the opposite sidewall at any point along the axis of flow. Such profiles have also been measured experimentally with traversing thermocouples.

Tenacity A term used in yarn and textile manufacturing to denote the strength of a yarn or filament of a given size. Numerically it is the newtons of breaking force per tex of lineal density (replacing the deprecated old unit, grams per denier). In testing tenacity, the yarn is usually pulled at the rate of 0.5 cm/s. To convert g/denier to N/tex, multiply by 0.0883.

Tensile Bar (dogbone specimen, dumbbell specimen) Any of several kinds of test specimens made for use in tests of tensile stress vs elongation. There are two main types, both having ends that have two to three times the cross section of the central, or *gauge-length* section, this geometry guaranteeing that, because the force is the same everywhere along the bar, the stress will be much lower in the end sections than in the gauge section. This geometry prevents failure at the grips and confines the plastic deformation, if any, to the gauge section. The first type is of rectangular cross section perpendicular to the length, and may be machined from sheet stock or molded. These are well defined in ASTM D 638, which also specifies certain geometries for specimens taken from tubing and round-rod stock. The round-rod specimens typify the second geometry, in which the transverse cross sections are circular. The flat type, but smaller, is also called for in the TENSILE-IMPACT TEST, w s.

Tensile Heat-Distortion Temperature An obsolete misnomer for DEFLECTION TEMPERATURE, w s.

Tensile-Impact Test An impact test that uses a pendulum striker to break a dogbone-shaped test specimen, described in ASTM D 1822 and 1822M. It differs from the Izod impact test in two important aspects: (1) the specimen is not notched and (2) it is broken in simple tension rather than bending. For these reasons, its interpretation is more straightforward.

Tensile Modulus Synonym for Young's modulus; see MODULUS OF ELASTICITY.

Tensile Product The product of tensile strength and elongation at break. In a Hooke's-law material (which few if any plastics are), the tensile product is twice the energy to rupture, so is related to toughness. In a material that exhibits a long, flat region of ductile elongation at constant stress during the tensile test, the tensile product is closely equal to the nominal energy to break.

Tensile-Shear Strength A measure of the shear strength of an adhesive bond in which two members are bonded in a LAP JOINT, w s, then pulled at both ends until the joint fails in shear. The strength is reported as the tensile force divided by the shear area (Pa). A double lap joint may be specified. Many tests of tensile-shear strength are listed among the ASTM Standards; five that are suitable for bonds made between rigid plastics, all in sec 16.06, are D 3163, D 3164, D 3165, D 3528, and D 3983.

Tensile Strength The maximum nominal stress sustained by a test specimen being pulled from both ends, at a specified temperature and at a specified rate of stretching. When the maximum nominal stress occurs at the YIELD POINT (w s), it shall be designated *tensile strength at yield*. When it occurs at break, it shall be designated *tensile strength at* *break*. The ASTM test for plastics is D 638 (metric, D 638M). The SI unit of tensile strength is the pascal (N/m^2), but trade publications in the US are still clinging to the pound (force) per square inch (psi). The strengths of commercial plastics that are neither plasticized nor fiber-reinforced range from about 14 to 140 MPa (2 to 20 kpsi).

Tensiometer An instrument, invented by P. L. du Noüy in 1919, for measuring SURFACE TENSION, w s, of liquids and interfacial tensions between immiscible liquids, consisting of a horizontal, platinum-wire ring suspended from the end of a slender cantilever beam whose movement is indicated by a pointer on a circular scale. To measure surface tension, one submerges the horizontal ring in the test liquid, then carefully raises it by turning the knob and pointer until the meniscus lifted by the ring breaks. The pointer indicates the surface tension in d/cm (= 10^{-3} N/m). Use of the instrument is described in ASTM Test D 1331 (sec 15.04).

Tentering Biaxial orientation of film or sheet by means of a TENTERING FRAME, w s.

Tentering Frame (tenter frame) A machine that continuously stretches, simultaneously in two perpendicular directions, a temperature-conditioned film or sheet, imparting biaxial orientation. Clamps attached to endless chains grip the sheet on both edges and, while accelerating in the direction of sheet travel, also move outward from the longitudinal centerline. Stretch ratio is about 3 to 4 in each direction, with about the same factor of increase in strength and modulus over those of the unoriented sheet. Tentering is usually done shortly downstream from the sheet extruder, but can also be done on film or thin sheet that has been extruded, cooled, and wound into coils for storage, then later reheated to be oriented.

TEP Abbreviation for TRIETHYL PHOSPHATE, w s.

Tera- (T) The SI prefix meaning $\times 10^{12}$.

Terephthalate Polyester Any polymeric ester of terephthalic acid (1,4-benzene dicarboxylic acid), but in particular the three commercially important thermoplastic resins, POLYETHYLENE TEREPHTHALATE, POLYBUTYLENE TEREPHTHALATE, and POLYCYCLOHEXYLENEDIMETHYLENE TEREPHTHALATE, w s.

Terephthaldehyde Resin See POLYESTER, SATURATED.

Terephthalic Acid (TPA, *para*phthalic acid, benzene-*p*-dicarboxylic acid) $C_6H_4(COOH)_2$. White crystals or powder, used in the production of alkyd resins and thermoplastic polyesters.

Termination (chain termination) The final phase of a polymerization in which chain growth ends through reaction of polymeric free radicals with each other or with smaller entities.

Terpene Resin See POLYTERPENE RESIN.

Terpolymer The product of simultaneous polymerization of three different monomers, or of the grafting of one monomer to the copolymer of two different monomers. An important commercial terpolymer is ABS resin, derived from acrylonitrile, butadiene, and styrene.

Terra Alba $CaSO_4 \cdot 2H_2O$. A finely powdered form of gypsum, used as a filler.

Terra Ponderosa Synonym for BARIUM SULFATE, w s.

Tertiary (adj) In organic chemistry, denoting a functional group in which three of its original hydrogen atoms have been replaced by other groups. Triphenylcarbinol, $(C_6H_5)_3COH$, and *t*-butyl alcohol, $(CH_3)_3$–COH, are examples of tertiary alcohols. Trimethylamine, $(CH_3)_3N$, is a tertiary amine.

Tesla (T) The SI unit of magnetic-flux density, equal to one weber per square meter (1 Wb/m^2). The older, now deprecated unit, the *gauss*, is equal to 10^{-4} T.

TETA Abbreviation for TRIETHYLENETETRAMINE, w s.

Tetrabasic Lead Fumarate $4PbO \cdot PbC_2H_2(COO)_2 \cdot 2H_2O$. A creamy-white powder used as a heat stabilizer for vinyl phonograph records, electrical-grade plastisols, and insulation. It is also used as a curing agent for chlorosulfonated polyethylene.

Tetrabromobisphenol A [4,4'-isopropylidene bis(2,6-dibromophenol)] $(CH_3)_2C(C_6H_2Br_2OH)_2$. (See structure at BISPHENOL A.) An off-white, crystalline solid, used as a flame retardant in epoxy resins, polyesters, and polycarbonates.

Tetrabromophthalic Anhydride Br_4C_6-2,3-$(CO)_2O$. A reactive intermediate containing 69% bromine, used as a flame retardant.

2,2',6,6'-Tetrabromo-3,3',5,5'-Tetramethyl-*p*,*p*'-Biphenol (TTB) An aromatic brominated flame retardant synthesized easily by a two-step process from 2,6-dimethylphenol. The unusual chemical structure of TTB enables its use as both a reactant and additive flame retardant. It has been used in high-impact polystyrene.

Tetrabutyl Titanate (TBT, butyl titanate, titanium butylate) $Ti(OC_4H_9)_4$. A catalyst for condensation and cross-linking reactions, also used to improve the adhesion of plastic compounds to metals.

Tetrachlorobisphenol A [4,4'-isopropylidene bis(2,6-dichlorophenol)] $(CH_3)_2C(C_6H_2Cl_2OH)_2$. (See structure at BISPHENOL A.) A monomer for flame-retardant epoxies, polyesters, and polycarbonates.

Tetrachloromethane Synonym for CARBON TETRACHLORIDE, w s.

Tetrachloroethane A strong solvent whose use has been curtailed because of toxicity and its suspected action on stratospheric ozone.

Tetrachlorophthalic Anhydride Cl_4C_6-2,3-$(CO)_2O$. A reactive intermediate for polyester resins, containing 50% chlorine, with flame-retardant action.

Tetraethylene Glycol Dicaprylate $(C_7H_{15}COOCH_2CH_2OCH_2CH_2)_2O$. A plasticizer for vinyl chloride polymers and copolymers.

Tetraethylene Glycol Di(2-ethylhexanoate) $(C_8H_{17}COOCH_2CH_2O–CH_2CH_2)_2O$. A secondary plasticizer for vinyl resins and a primary plasticizer for cellulosic plastics and synthetic rubbers. In vinyls, it is used at levels of 15 to 20 percent of the total plasticizer to impart good low-temperature flexibility. In nitrocellulose lacquer, it imparts cold-check resistance.

Tetraethylene Glycol Monostearate $C_{17}H_{35}COO(CH_2CH_2O)_4OH$. A plasticizer for ethyl cellulose and cellulose nitrate.

Tetrafluoroethylene (TFE, perfluoroethylene) $F_2C{=}CF_2$. A colorless gas derived by passing chlorodifluoromethane through a heated tube, the monomer from which polytetrafluoroethylene (Teflon®) is made.

Tetrafluoroethylene-Ethylene Copolymer See POLY(ETHYLENE-TETRAFLUOROETHYLENE).

Tetrafluoroethylene-Hexafluoropropylene Copolymer (FEP) See FLUORINATED ETHYLENE-PROPYLENE RESIN.

Tetrahydrofuran (THF, tetramethylene oxide) A colorless liquid obtained by the catalytic hydrogenation of furan, with the empirical formula C_4H_8O and the heterocyclic structure shown below.

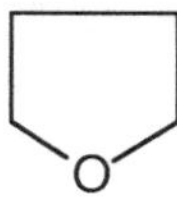

In addition to its many uses as an industrial intermediate, THF is a powerful solvent for PVC, polyvinylidene chloride, and many other polymers. It is often used as the carrier solvent for size-exclusion chromatography of polymers. Its presence in relatively small amounts increases the "bite" of vinyl printing inks, lacquers, and adhesives. THF has been polymerized to polytetramethylene ether glycol for use in the production of polyurethanes.

Tetrahydrofurfuryl Oleate $CH_3(CH_2)_7CH{=}CH(CH_2)_2C_4H_7O$. A plasticizer for polystyrene,and cellulosic, acrylic, and vinyl resins. In vinyls it is used as a secondary plasticizer, imparting resistance to low temperatures, and as a lubricant in stiff or highly filled calendering and extrusion compounds.

Tetrahydronaphthalene (1,2,3,4-tetrahydronaphthalene, Tetralin) $C_{10}H_{12}$. This bicyclic, semiaromatic hydrocarbon, produced by the partial hydrogenation of naphthalene, is an involatile solvent for rubbers, PVC, and natural resins.

Tetramer A polymer formed from four molecules of a monomer and/or made up of four mer units. See OLIGOMER.

1,2,4,5-Tetramethylbenzene Explicit name for DURENE, w s.

1,1,3,3-Tetramethylbutyl Peroxy-2-Ethylhexanoate A liquid peroxide superior to (solid) benzoÿl peroxide as a catalyst for polyesters and, because it is a liquid, easier to handle.

5,5'-Tetramethylene Di(1,3,4-dioxazol-2-one) (adiponitrile carbonate, ADNC) A white crystalline solid, capable of being reacted with diols and polyols to form light-stable urethane coatings, elastomers, and foams.

Tetramethylene Glycol See 1,4-BUTYLENE GLYCOL.

Tetramethylethylenediamine (TMEDA, *N,N,N',N'*-tetramethylethylenediamine) $(CH_3)_2NCH_2CH_2N(CH_3)_2$. An anhydrous, corrosive liquid used as a catalyst for urethane foams, coatings, and elastomers, and as a curing agent for epoxy resins.

Tetramethylthiuram Disulfide (TMTD) $[(CH_3)_2NCS_2-]_2$. A white, crystalline powder used as a fungicide, bacteriostat, and rodent repellent in vinyl compounds, and as a secondary accelerator in rubber curing.

Tetrapolymer Synonym for QUATERPOLYMER, w s.

Tex A convenient unit of lineal density of fibers: the mass in grams of one kilometer of fiber length. The SI equivalent is 10^{-6} kg/m (1 mg/m). Compare CUT, DENIER, and GREX NUMBER.

TFE Abbreviation for TETRAFLUOROETHYLENE, w s.

T_g Symbol for GLASS-TRANSITION TEMPERATURE, w s.

TGA Abbreviation for THERMOGRAVIMETRIC ANALYSIS, w s.

Theoretical Weight The weight of a part calculated from its specified dimensions and the density of the material.

Thermal Analysis Any of a broad group of techniques that all involve programed heating of a material and observing the accompanying physical and chemical reactions (e g, loss of water, decomposition), changes of state (T_g, T_m), or changes in physical or mechanical properties. See DIFFERENTIAL THERMAL ANALYSIS, DIFFERENTIAL

SCANNING CALORIMETER, DYNAMIC MECHANICAL ANALYZER, THERMOGRAVIMETRIC ANALYSIS.

Thermal Black See CARBON BLACK.

Thermal Capacity See HEAT CAPACITY and SPECIFIC HEAT.

Thermal Conductivity (k) The basic measure of steady heat-transfer rate within solid materials (and still fluids) by atomic or molecular contact and vibration. It derives from FOURIER'S LAW OF HEAT CONDUCTION, w s, and may be thought of as the rate of heat flow between two opposite faces of a unit cube whose other faces are perfectly insulated when the temperature at the warmer face is 1 K above that of the cooler face. The SI dimensions corresponding to this concept are $(J/s)/[m^2(K/m)]$, which reduce to $W/(m{\cdot}K)$. Some conversions from other units to SI are given in the Appendix. For plastics and other materials, k increases with rising temperature. One of the early ASTM tests, first issued in 1942, applicable to many types of solid materials and still included in section 08, is the guarded-hot-plate method of measuring thermal conductivity of flat-slab specimens, C 177. Method C 518 (sec 04.06) uses a calibrated heat-flow meter to determine k values. More recently, J Schroeder's evaporative-calorimetric method (1963) has become ASTM D 4351 for plastics.

Thermal Decomposition Breakdown of a plastic resulting from action by heat. With a given plastic, it occurs at a temperature at which some components of the material are separating, reacting together, or depolymerizing, with observable changes in micro- or macrostructure, mechanical and/or electrical properties, and, usually, reduction in molecular weight. Compare DEGRADATION.

Thermal Diffusivity (α) An important property for unsteady (transient) heat transfer, particularly in solid materials, equal to the thermal conductivity divided by the product of heat capacity and density, i e, $\alpha = k/(C{\cdot}\rho)$. For most solids, α increases slowly with rising temperature. Thermal diffusivity comes into play in the heating and cooling of thermoformable sheets, in the cooling of injection moldings in the mold, and the cooling of extrudates. The most-used method for measuring thermal diffusivity of solids, the thermal-pulse method, has so far been adopted by ASTM only for synthetic carbon and graphite (ASTM Test C 714, sec 15.01). The SI unit is m^2/s.

Thermal-Expansion Coefficient See COEFFICIENT OF THERMAL EXPANSION.

Thermal Fluid Any of many types of heat-stable, noncorrosive liquids such as glycols, silicone and other oils, and the eutectic mixture of biphenyl and biphenyl oxide ("Dowtherm A") that are used to transfer heat by convection. Water, saturated steam, and glycol solutions are also important. Examples of applications in the plastics industry are jacketed molds for rotational casting, heating of calenders, injection-mold chilling, controlling the temperatures of cored extrusion screws, and maintaining temperatures of liquids in storage tanks.

Thermal Gravimetric Analysis Synonym for THERMOGRAVIMETRIC ANALYSIS, w s.

Thermal History (heat history) The integrated product of time × temperature for a plastic, from the time it was first subjected to a high temperature to the present moment under consideration. Thermal history is an important consideration in processing (and reprocessing) heat-sensitive polymers such as rigid PVC and polypropylene.

Thermal-Impulse Sealing See IMPULSE SEALING.

Thermally Foamed Plastic A cellular plastic produced by applying heat to effect gas-generating decomposition or volatilization of a constituent (after ASTM D 883).

Thermally Stimulated Current (TSC) A technique useful in studying the transitions of amorphous, polar polymers with rising temperature. A preheated sample is electrically oriented by applying a strong electric field, then chilling the polymer with the field applied. The sample is removed from the field and reheated on a temperature ramp while the current generated by the release of dipoles is tracked. Current peaks relate to relaxation times of molecular motions within the polymer. The technique is often carried out at a number of different preheating temperatures and the results subjected to RELAXATION-MAP ANALYSIS (w s) in order to distinguish relaxations occurring at various molecular-weight levels in a typical polymer.

Thermal Polymerization A polymerization process performed solely by heating in the absence of a catalyst. Monomers such as styrene and methyl methacrylate are examples of those that can be thermally polymerized.

Thermal Properties Broadly, all properties of materials involving heat or changes with temperature. In section 08 of ASTM's Annual Book of Standards ("Plastics"), tests listed under "Thermal Properties" include many properties, from brittleness temperature, coefficient of expansion, deflection temperature, etc, to heat of fusion, glass-transition temperature, thermal conductivity, heat capacity, mold shrinkage, flammability, and many more.

Thermal Resistance (R-value) The ability of a material to retard the conductive passage of heat; the reciprocal of conductance. While plastics, with very few exceptions, are good thermal insulators, the property is of especial interest for cellular plastics, often used to provide thermal insulation. The R-value of a layer of thermal insulation is given by: $R = \Delta T/q$, the ratio of the temperature drop in the direction of heat flow from one surface of the layer to the other, divided by the rate of heat flow per unit of surface area (*heat flux*). English units are customary in the US. This is not as fundamental a property as THERMAL CONDUCTIVITY, w s, because the thickness of the layer is implicit in the resistance and should be separately stated.

Thermal Sealing (thermal heat sealing) A method of bonding two or more layers of plastics by pressing them between heated dies or tools that are maintained at a relatively constant temperature. See also HEAT SEALING.

Thermal Sensitivity See HEAT SENSITIVITY and THERMAL HISTORY.

Thermal Stability Synonym for HEAT STABILITY, w s.

Thermal Stabilizer See STABILIZER.

Thermal Stress Cracking (TSC) Crazing and cracking of some thermoplastics that results from overexposure to elevated temperatures.

Thermal Volatilization Analysis (pyrolysis analysis, TVA) Ramp heating of a plastic with passage of the evolved volatiles through one or more chemical detectors, sometimes with intervening, controlled-temperature, vapor-condensing traps. TVA is a powerful technique when coupled with THERMOGRAVIMETRIC ANALYSIS, w s.

Thermionic Emission Electron or ion emission due to the temperature of the emitter. The rate of emission increases rapidly with rising temperature, and is also very sensitive to the state of the emitting surface.

Thermistor A contraction of *thermal resistor*; a semiconductor whose resistance varies sharply and reproducibly with temperature, therefore useful for temperature measurement, and sometimes used for that purpose in plastics industry.

Thermoband Welding Trade name for a variant of hot-plate welding in which a metallic tape acting as a resistance element is adhered to the material to be welded. Low voltage is applied to heat the tape, and the adjacent plastic, to the plastic's melting range. Pressure may be applied to the joint while it cools.

Thermochromic (adj) Changing color with changing temperature, a characteristic of special materials useful as temperature indicators.

Thermocompression Bonding The joining together of two materials without an intermediate material by the application of pressure and heat in the absence of an electrical current.

Thermocouple (TC) A pair of connected, welded junctions formed by two wires of dissimilar metals such as iron and CONSTANTAN (w s). If a temperature difference exists between the two junctions, a weak emf, 40 to 50 μV/K, is generated in nearly linear proportion to the ΔT. Originally, one junction was placed in an ice bath to serve as a *reference*. Then, the emf developed in the circuit was simply convertible to temperature. Modern thermocouples employ a single junction and the instrument to which the TC is connected senses its own temperature and compensates the incoming signal for the difference between that temperature and 0°C. Plain wire TCs respond very rapidly to temperature changes but those used in plastics-processing equipment are always sheathed in a sturdy protective tube, so are slower. Metal compositions of commercial TC wires are so carefully controlled that, except for exacting laboratory work, it is usually unnecessary to calibrate thermocouples. In one type of hand-held instrument, called a PYROMETER, w s, the TC is integral with a microammeter whose needle moves across a temperature scale. Thermocouples are the most used temperature sensors in plastics processing because they are sturdy, simple, reliable, readily available, and cheap.

Thermodynamic Temperature See KELVIN.

Thermoelasticity Rubber-like elasticity exhibited by a rigid plastic and resulting from an increase in temperature. In this state, the plastic may be formable into a different shape. Retention of the desired shape may be achieved by cooling in place after forming. With thermosetting materials, prolonged heating may be necessary to effect cure in place.

Thermoform (v) To change the shape of a plastic rod, profile, tube, or sheet by first heating it to make it pliant, next, forming the desired shape, and finally, cooling the formed shape. SHEET THERMOFORMING, w s, is by far the most important class of these operations.

Thermoformability The ease with which a heat-softened plastic sheet (or rod, etc) can be given a new permanent shape, particularly by the techniques of SHEET THERMOFORMING, w s. Some attempts have been made to devise tests of sheet thermoformability, but none have been widely used or adopted by ASTM as of 1992.

Thermoforming (1) See SHEET THERMOFORMING. (2) Any process in which heat softening is used to assist in the forming or reshaping of a plastic rod, tube, bar, strip, or profile. When performed directly following extrusion of the profile, the term *postforming* is often used.

Thermogram A plot of the percent of original mass of a specimen remaining during a program of linearly rising temperature vs the specimen temperature. See THERMOGRAVIMETRIC ANALYSIS.

Thermographic Nondestructive Testing See INFRARED THERMOGRAPHY.

Thermographic-Transfer Process A modification of hot stamping wherein the design to be transferred is first printed (in reverse) on a film, from which it is transferred to the plastic part by means of heat and pressure.

Thermogravimetric Analysis (TGA) A testing procedure in which the diminishing mass of a specimen is recorded as the specimen's temperature is raised at a uniform rate (*ramped*). A typical apparatus consists of an analytical balance supporting (on a wire or rod) a platinum crucible containing the specimen, the crucible being situated inside an electric furnace, and means for recording and plotting the percent mass remaining vs temperature. Some TGA tests are conducted in air, others in controlled, successive atmospheres such as nitrogen in the first stage, followed by air in the second stage. Thermogravimetric curves so obtained (*thermograms*) provide useful information regarding polymerization and pyrolysis reactions, the efficiencies of stabilizers and activators, the thermal stability of final materials, and direct analysis.

Thermolastic See STYRENE-BUTADIENE THERMOPLASTIC.

Thermomechanical Spectrum A plot of a mechanical property such as tensile modulus or strength vs temperature.

Thermoplastic (n) A resin or plastic compound that, as a finished material, is capable of being repeatedly softened by heating and hardened by cooling. Examples of thermoplastics are: acetal, acrylic, cellulosic, chlorinated polyether, fluorocarbons, polyamides (nylons), polycarbonate, polyethylene, polypropylene, polystyrene, some types of polyurethanes, and vinyl resins. (adj) Of an organic material, capable of being repeatedly softened and hardened by heating and cooling.

Thermoplastic Elastomer (TPE) Any of a family of polymers that resemble elastomers in that they are highly resilient and can be repeatedly stretched to at least twice their initial lengths with full, rapid recovery, but are true thermoplastics and thus do not require curing or vulcanization as do most rubbers. See ELASTOMER for examples.

Thermoplastic Polyester Any of a class of linear terephthalate polyesters that are true thermoplastics. Commercially important are POLYETHYLENE TEREPHTHALATE, POLYBUTYLENE TEREPHTHALATE, and POLYCYCLOHEXYLENEDIMETHYLENE TEREPHTHALATE, w s. US sales of these resins in 1992 totaled 1.24 Tg (1.37×10^6 tons).

Thermoplastic Polyolefin (TPO) Any of a group of elastomers produced by either of two processes. In one, polypropylene is melt-blended with from 15 to 85% of terpolymer elastomer, ethylene-propylene rubber, or styrene-butadiene rubber. In the other, propylene is copolymerized with ethylene-propylene elastomer in a series of reactions. The smaller elastomeric domains obtained in the latter process are claimed to provide improved properties over the blended materials.

Thermoplastic Rubber See ELASTOMER.

Thermosetting (adj) Of a resin or plastic compound, brought to its final state by application of heat or by catalytic reaction, and incapable of being melted thereafter.

Thermosetting Plastic (thermoset) A resin or plastic compound that in its final state is substantially infusible and insoluble. Thermosetting resins are often liquids at some stage in their manufacture or processing, which are cured by heat, catalysis, or other chemical means. Much crosslinking occurs. After being fully cured, thermosets cannot be melted by reheating. Some plastics that are normally thermoplastic can be made unmeltable by means of crosslinking treatments or reactions. Some thermosetting plastics are alkyd, allyl, amino, epoxy, furane, phenolic, polyacrylic ester, polyester, and silicone resins.

Thermotropic Polymer See LIQUID-CRYSTAL POLYMER.

Theta Solvent A solvent for a particular polymer in dilute solution, and at a temperature called the THETA TEMPERATURE, w s, that allows the polymer chains to assume their unperturbed, random-coil configurations with theoretical root-mean-square distances between chain ends.

Theta State A term introduced by P. Flory to describe the condition in a polymer solution in which there is little interaction between the molecules of the solvent and those of the polymer, and in which the polymer molecules exist as statistically distributed coils.

Theta Temperature (θ, Flory temperature) See THETA SOLVENT. θ is also the temperature for the polymer solution at which the second virial coefficient in the equation giving molecular weight from osmotic pressure approaches zero.

THF Abbreviation for TETRAHYDROFURAN, w s.

Thickening Agent (antisag agent) A substance that increases the viscosity and/or thixotropy of fluid dispersions or solutions. Such agents are used widely in adhesives, coatings and paints to prevent flow and slumping while they are setting or drying to their final form. Examples of thickening agents are bentonite, calcium carbonates with high oil absorption, clays, chrysotile asbestos, hydrated siliceous minerals, magnesium oxide, soaps, stearates, and special organic waxes.

Thickness Gauging The thickness of many calendered, extruded, and cast products must be measured while they are produced in order to control the thickness within specified tolerance limits. The simplest methods, called *contact gauging*, use tools such as calipers, micrometers, and rolls that physically touch the product being measured. Today these are mostly used for checking and backup while *non-contact methods* make on-line measurements and transmit signals to computers, which in turn order process adjustments to be made. Non-contact gauging devices employ nuclear radiation (see BETA-RAY GAUGE), infrared radiation, ultrasound, air nozzles with means for measuring back pressure which varies with product thickness, electrical-capacitance sensors, and optical devices employing beams of light.

Thickness Variation The differences in thickness among different locations in a product of desired uniform thickness, such as film, sheet, pipe, wire coatings, laminates, bar stock, etc. In making sheet and film, one must be concerned with thickness variation in both the *machine direction* and *transverse direction*, since they usually have quite different causes and cures. The same is true of wall thickness in pipe, where the directions are axial and circumferential. (The thickness itself is radial.) Thickness variations are often of concern in molded and thermoformed products, even though major differences in thickness are there by design. In extruded products, when thickness variations are even a few percent of nominal, more material must be extruded per unit length of product in order to meet minimum thickness specifications, thus increasing unit cost and reducing profit.

Thin-Layer Chromatography See CHROMATOGRAPHY.

Thinner A liquid incapable of dissolving a resin but which can partly substitute for a solvent and, at the same time, reduce the viscosity of a paint, varnish, lacquer, or adhesive. See also DILUENT.

2,2'-Thiobis(4-*t*-octylphenolate)-*N*-Butylamine Nickel (Cyasorb UV 1084) An ultraviolet absorber used in polyolefins for items such as agricultural film wherein good weatherability is required.

Thiokol® Trade name of Thiokol Corp. for a line of polysulfides and similar materials. See POLYSULFIDE RUBBER.

Thiourea-Formaldehyde Resin A member of the amino plastics family, made by condensation of thiourea [$(NH_2)_2CS$] with formaldehyde.

Thixotropic Agent A chemical that imparts the property of THIXOTROPY, w s, to a solution or suspension. See also THICKENING AGENT.

Thixotropic Fluid See THIXOTROPY.

Thixotropy A time-dependent decrease in the viscosity of a liquid subjected to shearing or stirring, followed by a gradual recovery when the action is stopped. This is a desirable property in most paints because the painter wants the paint to spread easily but not to sag or slump when brushing, rolling, or spraying is stopped. The term is often mistakenly applied to PSEUDOPLASTIC FLUIDS, w s.

Threading up Actions taken in starting an extrusion operation after turning on the preheated extruder at low speed and beginning the flow of feedstock. As the first extrudate emerges from the die, it is gripped and moved by hand through the various parts of the downstream equipment until each of the driven elements is itself moving the extrudate. Extrusion rate is then gradually increased to the target level.

Thread Plug The male part of a mold that shapes an internal thread in the molding and must be unscrewed from the finished piece. Automatic unscrewing molds can do this mechanically for scores of bottle caps molded in a single shot.

Three-Dimensional Braid (3-D braid) A recent development in building reinforcement preforms for complex shapes that permits the placing of reinforcing fibers in three orthogonal (or nonorthogonal) directions so as to best support multidirectional stresses expected to act on the finished part in service.

Three-Plate Mold An injection mold with an intermediate, movable plate that permits center or offset gating of each cavity.

Three-Roll Stack (haul-off) The vertical array of three polished and cored, chrome-plated rolls that receives molten sheet from a sheet die and chills it while impressing the high finish of the rolls on the sheet itself. The position of the center roll is fixed, while the upper and lower rolls are pressed toward the center roll against adjustable stops by air cylinders exerting pressures of 300+ kPa per centimeter of roll width. Roll temperatures are separately controlled by warm water pumped through the shell annuli at high velocity. The molten sheet passes into the nip of the two upper rolls, whose speed, relative to the mean melt velocity in the die, determines how much the sheet is drawn down. Embossed sheet can be made by replacing the center roll with an embossing roll.

Threshold Limit Value Parts of vapor or gas per million parts of air by volume (ppm) or milligrams of *particulates* per cubic meter of air (mg/m^3) to which workers may safely be exposed for a stated time. Threshold limits for many materials have been published by the Environmental Protection Agency (EPA) and the the American Conference of Governmental Industrial Hygienists (ACGIH). Those limits are intended for use by persons trained in the field of industrial hygiene.

Throwing A textile term referring to the act of imparting twist to a yarn, especially while plying and twisting together a number of yarns. A person doing this is called a *throwster*.

Thrust Bearing Of an extruder, the heavy-duty bearing upon which the rear shoulder of the screw shank pushes as it transmits the rearward force due to the head pressure. (See THRUST LOAD.) Three types of thrust bearings are in use for single-screw extruders. In order of increasing merit they are: flat-roller bearings, tapered-roller bearings, and spherical-roller bearings. See also B-10 LIFE.

Thrust Load In an extruder or screw-injection molder, the rear-directed force in reaction to the forward buildup of pressure in the screw, culminating in the head pressure acting over the whole screw cross section, and equal to $\pi \cdot D^2 \cdot P/4$. (See also B-10 LIFE.) This force is much less in an injection molder of the same screw diameter because, while the screw is turning, the head pressure is low and when the screw stops, the injection rams take up the thrust. Therefore, their thrust bearings can be much smaller than those of equal-size extruders.

Ti Chemical symbol for the element titanium.

Tie Bar In a plastic molding press, one of two or four sturdy, cylindrical, steel posts that provide structural rigidity to the clamping mechanism and accurately guide platen movement.

Time-Temperature Equivalence Because an increase in temperature accelerates molecular motions, mechanical behavior of polymers at one temperature can be used to predict those at another by means of a shift factor equal to the ratio of relaxation time at the second temperature to that at the first. The principle can be used to combine measurements made at many temperatures into a single master curve for a reference temperature over many decades of time. See WILLIAMS-LANDELL-FERRY EQUATION.

Tin Stabilizer See ORGANOTIN STABILIZER.

TIOTM Abbreviation TRIISOOCTYL TRIMELLITATE, w s.

Titanate Coupler One of a family of organo-titanium compounds first developed by Kenrich Petrochemicals in 1978 and burgeoning since then. Types available include monoalkoxy, chelate, coordinate, and quaternary salts. They form molecular bridges between organic matrices (resins) and inorganic fillers and reinforcements. Conductivities of metal-filled plastics are increased by one to four orders of magnitude, while melt viscosities are reduced by factors of 0.3 to 0.1.

Titanium Dioxide (titanic anhydride, titanic acid anhydride, titanium white, titania) TiO_2. A white powder available in two crystalline forms, both tetragonal: *anatase* and *rutile*. Both are widely used as opacifying and brightening pigments in thermosets and thermoplastics, used alone when whites are desired or in conjunction with other pigments when tints are desired. They are essentially chemically inert, light-fast, resistant to migration and heat. The rutile form is denser and has the higher refractive indices (2.61 and 2.90 vs 2.55 and 2.45 for the anatase form), and thus has somewhat greater opacifying power for a given volume percent and particle-size distribution.

Titanium Trichloride $TiCl_3$. A catalyst for olefin polymerization.

TMDI See 2,2,4-TRIMETHYL-1,6-HEXANE DIISOCYANATE.

TMTD See TETRAMETHYLTHIURAM DISULFIDE.

***a*-Tocopherol** (ATP, vitamin E, 5,7,8-trimethyl tocol) $C_{29}H_{50}O_2$. An involatile liquid antioxidant, generally regarded as safe by the Food and Drug Administration. It has been shown to be a good heat stabilizer in polyolefins, providing protection at levels around 250 ppm. Both ATP and its breakdown products are environmentally safe.

TOF Abbreviation for TRIOCTYL PHOSPHATE, w s.

Toggle Action A mechanism that magnifies force exerted on a knee joint. It is used as a means of closing and locking press platens and also serves to apply pressure at the same time.

Tolerance Interval The specified allowance of variation of a dimension (or other quantity) above and below the nominal or target value in a production part or product. For a process whose average level and random variation are in control, symmetrical tolerance limits should be at least six process standard deviations apart in order to approach zero percent defective parts. Tolerances are better understood and getting much more attention now than a few decades ago when a New York molder was asked about tolerances on parts he was producing. His reply: "Hey, we got lotsa tolerance here! We hire people no matter what color or nationality they are."

Toluene (toluol, methylbenzene, methylbenzol) $H_3CC_6H_5$. A colorless, flammable liquid with a sharp, benzene-like odor, used as a solvent for cellulosics, vinyl organosols, and other resins. Toluene is also a synthesis intermediate for polyurethanes and polyesters.

Toluene-2,4-Diamine (TDA, *m*-tolylene diamine) $H_3CC_6H_3(NH_2)_2$. A colorless, crystalline material used in the production of toluene diisocyanate, a key material in the manufacture of polyurethanes.

Toluene-2,4-Diisocyanate (2,4-toluene diisocyanate, Br: tolylene diisocyanate, TDI) $H_3CC_6H_3(NCO)_2$. A water-white to pale yellow liquid with a sharp, pungent odor, produced by reacting toluene-2,4-diamine with phosgene. (Some of the 2,6-isomer is usually present.) It reacts with water to produce carbon dioxide. TDI is widely used in the production of polyurethane foams and elastomers, but is toxic and requires careful handling to keep its concentration in work-space air below the permissible threshold limits. See also DIISOCYANATE.

***p*-Toluenesulfonamide** (PTSA) $H_3CC_6H_4SO_2NH_2$. White leaflets, existing also in the *o*- form; both are used as solid plasticizers for ethyl cellulose, polyvinyl acetate, and rigid PVC.

p-Toluenesulfonylhydrazide A blowing agent similar to benzenesulfonylhydrazide, but having higher melting and decomposition temperatures.

p-Toluenesulfonylsemicarbazide A blowing agent with a high decomposition temperature (235°C) that makes it useful for foaming plastics that are processed at high temperatures, such as high-density polyethylene, polypropylene, rigid PVC, acrylonitrile-butadiene-styrene resins, polycarbonates, and nylons.

2,4-Tolylene Diisocyanate British synonym for TOLUENE-2,4-DIISOCYANATE, w s.

Toner A coloring agent in which all pigments and vehicles are organic. See also LAKE.

Top-Blowing (adj) Designating a blow-molding process in which air is injected into the parison at the top of the mold.

Torpedo (1) (spreader) A streamlined, conicylindrical block, supported by three ridges (*spider*) and placed in the path of flow of the plastic pellets within the heating cylinder of a plunger-type injection molder, to spread the stock into a thin annulus, thus providing intimate contact with the heating surfaces. (2) Years ago, some extruder screws ended in smooth torpedoes only a little less in diameter than the barrels, the idea being to provide final shear mixing and improve homogeneity of the melt arriving at the die. (3) The core of an in-line pipe die supported by spider legs is sometimes called a torpedo.

Torr The torr (for E. Torricelli, who invented the mercury barometer in 1643) was introduced a few decades ago when people grew weary of saying "millimeters of mercury", to which it was set equal. Now the torr, too, is deprecated. 1 torr = 133.322 Pa.

Torsion Engineering term for modes of shear stress and shear strain caused by twisting of bodies.

Torsional Braid Analysis A method of performing torsional tests on small amounts of materials in states in which they cannot support their own weight, e g, liquid thermosetting resins. A glass braid is impregnated with a solution of the material to be tested. After evaporation of the solvent, the impregnated braid is used as a specimen in an apparatus that measures motion of the oscillating braid as it is being heated at a programed rate in a controlled atmosphere. ASTM D 4065 provides information on torsional testing.

Torsional Modulus Shear modulus (G) as measured in a test in which the specimen is twisted. In ASTM D 1043, the test specimen is a flat rectangular bar with length about 7 times its width of 6.35 mm and thickness 1/3 the width. The apparent shear modulus is given by:

$G = k \cdot T \cdot L / (a \cdot b^3 \cdot \phi)$ where T is the torque exerted on the specimen, L, a, and b are its length, width and thickness, ϕ is the angle of twist, and k is a coefficient dependent on units and the ratio a/b.

Torsional Rigidity Of a fiber, wire, bar, tube, or profile shape subjected to twisting of one end relative to the other, the torque required to produce a twist of one radian. This rigidity is proportional to the shear modulus of the material and is strongly dependent on all dimensions, especially section thickness.

Torsional Test A test for determining shear properties of plastics, such as shear modulus, based on measuring the torques required to twist a specimen through a prescribed arc. ASTM 1043 is such a test.

Tortuosity Factor (1) The distance a permeating molecule must travel to pass through a film, divided by the thickness of the film. (2) The mean length of path of fluid molecules passing through a packed bed or porous medium (such as open-cell foamed plastic) divided by the thickness of the bed in the direction of the pressure gradient.

Total Shear A concept introduced around 1956 as an indicator of shear mixing in extruders, but also applicable to other processes in which shear is the principal flow mode. It is the integrated product of shear rate times time over the region in which a melt is undergoing shear flow. The ratio of initial to final *striation thickness* is proportional to total shear. See STRIATION (2).

TOTM Abbreviation for TRIOCTYL TRIMELLITATE, w s.

Toughness A term with a wide variety of meanings, no single precise mechanical definiton being generally recognized. Toughness generally inplies a lack of brittleness; having very substantial elongation to break accompanied by high tensile strength. One proposed definition for toughness is the energy per unit volume to break a material, equal to the area under the stress-strain curve. Toughness has also been equated to impact resistance, especially resistance to repeated impacts. Energy required to break a specimen in the TENSILE-IMPACT TEST, w s, divided by the gauge-length volume of the specimen, is a fairly straightforward toughness measure. Note, though, that energy to break in tension is definitely dependent on the time scale of the test. Toughness has also been equated to resistance to abrasion, and to resistance to penetration by points and cutting with sharp tools.

Tow A loose, untwisted rope that usually contains many thousands of filaments.

Towpreg A prepreg consisting of resin-impregnated TOW, w s, that may be braided or woven to form a reinforced-plastic structure.

Toxicity (poisonousness) Although most pure resins and polymers are relatively nontoxic, compounding additives such as stabilizers, colorants, and plasticizers must be carefully selected when products are to be used for food packaging or other applications involving human contact. Data are available on the safe limits of vapor concentration in the work place for most solvents and monomers. See NONTOXIC MATERIALS for a source of detailed information about the toxicity status of resins and additives.

TPA Abbreviation for TEREPHTHALIC ACID, w s.

TPE Abbreviation for THERMOPLASTIC ELASTOMER, w s.

TPO Abbreviation for THERMOPLASTIC POLYOLEFIN, w s.

TPP Abbreviation for TRIPHENYL PHOSPHATE, w s.

TPR Registered trade name of Uniroyal, Inc, for a family of thermoplastic rubbers based mainly on ethylene and propylene. Grades range in hardness (Shore A scale) from 65 to 90. Processable by the usual thermoplastics methods, these materials have the properties of vulcanized rubber.

TPS Abbreviation used by the British Standards Institution for "toughened polystyrene," equivalent to US high-impact polystyrene.

TPX Abbreviation for POLY(4-METHYLPENTENE-1), w s.

TR Abbreviation used by British Plastics Institution for *thio rubber*. See POLYSULFIDE RUBBER.

Tracking An electrical-breakdown phenomenon in which current caused by a large voltage difference between two conductors in contact with an insulating material gradually creates a conductive leakage path across the surface of the material by forming a carbonized track that appears as a thin, wriggly line between the electrodes. ASTM lists several tests of tracking resistance, all in section 10.02.

Trade Molder The British term for CUSTOM MOLDER, w s.

Trailing Flight Face (trailing flight) In an extruder screw, the rear side of any flight. The forward side is called the LEADING FLIGHT FACE, w s.

trans- An organic-chemistry prefix denoting an isomer in which certain atoms or groups are located on opposite sides of a plane of symmetry. Usually ignored when alphabetizing names of compounds. Compare *CIS-*.

Transducer A device that transforms the value of a physical variable into an electrical signal, usually voltage or current. Examples are: thermocouple, pressure transducer, linear variable differential transformer, (a motion transducer), tachometer generator, and force cell.

Transfer Coating A process for coating fabrics such as knits, which are extremely difficult to coat directly by conventional spread-coating methods. In a typical version of the process, a layer of plastisol is cast against a silicone-treated release paper. This first layer becomes the top coat or wear layer in the final product. After gelling of the first layer a second coating of urethane solution is applied to serve as an adhesive layer that bonds the wear layer to the fabric substrate. The composite is finally heated and pressed together. and the paper is stripped away. Many variations of this process are possible, e g, using vinyl or polyurethane foam, embossing, etc.

Transfer Molding A molding process used mainly for thermosetting resins and vulcanizable elastomers. The molding material, usually preheated, is placed in an open "pot" with a hole in its bottom atop the closed mold. The cross-sectional area of this pot is about 15% larger than the total projected area of all cavities and runners in the mold. A ram is placed in the pot above the material. Pressure, applied by a press platen to the ram, forces the molding material through the runners and gates, and into the cavities of the heated mold. Following a heating cycle in which the material is cured or vulcanized, the press is opened and the parts are ejected. In a variation called *plunger molding*, resembling injection molding, the plunger is more a part of the press rather than of the mold, and pressure is applied to the plunger by an auxiliary hydraulic ram. The compound flows faster in plunger molding and more frictional heat is developed, so molding cycles are generally shorter than in transfer molding.

Transfer-Molding Pressure The pressure applied to the cross-sectional area of the material pot or cylinder (MPa or kpsi).

Transition Section (transition zone, compression section) In a metering-type screw for a single-screw extruder, the section of decreasing channel volume per turn between the feed and metering sections, in which the plastic is changing state from a loosely packed bed of particles-*cum*-voids, to a void-free melt. The transition may be abrupt or gradual, the latter having been found more satisfactory for nearly all thermoplastics, and may be accomplished by increasing the root diameter of the screw or by reducing the lead angle or both. Reducing the root diameter has been by far the preferred method and may be done conically or helically. See CONICAL TRANSITION, HELICAL TRANSITION, and COMPRESSION RATIO.

Transition Temperature (1) The GLASS-TRANSITION TEMPERATURE, w s. (2) More generally, any temperature at which a polymer exhibits an abrupt change in phase or measurable property, or at which a property's rate of change with temperature changes abruptly (second-order transition).

Transmittance See LIGHT TRANSMITTANCE.

Transparent Plastic A plastic that transmits incident light with negligible scattering and little absorption, enabling objects to be seen clearly and brightly through it. At least nine basic classes of plastics are generally regarded as being permanently transparent in thick sections; some others possess near-transparency, at least for a limited period. The nine are: acrylics (foremost in usage for transparency), cellulosics, allyl diglycol carbonates, some epoxies, nylons, and plasticized PVCs, polycarbonate, polysulfone, crystal polystyrene, and polyphenyl sulfone. Some others, such as polyethylene and polypropylene, are highly transparent in thin films, but in thicker sections, are made translucent by their crystallites. The ASTM test for haze and luminous transmittance of transparent plastics is D 1003.

Transputer A type of process-control architecture in which computing elements are linked not only by a systems bus but also by additional links combining software and hardware, called "firmware" links. The gain in speed of handling control tasks (as of 11/91) is about a factor of 20, permitting the use of complex control algorithms for, say, injection molding, that couldn't be used with conventional microprocessor/bus systems.

Transverse Direction (1) In extruding sheet or film, the direction of the width, crosswise to the direction of extrusion. (2) In a uniaxially oriented plastic, either direction perpendicular to the direction of orientation (stretching). In a biaxially oriented sheet, the direction perpendicular to both axes (the plane) of orientation. (3) In a fiber-reinforced laminate, the thickness direction or, in a laminate with unidirectional reinforcement, either of the two directions perpendicular to the fiber lengths.

Trapped-Sheet Forming A process announced in 1956 for high-speed, pressure-thermoforming of thin, biaxially oriented polystyrene sheet into snap lids for dairy-product containers, so called because each circle in the sheet was edge-restrained during heating to prevent its shrinking and losing its oriented strength.

Treater (1) Equipment for preparing resin-impregnated reinforcements, including means for passing a continuous web or strand through a resin tank, controlling the amount of resin picked up, drying and/or partly curing the resin, and rewinding the impregnated reinforcement. (2) The equipment used in CORONA-DISCHARGE TREATMENT, w s.

Tree Formation The generation of a tree-like void structure in a transparent plastic by electron bombardment at a point on the surface. The effect in acrylic blocks is dramatically decorative. Similar breakdown structures form in dielectrics subjected to strong electric fields, eventually penetrating the dielectric and causing a short circuit.

Tremolite $Ca_2Mg_5Si_8O_{22}(OH, F)_2$. A silicate mineral similar to and sometimes sold as fibrous talc. It can be used in many applications in place of asbestos as a filler.

Triacetate A generic term for film or fibers of cellulose acetate in which at least 92% of the hydroxyl groups are acetylated. See also CELLULOSE TRIACETATE.

Triacetin Synonym for GLYCERYL TRIACETATE, w s.

Triallyl Cyanurate (TAC, 2,4,6-triallyloxy-1,3,5-triazine) $(CH_2=CHCH_2OC)_3N_3$. This heterocyclic compound, a solid below 110°C, is highly reactive and is used in copolymerizations with vinyl-type monomers to form ALLYL RESINS (w s). It is also used to crosslink unsaturated polyesters and raise their softening temperatures.

Triallyl Phosphate (TAP) $(CH_2=CHCH_2O)_3PO$. A monomer that can be polymerized with methyl methacrylate to produce flame-retardant copolymers.

Triaryl Phosphate A synthetic-ester-type plasticizer derived from isopropylphenol feedstock, useful as a flame-retarding plasticizer in vinyl plastisols.

Triaxial-Braid Preforming Any braiding technique for fibrous reinforcements by which three-dimensional preforms for reinforced-plastics structures are produced.

Triaxial Weaving Weaving in which the cloth is made from three yarns whose axes are 120° apart. When used as a reinforcing medium, the cloth yields a laminate whose properties in the plane are nearly isotropic.

Triazine Resin (NCNS resin) A class of thermosetting polymers prepared from primary and secondary biscyanamides with pendant arylsulfonyl groups. The biscyanamides are reacted together in solutions to form soluble prepolymers by addition polymerization. By refluxing these solutions, stable laminating varnishes are obtained. Alternatively, evaporation of solvents from the solutions yields the prepolymer in powder form for molding. Laminates prepared with triazine resins have good mechanical strength at high temperatures and are relatively fire-retardant.

Tribasic Pertaining to acids having three replaceable hydrogen atoms per molecule, e g, phosphoric acid, the most important one; or to acid salts in which two of three available hydrogens have been replaced by metals.

Tribasic Lead Maleate A yellowish-white crystalline powder, an effective heat stabilizer for vinyls and a curing agent for chlorosulfonated polyethylene.

Tribasic Lead Sulfate $3PbO \cdot PbSO_4 \cdot H_2O$. A heat stabilizer, especially for vinyl electrical-insulation compounds. It is very effective, has good electrical properties, and produces no gassing.

Tribology The engineering science that deals with friction, wear, and lubrication of surfaces sliding or rolling on one another, and the design of systems and components, such as gears and bearings, in which these actions are involved.

Tributoxyethyl Phosphate $(C_4H_9OC_2H_4O)_3PO$. A primary plasticizer for cellulosic, acrylic, and vinyl resins, imparting low-temperature flexibility and flame retardance. In vinyl plastisols, small amounts of tributoxyethyl phosphate markedly reduce viscosity. However, when used alone or in high percentages, this plasticizer causes inconveniently rapid gelation.

Tri-*n*-butyl Aconitate $C_3H_3(COOC_4H_9)_3$. A combination plasticizer and stabilizer for polyvinylidene chloride and synthetic rubbers.

Tributyl Borate (butyl borate) $(C_4H_9)_3BO_3$. A colorless liquid, used as an antiblocking agent in plastic films and sheets.

Tri-*n*-butyl Citrate $C_2H_4(OH)(COOC_4H_9)_3$. A nontoxic plasticizer for most thermoplastics, including cellulosics, polystyrene, and vinyls.

Tributyl Phosphate (TBP) $(C_4H_9O)_3PO$. A colorless liquid used as a primary plasticizer and solvent for cellulose acetate, chlorinated rubber and, in special applications, for vinyl resins. Its relatively high volatility limits its use as a plasticizer for vinyls.

Tri-*n*-butyl Phosphine $(C_4H_9)_3P$. A curing agent for epoxy resins, and a catalyst for vinyl and isocyanate polymerizations.

4,4,4-Trichloro-1,2-Butylene Oxide (TCBO) $Cl_3CCH_2CHOCH_2$. This highly reactive, liquid epoxide is used for modifying polyols to achieve fire retardance in polyurethane foams.

1,1,1-Trichloroethane (methyl chloroform) CH_3CCl_3. A nonflammable solvent that can be milled with resins to produce nonflammable adhesives. Less used than formerly because of the perceived need to minimize release of chlorinated compounds into the atmosphere.

Trichloroethylene (TCE, trichloroethene) $ClCH{=}CCl_2$. Until recently, a nonflammable solvent widely used for degreasing. It is a suspected cancer agent, mutagen, and stratospheric-ozone destroyer.

Trichlorofluoromethane (Freon 11, Genetron 11) CCl_3F. A chemically inert blowing agent used until recently with water in flexible polyurethane formulations to control foam density and load-bearing properties. It is also a refrigerant and former aerosol propellant. Freon 11 is being phased out of most applications by 1995 in the general push to reduce fluorocarbons in the atmosphere.

N-(Trichloromethylthio) Phthalimide A bacteriostatic agent used in vinyl fabrics for hospitals and households. It stops the growth of bacteria such as *staphylococcus aureus* that cause pink staining in white and pastel-colored PVCs.

Trichlorotrifluoroethane (Freon 113) $Cl_2FCCClF_2$. A colorless, nearly odorless solvent that boils at 47°C, formerly used as a blowing agent for integral-skin polyurethane foam, but now because of its suspected action on stratospheric ozone, being phased out.

Tricresyl Phosphate (TCP, TCF, tritolyl phosphate) $(CH_3C_6H_4O)_3{\equiv}PO$. One of the earliest commercial plasticizers for PVC, also suitable for cellulosics, alkyds, and polystyrene. Like other phosphate plasticizers, TCP imparts flame retardance and fungus resistance, even when used in amounts as low as 5% of the total plasticizer content. TCP and cresyldiphenyl phosphate are the plasticizers most widely used for these properties.

Tricresyl Phosphite $(CH_3C_6H_4O)_3P$. A colorless liquid used as a flame-retardant plasticizer and stabilizer for thermoplastics.

Tricyclohexyl Citrate $(C_6H_{11}OOCCH_2)_2C(OH)COOC_6H_{11}$. A nontoxic plasticizer for polystyrene, cellulosics, acrylics and vinyls.

Tridecyl Phosphite $(C_{10}H_{21}O)_3P$. A colorless liquid used as a stabilizer for polyolefins and PVC.

Tridimethylphenyl Phosphate (trixylenyl phosphate, TXP) A plasticizer for cellulosics and vinyl compounds in which a low-density, electrical-grade, flame-retardant plasticizer is required.

Tridirectional Laminate A reinforced-plastics material having reinforcing fibers running in three principal directions. If the directions are 120° apart and there are equal percentages of fiber lying in the three directions, the laminate properties will be almost isotropic in the laminate plane.

Triethyl Aluminum (aluminum triethyl, ATE) $(C_2H_5)_3Al$. A catalyst for the polymerization of olefins.

Triethyl Citrate (ethyl citrate) $(C_2H_5OOCCH_2)_2C(OH)COOC_2H_5$. An ester of citric acid, used as a plasticizer for many thermoplastics including vinyls, cellulosics, and polystyrene. It has won FDA approval for use in food packaging.

Triethylenediamine (1,4-diazabicyclo-2,2,2-octane, DABCO) This tertiary diamine, whose structure is shown below, is the most widely used amine catalyst for polyurethane foams, elastomers, and coatings. It is soluble in water and polyols.

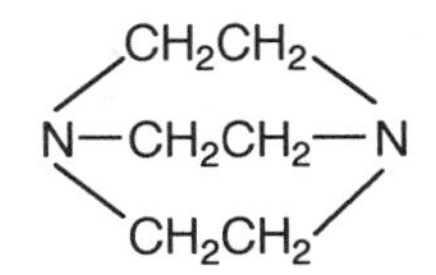

(Triethylenediamine)

Triethylene Glycol Diacetate $(-CH_2OCH_2CH_2OOCCH_3)_2$. A plasticizer for cellulosic plastics and some acrylic resins.

Triethylene Glycol Dibenzoate $(-CH_2OCH_2CH_2OOCC_6H_5)_2$. A secondary plasticizer, partly compatible with all common thermoplastics. In most resin systems it has a tendency to crystallize and bloom at higher concentrations, which property may be used to advantage to prevent BLOCKING, w s.

Triethylene Glycol Dicaprylate (triethylene glycol trioctanoate) $(-CH_2OCH_2CH_2OOCC_7H_{15})_2$. A plasticizer for ethyl cellulose and vinyl resins, with good low-temperature flexibility.

Triethylene Glycol Di(2-ethylbutyrate) $(-CH_2OCH_2CH_2OOC-C_5H_{11})_2$. A plasticizer for cellulosic, acrylic, and vinyl resins. Its largest use is in plasticizing polyvinyl butyral, the interlayer in automobile windshield glass.

Triethylene Glycol Di(2-ethylhexanoate) $(-CH_2OCH_2CH_2OOC-C_7H_{15})_2$. A plasticizer for cellulosic plastics, polymethyl methacrylate, PVC, and vinyl chloride-vinyl acetate copolymers. In vinyls it is usually used as a secondary plasticizer at 10 to 25 percent of the total plasticizer, to impart low-temperature flexibility.

Triethylene Glycol Dipelargonate $(-CH_2OCH_2CH_2OOCC_8H_{17})_2$. A plasticizer for vinyls and cellulosics.

Triethylene Glycol Dipropionate $(-CH_2OCH_2CH_2OOCC_2H_5)_2$. A plasticizer for cellulosic resins and polymethyl methacrylate.

Triethylenetetramine (TETA) $(-CH_2NHCH_2CH_2NH_2)_2$. A viscous, yellowish liquid, TETA is used as a room-temperature curing agent for epoxy resins.

Tri(2-ethylhexyl) Citrate A nontoxic plasticizer for PVC.

Tri(2-ethylhexyl) Phosphate See TRIOCTYL PHOSPHATE.

Triethyl Phosphate (TEP) $(C_2H_5O)_3PO$. A flame-retardant plasticizer for cellulosics, acrylics, some vinyl polymers, and unsaturated polyesters.

Trihydrazide Triazine A heterocyclic chemical blowing agent that decomposes at 250 to 260°C yielding, per gram, about 175 cm^3 of gas consisting mostly of ammonia and nitrogen. It is used in foaming polypropylene, acrylonitrile-butadiene-styrene resin, nylon, and other high-melting resins.

Triisodecyl Trimellitate $(C_{10}H_{21}OOC)_3C_6H_3$. An involatile plasticizer for PVC.

Triisooctyl Trimellitate (TIOTM) $(C_8H_{17}OOC)_3C_6H_3$. A plasticizer for cellulosic and vinyl plastics with low volatility, high resistance to soapy-water extraction, and essentially no marring effect on lacquered surfaces.

Trimer An oligomer formed by the union of three molecules of a monomer and/or made up of three mer units. See POLYMER and OLIGOMER.

Trimethyl Borate (methyl borate, trimethoxyborine) $(CH_3O)_3B$. A colorless liquid used as a flame retardant in plastics.

Trimethylene Glycol Synonym for 1,3-propanediol. See GLYCOL.

Trimethylene Glycol Di-*p*-aminobenzoate A diamine curing agent for polyurethanes, introduced in 1976 to replace the popular but toxic MOCA®, w s. It is very soluble in a variety of coating solvents.

2,2,4-Trimethyl-1,6-Hexane Diisocyanate (TMDI) $OCNCH_2C(CH_3)_2CH(CH_3)CH_2CH_2NCO$. A branched aliphatic isocyanate used in making polyurethanes.

Trimethylolethane Tribenzoate A solid plasticizer for PVC.

2,2,4-Trimethyl-1,3-Pentanediol (trimethylpentanediol, TMPD) $(CH_3)_2CH(OH)C(CH_3)_2CH_2OH$. One of the principal glycols used in making polyester resins, alkyd resins, and polyester plasticizers, containing one primary and one secondary hydroxyl group. It is made by the aldol condensation of isobutyraldehyde, yielding a water-insoluble white solid. TMPD is used in producing linear unsaturated polyesters and is particularly good for gel-coating resins.

2,2,4-Trimethyl-1,3-Pentanediol Monoisobutyrate Benzoate A plasticizer for PVC imparting good stain resistance.

Tri(*n*-Octyl-*n*-Decyl) Trimellitate (NODTM) A low-temperature plasticizer for vinyls and cellulosics, as permanent as polymeric plasticizers. It is used in heavy-duty applications such as truck seating, window channeling, film and sheeting subject to wide temperature ranges, and in baby wear.

Trioctyl Phosphate [TOP, TOF, tri-(2-ethylhexyl) phosphate] $[C_4H_9CH(C_2H_5)CH_2O]_3PO$. A plasticizer for PVC, imparting good low-temperature flexibility, resistance to water extraction, flame and fungus resistance, and minimum change in flexibility over a wide temperature range. It is also compatible with polyvinyl butyral, ethyl cellulose, and cellulose acetate-butyrate resins with a high butyral content.

Trioctyl Trimellitate (TOTM) $(C_8H_{17}OOC)_3C_6H_3$. A primary plasticizer for vinyls that hold up well at high temperatures. It combines the permanence of polymeric plasticizers with the low-temperature properties of monomerics. In vinyls, it is used for auto interior parts and for wire insulation good for temperatures to 105°C. It is also used with cellulosics and acrylics.

Triol A term sometimes used for *trihydric alcohol*, i e, an alcohol containing three hydroxyl (–OH) radicals.

1,3,5-Trioxane (*sym*-trioxane, triformol, trioxin,) $\overline{CH_2OCH_2OCH_2O}$. The stable, cyclic trimer of formaldehyde, a colorless, crystalline solid. It is easily depolymerized in the presence of acids to its monomer, or may be further polymerized to form ACETAL RESINs, w s. This trimer should not be confused with PARAFORMALDEHYDE, w s.

Triphenyl Phosphate (TPP) $(C_6H_5O)_3PO$. A crystalline powder, one of the original synthetic plasticizers for cellulose nitrate. It is also a flame retardant for vinyls, cellulosics, acrylics, and polystyrene.

Triple-Daylight Mold A mold having four plates: feed plate, floating cavity plate, stripper plate, and moving mold plate. When the mold opens, all the plates move apart making three openings among them.

Tris(2-Chloroethyl Phosphate) $(CH_2ClCH_2O)PO$. A plasticizer for polystyrene, cellulosics, and vinyls. It is also effective as a flame retardant for unsaturated polyesters and polyurethane foams.

Tris(2,3-Dibromopropyl) Phosphate $(CH_2BrCHBrCH_2O)_3PO$. A flame retardant for unsaturated polyesters, polyurethane foams, and other plastics.

Tris(1,2-Dichloroisopropyl) Phosphate $[CH_2ClC(CH_3)ClO]_3PO$. A flame retardant for unsaturated polyesters.

Tris(2,3-Dichloropropyl) Phosphate $(CH_2ClCHClCH_2O)_3PO$. A plasticizing flame retardant for many plastics including vinyls, cellulosics, acrylics, polyolefins, phenolics, polyesters, and polyurethane foams.

Trisnonylphenyl Phosphite (TNPP) An FDA-sanctioned heat stabilizer and antioxidant used in styrene-butadiene copolymers.

Tritactic Polymer An isotactic or syndiotactic polymer that is also of the *cis-* or *trans* form because the molecules are unsaturated and have double bonds.

Tritolyl Phosphate Synonym for TRICRESYL PHOSPHATE, w s.

Trixylenyl Phosphate (TXP) See TRIDIMETHYLPHENYL PHOSPHATE.

Trommsdorff Effect See AUTOACCELERATION.

True Density (solid density) Of a specimen of porous material, the mass of the specimen divided by (volume of the specimen less the volume of its voids), i e, $\rho_t = M/[V(1 - \varepsilon)]$, where ε is the void fraction and V is the specimen volume. Compare APPARENT DENSITY.

True Strain In a tensile test, the integral of the differential increase in length divided by the length at that point in the test, i e,

$$\text{True strain} = \int_{L_O}^{L_f} \frac{dL}{L} = \ln\left(L_f / L_O\right)$$

In this equation, L_O is the original gauge length of the specimen and L_f is the gauge length at the stress-strain point of interest. True strain is always less than nominal strain (see STRAIN), but the difference is small unless the strain is large. For example, a nominal strain of 0.5 (50%) corresponds to a true strain of 0.41 (41%).

True Stress The quotient of force divided by true cross-sectional area in the specimen's gauge length, at any point during a tensile or compressive test. True stress is always larger than nominal stress (see STRESS), but the difference is negligible at low strains. In a ductile plastic having a Poisson's ratio of 0.4, for example, if the true strain were 0.41 (41%), the cross-sectional area would be 70% of the initial area, so true stress would be 1.43 times nominal.

TSC Abbreviation for THERMAL STRESS CRACKING or THERMALLY STIMULATED CURRENT, either of which see.

T-Slot Die See MANIFOLD (C).

TTB See 2,2',6,6'-TETRABROMO-3,3',5,5'-TETRAMETHYL-*p,p*'-BIPHENOL.

Tubing (1) Any of a wide range of continuous extrusions, usually having circular annular cross sections, and flexible enough to be wound on a core or coiled. Some large-diameter tubing has such thin walls that it is flattened before winding. Some stiff but coilable, small-diameter tubing, such as saran and nylon tubing used for automotive fuel lines and compressed-air service, has relatively thick walls. Unlike pipe, tubing is not usually cut to standard lengths nor is it expected to be nearly straight. With tubing, the internal diameter is usually more carefully controlled than the OD, whereas with pipe the reverse is true. (2) Cylindrical fabric made by braiding, weaving, or knitting. The term *sleeving* is applied to such tubing less than 10 cm in diameter.

Tubing Die A die with an annular opening used to extrude plastics tubing. The core (*mandrel*) of the die may be fitted with a water-cooled extension that aids in chilling the extrudate and bringing its internal diameter within tolerances. See also PIPE DIE.

Tubular Film See FILM BLOWING.

Tumbling (tumble finishing, barreling) A finishing operation for small plastic articles in which gates, flash, and fins are removed and/or surfaces are polished, by rotating them in a barrel together with wooden pegs, sawdust, and (sometimes) polishing compounds. The barrels are usually of octagonal shape with alternate open and closed panels, the open panels covered with screen to permit fragments of removed material to fall out. Blocks of dry ice may be added to the tumbling medium to embrittle the parts and thus facilitate cleaner break-off of flash; but in this case the barrel must be closed to retard evaporation of the dry ice.

Tumbling Agitator (tumble mixer) A cylindrical or conical vessel rotating about a horizontal or inclined axis, with internal ribs that lift the material and then let it tumble back into the charge. They are used mainly for blending dry materials, e g, color concentrates and resin pellets.

Tungsten (W) A hard, dense metallic element that has occasionally been used in powder form as a plastics filler to increase density. Tungsten carbide, extremely hard and widely used in metal-cutting tools, has also been used to impart abrasion resistance in plastics.

Tunnel Gate Synonym for SUBMARINE GATE, w s.

Tunneling A condition occurring in incompletely bonded laminates, characterized by release of longitudinal portions of the substrate and deformation of these portions to form tunnel-like structures.

Turbulent Flow Flow for which the stream REYNOLDS NUMBER (w s) exceeds about 2100 to 4000. Polymer solutions, particularly dilute ones, can easily experience turbulent flow but polymer melts rarely (if ever) do because of their extremely high viscosities.

Turnkey (adj) Describing a complete manufacturing system or facility, such as a polymerization plant of sheet-extrusion line, delivered to the customer in ready-to-operate condition. ("He only has to turn the key.")

Twill Weave A cloth weave in which the warp yarn runs alternately over two fill yarns and under one fill yarn. The twos and ones may be staggered with adjacent or regularly spaced warp yarns to produce a diagonal effect. In dutch twill, wires of different sizes are used to produce a very dense screen with tortuous passages (light doesn't pass directly through) and an effective MESH NUMBER (w s) of 1200 or more. Twilled screens have sometimes been used in fiber extrusion and twill-woven cloths are used in some reinforced-plastics structures.

Twinning A movement of planes of atoms in the crystal lattice parallel to a specific (twinning) plane so that the lattice is divided into two symmetrical parts which are differently oriented. The amount of movement of each plane is proportional to its distance from the twinning plane.

Twin-Screw Extruder See EXTRUDER, TWIN-SCREW.

Twin-Shell Forming A high-speed thermoforming process for producing bottles and other hollow objects. Two thermoplastic sheets from separate roll-unwind stands are conveyed through heating apparatus, then positioned between facing halves of vacuum-forming molds. After closing the molds, vacuum is drawn on each half to simultaneously form the articles and seal the edges. When molds are arranged on a pair of endless conveyors, the process becomes continuous, sheets being fed into one end and a continuous web of formed products emerging from the other end, ready for separation from the waste portion of the sheets. The trim is ground and returned to the sheet-extrusion operation.

Twist (1) A textile term, the number of turns (360°) per unit length that a multifilament yarn, staple yarn, or other structure is turned or twisted around its longitudinal axis into a stable structure. (2) The unintended, progressive spiraling seen in some pultruded products.

Twist Direction The direction in which a yarn, held vertically, spirals in the downward direction about its own axis. If the direction is counterclockwise, the yarn is said to have S-twist; if clockwise, Z-twist.

Two-Color Molding See DOUBLE-SHOT MOLDING.

Two-Component Spraygun A spray head with fittings for attachment to two separate feed lines, each carrying one component of a reactive resin mixture, e g, as in urethane foam-in-place molding or polyester gel-coating.

Two-Level Mold (double-decker mold) An injection mold having two layers of cavities, used in making low-mass articles with large projected areas, so as to make best use of both plasticating and clamping capacities.

Two-Roll Mill See ROLL MILL.

Two-Shot Injection Molding Confusingly, this term has been used in the literature for two processes that are distinctly different. One is described under DOUBLE-SHOT MOLDING, w s. In the other process, one first injects a metered amount of one material into a single-cavity mold. As this material just begins to chill against the cold mold surfaces, a second material is injected. This fills the interior and forces the first material outward to the cavity surfaces. The second polymer, usually a reclaimed material, forms the interior of the finished article, while the virgin material first injected forms the outer shell and surface of the article.

Two-Shot Molding See DOUBLE-SHOT MOLDING.

TXP Abbreviation for *trixylenyl phosphate*, synonym for TRIDIMETHYLPHENYL PHOSPHATE, w s.

Type 8 Nylon Not to be confused with NYLON 8, w s, Type 8 nylon is a chemically treated nylon 6/6. It is thermoplastic, light yellow, with a leathery flexibility and excellent resistance to common solvents and abrasion. It has been used to impart abrasion resistance to the denim knees of jeans and overalls by impregnation of the cloth, and as an abrasion- and solvent-resistant coating for work gloves.

Type J Thermocouple A THERMOCOUPLE, w s, made up of one or two welded junctions of iron and CONSTANTAN (w s) wires and widely used in measuring temperatures in plastics-processing equipment.

Type K Thermocouple A THERMOCOUPLE, w s, made up of one or two welded junctions of CHROMEL AND ALUMEL (w s) wires, widely used in measuring high temperatures in oxidizing atmospheres, and enjoying considerable use in plastics processing.

Type T Thermocouple A THERMOCOUPLE, w s, made up of one or two welded junctions of copper and CONSTANTAN (w s) wires, used more in the measurement of low temperatures than high ones.

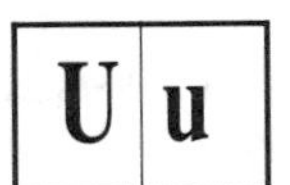

U (1) Chemical symbol for the element uranium. (2) In heat-transfer engineering, symbol for OVERALL CONDUCTANCE, w s.

Ubbelohde Viscometer (Cannon-Ubbelohde viscometer) An instrument made of Pyrex glass and consisting of an upper reservoir that drains through a marked capillary to a lower, vented chamber and thence to a second reservoir. The time taken by the test liquid to drain through the capillary is proportional to the viscosity (with slight corrections). This is one of several similar types used to measure viscosities of polymer solutions. Three relevant ASTM Methods are D 1243, D 1601, and D 2857. See also VISCOSITY.

UF Abbreviation for *urea-formaldehyde* resin. See AMINO RESIN.

UHMWPE Abbreviation for ULTRA-HIGH-MOLECULAR-WEIGHT POLYETHYLENE, w s.

Ultimate Elongation In a tensile test, the nominal elongation at rupture. See ELONGATION.

Ultimate Strength The maximum nominal stress a material can withstand when subjected to an applied tensile, compressive, or shear load. If the mode of loading is not specified, it is assumed to be tensile. In materials that exhibit a definite yield strength, ultimate strength will usually mean the nominal stress at break, which can be less than the maximum.

Ultracentrifuge A small centrifuge capable of rotational speeds to 100,000 rpm, creating sedimentation forces up to a million times gravity. At lower speeds and long sedimentation times, an equilibrium radial distribution of larger and smaller molecules is attained. As the centrifuge spins, a narrow, collimated light beam is passed through a cell containing a polymer solution. By measuring refractive index or light absorption at different radii (i e, depths in the cell), and using complex data-reduction methods, the number- and weight-average molecular weights of the polymer can be estimated and a graph of the molecular-weight distribution drawn. A second method uses the same techniques to measure sedimentation velocities at much higher speeds and over shorter time periods. Ultracentrifugation has been most useful with biological polymers, which tend to be all one size (*monodisperse*), making the data analysis less complicated.

Ultra-High-Molecular-Weight Polyethylene (UHMWPE) Any polyethylene having an average molecular weight (M_w) in the range from 1 million to 5 million g/mol. Density is about 0.94 g/cm^3. These materials, like polytetrafluoroethylene, do not truly melt and are processed by compression and sintering, and related methods. Small amounts of PE resins having somewhat lower molecular weight may be added as processing aids.

Ultra-Low-Density-Polyethylene (ULDPE) Any linear polyethylene with density less than 0.90 g/cm^3, and possibly as low as 0.86 g/cm^3. ULDPE films have better optical properties and better resistance to puncture, impact, and tearing than conventional linear low-density PEs.

Ultramarine-Blue Pigment A pigment family comprising a complex of double silicates of sodium and aluminum in combination with sodium polysulfide. They produce bright, clean tones even in combination with white pigments, and are resistant to the high temperatures employed in processing thermoplastics.

Ultrasonic Cleaning A method used for thoroughly cleaning molded plastics for electrical components and mechanical parts. A piezoelectric transducer (e g, a crystal of barium titanate), mounted on the side or bottom of a cleaning tank, is excited by an ultrasonic generator to produce high-frequency vibrations in the cleaning medium. These vibrations cause intense cavitation in the liquid and dislodge contaminants from crevices, and even from blind holes, that normal cleaning methods would not remove.

Ultrasonic C-Scan A nondestructive inspection technique for reinforced plastics in which the energy absorbed from a short ultrasonic pulse is measured. This is quantitatively different for samples containing delaminations, voids, or too little or too much reinforcement, than for solid samples of the correct composition.

Ultrasonic Degating A degating method used for small plastic parts produced by a family mold. The molding-machine operator removes the runner system and attached parts from the mold and loads the branched structure into the degating machine, where the runner makes contact with an ultrasonic horn. High-frequency-sound vibrations are transmitted through the runners to the narrow gates, causing them to melt, and the parts drop through holes into a sorting system.

Ultrasonic Frequency A sound frequency above the limit of human audibility, approximately 18 kHz. Most ultrasonic devices operate well above this level.

Ultrasonic Inserting A method of incorporating metallic inserts into plastics articles by means of ultrasonic heating. A plain cylindrical hole or well is molded in the plastic article by means of a core pin, the hole diameter being slightly less than that of the insert. Ultrasonic vibration and light pressure are applied as the metal part is being inserted, melting the plastic within a small radial distance of the inside hole surface. The displaced melt flows into the knurls, flutes or undercuts in the insert's outer surface and freezes, locking the insert into position.

Ultrasonic Prepreg Cutting A recently introduced method for cutting PREPREGs (w s) that uses an intense and extremely narrow beam of ultrasonic energy.

Ultrasonic Staking The process of forming a head on a protruding peg in a plastic article for the purpose of holding a surrounding part in position (see STAKING), utilizing ultrasonic heating and pressure to melt the tip of the protrusion and form it into a head. The process is very fast, usually taking only a fraction of a second.

Ultrasonic Welding (ultrasonic sealing) A method of welding or sealing thermoplastics in which heating is accomplished with vibratory mechanical pressure at ultrasonic frequencies (20 to 40 kHz). Electrical energy is converted to ultrasonic vibration by a transducer, directed to the area to be welded by a horn, and localized heat is generated by friction and impact at the surfaces to be joined. Other parts of the assembly are not heated. The process is most effective for rigid and semirigid plastics, since the energy is rapidly dissipated in soft, flexible materials. It will work with dissimilar plastics if they are melt-compatible and melt in the same temperature range.

Ultraviolet (UV) The region of the electromagnetic spectrum between the X-ray region and the violet end of the visible-light range, including wavelengths from about 3 to 200 nm. Photons of radiation in the UV region have sufficient energy to initiate some chemical reactions and to degrade many neat resins.

Ultraviolet Curing The process by which certain polymers or coatings are cured, with the aid of a photoinitiator, by exposure to ultraviolet radiation. One such polymer system is the oligomer tris(2-hydroxymethyl)isocyanurate triacrylate with an initiator and an acrylic monomer such as 2-phenoxyethyl acrylate in ratios from 10 to 100 parts per 100 parts of the triacrylate (USP 4,812,489).

Ultraviolet Degradation One of the most serious degradation threats to plastics being used outdoors. Changes in plastics such as crazing, chalking, dulling of the surface, discoloration, and fading; changes in electrical properties; lowering of strength and toughness; and even disintegration, have been caused by exposure to UV radiation, particularly the longer wavelengths near the visible violet. The presence of oxygen can exacerbate the process. Chain scission is the major mechanism.

Ultraviolet Printing (UV printing, UV-curing decoration) The process of printing or decorating with inks that cure rapidly by exposure to ultraviolet light. The inks are solvent-free, thus avoiding problems with air-pollution regulations, and contain a UV-sensitive catalyst that cures the ink in as little as one second. Mercury-vapor lamps can be used as the source of UV light. For polymers that are subject to degradation by UV-induced oxidation, equipment has been designed to blanket the exposed area with nitrogen.

Ultraviolet Spectrophotometry A method of chemical analysis similar to INFRARED SPECTROPHOTOMETRY (w s) except that the spectrum is obtained with ultraviolet light. It is somewhat less sensitive than the IR method for polymer analysis, but is useful for identifying and measuring plasticizers and antioxidants.

Ultraviolet Stabilizer An additive that protects plastics against ULTRAVIOLET DEGRADATION, w s, and that may accomplish its protection in various ways. An additive that preferentially absorbs UV radiation and dissipates the associated energy in a harmless manner is sometimes called an *ultraviolet absorber* or *ultraviolet screening agent*. Additives that do not actually absorb UV radiation but protect the polymer in some other manner are called *ultraviolet stabilizers* or other names indicative of the mechanism of stabilization. For example, products that remove the energy absorbed by the polymer before photochemical degradation can occur are called *energy-transfer agents* or *excited-state quenchers*. Other modes of UV stabilization are singlet-oxygen quenching, free-radical scavenging, and hydroperoxide decomposition. Classes of such stabilizers in use are the benzophenones, benzotriazoles, substituted acrylates, aryl esters, and compounds containing nickel or cobalt. Still another group consists of dispersed pigments, such as the very effective carbon black, which, of course, color the plastics and render them opaque.

Ultra-Viscoson Combustion Engineering's trade name for a somewhat portable instrument that consists of a small metal reed vibrated lengthwise at ultrasonic frequency (≈ 28 kHz), together with a computer. To measure the viscosity of a liquid, one immerses the reed in the liquid with the oscillator turned on. The computer measures the damping coefficient of the reed vibration and from that infers the product of viscosity times density. Dividing the reading by density gives the viscosity. The range is from 0.01 to 50 Pa·s for densities near 1 g/cm^3. Note that, with polymer melts, the viscosity determined at this high frequency may differ substantially from that obtained in steady-flow capillary rheometers or low-frequency cone-and-plate instruments.

Undercure A condition of inadequate mechanical properties in a thermosetting resin or elastomer resulting from too little time and/or too low temperature during the curing cycle.

Undercut (n) A lateral indentation in a molded part (or protuberance in a mold) that tends to impede withdrawal of a molded part from the mold. Articles of flexible materials such as plasticized vinyls can often be removed without difficulty from molds with severe undercuts, but undercuts must either be avoided in rigid materials or, where they must be part of the design, the mold must have movable parts that withdraw (*side draws*) before the part is to be ejected. Slight undercuts are sometimes deliberately designed into one half of a mold to cause a part to remain in that half when the mold opens. See also SIDE-DRAW PIN.

Underwater Pelletizing A system, used mostly with 15-cm-diameter and larger, high-output extruders in resin-manufacturing plants, in which a circular, heated die plate containing up to several hundred strand holes is enclosed within a water-tight casing and sprayed with water, while a rapidly spinning fly-knife slices the emerging strands of melt into lengths about equal to their diameters and flings the warm globs into the water spray. The water falling to the bottom carries the pellets out of the casing and onto a dewatering screen and drying conveyor. A delicate balance must be reached among several requirements: (1) cutting the pellets cleanly and chilling them instantly so that they do not grow "tails" nor form doubles and larger clusters;

(2) maintaining a uniform metal temperature of the die over its entire face so that melt doesn't freeze in the holes and the extrusion rate is the same in all; (3) matching the knife speed with the extrusion rate to get the desired pellet length; and (4) draining the emerging pellets in a way that uses their remaining sensible heat to fully dry them but cools them sufficiently so that they do not soften and fuse together in the collecting bin. Surface tension in the molten particles rounds them into nearly spherical or ellipsoidal final shapes.

Uniaxial Load A condition whereby a test sample or structural member is stressed in only one direction.

Uniaxial Orientation An orientation process that stretches the product in only one direction, as in manufacture of staple fiber, monofilaments, and melt-cast film.

Uniaxial Strain Tensile or compressive strain in a single direction—the usual testing mode—and typically in the length direction of a test specimen or structural member. See also STRAIN and TRUE STRAIN.

Uniaxial Stress Tensile or compressive stress in a single direction, usually the lengthwise direction of a test specimen or structural member. See also STRESS and TRUE STRESS.

Unicellular Plastic A term that has sometimes been used for CLOSED-CELL FOAMED PLASTIC, w s.

Unidirectional Laminate A reinforced-plastic structure in which substantially all of the fibers are parallel. The modulus of elasticity (E) and strength of such a laminate in the direction of reinforcement will be somewhat more than the product of the volume fraction of reinforcing fiber times its corresponding properties. (See LAW OF MIXTURES.) However, properties in the transverse directions are essentially those of the matrix resin. See also BI- and TRIDIRECTIONAL LAMINATE.

Unit Elongation Synonym for ELONGATION, w s.

Unit Mold A mold designed for quick changing of interchangeable cavities or cavity parts.

Unloading Valve A valve that limits the maximum pressure in a hydraulic line (or other fluid space) to a desired value by diverting the flow of fluid from a pump to a bypass line. See also RUPTURE DISK.

Uns- (*unsym-*) Abbreviation for *unsymmetrical*, a prefix denoting unsymmetrical disposition of substituents of organic compounds with respect to the carbon skeleton or a functional group. It is usually ignored in alphabetization of compound names.

Unsaturated Compound An organic compound having one or more incidences of two or three bonds between two adjacent atoms, usually carbon or nitrogen atoms, and capable of adding other atoms at such points to reduce it to a single bond, thus becoming saturated. Multiple unsaturation is common, as in dienes, aromatics, and oils and fats.

Unsaturated Polyester See POLYESTER, UNSATURATED.

Unwind Unit (unwind) (1) In plastics coating and film laminating, a stand or a driven machine holding a roll of the substrate to be coated or laminated and supplying the web to the coating equipment at the rate and tension needed. (2) In molding from continuous prepreg, a similar unit as in (1) but sometimes including additional functions.

Unzipping The fast reversal of polymerization, with release of monomer, that can occur in addition homopolymers once a stable end group has been removed. Copolymerization helps to minimize unzipping.

UP Abbreviation used by the British Standards Institution for *unsaturated polyester*. See POLYESTER, UNSATURATED.

Up-and-Down Method (staircase method) A testing protocol sometimes used to estimate the average value of a property that, with limited resources for testing, must be tested at discrete levels of experimental factors, such as temperature or impact energy, that are strong determinants of the property. One specimen is tested at level A of the factor and passes or fails. If it passes, the factor intensity is raised to level B, significantly higher than A, and the test is repeated with a new specimen. (If it fails at level A, intensity is lowered.) If it passes at level B, intensity is again raised, to level C. When it eventually fails, the level is reduced one level, and so on, until the allotted number of specimens has been tested. This procedure has been used in determining BRITTLENESS TEMPERATURE, w s, and in several impact tests. The method is fairly efficient at determining the average sought, but at the expense of standard deviation.

UPC (Universal Product Code) This is a 10-digit numeric code that uniquely identifies products. The first five digits identify the manufacturer or organization controlling the label of the product, while the last five, called the item code, identify individual products within those companies.

Upper Newtonian Viscosity (μ_∞) The limiting viscosity, independent of shear rate, observed at extremely high shear rates with some polymer solutions, and a parameter of some rheological models for nonNewtonian flow. See, for example, POWELL-EYRING MODEL.

Upstroke Press A hydraulic press in which the main ram is situated below the moving table, pressure being applied by an upward movement of the ram.

Urea (carbamide) $H_2NC(O)NH_2$. A white, crystalline powder derived from the decomposition of ammonium carbonate. It is used in the manufacture of urea-formaldehyde resins. See AMINO RESIN.

Urea-Formaldehyde Foam A foam produced by combining a urea-formaldehyde resin with a detergent-type foaming agent under pressure. Upon release of pressure, a foam of about the consistency of shaving cream emerges and cold-cures within 2 to 4 hours. The foam is of low density, is noncombustible, and dries within 1 to 2 days. The dried foam has some resiliency, good thermal-insulation qualities, and is sound-absorbent. Although not recommended for continuous exposure to temperatures above 100°C, the material does not decompose and release gases until heated to a much higher temperature.

Urea-Formaldehyde Resin See AMINO RESIN.

Urea Plastic See AMINO RESIN.

Urethane (1) (ethyl carbamate) $H_2NCOOC_2H_5$. This compound may be thought of as urea in which one $-NH_2$ group has been replaced by an ethoxy group, $-OC_2H_5$. Curiously, urethane itself has no direct application in making polyurethanes. (2) A compound of the general structure RHNCOOR', formed by reaction of an alcohol with an isocyanate, which is (1), above, with one of the amino hydrogens replaced by R and the ethyl of ethoxy broadened to include other alkyl or aryl radicals. (3) A chain unit in polyurethanes, –RHNCOOR'–, which is formed by the reaction of a diol and an isocyanate. (4) Shorthand substitute for POLYURETHANE, w s.

Urethane Coatings (polyurethane coatings) ASTM has designated five types of urethane coatings. Type I is a one-component system modified with a drying oil such as linseed or soya, which reacts with oxygen from the air to effect cure, used in wood finishes. Type II is based on an isocyanate-terminated prepolymer in a solvent that dries by evaporation and cures by reaction with moisture in the air. It is used for coating wood, rubber, and leather. Type II is based on a blocked isocyanate, a polyester, a curing agent, and a suitable solvent. The applied coating is heated to effect curing. It is used for wire covering and industrial finishes. Type IV is a two-component system, one being a prepolymer made from a diisocyanate and a polyol, the other being a catalyst such as a tertiary amine. It is used for heavy-duty industrial finishes with good resistance to chemicals, abrasion, and corrosion. Type V is also a two-component system, one being a polyisocyanate (usually an adduct of a diisocyanate and trimethyolpropane) and the other being a polyol, in solvents that evaporate after application of the coating. The reaction proceeds at ambient temperatures without the aid of a catalyst. It is used as a high-performance industrial coating.

Urethane Foam See POLYURETHANE FOAM.

Urethane/Imide Modified Foam See POLYURETHANE/IMIDE MODIFIED FOAM.

Urethane Plastic See POLYURETHANE and POLYURETHANE FOAM.

Usable Life See POT LIFE.

UV Abbreviation for ULTRAVIOLET, w s.

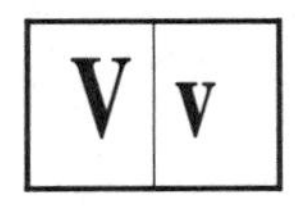

v (1) Symbol for velocity.

V (1) SI abbreviation for VOLT, w s. (2) Chemical symbol for the element vanadium. (3) Symbol for system volume.

Vacuum-Bag Molding See BAG MOLDING.

Vacuum Calibration Synonym for VACUUM SIZING, w s.

Vacuum Casting A method used for casting fluid thermosetting resins to avoid inclusions of air bubbles. The mold is placed in a vacuum chamber and filled with resin from an external hopper. Vacuum is applied to pull out bubbles, held until they have all risen to the surface, then released. Curing follows.

Vacuum Deposition See VACUUM METALIZING.

Vacuum Forming (straight vacuum forming) The simplest, original technique of SHEET THERMOFORMING, w s.

Vacuum Impregnating The process of impregnating electrical components by subjecting the parts to a moderate vacuum to remove air and other volatiles, introducing the impregnant to penetrate the parts, then releasing the vacuum and curing. Epoxy, phenolic, and polyester resins are often used. See also POTTING and ENCAPSULATION.

Vacuum Metalizing A decorating process used to make plastic objects resemble shiny metals. The article to be coated is first thoroughly cleaned to remove grease, dirt, or mold release, then is coated with a lacquer by dipping or spraying, and thoroughly dried. The prepared articles are placed on racks inside a chamber, a very high vacuum (0.067 Pa absolute) is created inside the chamber, then loops of metal wire, usually aluminum, are vaporized on hot tungsten wires and the metallic vapor condenses on the surfaces of the articles. A top coat of protective lacquer, which may be tinted to produce a gold or copper color, is then applied to the metal coating. In a variation called the *cathode sputtering process*, the metal is vaporized by an electrical discharge between electrodes in the vacuum chamber. This process gives a somewhat more uniform film. When the metallic coating is applied to the surface of the article that is normally viewed, it is called *first-surface metalizing*. Transparent plastics such as polystyrene and acrylics can be decorated on the back side, thereby protecting the extremely thin metal coating with the plastic itself, in which case the process is called *second-surface metalizing*.

Vacuum Sizing In extruding pipe or tubing, the passing of the extrudate through flexible seals into a tank under mild vacuum, causing the still pliant tube to expand against water-cooled sizing rings and bringing the final outside diameter within specified tolerances. The tank is sometimes called a *forming box*. See also SIZING PLATE and SIZING RING.

Vacuum Snap-Back Forming (snap-back forming) A variation of SHEET THERMOFORMING, w s, in which the heated plastic sheet, positioned above a vacuum box and below a male plug, is first drawn into a concave shape by means of vacuum applied to the bottom of the vacuum box, then pulled upward or "snapped back" against the surface of the plug by means of vacuum drawn through tiny holes in the plug. The process provides uniform thickness for deeply drawn products.

Vacuum Venting The drawing of a vacuum on the cavity of an injection (or other) mold in order to eliminate molding defects such as short shots, voids and, particularly, burned spots (*dieseling*). The vacuum may be drawn by means of tubes leading to vents in sharp corners, blind holes, etc. In one implementation of the concept, the entire mold is enclosed in a vacuum-tight box with a parting line coplanar with that of the mold, its mating surfaces sealed by O-rings.

Valence In ionic chemical bonding, the property of an element that is measured by the number of atoms of hydrogen (or its equivalent) that one atom of the element can hold in combination if negative, or can displace in a reaction if positive. Many elements have more than one valence, corresponding to lower and higher states of oxidation or reduction. In covalent bonding, the number of outermost-shell electrons that an element has available for sharing with other elements.

Valence Electron An outermost-shell electron that is gained, lost, or shared in a chemical reaction.

Valley Printing (inlay printing) A printing process for flat plastic surfaces in which ink is applied to the raised portions of an embossing roll that simultaneously embosses the plastic surface and deposits ink in the valleys of the embossed surface. The plastic surface to be valley printed must be warm enough for controlled deformation by the embossing roll. It can be a freshly extruded or calendered sheet, or a cold sheet that has been heated by contact with hot rolls or by radiant heaters. The process is similar to flexographic heating in that both print from raised portions of a cylinder. However, in flexographic printing the roll is flexible while in valley printing it is rigid.

Value (color value) The *lightness* of a color. A color may be classified as equivalent to some member of a series of shades ranging from black (the zero-value member) to white. The other two fundamental characterizers of color are *hue* and *saturation*.

Valved Extrusion An extrusion operation in which melt pressure and, to a lesser extent, throughput are controlled by an adjustable valve. For example, when a screen pack is used to remove foreign matter from the melt stream, a valve may be inserted between the screen pack and the die. When the screen is newly installed and clean, the valve will be partly closed to adjust head pressure, melt temperature, and extrusion rate to the desired levels. As the screen becomes blocked by collected solids and develops more resistance to flow, the valve is gradually opened to maintain the original conditions. Valving mechanisms of various types have also been placed at the end of the first stage of a two-stage screw used in vented operation, with the purpose of increasing the working and the melt temperature in the rear stage and thereby improving extraction in the vented section; and also to provide assistance in equalizing the flow rates in the two pumping sections.

Van der Waals Force (secondary valence force, intermolecular force) An attractive force, much weaker than primary covalence bonds, between molecules of a substance in which all the primary valences are saturated. They are believed to arise mainly from the *dispersion effect*, in which temporary dipoles induce other dipoles in phase with themselves.

Vapor As most frequently used, the term vapor means a substance that, although present in the gaseous phase, generally has a stable liquid or solid state at ambient temperatures. *Gas*, on the other hand, is used for substances that do not have stable liquid or solid states at ambient conditions. Thus, we speak of *water vapor*, but *oxygen* and *nitrogen gases* in the atmosphere.

Vapor Barrier A layer of material through which water (or other) vapor will pass only slowly, or not at all.

Vapor Degreasing A process of removing grease from parts and equipment components by suspending them with a closed chamber over a pool of boiling, nonflammable solvent such as a mixture of chlorofluorinated hydrocarbons.

Vapor-Liquid Chromatography See CHROMATOGRAPHY.

Vapor Pressure The pressure of the vapor phase when a solid or liquid is in equilibrium with its vapor. Vapor pressure increases exponentially with absolute temperature at a rate that is unique for each pure substance and closely related to its heat of vaporization. In homologous organic compounds, vapor pressure at any temperature decreases with increasing molecular weight. The term has little meaning for plastics because high polymers decompose before evaporating.

Vapor Transmission If the vapor is not otherwise identified, this phrase is understood to mean WATER-VAPOR-TRANSMISSION RATE, w s.

Variance (1) In statistical quality control, the variance of a sample of n process measurements x_i is given by the equation

$$s^2 = \frac{\sum_{i=1}^{n}(x_i - \bar{x})^2}{n-1}$$

where $\bar{x}$ is the ARITHMETIC MEAN (w s) of the sample. The square root of the variance s is the sample *standard deviation.* s^2 estimates, efficiently and without bias, the population variance, σ^2. (2) An agreed-upon difference between a specification for a product or property and the actual average value achieved in a lot of product.

VCM Abbreviation for *vinyl chloride monomer.* See VINYL CHLORIDE.

Veil A thin mat of very fine, relatively long fibers used at the outermost layer of a composite in order to improve surface appearance and smoothness.

Velocity Profile A graph of the fluid velocity in a stream at various points along a coordinate direction perpendicular to the flow and in which direction the velocity is changing most sharply. This graph has the form of a parabola for a Newtonian liquid in laminar flow through a circular tube. For pseudoplastic liquids (polymer melts), the curve is a parabola of higher degree, usually 2.25 to 4 (instead of 2.0), rising more rapidly near the tube wall and flattening near the center. For a pure drag flow between parallel surfaces, the profile is linear regardless of the type of fluid.

Vent See AIR VENT.

Vented Extruder See EXTRUDER, VENTED.

Venturi Cooling Ring A design of air-cooling ring for blown-film extrusion in which a slot around the inside of the ring and near the bottom injects air vertically upward at high velocity. This not only cools the bubble but also, by lowering air pressure between the bubble and the annular jet, helps to quickly expand the bubble and stabilize its position and movement.

Venturi Dispersion Plug In injection molding, a plate having an orifice with a conical relief drilled therein which is fitted in the nozzle to aid in the dispersion of colorants in a resin.

Vermiculite A complex, mica-like mineral containing water, which causes the mineral to expand from six to twenty times in volume when heated to a high temperature in its softening range. The expanded material is sometimes used as a density-reducing filler in plastics. Compare PERLITE.

Versamid® General Mills' trade name for FATTY POLYAMIDEs, w s.

Vertical Extruder An extruder arranged so that the barrel is vertical and extrusion is usually downward or upward.

Vertical Flash Ring (1) The clearance between the force plug and the vertical wall of the cavity in a positive or semipositive compression mold. (2) The ring of excess material that escapes from the cavity into this clearance space.

Very-Low-Density Polyethylene (VLDPE) Any polymer of ethylene with some higher-olefin content, in the density range from 0.90 to 0.915 g/cm^3, produced by the Union Carbide gas-phase process (Flexomer®). The materials have low moduli, with properties between those of low-density polyethylene and ethylene-propylene rubbers. They are useful for stretchable films. See also ULTRA-LOW-DENSITY POLYETHYLENE.

VF_2/HFP Abbreviation for VINYLIDENE FLUORIDE-HEXAFLUOROPROPYLENE COPOLYMER, w s.

v_i Symbol for volume fraction of component *i* in a blend or composite; or for a velocity component in the *i*-direction.

Vibration Welding A joining method in which two plastic parts are pressed together and one is vibrated through a small angular displacement in the plane of the joint. The frictional heat so generated melts the plastic at the interface. Vibration is stopped and pressure and alignment are maintained until the joint freezes. Recommended ranges for the four controllable factors are: amplitude of vibration, 3–4 mm; pressure, 70–170 kPa; weld time, 2–3 s; and hold time—geometry- and material-dependent—about half the weld time. See also SPIN WELDING.

Vibratory Feeder A device for conveying dry materials from storage hoppers to processing machines, comprising a tray or tube vibrated by mechanical or electrical pulses. The frequency and amplitude of vibration control the rate of transport.

Vibratory Mill (Vibro-Energy® mill) A top-loading, cylindrical, grinding device mounted on springs and employing ball-type grinding media, driven by an eccentrically weighted motor that subjects the working chamber to three-dimensional vibration. These mills provide faster size reduction and are much more energy-efficient than conventional BALL MILLs, w s.

Vicat Softening Point The temperature at which a flat-nosed needle of 1-mm^2 circular cross section penetrates a thermoplastic specimen to a depth of 1 mm under a specified load using a uniform rate of temperature rise (ASTM D 1525). This test is used for thermoplastics such as vinyls, polystyrene, acrylic, and cellulosics that have no definite melting ranges.

Vickers Hardness A test similar to that of BRINELL HARDNESS, w s, using an indenter in the form of a square-based diamond pyramid, with a vertex angle of 136° between the opposite faces. The result is expressed as the applied load divided by the projected area of the impression.

Vinal Generic name for a manufactured fiber in which the fiber-forming substance in any long-chain synthetic polymer is composed of at least 50% by weight of vinyl alcohol units ($-CH_2CHOH-$), and in which the total of the vinyl alcohol units and any one or more of the various acetal units is at least 85% by weight of the fiber (Federal Trade Commission).

Vinyl (ethenyl) The unsaturated group $CH_2{=}CH-$ which is the basis for all vinyl plastics. The name vinyl is used when the open bond is filled by anything but H (ethene) or a hydrocarbon radical (olefin). The term also includes the much broader compound class: (AB)C=C(XY).

Vinyl Acetate $H_2C{=}COOCH_3$. A colorless liquid obtained by the reaction of acetylene and acetic acid in the presence of a catalyst such as mercuric oxide. It is the monomer for polyvinyl acetate, and a comonomer and intermediate for many members of the vinyl plastics family.

Vinyl Alcohol (ethenol) $H_2C{=}CHOH$. A conceptual compound, the theoretical monomer of POLYVINYL ALCOHOL, w s, but unknown in the free state. All attempts to synthesize it have led instead to its tautomer, acetaldehyde (CH_3CHO).

Vinyl Alcohol Plastic See POLYVINYL ALCOHOL.

Vinylation The process of forming a vinyl derivative by reaction of alcohols, amines, or phenols with acetylene. Such derivatives are intermediates for polymers.

Vinylbenzene A synonym for STYRENE, w s.

Vinyl Butyrate $CH_2{=}CHOOC_3H_7$. A volatile liquid monomer for polymers used in water-based paints.

9-Vinylcarbazole (*N*-vinylcarbazole) A tricyclic tertiary amine, $H_2C{=}CHN(C_6H_4)_2$, with the structure shown below.

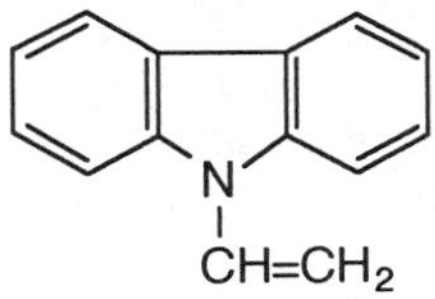

CH=CH$_2$

This monomer, derived from acetylene and carbazole, is used in the production of POLY(*N*-VINYLCARBAZOLE), w s.

Vinyl Chloride (chloroethylene, chloroethene, VC) $H_2C{=}CCl$. A colorless gas at normal temperatures and pressures that boils at −13.9°C, made by reacting ethylene with chlorine or hydrogen chloride to obtain ethylene dichloride, which is cracked to form vinyl chloride. VC is usually handled at low temperatures in its liquid state, and is the monomer for POLYVINYL CHLORIDE, w s. There is strong evidence that VC (but not its polymer) causes liver cancer in humans, so it must be scrupulously kept out of the work environment.

Vinyl Chloride-Ethylene Copolymer Any copolymer of vinyl chloride with small percentages of ethylene. These resins possess superior heat stability and hot strength, and require lesser amounts of impact modifiers to achieve satisfactory impact strength than does straight PVC homopolymer. They are useful in producing films and bottles for packaging, since their better heat stability provides more latitude in selecting nontoxic stabilizers.

Vinyl Chloride Plastic See POLYVINYL CHLORIDE.

Vinyl Cyanide A synonym for ACRYLONITRILE, w s.

Vinyl Ester Resin Any of several epoxy-related resins in which the epoxide groups have been replaced by ester groups, typically acrylic. When cured, they have excellent resistance to strong chemicals such as chlorine and caustics.

Vinyl Ether See VINYLETHYL ETHER, VINYLISOBUTYL ETHER, and VINYLMETHYL ETHER.

Vinylethylene Synonym for BUTADIENE, w s.

Vinylethyl Ether (EVE, ethylvinyl ether) $H_2C{=}CHOC_2H_5$. A colorless monomer that can be polymerized either in the liquid or gaseous state. In plastics, it is used as a comonomer and intermediate.

Vinyl Fluoride (fluoroethylene, fluoroethene) $H_2C{=}CHF$. A colorless gas, the monomer for POLYVINYL FLUORIDE, w s.

Vinyl Foam Although cellular vinyls can be produced by many methods, including mechanical frothing and leaching-out of soluble additives, the most widely used procedure is chemical blowing. From 1 to 2% of a blowing agent such as AZOBISFORMAMIDE, w s, is incorporated in a vinyl compound or dispersion, remaining inert until it is decomposed by processing heat to release a gas. Such compounds are processed by calendering, extrusion, injection molding, compression molding, slush casting, and rotational casting. Vinyl foams are widely used in clothing, flooring, footwear, furniture, and packaging.

Vinylformic Acid A synonym for ACRYLIC ACID, w s.

Vinylidene Indicating a bisubstituted ethylene in which both hydrogens on one carbon atom have been replaced, i e, $H_2C{=}CXY$, or the vinylidene group, $H_2C{=}C{=}$. X and Y are usually the same element.

Vinylidene Chloride (1,1-dichloroethylene) $H_2C{=}CCl_2$. A colorless, volatile liquid that is produced by the dehydrochlorination of 1,1,2-trichloroethane. It is the monomer for POLYVINYLIDENE CHLORIDE, w s, and is a comonomer with vinyl chloride (see SARAN) and other monomers such as acrylonitrile.

Vinylidene Chloride-Acrylonitrile Copolymer Any VC copolymer containing 5 to 15% acrylonitrile, and mainly used as coatings for cellophane, paper, and films of other polymers. They are comparable with saran in their low permeability to oxygen and carbon dioxide, and have good chemical resistance, toughness, transparency, and heat-sealability. They have also found some applications as low-flammability fibers.

Vinylidene Chloride Plastics See POLYVINYLIDENE CHLORIDE and SARAN.

Vinylidene Fluoride (1,1-difluoroethylene) $H_2C{=}CF_2$. A colorless, nearly odorless gas prepared by the dehydrohalogenation of 1-chloro-1,1-difluoroethane, or by the dehalogenation of 1,2-dichloro-1,1-difluoroethane. It polymerizes readily in the presence of free-radical initiators to produce the homopolymer polyvinylidene fluoride, and is also copolymerized with olefins and other fluorocarbon monomers to make fluorocarbon elastomers. See POLYVINYLIDENE FLUORIDE.

Vinylidene Fluoride-Hexafluoropropylene Copolymer (VF_2/HFP) Any of a family of chemical- and heat-resistant, vulcanizable elastomers containing 60 to 85% VF_2 (Du Pont's Viton® A). Terpolymers with small amounts of tetrafluoroethylene are also available.

Vinylisobutyl Ether (isobutylvinyl ether, IVE) $H_2C{=}CHOCH_2{-}CH(CH_3)_2$. A colorless, flammable liquid used to make polymers and copolymers used in coatings, adhesives, and lacquers; and modifiers for alkyd resins and polystyrene. See POLYISOBUTYLVINYL ETHER.

Vinylmethyl Ether (methyl vinyl ether) $H_2C{=}CHOCH_3$. A low-boiling liquid (6°C) or gas, polymerizable to POLY(VINYLMETHYL ETHER), w s. It is also used as a modifer for alkyd resins and polystyrene.

Vinyl Plastic See VINYL RESIN.

Vinyl Propionate $H_2C{=}CHOOCC_2H_5$. A volatile liquid, the monomer for emulsion-paint polymers.

1-Vinyl-2-Pyrrolidone (*N*-vinyl-2-pyrrolidone) A cyclic monomer derived from acetylene and formaldehyde, with the structure below.

N O

$HC{=}CH_2$

See POLY(1-VINYLPYRROLIDONE).

Vinyl Resin According to strict chemical nomenclature, this term includes all resins and polymers made from monomers containing the vinyl group, $H_2C{=}CHX$. Thus, in the chemical literature, polystyrene, polyolefins, polymethyl methacrylate, and many other styrenic, ethenic, and acrylic copolymers are classified as vinyls. However, in the plastics literature, the above materials are given their own classifications and the term vinyl is restricted to compounds in which X, above, is not H, a hydrocarbon radical, nor an acrylic-type ester. In daily use, the term *vinyl plastics* refers primarily to POLYVINYL CHLORIDE, w s, and its copolymers, and secondarily to the following: POLYVINYL ACETAL, POLYVINYL ACETATE, POLYVINYL ALCOHOL, POLYVINYL BUTYRAL, POLY(*N*-VINYLCARBAZOLE), POLYVINYL DICHLORIDE, POLYVINYL FORMAL, POLYVINYLIDENE CHLORIDE, POLYISOBUTYLVINYL ETHER, and POLY(1-VINYLPYRROLIDONE).

Vinyl Stearate $H_2C{=}CHOOCC_{17}H_{35}$. A white, waxy solid, used as an internal plasticizer by means of copolymerization at low levels.

Vinylstyrene Synonym for DIVINYLBENZENE, w s.

Vinyltoluene $H_2C{=}CHC_6H_4CH_3$. A colorless liquid, the commercial forms comprising a 60:40 mixture of the *m*- and *p*- isomers, used as a solvent and as a polymerizable monomer in place of styrene in the production of polyester resins.

Vinyltrichlorosilane $H_2C{=}CHSiCl_3$. A coupling agent used in glass-reinforced polyesters.

Vinyltriethoxysilane $H_2C{=}CHSi(OC_2H_5)_3$. A coupling agent used in glass-reinforced polyesters, polyethylene, and polypropylene.

Vinyl-tris(β-Methoxyethoxy)silane $H_2C{=}CHSi(OCH_3OC_2H_5)_3$. A silane coupling agent used in glass-reinforced-polyester and -epoxy structures.

Vinyon Generic name for a manufactured fiber in which the fiber-forming substance is any long-chain synthetic polymer composed of at least 85% by weight of vinyl chloride units, $-CH_2CHCl-$ (Federal Trade Commission).

Virgin Material Any plastic compound or resin that has not been subjected to use or processing other than that required for its original manufacture.

Viscoelastic Having both viscous and elastic properties.

Viscoelasticity The tendency of a plastic to respond to stress (or strain) as if it were a combination of an elastic solid and a viscous liquid. This property, possessed by all plastics to some degree, dictates that while plastics have solid-like characteristics such as elasticity, strength, and form stability, they also have liquid-like characteristics such as flow over time that depend on temperature, pressure, and stress. As a result, the response to stress depends on both the rate of application of the stress and the time for which it is maintained. With many—probably most—plastics, strengths and moduli are higher when stress is applied within a fraction of a second than when it is applied over a period of hours. Viscoelasticity is responsible for the phenomenon of CREEP, w s, in plastics, an important design consideration for plastics structures that will bear sustained loads. Conceptually, viscoelasticity has been portrayed by simple mechanical models and combinations of them. See MAXWELL MODEL, VOIGT MODEL, and TIME-TEMPERATURE EQUIVALENCE.

Viscometer (viscosimeter) An instrument used for measuring the viscosity and sometimes other flow properties of fluids having low to moderate viscosities. (Instruments used with highly viscous materials, such as polymer melts, are usually called *rheometers.)* A widely used type of viscometer is the rotational type, in which a rotor turns within a cup containing the liquid sample and the torque required to turn the rotor or to hold the cup stationary is measured, along with speed of rotation. Of the many other types, some measure the rate at which a bubble rises through the liquid, or a ball falls through it; others measure the time required for a known quantity of the liquid to drain by gravity through an orifice at the bottom of a cup; still others measure the rate of laminar flow of the liquid through a capillary tube of known dimensions operating with a measured pressure differential. In the plastics industry, viscometers are used mostly to measure the viscosities of polymer solutions and slurries. See AIR-BUBBLE VISCOMETER, BROOKFIELD VISCOMETER, CAPILLARY VISCOMETER, FORD VISCOSITY CUPS, STORMER VISCOMETER, VISCOSITY and ZAHN VISCOSITY CUP.

Viscose A solution of xanthated cellulose in dilute sodium hydroxide from which rayon fibers and cellophane films are formed. The xanthated cellulose is produced by reacting alkali cellulose, i e, wood fibers or cotton linters treated with sodium hydroxide, with oxygen and carbon disulfide. Rayon produced by this method is known as *viscose rayon*. See also RAYON.

Viscosity (viscosity coefficient, μ or η) Resistance to flow, a fundamental property of fluids, first quantitatively defined by I. Newton in his *Principia*. A modern version of his equation of viscosity is:

$$\tau_{xz} = -\mu \frac{dv_z}{dx}$$

in which dv_z/dx is the rate of change of z-directed velocity at the coordinate location x in the fluid, x being directed perpendicular to z; μ is the viscosity, a function of temperature and, more weakly, static pressure; and τ_{xz} is the z-directed *shear-stress* component acting on an imagined element of fluid surface normal to the x-direction. dv_z/dx, the *velocity gradient*, often written without the subscript, is also called the *shear rate*, usually symbolized as $\dot{\gamma}$. (The negative sign indicates that the stress in the fluid opposes the velocity.) In the conceptually simplest case, that of steady flow between parallel plates separated by a distance h, one of which is moving with velocity v relative to the other, the stress is just the force F required to drag the moving plate divided by the plate area A, and Newton's equation reduces to: $F/A = -\mu\ (V/h)$. For the circular-tube geometry of orifice-type rheometers, the shear rate and stress vary from zero at the tube axis to a maximum at its inside surface, while the velocity does just the reverse. Since for ordinary fluids, Newton's law holds throughout the fluid, it also holds at the tube wall (unless there is slip, rarely proved with polymeric solutions and melts). Rheologists have found it convenient to report their measurements in terms of the shear stress at the tube wall, $\Delta P \cdot R/2L$, and the Newtonian (or *apparent*) shear rate at the wall, $4Q/\pi R^3$ (= $4V/R$), which contain all the quantities they actually measure, i e, the pressure drop from entrance to exit, the tube radius, and the steady flow rate, Q, which is equal to πR^2 times the average velocity V. By Newton's law, the viscosity is given by the quotient of the shear stress and the shear rate, i e, $\pi R^4 \cdot \Delta P/(8Q \cdot L)$. This expression is just an inversion of the Hagen-Poiseuille equation.

Newton's equation applies accurately to ordinary fluids such as water, pure organic liquids, familiar oils and honey, even to dilute solutions of polymers, but not to concentrated polymer solutions or most molten plastics. Measurements show that these latter materials deviate from Newton's law in that their viscosities, as given by the quotient of shear stress/shear rate, diminish with rising shear rate (and stress). That is, they are nonNewtonian and pseudoplastic. For such materials, the viscosity symbol μ is replaced by η to remind readers of its dependence on shear.

Because some chemical engineers like to think of shear stress in its alternate identity, *momentum flux*, because many instruments were developed to measure viscosities related to specific industrial uses, and because the range of viscosities over all fluids and conditions is tremendous (more than 15 orders of magnitude), a bewildering plethora of viscosity units has evolved. Some useful conversions are given in the Appendix. We can hope that all others will eventually give way to the SI unit of viscosity, the pascal-second (Pa·s). If, in the (not very practical) parallel-plate setup described above, the plate areas were 1 m^2, their separation 1 m, their relative velocity 1 m/s, and the drag force 1 N, the viscosity would be exactly 1 Pa·s. For many years scientists and engineers used the *poise*, equal to 0.1 Pa·s, and the centipoise, as their working units. Plastics engineers have also used the psi·s ($lb_f \cdot s/in^2$, = 6895 Pa·s). The Pa·s may also be viewed, through the momentum-flux perspective, as 1 kg/(m·s).

Many early instruments tried to gauge viscosity by the time required for a vessel full of liquid to drain through a short tube in its bottom. These, with suitable corrections, provide estimates of the KINEMATIC VISCOSITY, w s. If mass-based viscosity units are used, kinematic viscosity will have the dimensions (length)/time, or m^2/s in SI. Clearly, one must be careful, in using reported viscosities, to identify unambiguously the units used. In the older literature, and even today, alas! the pound and kilogram may be either force or mass units, though in SI the kilogram, one of the seven base units, is strictly assigned to mass. To convert a kinematic viscosity to an equivalent absolute viscosity, one must know the liquid's density at the stated temperature. The old scientific unit, the Stokes, equal to 1 cm^2/s, is closely related to the poise.

All liquid viscosities decrease with rising temperature, some much more steeply than others. The range of polymer-connected viscosities is very wide, from 0.001 Pa·s for very dilute solutions at room temperature to 100–5000 Pa·s for molten plastics, and many times more for cooler amorphous polymers. ASTM tests for flow properties of plastics include D 789, 1238, 1243, 1601, 1823, 1824, 2393, 2857, 3749, and 4440. See also the following viscosity-related terms.

BINGHAM PLASTIC
BROOKFIELD VISCOMETER
CAPILLARY RHEOMETER
CAPILLARY VISCOMETER
CONSISTENCY
CUP-FLOW TEST
DILATANCY
DILUTE-SOLUTION VISCOSITY
EXTRUSION PLASTOMETER
HAGEN-POISEUILLE EQUATION
INHERENT VISCOSITY
INITIAL VISCOSITY
INTRINSIC VISCOSITY
KINEMATIC VISCOSITY
LAMINAR FLOW
MELT-FLOW INDEX
NEWTONIAN FLOW
PSEUDOPLASTIC FLUID
REDUCED VISCOSITY
RELATIVE VISCOSITY
RHEOLOGY
RHEOMETER
RHEOPEXY
SAYBOLT VISCOSITY

SPECIFIC VISCOSITY
STORMER VISCOMETER
THIXOTROPY
UBBELOHDE VISCOMETER
ULTRA-VISCOSON
VISCOELASTICITY
VISCOMETER
VISCOUS FLOW
YIELD VALUE
WEISSENBERG RHEOGONIOMETER
ZAHN VISCOSITY CUP

Viscosity-Average Molecular Weight (M_v, $\overline{M}_v$) An averaged molecular weight for high polymers that relates most closely to measurements of dilute-solution viscosities of polymers. The defining equation is

$$M_v = \left[\frac{\sum_{i=1}^{\infty} N_i M_i^{1+a}}{\sum_{i=1}^{\infty} N_i M_i} \right]^{1/a}$$

where N_i is the number of individual molecules having the molecular mass M_i. The exponent a, between 0.6 and 0.8 for many polymer/solvent systems, is best evaluated from measured viscosities of dilute solutions of narrow-molecular-mass fractions of polymers, determining the intrinsic viscosity $[\eta]$ for each fraction, then fitting the following equation to the data: $[\eta] = K' M^a$. For $a = 1$, occasionally seen, M_v = the weight-average molecular weight, M_w. Tables of K' and a are given in *Polymer Handbook*, J. Brandrup and E. H. Immergut, Editors, Wiley-Interscience, second and third editions.

Viscosity Coefficient The quantitative value of a fluid's viscosity at particular conditions, as contrasted with the qualitative concept of viscosity as resistance to flow. In recent decades, the "coefficient" has gone out of usage and viscosity coefficient and "coefficient of viscosity" have become, simply, VISCOSITY, w s.

Viscosity Depressant A substance that, when added in a relatively minor amount to a liquid, lowers its viscosity. Such materials, e g, ethoxylated fatty acids, are often incorporated in vinyl plastisols to lower their viscosities without increasing plasticizer levels. The viscosity depressant may be incorporated in the plastisol during the original mixing or may be added later to offset viscosity increase expected after prolonged storage.

Viscosity Number The IUPAC term for REDUCED VISCOSITY, w s.

Viscosity Ratio The IUPAC term for RELATIVE VISCOSITY, w s.

Viscous A qualitative term denoting that the material to which it is applied is "thick" and flows sluggishly, rather than being "thin" and flowing freely. The transition region between "free-flowing" and "viscous" corresponds roughly to viscosities from 1 to 30 Pa·s.

Viscous Dissipation (viscous-heat generation) In melt processing, wherever there is flow, the resistance of molecules to flow, i e, VISCOSITY, w s, causes heat to be generated within the melt. The rate of dissipation equals the product of shear stress times shear rate, or viscosity times the square of the shear rate. Because both the viscosity and shear rate are high in processes such as extrusion, injection and transfer molding, and intensive mixing, viscous dissipation is a principal mechanism of heating plastics in those processes.

Viscous Flow (laminar flow) A type of fluid motion in which all molecules of the fluid flow along straight lines parallel to the axis of a containing pipe or channel. (This definition arises from O. Reynolds' classic experimental demonstration of the transition from viscous to turbulent flow, which is described in most elementary texts on fluid flow.) See also LAMINAR FLOW.

Visible Light The narrow band in the electromagnetic spectrum that the human eye perceives, from about 380 nm (violet) to 760 nm (red).

Viton® Du Pont's trade name for a family of copolymer fluoroelastomers with a wide range of properties among them, but, in particular, good resistance to high temperatures and chemicals. They are used for gaskets, O-rings, oil seals, diaphragms, pump and valve linings, hose, tubing, and coating fabrics.

VLDPE Abbreviation for VERY-LOW-DENSITY POLYETHYLENE, w s.

Void (1) In a solid plastic article or laminate, an unfilled space within the article large enough to scatter light (in transparent materials) or other radiant energy that might be used to detect such spaces. (2) In cellular plastics, a cavity unintentionally formed and substantially larger than the characteristic individual cells (ISO). (3) An empty volume within any material or liquid medium. See also BLISTER.

Void Fraction The fraction or percentage of the volume of an article or material sample, such as powder or foam, that is within the material and contains only vacuum or gas. Notwithstanding the special definition of VOID (2), above, "void fraction", when applied to cellular plastics, includes and is mainly attributable to the volume of the cells. In a reinforced-plastic article or laminate, void fraction = 1 − volume fraction of polymer (resin) − volume fraction of reinforcement − volume fraction of filler. Expressed in terms of mass fractions and measured (true) densities of the components,

$$\text{Percent voids} = 100[1 - \rho_l(m_p/\rho_p + m_r/\rho_r + m_f/\rho_f)]$$

where ρ_l is the measured density of the laminate.

Voigt Model A conceptual, mechanical model useful as an analogy to the deformation behavior of viscoelastic materials. It consists of, side-by-side (in parallel), an elastic coil spring and a viscous dashpot rigidly connected at each end. When the ends are pulled apart, they will separate gradually and ever more slowly until the spring is stretched to a length corresponding to the pulling force divided by the spring stiffness (spring constant), when motion stops. Compare this with the MAXWELL MODEL.

Volatile Easily evaporated or vaporized at temperatures below about 40°C; low-boiling.

Volatile Loss The loss in weight, usually unintended, of a substance due to evaporation of one or more constituents.

Volatiles Content The percent weight loss, through loss of volatiles, from a plastic or impregnated reinforcement held at a specified temperature for a specified time (sometimes, under vacuum).

Volatility Of liquids and some solids, the tendency to evaporate when exposed to the atmosphere. This qualitative idea is closely related to VAPOR PRESSURE, w s. The *relative volatility* of two liquids at any temperature is the ratio of their vapor pressures at that temperature.

Volt (V) The SI unit of electromotive force (emf), equal to the difference in electric potential between two points of a conductor carrying a constant current of one ampere when the power dissipated between these points equals one watt (i e, in SI, 1 V ≡ 1 W/A). It is also the potential difference required to cause a steady current of one ampere to flow through a conductor whose resistance is one ohm.

Volume The space occupied by an article or sample of material, including any voids, within the defining surfaces. The SI unit of volume is the cubic meter, m^3, known in the past by the name *stere*, now deprecated (but alive and well in crossword puzzles). SI also allows the use of convenient subvolumes, e g, mm^3, cm^3. The exponent also operates on the abbreviated prefix in each case [i e, 1 cm^3 = 1 $(cm)^3$ = 10^{-6} m^3, not 0.01 m^3]. The special name *litre* (*liter* in the US) has been approved for the cubic decimeter (dm^3) but is to be used only for volumetric capacity and dry and liquid measure. No prefix other than milli- (m) or micro- (μ) should be used with liter. Some conversions of other volume units to SI are given in the Appendix.

Volume Coefficient of Thermal Expansion (cubical expansion coefficient) The rate of change of volume of a material with rising temperature, divided by the volume, i e,

$$\frac{\left(\frac{\partial V}{\partial T}\right)_P}{V}$$

the P subscript reminding one that the coefficient, which is mildly pressure-sensitive, is tied to the particular pressure at which it has been measured (most often, atmospheric). The SI unit is K^{-1}. Over a short interval of temperature, the expression above is very nearly equal to the change in volume divided by the change in temperature, i e, $\Delta V/\Delta T$. In isotropic materials, the cubical expansion coefficient is three times the linear expansion coefficient, the quantity usually meant by COEFFICIENT OF THERMAL EXPANSION, w s. See also DILATOMETER.

Volume Expansion The change in volume of a test specimen under specified test conditions (ISO). This is usually expressed as a percentage of the original volume. See also SWELLING.

Volume Resistivity (specific insulation resistance) The ratio of the potential gradient parallel to the current in a material to the current density. In SI, volume resistivity is numerically equal to the direct-current resistance between opposite faces of a one-meter cube of the material, with the unit ohm-meters (Ω·m). The smaller cgs unit, Ω·cm, is still widely used.

Vulcanized Fiber Cellulosic material that has been partly gelatinized by action of a chemical (usually zinc chloride solution), then heavily compressed or rolled to the required thickness, leached free of the zinc chloride, and dried. It has been used for electrical insulation, luggage, and materials-handling equipment.

Vulcanizing (vulcanization) The chemical reaction, usually accompanied by crosslinking, that induces extensive changes in the physical properties of a rubber or elastomer, brought about by reacting the material with sulfur and an accelerator. The discovery of vulcanization by C. Goodyear in 1839 was the beginning of a practical rubber-products industry. The changes brought about by vulcanizing include decreased plastic flow, reduced surface tackiness, increased moduli and resilience, much greater tensile strength, and considerably reduced solubility. Some thermoplastics, such as polyethylene, can be modified to be vulcanizable. The associated crosslinking causes the final product to resist flow and deformation at temperatures above the melting range of the original polymer.

W w

w Symbol for width or, in thermodynamics, work done by a system.

W (1) SI abbreviation for WATT, w s. (2) Chemical symbol for the element tungsten (from *wolframite*, the mineral in which it was first recognized).

Wall Stress (1) In a filament-wound pressure vessel, the stress calculated from the pressure or load divided by the entire cross-sectional area of the wall (not just that of the reinforcement). (2) In a fluid flowing through a channel (such as a die, tube, or extruder-screw channel), the shear stress at any channel surface. In the simplest case of a fluid in steady flow through a circular tube of radius R and length L, the wall stress is given by $\Delta P \cdot R/2L$, where ΔP is the pressure drop from inlet to outlet.

Walnut-Shell Flour See NUTSHELL FLOUR.

Warm Forging The process of forming thermoplastic sheets or billets into desired shapes by pressing them between dies in a press, when the material and/or the dies have been preheated. The blanks may be billets formed by extrusion, or may be die-cut from sheets. Forging dies may be inexpensive castings of the matched-metal type. The process is usually employed for relatively thick parts of polyolefins, including glass-reinforced compounds. One method of performing the process employs two shaped metal punches with a common die ring located between the upper and lower punches, the ring containing a preheated plastic billet. Another method uses a rubber pad on one platen to force the plastic sheet or billet into conformity with a metal die on the other platen. Warm forgings accurately reproduce mold detail, and exhibit low mold shrinkage, good impact strength, creep resistance, and thermal stability of dimensions. See also COLD FORMING and SOLID-PHASE FORMING.

Warp (1, n) In the textile industry, those threads of a cloth that are parallel to the selvage, i e, running lengthwise in the loom. (2, v) To change shape spontaneously, seen particularly in flat surfaces such as sheet and sides of boxy shapes. Such changes are often traceable to creep caused by stresses generated during molding or forming, by uneven absorption of water or a solvent, by uneven heating, or, in fiber-reinforced thermosets, by unequal curing in thin and thick sections.

Warpage Distortion caused by nonuniform change of internal stresses (ASTM D 883). See also DISHED.

Wash In reinforced-plastics molding, an area where the reinforcement placed in the mold has moved during closing of the mold, resulting in a resin-rich (and reinforcement-poor) region.

Water Absorption The percentage increase in weight of a plastic article when immersed in water for a stipulated time and at a specified temperature (usually room temperature). ASTM tests for plastics, rigid cellular plastics, and certain thermosets are D 570, D 2842, and D 3419, respectively. Most plastics absorb water to some extent, varying from almost zero in the case of polytetrafluoroethylene and polyolefins, to complete dissolution for some types of polyvinyl alcohol and polyethylene oxide. Water absorption can cause swelling, leaching of additives, plasticizing and hydrolysis, which in turn can cause discoloration, embrittlement, stress cracking, lowering of mechanical and electrical properties, and reduced resistance to heat and weathering. However, the amount of water absorbed does not correlate with the extent of harmful effects that might result. As a contrary example, the Izod impact strength of nylon 6/6 is doubled by water absorption at 23°C and 50% RH over that of the bone-dry, as-molded resin.

Water-Extended Polyester (WEP) An unsaturated polyester resin that has been "filled" with water rather than conventional fillers. In the process usually employed, a liquid polyester resin containing a promoter is mixed, in a blending machine, with water containing methyl ethyl ketone peroxide as a catalyst. The product is a low-viscosity emulsion that is easily poured into open molds, and that cures without additional heat by its own exothermic reaction. When the process conditions are properly controlled, the water (except that at the surface of the casting) is retained indefinitely as microscopic droplets in the cured casting. The cured material has some of the characteristics of woods, and WEP castings can be nailed, stapled, and finished in much the same way as woods. However, variations in physical properties such as shrinkage and warpage upon long-term loss of water, are greater than those of woods. Within these limitations, the working properties of WEP combined with its economy and ease of processing have stimulated interest in its use as a replacement for woods, plaster, and other materials in some furniture parts, and in decorative applications such as frames, lamp bases, and statuary.

Water-Soluble Resin Any of several resin types that are produced by polymerization reactions in which the chain growth results from breaking of ring structure or double bonds of the monomers. Examples are alkyl- and hydroxyalkyl cellulose derivatives, carboxymethyl cellulose, polyvinyl alcohol, polyvinyl pyrrolidone, polyacrylic acid, polyacrylamide, polyethylene oxide, and polyethylene-imide.

Water-Vapor-Transmission Rate (WVTR, WVT) The amount of water vapor passing through a stated area and thickness of a plastic sheet or film in a given time, when the sheet or film is maintained at a constant temperature and the difference in relative humidity between the two faces of the sheet is 50%. Typical units are g·mil/100 sq in/24 h or g·mm/m^2/24 h, both of which are ambiguous because of the repeated divisions. The latter can be rewritten, removing the ambiguity, as g·mm/(m^2·d). The 50% humidity difference is understood but not expressed. Thus, a WVTR is not a PERMEABILITY, w s. The two temperatures most often reported are 23° and 37.8°C. Since 50% of the vapor pressure of water at 23° is 1.403 kPa and at 37.8° is 3.274 kPa, the WVTRs at the two temperatures are based on different driving forces and are therefore not comparable in the way that permeabilities are. The ASTM test most used for WVTR of plastic films and sheets is E 96; others, all in section 15.09, are D 895, D 3079, and F 372.

Watt The SI unit of power, equal to 1 J/s (= 1 m·N/s), expressing the rate at which work is done or energy expended. In the case of alternating electric current, the watt is computed as the product of voltage across the circuit, times the current flow in amperes times the cosine of the phase angle between the current and the impressed voltage. In purely resistive DC circuits, watts = volts × amperes. Some conversions of other power units to SI are given in the Appendix.

Wax Solid, low-melting substances that may be of plant, animal, mineral, or synthetic origin. Waxes are generally slippery (though beeswax is somewhat sticky), plastic when warm, and, because their molecular weights are rather low, fluid when melted. See PARAFFIN WAX.

WAXS Abbreviation for WIDE-ANGLE-X-RAY SCATTERING, w s.

Wb SI abbreviation for WEBER, w s.

Weathering A broad term encompassing exposure of plastics to solar or ultraviolet light, temperature, oxygen, humidity, snow, wind, and air-borne dust and biological or chemical agents, such as smog. ASTM D 1435 prescribes procedures for testing the effects of outdoor weathering on plastics. See also ARTIFICIAL WEATHERING.

Weather-O-Meter Trade name of an apparatus for subjecting articles to accelerated weathering conditions, including intense ultraviolet radiation, water sprays, and elevated temperatures to 50°C.

Web (1) A continuous film or fabric in process in a machine. In extrusion coating, the *molten web* is that which issues from the die and becomes the coating, and the *substrate* (*web*) is the material being coated. (2) A continuous length of sheet material handled in roll form as contrasted with the same material cut into short lengths.

Web Coating Any of a number of processes by which coatings are applied to continuous substrates such as papers, cloths, and metal foils, including CALENDER COATING, EXTRUSION COATING, FLOW COATING, GRAVURE COATING, ROLLER COATING, SPREAD COATING, w s, plus others listed at COATING METHODS.

Weber (Wb) The SI unit of magnetic flux, defined as the magnetic flux which, linking a circuit of one turn, produces in it an electromotive force of one volt as the flux is reduced to zero at a uniform rate in one second. Therefore, 1 weber = 1 volt·second. The former unit of flux, the maxwell, part of the so-called "absolute" system of electrical units, equals 10^{-8} Wb.

Web Gate Synonym for DIAPHRAGM GATE, w s.

Weft (woof, fill, filler yarn) In the textile industry, the transverse threads or fibers in a woven fabric; those fibers running crosswise to the WARP (w s).

Weigh Feeder Synonym for GRAVIMETRIC FEEDER, w s.

Weight The force with which a body on or near the earth's surface is attracted to the earth, equal to the body's mass × g/g_c, where g is the acceleration due to gravity and $1/g_c$ is the proportionality constant in Newton's second law of motion (the law of momentum change). In the SI system, g_c = 1.00000 kg·m/(N·s^2) while g varies slightly (at sea level) from 9.78039 m/s^2 at 0° latitude to 9.83217 m/s^2 at 90° latitude because of the earth's spin and consequent flattening at the poles, with its "standard value", at about 45° N latitude, being 9.80655 m^2/s. Thus the abhorred kg_f is equated to 9.80655 N. See also FORCE.

Weight-Average Molecular Weight (M_w, $\overline{M}_w$) For a sample with distributed molecular weights (all commercial polymers), the defining equation is

$$M_w = \frac{\sum_{i=1}^{\infty}(N_i M_i)(M_i)}{\sum_{i=1}^{\infty} N_i M_i} = \frac{\sum_{i=1}^{\infty} N_i M_i^2}{\sum_{i=1}^{\infty} N_i M_i}$$

where N_i is the number of individual molecules having molecular mass M_i and N_iM_i equals the mass of the N_i molecules in the sample with molecular mass M_i. The numerator and denominator quantities are also known as the *second* and *first original moments* of the distribution. M_w may be determined from measurements of LIGHT SCATTERING or SIZE-EXCLUSION CHROMATOGRAPHY, w s. And it is M_w upon which melt viscosity in thermoplastics is strongly dependent. See also MOLECULAR WEIGHT, MOLECULAR-WEIGHT DISTRIBUTION, NUMBER-AVERAGE MOLECULAR WEIGHT, and VISCOSITY-AVERAGE MOLECULAR WEIGHT.

Weight Loss The loss of mass of an article or test sample as a result of exposure to particular conditions as, for example, a few hours at 90°C in a vacuum oven, or two years outdoors on a weathering panel, often expressed as a percentage of the original mass. See also VOLATILES CONTENT, which may be the same as weight loss when the latter is due exclusively to loss of volatiles.

Weir A simple device for controlling flow in open channels, consisting of a plate serving as a dam and having at its top center a V-notch or adjustable, rectangular, vertically sliding gate. Weirs can also be calibrated for flow measurement.

Weissenberg Effect A phenomenon sometimes encountered in rotational-viscometry studies of polymer melts and solutions at high speeds, characterized by the tendency of the polymer solution to climb the wall of the cup or the shaft of the rotor immersed in it.

Weissenberg Number (N_{We}) In the flow of viscoelastic liquids, the dimensionless Weissenberg number represents the ratio of the viscoelastic force to the viscous force and has sometimes been equated to $N_1/2\tau$, where N_1 = the first normal stress in a viscoelastic fluid flowing in simple shear and τ = the shear stress.

Weissenberg Rheogoniometer A vertical cone-and-plate rheometer (K. Weissenberg, 1948, and improvers since) for viscoelastic liquids in which the cone can twist through a measured angle while the plate is rotated at speeds providing shear rates to 100 s^{-1}. The cone and plate are enclosed in an oven that provides control of the sample temperature. The plate shaft rests on a ball bearing at its bottom which in turn rests on a force transducer. Thus it is possible to measure both the restoring torque on the cone and the normal force on the plate. The first provides an estimate of the liquid's viscosity while the normal force gives an estimate of the normal stress. See also MECHANICAL SPECTROMETER.

Weldbonding A process developed in the former USSR and introduced in the US by the Air Force Materials Laboratory. Developed for the aerospace industry, the process combines spot welding with adhesive bonding of aluminum structures. It has provided an economical and efficient means of laying up and oven curing large, epoxy-bonded assemblies.

Welding The joining of two or more pieces of thermoplastic by fusion at adjoining areas, either with or without addition of plastic from another source (such as welding rod). The term includes *heat sealing*, with which it is synonymous in some countries, but in the US the term heat sealing is limited to film and sheeting. Welding is almost always done with two (or more) pieces of the same plastic, but it can be done with compatible plastics that melt in the same temperature range. The various welding methods are described at the entries listed below.

BUTT FUSION
DIELECTRIC HEAT SEALING
EXTRUDED-BEAD SEALING
FRICTION WELDING
HEAT SEALING
HIGH-FREQUENCY WELDING
HOT-GAS WELDING
HOT-PLATE WELDING
IMPULSE SEALING
INDUCTION WELDING
JIG WELDING
SPIN WELDING
STITCHING
THERMOBAND WELDING
ULTRASONIC WELDING
VIBRATION WELDING

Weld Line (weld mark, flow line) (1) A flaw on a molded plastic article marking the meeting of two flow fronts within the mold. Because the two fronts may have cooled and skinned over before meeting, or had too little time in the molten state for interdiffusion of molecular segments across the interface, the weld may be imperfect and weak. This can be a serious problem with fiber- or mica-reinforced thermoplastics because of failure of fibers or flakes to penetrate the weld interface. (2) In extrusion of pipe, tubing, and some profiles from end-fed dies in which the cores are supported by spiders, a line parallel to the product axis where the flow front was split by a spider leg and subsequently reunited downstream. Weakness at such weld lines can depress the hoop strength of the pipe or tubing.

Wet Flexural Strength The flexural strength measured after boiling a test specimen in water, usually less than the strength of the original, dry specimen. See also FLEXURAL STRENGTH.

Wet Layup In the reinforced plastics molding, the process of forming an article by first applying a liquid resin (sometimes a special GEL COAT, w s) to the mold surface, then applying a reinforcing backing layer with more resin.

Wet-out The degree to which an impregnating resin has filled the voids among the filaments being impregnated. This may be expressed quantitatively as $100 \times (1 - v_r - VF)/(1 - v_r)$ in which v_r = the volume fraction of reinforcing fiber in the laminate and VF = the final VOID FRACTION, w s.

Wet-out Time The time required for a resin to completely fill the interstices of a reinforcement material and wet the surfaces of the fibers, usually determined by an optical or light-transmission method.

Wet Spinning See SPINNING.

Wet Strength (1) The strength of paper (or other material) when saturated with water, a term often used in discussions of processes wherein the wet strength of paper is increased by treatment with resins. (2) The strength of an adhesive joint determined immediately after removal from a liquid (usually water) in which it has been immersed under specified conditions of time, temperature and pressure.

Wettability The ability of a solid surface to accept contact of and by a liquid, allowing it to spread freely and completely cover the surface. Wettability is closely linked to the equality of components of surface energy and SURFACE TENSION (w s) of the solid and liquid involved. If the surface is wettable by the liquid, the contact angle of a droplet on the surface will be less than 10°.

Wetted out The condition of an impregnated reinforcement where in substantially all voids between the sized strands and filaments are filled with resin; 100% WET-OUT, w s.

Wetting Agent A compound that causes a liquid to penetrate more easily into, or to spread over the surface of, another material, usually by reducing the liquid's surface tension. Common wetting agents are soaps, detergents, and surfactants. They are widely used in polymerization reactions and in preparing emulsions of plastics.

Wet Winding A filament-winding process wherein the strand is impregnated with resin just prior to contact with the mandrel.

Whiskers A colloquial term used for nearly perfect, single-crystal fibers produced synthetically under controlled conditions from inorganic materials such as aluminum oxide, beryllium oxide, boron, boron carbide, graphite, magnesium oxide, metals, quartz, silicon carbide, and silicon nitride. They range in diameter from 0.5 to 30 µm, and in length from 1 µm to several mm. Whiskers are available as loose fibers, mats, and felts. Having tensile strengths and moduli from 5 to 10 times those of glass, they impart extremely high strength and stiffness to reinforced-plastics structures. Ceramic whiskers are also costly: 500 to 25,000 $/kg, but prices should drop in the future.

White Bole (bolus alba) A synonym for KAOLIN, w s.

Whiting A finely divided form of CALCIUM CARBONATE, w s, obtained by milling high-calcium limestone, marble, shell, or chemically precipitated calcium carbonate.

Wide-Angle X-ray Scattering (WAXS) A technique for determining the amount of crystallinity and the sizes and perfection of crystals in polymers, in which diffraction patterns of X rays scattered at 20° to 50° from the incident beam are recorded on film and measured. X rays of wavelength from 0.1 to 0.3 nm are used to elucidate structural features with sizes from 0.1 to 2 nm. Compare SMALL-ANGLE X-RAY SCATTERING.

Williams-Landell-Ferry Equation (WLF equation) An empirical equation for the TIME-TEMPERATURE EQUIVALENCE (w s) of creep and other properties, that has been successful with many plastics. It is

$$\log_{10} a_T = \frac{-17.44(T - T_g)}{51.6 + T - T_g}$$

where T and T_g = the temperature of interest and the glass-transition temperature of the polymer, K, and a_T is the SHIFT FACTOR (w s), i e, the ratio of the viscosities at the two temperatures. The equation holds over the range from T_g to about T_g + 100 K.

Winder (windup) Any device that collects a continuously extruded, calendered, or coated product onto a central core in progressive layers. The range of winder designs, sizes, and speeds is huge, corresponding to the equally wide range of products that are wound—from multi-bobbin winders for tiny reels of fishing leader, to dual, quick-change winders for coated paper, film, and wire-coating lines, to large, slowly turning winders of coated, multistrand cable and undersea cable.

Window A globule of incompletely plasticated material in a thermoplastic film, sheet, or molding that is visible when viewed by transmitted light. It is equivalent to FISHEYE, w s, except that the term window is usually employed to indicate a clear spot in an otherwise colored or opaque material.

Wire Coating The application of a plastic, rubber, or enamel coating to a single- or multi-strand wire, or to a cable of many previously coated single wires. Most wire coating is done by extrusion from the melt, but some, such as magnet wire for electric motors, has been done by passing the wire through a solution of thermosetting resin, then evaporating the solvent and curing the resin in an oven. Lineal rates on extrusion-coating lines range from 0.5 m/s on a line overcoating a large cable containing hundreds of wires to 30 m/s on a line coating hook-up wire. Over 500 Gg (0.55×10^6 tons) of leading thermoplastics were used to coat wire and cable in 1992.

Wire Gauge (wire gage) (1) Any of several shorthand systems of consecutive numbers, each number relating inversely to a particular wire diameter. Steel producers in the US use the Steel Wire Gauge, ranging from 7/0 (0000000), = 0.4900 in., to 0, = 0.3065 in., to 50, = 0.0044 in. In Britain, the British Standard Wire Gauge (Imperial Wire Gauge) has long been used, with diameters close to those of the Steel Wire Gauge. (This may be changing to metric.) Copper and aluminum wires, formerly given in Brown & Sharpe (B&S) wire gauge, are now specified in decimal-fractional inches. Contrarily, music- (piano-) wire sizes <u>increase</u> with their gauge numbers. The Standard for Metric Practice, ASTM E 380, has strangely omitted this important area of measurement. Presumably, in SI, there are no "gauges", and wire sizes are given in millimeters, as are screen sizes. (2) A metal plate perforated with graduated and labeled holes with which one may determine the size of a wire or drill bit by identifying the smallest hole through which the wire will pass.

Wire Train (wire-coating line) The entire assembly that is utilized to produce a resin-coated wire, normally consisting of a feed or let-off spool, a wire preheater, an extruder, a crosshead and die, cooling means, a coating-thickness gauger, a spark tester for pinholes, and a windup.

WLF Equation See WILLIAMS-LANDELL-FERRY EQUATION.

Wollastonite The mineral CALCIUM SILICATE, w s.

Wood Alcohol An impure alcohol historically obtained from the destructive distillation (pyrolysis in the absence of air) of wood, whose main constituent was METHANOL (w s). Today, methanol is synthesized from carbon monoxide and hydrogen. It is toxic and is used as a denaturant in ethanol to make it impotable.

Wood Flour (wood meal) Finely pulverized, dried wood, used as a filler in thermosetting molding compounds such as phenolics and ureas. The woods are resin-free softwoods such as pine, fir, and spruce, and, to a lesser extent hardwoods. Woods shredded to fibrous form are also used as a reinforcement.

Woodgraining A group of processes used to impart wood-like appearance to sheets or shaped articles. The substrates may be of plastic, wood, steel, or any other material. Among the processes used are conventional laminating techniques, multiple-coat painting, hot stamping, and introduction of several colors into the melt during molding.

Woof Synonym for WEFT, w s.

Work (n) The action of a force through a distance; the product of the force times the distance. Also, the action of a torque through an angular displacement; the product of the torque times the displacement in radians. The SI unit of work, the same as that of energy, is the joule (J), equal to 1 N·m.

Working Life (1) Synonym for POT LIFE, w s. (2) Synonym for SERVICE LIFE, w s.

Working Stress See ALLOWABLE STRESS.

Wrinkle (n) (1) An imperfection in reinforced plastics that has the appearance of a wave molded into one or more plies of fabric or other reinforcing material (ASTM D 883). (2) In a plastic film or coated cloth, an inadvertent crease.

WVTR Abbreviation for WATER-VAPOR-TRANSMISSION RATE, w s.

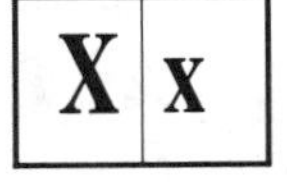

x (1) Symbol for mole fraction, usually subscripted to indicate the component of interest. (2) Symbol for general variable, or independent variable, and the horizontal graphing axis, or abscissa.

X Symbol, preceded by a number, for power of magnification and, closely related, sometimes used in place of ×, the multiplication operator.

Xaloy® Trade name of Xaloy, Inc, originator and leading supplier of bimetallic cylinders, used in extruders and injection molders, in which the inside layer is one of several extremely hard alloys. Number 101 has been the most used for general extrusion service, number 306 for corrosion protection. See also BIMETALLIC CYLINDER.

Xanthate A sodium salt of a dithiocarbonic acid ester, in particular the one formed in the viscose-rayon process by the reaction between sodium hydroxide cellulose and carbon disulfide and having the structure shown below, called cellulose xanthate or *viscose*. The viscose is subsequently precipitated, filtered, extruded as filaments into dilute sulfuric acid, washed, and dried to make viscose rayon.

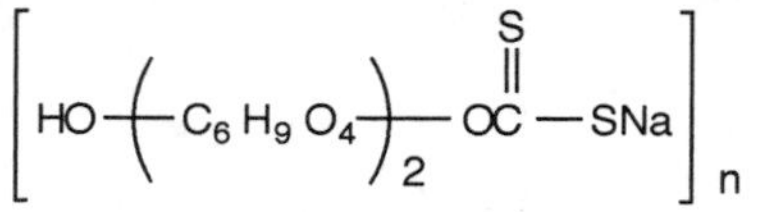

$$\left[HO\left(C_6H_9O_4 \right)_2 OC(=S)-SNa \right]_n$$

See CELLULOSE for the interesting structure of the cellulose mer in parentheses in the diagram.

Xenon-Arc Aging A test for evaluating the light stability of plastics, employing a xenon-gas-discharge lamp of special design that emits radiation duplicating the spectrum of natural sunlight more closely than most artificial sources. ASTM lists two such tests for plastics, D 4459 with dry specimens and G 26, in which the specimens may or may not be sprayed with water.

XLPE Abbreviation for *crosslinked polyethylene*. See RADIATION CROSSLINKING.

XPS Abbreviation for *expandable* or *expanded polystyrene*. See POLYSTYRENE FOAM.

X Ray Any electromagnetic radiation with wavelength in the range from 0.003 to 3 nm, produced by the bombardment of a target with cathode rays. Those with the shorter wavelengths are more energetic and are called *hard X rays*, while those with the longer wavelengths are called *soft X rays*.

X-Ray Diffraction Crystals, whose interatomic spacings are commensurable with the wavelengths of some X rays, can act as diffraction gratings for X rays. When X rays are directed obliquely at a crystal surface, and the resulting radiation is captured on photographic film, a symmetrical pattern of spots is observed that is related to the positioning of atoms in the crystal. This 80-year-old technique has been useful in studying crystalline structure in polymers. See also SMALL-ANGLE X-RAY SCATTERING and WIDE-ANGLE X-RAY SCATTERING.

X-Ray Microscopy The conversion of X-ray-diffraction patterns into pictures showing the positioning of atoms in crystals as if they were being viewed through a very powerful microscope. In *point-projection*

X-ray microscopy, the enlarged picture is obtained from X rays emitted from a pinhole source. The technique is useful for studying the structure of materials such as foamed plastics, laminates, and fibers.

Xylene $C_6H_4(CH_3)_2$ A commercial mixture of the three isomers, *o*-, *m*-, and *p*-xylene, used as a solvent for alkyd resins, polystyrene, natural resins, rubber, and polyisobutylene.

***o*-Xylene** (1,2-dimethylbenzene) 1,2-$C_6H_4(CH_3)_2$. A colorless liquid used as a feedstock in the production of phthalic anhydride. It can be extracted from the mixed isomers (see XYLENE) by distillation and can be isomerized to *p*-xylene.

***p*-Xylene** (1,4-dimethylbenzene) 1,4-$C_6H_4(CH_3)_2$. A colorless liquid used in the synthesis of terephthalic acid and dimethylterephthalate, both of which are intermediates for polyester fibers and films.

***p*-Xylene-α,α'-Diol** $C_6H_4(CH_2OH)_2$. A white crystalline solid, used as a crosslinking agent in polyurethanes, and in the production of polyesters and polycarbonates.

Xylenol Resin A phenolic-type resin produced by condensing xylenol (3,5-dimethylphenol) with an aldehyde. POLYPHENYLENE OXIDE (w s) is made from 2,6-xylenol.

Xylox Resin® Trade name for a family of heat-resistant thermosetting resins made by the condensation of aralkyl ethers and phenols, resulting in hydroxyphenylene-*p*-xylene prepolymers that can be cured to hard, intractable resins by reaction with hexamethylenetetramine or epoxy compounds. These thermosetting resins have the good qualities of phenolics and epoxies, with superior mechanical and electrical properties at elevated temperatures.

***p*-Xylylene** (PX) $H_2C{=}C_6H_4{=}CH_2$. A highly reactive monomer from which parylene polymer is formed, $(-CH_2C_6H_4CH_2-)_n$. The gaseous monomer readily forms a stable solid dimer, convenient for shipping, from which the monomer is easily regenerated by heating. PX polymerizes spontaneously in vacuum on any cool surface to form tough, uniform, impervious films. Dimers with ring-substituted chlorine are also available. See also DI-*p*-XYLYLENE and PARYLENE.

***m*-Xylylenediamine** A solid diamine useful as curing agent for epoxy resins.

Xylylene Diisocyanate A mixture of the *m*- and *p*-isomers, used in the production of polyurethane coatings.

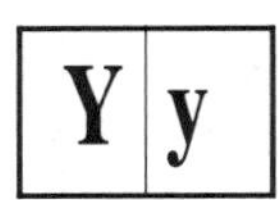

y (1) (yr) Abbreviation for year. (2) Symbol for general dependent variable and the ordinate (vertical) axis in two-dimensional graphing.

Y Chemical symbol for the element yttrium, which has been used as a catalyst for ethylene polymerization.

Yarn Generic term for spun strands of fibers or filaments in a form suitable for weaving or otherwise intertwining to form a fabric. A yarn may contain from a dozen to more than 100 individual fibers.

Yellowness Index (YI) A measure of the yellowing of a plastic, such as might occur after lengthy exposure to light. It is determined according to ASTM D 1925, and is therein defined as the deviation in chroma from whiteness or water-whiteness in the dominant wavelength range from 570 to 580 nm. The index is computed from the three tristimulus values measured with a spectrophotometer, relative to a magnesium oxide standard.

Yield Point In tensile testing, the first point on the stress-strain curve at which an increase in strain occurs without an increase in stress. This is the point at which permanent (plastic) deformation of the specimen begins. Many plastics do not exhibit an identifiable yield point.

Yield Strength The stress at which a material exhibits a specified limiting deviation from the proportionality of stress to strain. Unless otherwise specified, this stress will be the stress at the yield point (ASTM D 638 and D 638M). See also OFFSET YIELD STRENGTH.

Yield Value In rheology, the finite shear stress that must be applied to a semifluid before flow will start. If, at stresses greater than the yield value, shear rate is proportional to stress minus yield value, the material is called a BINGHAM PLASTIC, w s. A familiar example of a semifluid with a yield value is ordinary toothpaste.

Young's Modulus (E) See MODULUS OF ELASTICITY.

Z z

z (1) In rectangular and cylindrical coordinate systems, the symbol for the vertical (axial) coordinate. (2) Symbol for STANDARD NORMAL DEVIATE, w s.

Z Symbol for atomic number, electrical impedance, or SECTION MODULUS, w s.

Zahn Viscosity Cup One of several similar, one-shot devices for obtaining "quick and dirty" measurements related, in a roughly linear way, to the KINEMATIC VISCOSITY, w s, of the test liquid, typically a free-flowing oil. The cupful of test liquid is brought to the desired temperature in a bath of heating medium, then held over a collection vessel and allowed to empty through a short tube in its bottom. Measurements are reported as Zahn Number-*X* seconds, *X* indicating which of five different cups was used. Compare SAYBOLT VISCOSITY and see also VISCOSITY.

Z-Average Molecular Weight (M_Z, $\overline{M}_z$) A higher-degree average than weight average or viscosity average, but closer to the former, and defined by the equation

$$M_z = \frac{\sum_{i=1}^{\infty} N_i M_i^3}{\sum_{i=1}^{\infty} N_i M_i^2}$$

M_Z is more sensitive than the other averages to the largest molecules present in the sample. The sums in the numerator and denominator are also known as the *third* and *second original moments* of the molecular-weight distribution.

Z-Calender A calender with four rolls arranged so that, as the web of material passes through them, the cross section of its path has a shape resembling the letter Z.

Zein A naturally occurring, high-molecular-weight protein, a polymer of amino acids linked by peptide bonds, derived from corn. It is considered to be a member of the protein family of plastics, the main member of which is casein plastic. Zein resins, rarely seen today, years ago were used for fibers (e g, Vicara), films, and paper coating.

Zero Defects (ZD) The state of quality of manufactured product in which every finished item is within the preset specification limits for all its quality factors. This has long been a goal of certain critical industries and is now spreading into others. It is most likely to be achieved by an ongoing program of total quality management (TQM) that makes use of STATISTICAL PROCESS CONTROL, w s.

Ziegler Catalyst Any of a large group of catalysts made by reacting a compound of a transition metal chosen from groups IV through VIII of the periodic table with an alkyl, hydride, or other compound of a metal from groups I through III. A typical example is the reaction product of an aluminum alkyl with titanium tetrachloride or titanium trichloride. These catalysts were first discovered by the German chemist K. Ziegler (late 1940s) for the low-pressure polymerization of ethylene. Subsequent work by G. Natta (early 1950s) showed that these and similar catalysts are useful for preparing stereoregular polyolefins; thus, the family of catalysts is sometimes called *Ziegler-Natta catalysts.*

Zimate Trade name of the R. T. Vanderbilt Co. for a group of dialkyl dithiocarbamates, useful as accelerators for curing rubber.

Zinc Baryta White Synonym for LITHOPONE, w s.

Zinc Borates White, amorphous powders of indefinite composition, containing various amounts of zinc oxide and boric oxide. They are used as flame retardants in PVC, polyvinylidene chloride, polyesters, and polyolefins, often in combination with antimony trioxide.

Zinc Oxide (Chinese white, flowers of zinc, zinc white) ZnO. An amorphous white or yellowish powder, used as a pigment in plastics. It is said to have the greatest power to absorb ultraviolet light of all commercially available pigments. Mixtures of zinc oxide and secondary organic additives such as butyl Zimate act in strong synergism and increase the protection against UV light as much as 66 times that of ZnO alone. See ZIMATE.

Zinc Palmitate $Zn(OOCC_{15}H_{31})_2$. An amorphous white powder used as a lubricant in plastics.

Zinc Ricinoleate $Zn[OOC(CH_2)_7CH{=}CHCH_2CH(OH)C_6H_{13}]_2$. An amorphous white powder used as a stabilizer in vinyl plastics.

Zinc Stabilizer Any of a group of zinc soaps of fatty acids, usually formulated in combination with barium and calcium soaps and organic phosphites in plasticized PVC compounds. The new interest in zinc compounds is driven by their greater environmental safety vis-a-vis traditional stabilizers based on heavy metals such as cadmium and lead. The levels required for equal stabilization are about 40% higher—3 to 4 phr—with an associated incremental cost of compound.

Zinc Stearate $Zn(OOCC_{17}H_{35})_2$. A white powder used as a lubricant and stabilizer in vinyl compounds.

Zirconia (zirconium oxide) ZrO_2. A white monoclinic powder used as a pigment when good electrical properties are required. Though zirconia is refractory (melting point = 2715°C), it has been spun into fibers. Their strength is boosted considerably by coating them with boron nitride or silicon carbide.

ZMBT Abbreviation for the zinc salt of 2-mercaptobenzothiazole, a light-cream-colored powder used as an accelerator in curing natural rubber and butadiene-styrene copolymers.

Zn Chemical symbol for the element zinc.

Z-Nickel An obsolete designation for a corrosion-resistant alloy that was specified some decades back for polyvinylidene-chloride extruders in the early years of that resin. Its composition is believed to have been close to what is now known as *duranickel 301*, an alloy containing 94% nickel, 4.4% aluminum, 0.5% each of silicon and titanium, 0.3% iron, and traces of carbon, manganese, copper, and sulfur.

Zone Melting A method of polymer fractionation in which a sample of polymer is placed on top of a vertical tube filled with frozen solvent. A small zone at the top is melted, dissolving the polymer, whose molecules tend to migrate through the solvent at size-dependent velocities. The next zone is melted while the top one is refrozen, and so on, down the tube. When all zones have been refrozen, the column is cut into segments from which the polymer fractions are captured by evaporating the frozen solvent.

Zr Chemical symbol for the element zirconium.

APPENDIX

Conversion Factors for Selected Units

The reader is directed to ASTM *Practice for Use of the International System of Units*, E 380-89, whence most of the conversion factors listed here came. However, E 380's conversion factors are stated to as many as seven significant figures, several more than are needed for everyday work in the plastics industry or, for that matter, in polymer science, so in this list they will be given to four significant figures. If the first digit of the factor is in the range $1 \le$ first digit < 4, five actual digits are given; others have four. Thus, since the error in each should be no more than 0.5 unit of the last digit, all should be accurate to 0.01% or better. If fewer than four digits are given, the conversion is exact. For example, 1 inch $\equiv$ 2.54 cm. Also, E 380 states every factor in standard scientific format, e g, the factor by which a viscosity in $lb_f \cdot s/in^2$ must be multiplied to convert it to Pa·s is stated as 6.894,757 E+03, whereas here it will be given more readably as 6.895 kPa. Only SI-sanctioned prefixes are to be used with the base units. They are:

exa (E)	$= 10^{18}$	deci (d)	$= 10^{-1}$
peta (P)	$= 10^{15}$	centi (c)	$= 10^{-2}$
tera (T)	$= 10^{12}$	milli (m)	$= 10^{-3}$
giga (G)	$= 10^{9}$	micro (μ)	$= 10^{-6}$
mega (M)	$= 10^{6}$	nano (n)	$= 10^{-9}$
kilo (k)	$= 10^{3}$	pico (p)	$= 10^{-12}$
hecto (h)	$= 10^{2}$	femto (f)	$= 10^{-15}$
deka (da)	$= 10^{1}$	atto (a)	$= 10^{-18}$

The prefixes giga and pico are pronounced jig'a and peek'o. The conversions given here are all from non-SI units to SI units. Each mass unit that has been used as a force unit is so identifed by "(force)" in its name and by the subscript "f" in its abbreviation. The *liter* (L) is approved in SI as an alternate name for the volume measure, cubic decimeter (dm^3). Note that the powers applied to prefixed units are applied to the prefixes, too. Example: 1 dm^3 = 1 $(dm)^3$, not 0.1 m^3.

Should you wish to convert from one non-SI unit (A) to another (B), multiply the quantity in A units by F_A/F_B, where F_A and F_B are the listed equivalents for A and B in the common SI unit. Be sure to include the powers of ten given by the attached abbreviated prefixes. For example, suppose that you wish to convert a modulus given as 7634 kg_f/mm^2 to psi (lb_f/in^2). The new modulus value = 7634 $\times$ $(9.807 \cdot 10^6/6.895 \cdot 10^3) = 1.0858 \cdot 10^7$ psi, or 10.858 Mpsi.

Length, Area, and Volume

Length

angstrom (Å) =	0.1 nm
finger	2.1090 cm
foot (ft)	0.30480 m
inch (in)	2.5400 cm
microinch (μin)	25.400 nm
mil	25.4 μm
mile (US statute) (mi)	1.6093 km
yard (yd)	0.9144 m

Area

acre (ac)	4047 m^2
are (pron. like *bar*)	100 m^2
circular mil (wire)	506.7 μm^2
hectare	10,000 m^2
square foot (ft^2)	9.290 dm^2
square inch (in^2)	6.452 cm^2
square mile	2.590 km^2
square yard	0.8361 m^2

Volume

barrel (US, 31 gal)	119.24 L (dm^3)
barrel (oil, 42 gal)	158.99 L
cubic foot (ft^3)	28.317 L
cubic inch (in^3)	16.387 cm^3
cup (8 fl oz)	0.23659 L
fluid ounce (US) (fl oz)	29.574 cm^3
gallon (US liquid) (gal)	3.7854 L
quart (US liquid) (qt)	0.94635 L
teaspoon (tsp)	4.9289 cm^3

Mass

grain (gr)	64.799 mg
hundredweight (short, 100 lb) (cwt)	45.359 kg
ounce (avoirdupois) (oz)	28.350 g
ounce (troy)	31.103 g
pound (avoirdupois) (lb)	0.45359 kg
ton (US, short) (T)	907.18 kg
tonne	1000 kg

Elapsed Time

day (mean solar) (d, da)	8.64·104 s
hour (h, hr)	3600 s
year (365 days) (yr)	$3{,}1536 \cdot 10^7$ s

Temperature and Temperature Interval

Δt, °Celsius (or °Centigrade) (°C)	1.0 K
t, °Celsius (°C)	$T_K = t_C + 273.15$, K
Δt, °Fahrenheit (°F)	0.5556 K
t, °Fahrenheit (°F)	$T_K = 255.37 + 0.5556\ t_F$, K

Density, Lineal Density, Concentration

Density

gram per cubic centimeter (g/cm^3)	1000 kg/m^3
pound (avoir) per cubic inch (lb/in^3)	27,680 kg/m^3
pound (avoir) per cubic foot (lb/ft^3)	16.018 kg/m^3
ton (short) per cubic yard (T/yd^3)	1186.6 kg/m^3

Lineal Density

denier	0.11111 mg/m
milligram per foot (mg/ft)	3.2808 mg/m
tex	1.0 mg/m

Concentration

grain per gallon (gr/gal)	17.118 kg/m^3
grain per cubic foot (gr/ft^3)	2.2884 g/m^3
pound-mole per cubic foot ($lb\text{-}mol/ft^3$)	16.018 mol/L

Velocity (Speed), Angular Velocity, Frequency

Velocity

foot per second (ft/s)	0.3048 m/s
foot per minute (ft/min)	5.08 mm/s
kilometer per hour (km/h)	0.27778 m/s
mile per hour (mi/h, mph)	0.4470 m/s

Angular Velocity

revolutions per minute (rpm)	0.10458 rad/s
revolutions per second (rps)	6.275 rad/s

Frequency

cycle per minute	0.016667 Hz
cycle per second	1.0 Hz
vibrations per second	1.0 Hz

Force, Weight, Torque

Force

dyne (d)	10 μN
gram(force) (g_f)	9.807 mN
kilogram(force) (kg_f)	9.807 N
kilopond (kp)	9.807 N
ounce(force) (oz_f)	0.27801 N
pound(force) (lb_f)	4.448 N
ton(force) (= 2000 lb_f) (T_f)	8896 N

Torque, Bending Moment

dyne·centimeter (d·cm)	$1.0 \cdot 10^{-7}$ N·m
kilogram(force)·meter ($kg_f \cdot m$)	9.807 N·m
ounce(force)·inch ($oz_f \cdot in$)	7.061 mN·m
pound(force)·foot ($lb_f \cdot ft$)	1.3558 N·m
pound(force)·inch ($lb_f \cdot in$)	0.11298 N·m

Pressure, Stress, Strength

atmosphere (standard) (atm)	101.32 kPa
bar	100 kPa
gram(force) per denier (fiber strength)	88.26 N/(g/m)
inch of mercury, 60°F (in Hg)	3.3768 kPa
inch of water, 60°F (in H_2O)	248.84 Pa
kilogram(force) per square centimeter (kg_f/cm^2)	98.07 kPa
kilogram(force) per square millimeter (kg_f/mm^2)	9.807 MPa
kilopound(force) per square inch (kpsi)	6.895 Mpa
millimeter of mercury, 0°C (mm Hg)	133.32 Pa
ounce(force) per square inch (oz_f/in^2)	430.9 Pa
pound(force) per square foot (lb_f/ft^2, psf)	47.88 Pa
pound(force) per square inch (lb_f/in^2, psi)	6.895 kPa
torr, 0°C	133.32 Pa

Flow Rate, Throughput

cubic foot per minute (ft^3/min, cfm)	0.4719 L/s
cubic inch per second (in^3/s)	16.387 cm^3/s
gallon (US) per minute (gal/min, gpm)	0.06309 L/s
kilogram per hour (kg/h)	0.27778 g/s
pound per hour (lb/h, pph)	0.12560 g/s

Energy, Work, Heat

Note: British thermal units (Btu) and calories in this Appendix are the "International Table" values, 1055.056 J and 4.186,800 J, respectively. See main-section entries BRITISH THERMAL UNIT and CALORIE for discussion of other types of Btu and calorie.

British thermal unit (Btu)	1.0551 kJ
calorie (cal)	4.187 J
cubic foot·atmosphere (ft^3·atm)	2.8692 kJ
erg	0.1 μJ
foot·pound(force) (ft·lb_f)	1.3558 J
horsepower (550 ft·lb_f/s)·hour (hp·h)	2.6845 MJ
kilocalorie ("large calorie") (kcal)	4.187 kJ
kilowatt·hour (kW·h)	3.6 MJ
liter·atmosphere (L·atm)	101.33 J
meter·kilogram(force) (m·kg_f)	9.807 J
watt·hour (W·h)	3.6 kJ

Power, Energy-Dissipation Rate

British thermal unit per hour (Btu/h)	0.29307 W
Foot·pound(force) per second (ft·lb_f/s)	1.3558 W
horsepower (550 ft·lb_f/s) (hp)	745.70 W
kilocalorie per hour (kcal/h)	1.1630 W
ton of refrigeration (12,000 Btu/h)	3.5168 kW

Heat Capacity

British thermal unit per pound degree Fahrenheit [Btu/(lb·°F)]	4.187 kJ/(kg·K)
calorie per gram·degree Celsius [cal/(g·°C)]	4.187 kJ/(kg·K)
kilocalorie per kilogram·degree Celsius [kcal/(kg·°C)]	4.187 kJ/(kg·K)

Heat-Transfer Quantities

Heat Flux

$Btu/(h \cdot ft^2)$	$3.1546\ W/m^2$
$Btu/(s \cdot ft^2)$	$1.1357\ kW/m^2$
$calorie/(s \cdot cm^2)$	$41.868\ kW/m^2$
$kcal/(min \cdot cm^2)$	$6.978\ kW/m^2$

Thermal Conductivity

$Btu/[h \cdot ft^2 \cdot (°F/ft)] = Btu/(h \cdot ft \cdot °F)$	$1.7307\ W/(m \cdot K)$
$Btu/[h \cdot ft^2 \cdot (°F/in)] = Btu \cdot in/(h \cdot ft^2 \cdot °F)$	$0.14423\ W/(m \cdot K)$
$Btu/[s \cdot ft^2 \cdot (°F/ft)] = Btu/(s \cdot ft \cdot °F)$	$6.231\ kW/(m \cdot K)$
$calorie/[s \cdot cm^2 \cdot (°C/cm)] = cal/(s \cdot cm \cdot °C)$	$418.68\ W/(m \cdot K)$

Film Coefficient, Overall Heat-Transfer Coefficient

$Btu/(h \cdot ft^2 \cdot °F)$	$5.678\ W/(m^2 \cdot K)$
$cal/(s \cdot cm^2 \cdot °C)$	$41.868\ kW/(m^2 \cdot K)$

Thermal Diffusivity

square centimeter per second (cm^2/s)	$1.0 \cdot 10^{-4}\ m^2/s$
square foot per hour (ft^2/h)	$2.5806 \cdot 10^{-5}\ m^2/s$
square foot per second (ft^2/s)	$0.09290\ m^2/s$

Viscosity, Kinematic Viscosity

Viscosity (Absolute Viscosity, Dynamic Viscosity)

centipoise (cp)	1.0 mPa·s
kilopoise	100 Pa·s
poise	0.1 Pa·s
pound(force)·second per square inch ($lb_f \cdot s/in^2$, psi·s)	6.895 kPa·s
pound(force)·second per square foot ($lb_f \cdot s/ft^2$, psf·s)	47.88 Pa·s
pound per foot·second [lb/(ft·s)]	1.4882 Pa·s
pound per foot·hour [lb/(ft·h)]	0.41338 mPa·s

Notes: (1) 1 pascal·second (Pa·s) may be interpreted either as the quotient of shear stress and shear rate, i e, $1\ (N/m^2)/(s^{-1}) = 1\ N \cdot s/m^2$, or as the quotient of momentum flux and velocity gradient, i e, $1\ [(kg \cdot m/s)/(m^2 \cdot s)]/[(m/s)/m] = 1\ kg/(m \cdot s)$. (2) Viscosity units tied to particular equipment, e g, Saybolt seconds, are discussed at entries in main section.

Kinematic Viscosity

centistokes	1.0 mm^2/s
square centimeter per second (cm^2/s)	$1.0 \cdot 10^{-4}$ m^2/s
square foot per hour (ft^2/h)	25.806 mm^2/s
square foot per second (ft^2/s)	0.09290 m^2/s
square meter per hour (m^2/h)	$2.7778 \cdot 10^{-4}$ m^2/s
stokes	$1.0 \cdot 10^{-4}$ m^2/s

Permeance and Permeability

Note: See PERMEANCE and PERMEABILITY in main section. For a homogeneous material, permeability equals the product of permeance times film or sheet thickness. Although permeability is theoretically independent of thickness, actual measurements have sometimes found it to be thickness-dependent. Refer to ASTM D 1434 for further information.

Permeance

barrer: 10^{-10}mL(STP)/ [$cm^2 \cdot s \cdot$(cm Hg)]	33.49 fmol/($m^2 \cdot s \cdot Pa$)
inch-pound: mL(STP)/ [(100 in^2)·d·atm]	79.00 fmol/($m^2 \cdot s \cdot Pa$)
metric: mL(STP)/(m^2·d·atm)	5.097 fmol/($m^2 \cdot s \cdot Pa$)

Permeability (of diffusing substance through film or sheet)

barrer: 10^{-10}mL(STP)/ [cm·s·(cm Hg)]	334.9 amol/(m·s·Pa)
cm^3·mm/[m^2·(24 h)·atm]	5.096 amol/(m·s·Pa)
inch-pound: mL(STP·mil/ [(100 in^2)·d·atm]	2.0064 amol/(m·s·Pa)
metric: mL(STP)·mil/ (m^2·d·atm)	0.12945 amol/(m·s·Pa)

Permeability (of water through porous medium)

perm = 1 (ft^3/d)·cp/{ft^2·[(lb_f/ft^2)/ft]}	0.15596 μm^2

Note: See PERMEABILITY (2) in main section. For a liquid other than water, the perm, in μm^2, = 0.15596 × (liquid viscosity, mPa·s).

Gas-Transmission Rate

barrer: 10^{-10}mL(STP)/(cm^2·s)	0.04462 nmol/(m^2·s)
inch-pound: mL(STP)/ [(100 in^2)·d]	8.004 nmol/(m^2·s)
metric: mL(STP)/(m^2·d)	0.5164 nmol/(m^2·s)